POPULAR BELIEFS
AND
FOLKLORE TRADITION
IN SIBERIA

Indiana University Publications
Uralic and Altaic Series

Editor: Thomas A. Sebeok
Volume 57

Popular Beliefs and Folklore Tradition in Siberia

Edited by

V. Diószegi
Budapest

Indiana University, Bloomington
Mouton & Co., The Hague, The Netherlands, 1968

English translation revised
by
Stephen P. Dunn,
Center for Slavic Studies,
University of California, Berkeley

Library of Congress Catalog Card Number: 66—64654

Printed in Hungary

Contents

IN MEMORIAM ANTAL REGULY
(1819–1858)

Reguly Antal

Introduction

In 1850 a modestly produced book appeared in Hungary. It was entitled *Reguly-Album* and issued by the Reguly Society. The volume contained contributions by distinguished authors of contemporary Hungarian literature, short stories, poems, plays, and also a few scholarly papers. On the faded pages the readers found treatises about Hungarian prehistory, and became acquainted with problems of Hungarian heroic saga; Ferenc Toldy devoted a rather extensive study to the actual results of research concerning ethnic and linguistic affinities between the Finno-Ugrian peoples, as well as Antal Reguly's travels in Finland, Russia and Siberia, treated from the point of view of the scientific objective this great researcher had set for himself. With this volume the scholars and authors wished to express their deep respect for Reguly whose health was by that time seriously impaired.

After finishing his university courses in Pest, and his preliminary studies in various libraries of Germany, Sweden, Finland and Russia, Reguly set out on journeys in the Finnish, Lappish, Vogul and Ostyak areas, and spent longer periods of time with the Samoyeds, the Zyrian, Cheremiss, Votyak, Chuvash and other peoples. His primary interest lay in Finno-Ugrian linguistic affinities, in the relationships of the Hungarian language with the other Finno-Ugrian languages. The scholars in Stockholm and Helsinki all backed him, and the pride of Finnish science, M. A. Castrén, gave him valuable instructions; and also the *Kalevala* published in 1835 had a stimulating effect on him. In the centre of Russian education, he perfected his knowledge of Russian, but he also learned some other languages, such as Komi, Mari, Mordvinian, Chuvash and Turkish. He endeavoured further to get acquainted with the ethnography, geography, anthropology and history of the Asiatic peoples, in which effort he extended his interest to China, as well as to the Mongolian and Iranian territories.

An essay of his, is evidence enough of his having been well versed in the history and language of the Mongols.[1] He gave due consideration to the cultural interrelations between the various ethnic groups, and the mutual influence that may exist between the various languages. He paid particular attention to Asiatic religion since he was deeply aware of the great influence that the religious beliefs had exercised on the formation of the Asiatic mind and intellect. While dealing with studies like these, he never lost sight of his main object which was to investigate the languages, traditions and way of life of the Finno-Ugrian nations. The Hungarian Academy of Sciences and his Russian friends supplied him with the means that enabled him to begin the series of field-trips to the various Finno-Ugrian and Turk-

[1] A dzungárokról és azoknak a magyarokkal állított fajrokonságáról (About the Dzungars and their alleged relationship with the Hungarians), in *Magyar Akadémia Értesítő* (Bulletin of the Hungarian Academy) X, Pest 1850, pp. 155–68.

ish peoples which took him three years to finish. Antal Reguly's field research in the Vogul and Ostyak regions is of greatest importance not only for its linguistic value but also for the material it had contributed to the ethnography and geography of the area concerned. One of the results of his geographical observations was a map which he published in St. Petersburg in 1846 under the title *Ethnographisch-geographische Karte des nördlichen Uralgebietes, entworfen auf einer Reise in den Jahren 1844 und 1845 von Anton Reguly.* The map was executed according to the cartographical methods of the time.[2] Reguly marked the boundaries of regions where tillage, animal-breeding, or forest husbandry were carried on. In the Vogul-inhabited areas he indicated the place of each hunting hut, the yurts of the nomadic reindeer-breeders, the names of owners, the strip of land on which they usually wandered throughout the summer periods of pasturing, and he also indicated the water courses for the boats.

When Antal Reguly returned from Siberia, the learned circles in Hungary – and primarily the Hungarian Academy of Sciences – were looking forward to the publication of his results. But Reguly could never again recover strength enough to publish his invaluable collection. The material he brought home had to wait for some time to be worked up and published by other researchers of the Finno-Ugrian languages and ethnography. The first work of considerable scope and importance in this line was that of Pál Hunfalvy (*A vogul föld és nép* [The Vogul Territory and People]), a monograph on the ethnography of this people, based on Antal Reguly's notes and published in 1864, at Budapest. The extremely copious material of folk poetry in Reguly's collection was successively published by Bernát Munkácsi, partly by József Pápay and subsequently by Miklós Zsirai. For the elaboration of finds thanks are due to József Budenz.[3] A considerable part of Reguly's manuscripts are still in need of further elaboration.

It is a matter of general knowledge that Reguly had cherished far-reaching plans. He had envisaged, amongst other schemes, the publication of a large-scale comprehensive work dealing (1) with earlier history of Finno-Ugrian research, (2) with his own field research; (3) he wanted to publish his Vogul, Ostyak, Cheremiss etc. texts, together with the necessary explana-

[2] A detailed discussion of the Map appeared in No. 20 of St. Petersburger Zeitung, 1847. Later on, J. Pápay entered into thoroughgoing discussion of the Map in a treatise: Reguly Antal uráli térképe (Antal Reguly's Map of Ural), in *Földrajzi Közlemények* (Geographical Bulletin) XXXV, Budapest 1906.

[3] Of the relevant literature, the following standard works are worthy of mention: B. Munkácsi, *Vogul népköltési gyűjtemény* (Collection of Vogul Folk Poetry), Vols I, II, and IV, Budapest 1892–1921; J. Pápay, *Osztják népköltési gyűjtemény* (Collection of Ostyak Folk Poetry); *Dritte asiatische Forschungsreise des Grafen Eugen Zichy,* Vol. 5. Budapest–Leipzig 1905: J. Pápay, Die Ostjakischen Heldenlieder Regulys. *Journal de la Société Finno-Ougrienne* XXX, 36, Helsinki 1913–1918; J. Pápay, *Északi osztják medveénekek* (Bear Songs of the Northern Ostyaks), 1934; M. Zsirai, *Osztják (chanti) hősénekek* (Ostyak [Khanti] Heroic Songs). From the unpublished material of A. Reguly and J. Pápay, Vol. I. Budapest 1944, Vol. II. Budapest 1951; J. Budenz, Cseremisz tanulmányok (Cheremiss Studies), *Nyelvtudományi Közlemények* (Linguistic Bulletin) Vol. III, Pest 1864, pp. 397–470, Vol. IV, Pest 1865, pp. 48–105; J. Budenz, Erdei és hegyi cseremisz szótár (Forest- and Mountain-Cheremiss Vocabulary), *Nyelvtudományi Közlemények* (Linguistic Bulletin), Vol. 4, Pest 1865, pp. 48–105; J. Budenz, Mordvin közlések (Mordvinian Papers), *Nyelvtudományi Közlemények* (Linguistic Bulletin), Vol. 5. Pest, pp. 81–238.

tory notes and studies, as well as the dictionaries and grammars of these languages; (4) it was also his intention to write a handbook on the history, geography and ethnography of the eastern Finno-Ugrian peoples.[4]

Antal Reguly did not have time to carry out his plans and schemes, nevertheless it is owing to no small measure to his activity that so many worthy scholars of Hungarian linguistics and ethnography focussed attention on the Finno-Ugrian peoples. It was in his wake that Bernát Munkácsi went on research expeditions to the Votyak and Vogul areas, and Károly Pápai, József Pápay and János Jankó visited the Ostyak settlements. Incited by Antal Reguly, Finno-Ugrian research in Hungary arrived at a flourishing stage where the study of these peoples were advanced by such outstanding scholars as — to continue the list — József Szinnyei, Ödön Beke, Dávid Fokos-Fuchs, Irén Sebestyén, etc.

The present volume is dedicated by the Hungarian Academy of Sciences to the memory of Antal Reguly. It should not only remind us of the genius and achievement of the great scholar but it is also meant to reveal the close connection that exists between Finno-Ugrian research, on the one hand, and the exposition of folk culture throughout northern Eurasia, on the other. The countless problems of folklore of the various ethnic groups ranging from the westernmost Lapps to the numerous East-Siberian tribes are so intricately entangled, so closely linked up with one another, that a true solution can only be obtained by a complex method of research, which—instead of narrowing down—strives to broaden our approach to the fullest possible extent. Recent investigations into the ethnography of the Finno-Ugrian peoples and Finno-Ugrian linguistics readily support this outlook: they all bear witness to the fact that the present-day cultural circumstances of the Ob-Ugrians can only be clarified in the light of results concerning the folklore of the Paleo-Siberian peoples.

Our volume contains papers by researchers of several countries. The methods are varied with which they elucidate the problems of North-Eurasian shamanism, totemism, burial rites and other phenomena approached, in particular cases, from the angle of ethnography, ethnology, folklore, linguistics, history, archeology, etc. However, the understanding maintained by all contributors is uniform in so far as they are convinced that it is only through international cooperation that the solution of the relevant problems can be promoted. It is very gratifying to see the experts of several nations contributing to this volume in commemoration of Antal Reguly, and setting such an example of international cooperation.

The topics dealt with by the various authors are essentially connected with each other. It has been considered more serviceable to issue such a book than a collection of thematically diverse essays. Questions of shamanism, bear-cult, totemism, sacrificial rites, customs, hunters' rites stand in the foreground of ethnological research, and the authors have contributed by valuable new data, discoveries or theories to what we have already known from the literature. In all likelihood, a great number of particular data, such as are contained in this volume, will have to be published before so substantially comprehensive up-to-date works can be issued

[4] M. Zsirai, *Finnugor rokonságunk* (Our Finno-Ugrian Relations), Budapest 1937, 519 p.

on the theme as were the books of U. Harva, D. Zelenin, M. G. Levin, and L. P. Potapov at the time of their publication.[5] May this memorial volume to Reguly be a contribution to further syntheses.

The difficult task of editing has been undertaken by Vilmos Diószegi to whom, also in the name of the Ethnological Commission of the Hungarian Academy of Sciences, we herewith express our sincerest thanks. Our gratitude is also due to the collaborators whose contributions from Leningrad, Berlin, Stockholm, etc. has made it possible for us to offer this volume *in memoriam Antal Reguly* with such a rich content to the benefit of ethnology.

Béla Gunda

[5] U. Harva, Die Religiösen Vorstellungen der altaischen Völker, in *FFC*, 125, Helsinki 1938; (D. Zelenin) Д. Зеленин: *Культ онгонов в Сибири* (Idol Cult in Siberia), Moscow–Leningrad 1936, French translation: D. Zélénine: Le culte des Idoles en Sibérie, Paris 1952; (M. G. Levin and L. P. Potapov) М. Г. Левин–Л. П. Потапов: *Народы Сибири* (The Peoples of Siberia), Moscow–Leningrad 1956, English translation: M.G. Levin and L.P. Potapov: The Peoples of Siberia, Chicago and London 1964.

Antal Reguly

by

J. Kodolányi Jr., Budapest

From the time of Friar Julian's[1] oriental journey Hungarian scholars have shown an ever-increasing interest in discovering the kindred peoples of the Magyars, and trying to prove the relationship. Many zealous investigators ransacked Hungarian history in order to set up theories—usually without any solid basis—concerning the Magyars' racial or cultural connections with various ethnic groups, but there were few who could support their hypothesis with acceptable evidence. Lack of adequate knowledge and thorough preparatory studies, as well as the false pride that manifested itself in refusing kinship with 'unworthy' peoples gave rise to obscure propositions and untenable, though fanciful theories, some of which gravely obstructed the progress of more serious research conducted by investigators who happened to hit upon the right road.

Although Hungarian research produced famous pioneers such as Sándor Kőrösi Csoma (who was not genuinely interested in the affinity of the Finno-Ugrian languages), Mátyás Bél, András Huszti, János Sajnovics and Sámuel Gyarmathi, the discovery of the relationship between the Hungarian and the other Finno-Ugrian languages is due to West- and North-European travellers and scholars. Nevertheless, the contributions of the Hungarian scholars Sajnovics and Gyarmathi were most significant in this respect. They used methods of comparative linguistics in proving the relationship. Unfortunately, however, they did not have sufficient knowledge about the peoples concerned.

Antal Reguly[2] was the first Hungarian linguist to base his studies on the material he collected for himself among the Finno-Ugrian peoples. His aim

[1] Friar Julian was a Dominican who, during the reign of King Andrew II of Hungary, not long before the Mongol invasion in the first half of the 13th century, set out with three other members of the same order to discover those tribes of the Magyars who were left behind in the Eastern European homeland. After a long and perilous journey he alone reached a region near the Volga where he found a people speaking a language the sound of which made him feel that he discovered the Magyars who had been separated from the Hungarians of the Carpathian Basin at the dawn of their written history. He returned to Hungary towards the end of 1236 and soon afterwards submitted a report on his journey to Rome. In 1237 he set out again, but turned back because information had reached him that the advancing Mongolian hordes had devastated the region he was planning to go to. A description of his first journey is known from a report sent to Pope Gregory IX by the Dominican Friar Richard, a copy of which has come down to us in a Vatican codex (*Liber Censuum*) dating back to the end of the 13th century.

[2] Antal Reguly, born on 13th July, 1819 at Zirc (Veszprém county), attended secondary school at Székesfehérvár from 1828 till 1834, then studied for law at the Academy of Győr from 1834 till 1836 and at the University of Pest from 1836 till 1839. When leaving for Germany in 1838, he originally wanted to continue his legal studies. Attracted by Scandinavian culture, he proceeded still farther to the north; studying in the Royal Library of Stockholm and certain acquaintances from the scientific world directed his interest to the Finno-Ugrian peoples and languages.

was (a) to gather evidence which would support their linguistic relationship, (b) to clarify the origin of the Hungarian people and (c) to complete the work which had been conducted before him. He failed during his life-time to reach any decisive conclusions, because Reguly like so many other philologists of his time lacked appropriate scientific training. An even greater handicap was illness, which often caused delays, prevented him from writing down and arranging his material and eventually caused his untimely death.

As a young man, who prepared for the career of a statesman and studied law, he did not acquire the qualifications of an ethnographer or a linguist and did not know much about anthropology and cartography. But the study of history left a deep impression on him. He travelled extensively in Northern Hungary, Austria, Moravia and Galicia and, after finishing his studies, made a four-month tour of Western Europe. He visited Vienna and Prague, then went to Saxony, Prussia and the Island of Rügen and, through Rostock, to Hamburg. A romantic imagination drove him on to Copenhagen and Stockholm; eager to get acquainted with the rugged world of the North, to see the aurora borealis and to learn more about the progressive institutions of Scandinavia, he continued his journey.

Reguly spent a month in Sweden, from October 6th to November 5th, 1839. Brief as his stay was, it became a turning-point in his life. There he met and had exhaustive conversations with the Finnish poet Arvidson, who aroused his ambition to search for the origin of the Hungarians and investigate their relationship to other Finno-Ugrian peoples. Faced with hundreds of problems, he decided then and there to devote his life's work to the solution of these problems. There was no turning back now: he travelled to Finland, made long tours among the Finns and the Saams (Lapps), acquired a thorough mastery of the Finnish language and undertook profuse linguistic and ethnographic studies, under the guidance of Castrén.

In a letter dated October 20th, 1840, he disclosed his plans in the following words: 'It means that I want to spend next winter in Lapland: this will be the practical part of my studies. I hope to sail on October 20th up to Umeå through the Gulf of Bothnia and from there to the Swedish, Norwegian and Finnish Lapps in Kola, then to the Arkhangelsk and Olonets provinces and the Lagoda region, and finally back to Helsingfors. This will permit me to study the influence of Russian on the Finnish language.' Further on he writes: 'But to give a complete sketch of my plan, which I owe you, let me say that in April I intend to go—provided I have the means—to St. Petersburg, a city I regard as the point of departure for any journey to the Eastern Finns. I shall have to spend some time there in order to become acquainted with Russian scholars and obtain instructions from them and means from the sources available there for my further undertakings. It would be a good thing if I could spend some time in the region of Moscow in order to continue my sorely needed Russian studies. Then in the summer I would visit Nizhni Novgorod to whose markets people swarm from all parts of Russia, and from Central Asia, too. There I would gain experience which several months of travel elsewhere could not provide me. My trip to Southern Russia would be undertaken on the basis of information thus obtained. Next winter I would be busy studying Turkish in Turkey—again if I can afford it—and the whole trip would be wound up by a visit to the Csángós of Moldavia and the Székely population of Transylvania. Of course, the larger the number of languages covered, the more fruitful any studies in the field of comparative philology would be. Just as my knowledge of Finnish is of use to me in my studies in Lappish, so also will these two languages will help me to learn Turkish, for today no one doubts the relationship between the Turkic and the Finnic tongues.'[3]

Reguly's interest in linguistics is stressed again in another letter addressed to his parents: 'At present the investigation of languages attracts me most'.[4]

His interest went far beyond linguistics, as can be seen not only from his life-work but also from his research program he outlined in a report addressed to the Hungarian Academy of Sciences on May 2nd, 1842.

'The first part is the groundwork, which should therefore be examined and defined with mature deliberation if we wish to see the problem thoroughly investigated, and if we do not wish to work in vain. The aim being a linguistic one, I must approach it by means of linguistic method. Since peoples speaking related languages must also be ethnically related, this method, which is the only secure and infallible one for ascertaining the kinship of peoples long since separated, will at the same time establish a basis for historical analysis which will frame the second sphere of our tasks concerning the problems of relationship of these peoples. Once the linguistic part has been clarified and defined, we must deal with the ethnographic and historical aspects in order to provide a full discussion of the question. Hence, I have made it my duty to observe during my journey everything related to the looks, work, character, way of living, dress, customs, superstitions, mythology and material culture of the peoples concerned.'[5]

If we examine the above quotation in the light of Reguly's life-work, we can, and should, emphasize three significant features, which, in our opinion, have failed so far to receive the attention they deserve. These are: first of all his interest in ethnography, secondly the importance he attached to the exploration of the mythology of the Ob-Ugrian peoples and thirdly his recognition of the two corner-stones of Hungarian ethnogenetic research — Finno-Ugric and Turkic. Considering his possibilities and the means at his disposal, we can hardly reproach him for his not having carried out all the research tasks he had outlined.

This bold pioneer started the execution of his vast plan by visiting the Baltic Finns. Later, in 1841, he went to St. Petersburg. The letters of introduction with which he was provided, and his engaging manners soon won for him many friends and patrons among the scholars and officials in the Russian capital. It was with their aid that he studied and prepared himself for the long journey and self-imposed task of seeking out the East-European, and particularly the Trans-Uralian peoples related to the Magyars. He never stopped studying but went on without respite even when illness and exhaustion would have made a rest imperative. He became thoroughly familiar with anthropology, ethnology, history, archaeology, geography and above all, of course, with linguistics. An inextinguishable thirst for knowledge, a definitive trait of his mind, drove him further and further, until he was able to understand the profound unity of things he had sensed at the

[3] Toldy (1850) pp. XXIX–XXX.
[4] Toldy (1850) p. XXXV.
[5] Pápay (1905) pp. 4–5.

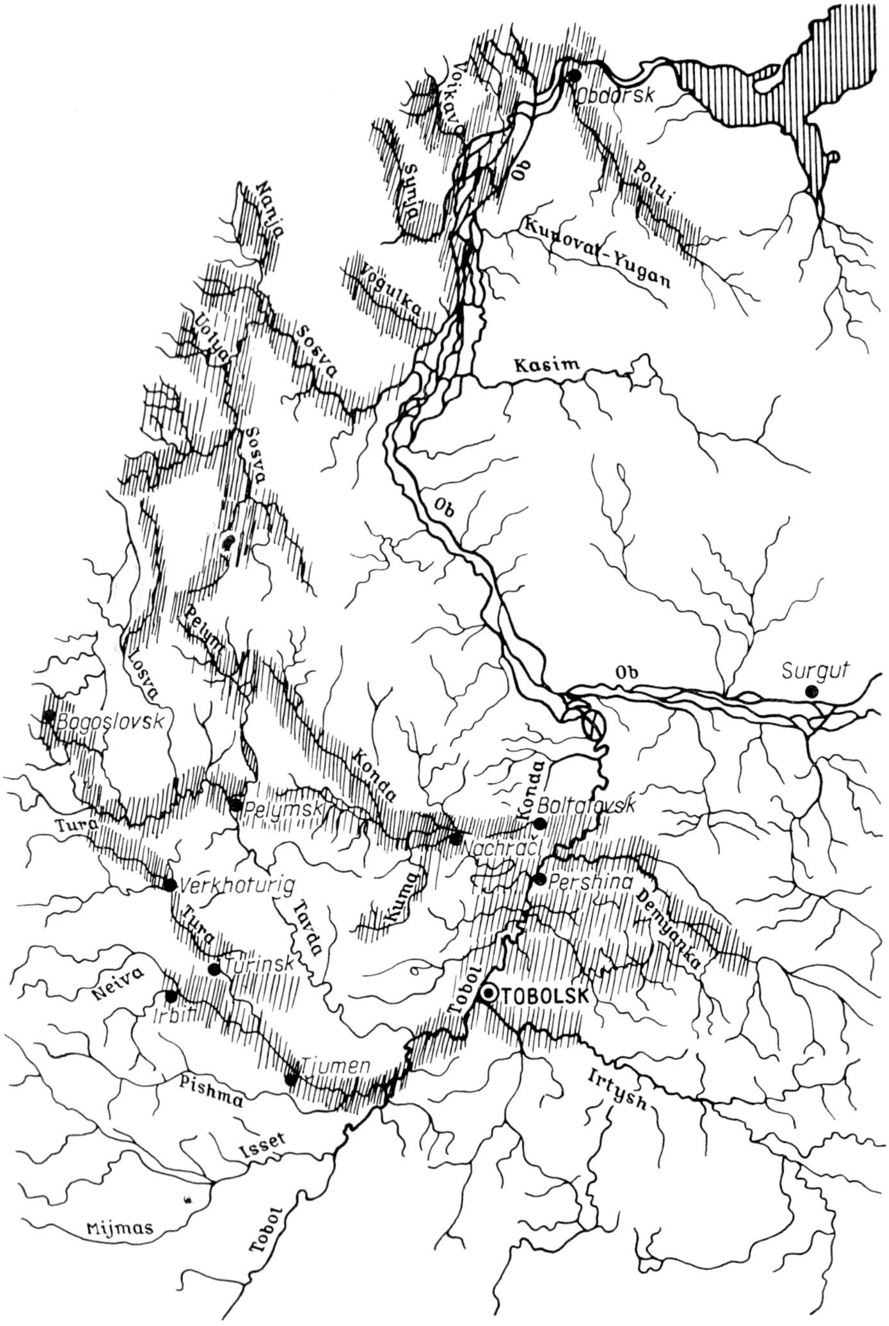

Fig. 1. The area in which Antal Reguly made his collections

bottom of the vast mass of details. His work would perhaps have been more efficient if he had restricted his field of interest to the study of some well-defined actual research problems of his time. But Reguly could not be content with that: he undertook responsibility for others. He realized that the valuable material still available for the collectors of his epoch was rapidly perishing, and was very much aware of the fact that a long time might pass before another Hungarian researcher would be able to access the geographical regions of the distant Finno-Ugrian relatives.

In the beginning his parents helped him financially, but later Reguly had to look for sponsors. A former history teacher of his appealed to the Secretary of the Hungarian Academy of Sciences, saying that in order to encourage him in his work he should be made a corresponding member. On February 8th, 1841, the Academy sent him a grant of two hundred forints, urged him to pursue his studies in Finnish and promised more help if his reports would prove to be of some value. Of course, this modest sum of money meant no real help and did not permit Reguly to start his journey to Siberia. While he was busy at his preparatory work in St. Petersburg, struggling against a double handicap of poverty and illness, his affairs at the Academy advanced only very slowly. Months went by, with hope and despair alternating. At last, towards the end of 1842, thanks to István Széchenyi who spoke in his favour, he received a grant of one thousand forints. Announcing this to his parents, Reguly writes: 'This letter has infused new life into me. My whole being is perturbed and I am hardly able to control my thoughts. So now, by a magic touch of the Academy, I have changed from a would-be Russian traveller into a Hungarian traveller.'[6] By July, 1843, Reguly had packed and was ready to leave but the promised money had not yet arrived. Since he considered winter the most suitable season for the journey, he left on October 9th with a loan from his friend K. E. von Baer. Had he but waited a little longer, the news that he had been elected a corresponding member of the Hungarian Academy, together with the promised money would have reached him in St. Petersburg.

Reguly was harassed by financial difficulties during the whole journey. He was often ill and found himself in desperate situations. But his tremendous will-power helped him to overcome all physical trouble. Finally he completed his Siberian field work in spite of obstacles which, at times, seemed to be insurmountable, collecting much valuable material of folklore and linguistics.

He reached Nizhni Novgorod via Moscow on October 21st, and on October 27th went from there to Kazan through the Mari (Cheremiss), Chuvash and Tatar areas. He spent two weeks in Kazan, studying the Mari and the Chuvash languages and establishing contacts with local scholars. On November 8th, he moved through a Tatar-inhabited region to Malmyz, and from there, crossing the province of Vyatka, he came to the border of Perm, and reached the land of the Udmurts (Votyaks). He stayed for a week in the town of Perm and studied the geography of that province. He went on through Solikamsk and finally crossed the Ural mountains on December 4th. He had hoped to continue his journey along the Lozva river. But he made his first stop in Verkhotur'je, arriving at Vsevolodo-Blagodatski on the 15th.

[6] Toldy (1850) p. LV.

As he passed through Mari, Chuvash, Tatar and Udmurt districts he took every opportunity to improve or extend his ethnographic knowledge of these peoples. This is clearly evident from the items he collected there.

In Vsevolodo-Blagodatski he was lucky enough to get two Mansi (Vogul) interpreters and after a short rest he began to study their language. An unusually warm friendship developed between Reguly and his interpreters who, later, revealed their innermost thoughts to him. It was through them that he obtained his first true picture of the Mansi people. 'There are some'—he wrote to Baer—'who have painted a terrifying picture of the Voguls; but they lack judgement and are badly prejudiced. They see only the strange outward form unpleasant to their eyes but don't care for what it contains; hence, they do not even suspect that there is a more human and less spoilt nature than their own at work here.'[7] Reguly found that the Mansi were dying out and regarded only the inhabitants of the banks of the Lozva and the Pelim as the true representatives of this ethnic group. He believed that the place names of the region were of Permian origin and hence concluded that the Mansi had immigrated into the area. He supported his view by referring to the tradition of the Khanty (Ostyak) people, according to which they, too, had made their way from the south-west to their present place of inhabitance.

After a short stay in Vsevolodo-Blagodatski, the Hungarian traveller soon resumed his journey and by March 3rd, 1845 had toured all the Mansi and Khanty settlements, the regions inhabited by the Nenets (Yurak Samoyeds) and the border areas in the northernmost parts of the Ural, adjoining the Komi (Zyrian) territory. Following the instructions of the Hungarian Academy of Sciences, he visited, on the way back, the Mari, Chuvash and Mordva areas in the Volga region, in spite of his recurring fits of illness. At last he returned to St. Petersburg with a rich haul of material on August 25th, 1846.

From Russia Reguly wished to go to Germany because he believed that he could do better work in arranging and classifying his collection under the guidance of German experts. Before leaving, he published his *Ethnographisch-geographische Karte des nördlichen Uralgebietes* which remained for a long time the only reliable source for later investigators—although they rarely acknowledged it. This work was given a most favourable reception by the Russian reviewers.

Unfortunately, Reguly could not carry out his plans. Wishing to recover entirely from the illness into which he had so often relapsed—sometimes very seriously—he spent some time at various hydrotherapeutic institutions. He tried to resume work whenever his health showed a slight improvement. Between two spells of treatment he came home to Hungary for a short visit in 1847. The following year he was appointed First Custodian of the University Library, but, because of the tragic end of the War of Independence, was not able to occupy this position until 1850. Even then his serious illness did not permit him to work very much. Encouraged by Ferenc Toldy and Pál Hunfalvy, he occasionally tried to concentrate on the vast material he had collected, but most of it remained as it was for decades after his death. In fact, some of his manuscripts have to this day

[7] Toldy (1850) pp. LXXI–LXXII.

not been prepared for publication. The first edited material was given to the public by Pál Hunfalvy in a work entitled *A vogul föld és nép* (The Vogul Territory and People). Reguly, however, did not live to see the publication of this book. His death in August 1858 put an end to his scientific work and delayed the submission of the results of his journey to the researchers waiting for them all over the world.

*

As we have already mentioned, Reguly's interests were not restricted to a special branch of science or to a single topic, but ranged over a number of fields in an attempt to collect as much material as possible in order to throw light on the origin of the Magyars and to prove his hypothesis of the Finno-Ugrian kinship. There is no doubt, of course, that he gave priority to linguistics, but he was also fully aware of the importance of the cognate fields. His effort to gather the largest possible amount of data for the benefit of comparative linguistics resulted, at the same time, in the collection of an immense store of ethnographic material. During his field work, which involved several years travelling in the terrain, he recorded a great deal of linguistic material and folk poetry, listed important ethnographic and anthropological observations and collected the related objects of material culture. He also drew maps of all parts of the Ural region he had visited.

In a letter written to Baer on January 24th, 1844, he characterized the Mansi folk poetry as follows: 'Their poetry allows a division into groups of heroic songs, bear songs and lyrics; hymns and prayers constitute a fourth group. Their heroes go about in an armour of iron, wear iron or chain shirts and helmets and carry lances, etc. They sing their bear songs at bear festivities which are usually held three days after the killing of a bear. I took down six songs of this kind, two of which run into some 30 pages; the seventh and last, a burial song, unknown to Baktiar, I hope to record with Yurkina's help during my trip to the Lozva.'[8] His recordings of folk poetry amount to many thousands of lines, 17,102 lines of Khanty text alone. The publishing of this material was started by Pál Hunfalvy and continued by József Budenz. Later Bernát Munkácsi, József Pápay and Miklós Zsirai furthered this work. Although the approach of these scholars is marked by a linguistic interest, the works they published are of great ethnographic importance, too.

When compared to this bulk of folklore records, Reguly's collection of ethnographic objects seems insignificant. Nevertheless, the impetus they gave to the development of Hungarian ethnography must be emphasized here. Reguly's interest in this line is eloquently proved by the fact that, in spite of his illness and the great exertion such work demanded, he still devoted much time to the collection of objects. Nor was his work fruitless, for these objects are very valuable today, reflecting the culture of the peoples in the period, more than a century ago.

In 1850 Ferenc Toldy gave the following description of Reguly's ethnographic objects: 'At a well-attended meeting in November, 1847, the Academy received a comprehensive report from the Secretary on Reguly's travels and their general results, with all the attention they deserved. On this occasion, also the collection of ethnographic pieces was exhibited

[8] Pápay (1905) p. 12.

which Reguly had been gathering with so much care and at such great costs. It consisted of various items such as: (1) Clothing, especially Lappish, Samoyed, Vogul, Ostyak and Mordvin furs; men's and women's garments and undergarments made of linen, wool and nettle cloth; headgear, aprons, shoes and a wide assortment of jewellery. Many of these items, particularly the women's shirts, were decorative, and embroidered with great care and skill; the combination of colour and design often displayed good taste. Some of the Lappish and Vogul furlined clothes had been worn by Reguly himself. (2) Puppets, i.e. male and female figures of Samoyed and Cheremiss origin. (3) Models, such as Vogul, Cheremiss, Ostyak and Russian figurines, sleds, skates, bows and nets. (4) Weapons; a representative collection of arrows and similar equipments. (5) A genuine Vogul lute (*sangur*) and a genuine Ostyak idol.

Reguly donated this collection to the Academy, saying that if it did not wish to exhibit the objects, they should be deposited in the National Museum. 'I need hardly mention that this collection, though not very large, is the only one of its kind in Europe. I consider it well worth-while to have all the objects drawn and published, partly in colour prints, either in Reguly's travel book or in the Annals of the Academy.'[9]

After a preliminary exhibition at the Hungarian Academy of Sciences, Reguly's collection was passed over to the National Museum, where it constituted the basic material for the collections of the Ethnographic Museum which was established much later. Unfortunately, of the whole collection only 76 pieces found their way there, the rest having been damaged in the meantime by moths or destroyed in some other way.

Reguly's ethnographic work covered Finland, the Baltic Finns, the settlements of the Volga Finns, the Tatars and the Chuvashes, but most of his vast legacy concerns the Ob-Ugrians.

Miklós Zsirai said: 'It was mainly thanks to the manuscripts he left behind that a school of comparative linguistics could develop in this country and that the great problem of the origin of the Hungarian language was solved largely through Hungarian contributions'.[10]

The same may be said if we examine Reguly's influence on ethnography: the two linguists who followed in his footsteps, Munkácsi and Pápay, really delivered an astonishing wealth of ethnographic material. The ethnographers Károly Pápai and János Jankó started in the same way as Reguly with the avowed intention of doing ethnographic research work. Even their biographies show a strange resemblance to Reguly's, for both died young without having had time to elaborate the material they had collected. János Jankó reached the Ob region starting from Finland and he, too, was interested in the Volga-Finnic and the Turkic peoples.

There is little doubt that Reguly's influence on Siberian ethnography is as important as his contribution to Finno-Ugrian linguistics. Nor must we forget the role Reguly played in the study of the religious life of the Siberian peoples, chiefly of the Ob-Ugrians. Hunfalvy was the first to describe the religious life of the Mansi and he used Reguly's records on folk poetry and his linguistic data. It is true that Reguly collected his material for a different

[9] Toldy (1850) pp. CXXII–CXXIII.
[10] Zsirai (1939) p. 1.

purpose, but it is no less true that his interests extended to religious life. We read in a letter he wrote to Baer: 'I cannot praise my informant Baktiar warmly enough. Sometimes he sits with me from morning to night, and he is so attached to me and so intimate with me that there is perhaps no phase in their religion and rites about which he has failed to tell me. And he . . . is also well known among his people, since he is a singer and often takes part as a priest in their annual horse sacrifices. These two capacities of his are again of scholarly use to me, for they offer me a rich source of material. So far I have recorded some 600 pages of various songs, prayers and melodies from him, which are of great interest even if their reference to the life and intellectual stage of the people are left out of consideration.'[11]

In another letter addressed to Baer, he describes a rite: 'Having returned to Verkh-Pelim (Upper Pelim), we prepared ourselves for a Vogul religious rite. *Pălm tărom* "deity of Pelim" listens with kindness to the suffering and supplicant. They pray and make pledges to him not only near Pelim, but also along the Lozva and the Konda and even the Northern Sosva river. My own Baktiar was indebted to him for saving his life when he was seriously ill, two years ago. At that time he had promised him the sacrifice of a horse; so far, however, poverty has prevented him from keeping his promise. Therefore I purchased a colt (after all a colt is a horse) from a Vogul and on March 29th (according to the Russian calendar) we started out to Pegunpaul to see the old *ńait* ("magician, priest", Finnish *noita*). *Pălm tărom*'s actual residence is in the *tumen*, a birch grove, near the village of Tarla on the bank of the Pelim, where he stands during the sacrifice on an unusually thick branch of birch almost thicker than the trunk itself. This is where they used to bring sacrifices to him in the past. Some years ago, however, this religious rite was transplanted to the village of Pegun, at a distance of two versts from here, because the road to Pelim passes the grove and profane passers by might disturb the performance of the rite. After having played a hymn on the *sangur* (a Vogul musical instrument) to invoke the deity, they performed the ceremony of killing the animal. Then they cooked it and ate of its meat, and so completed the sacrifice. The next morning I arrived with my sanctified and happy Baktiar at Pelim.'[12]

Further facts, provided by Reguly about the religious life of the Ob-Ugrian people are connected with their bear cult. In a letter sent to his friend Baer on January 24th, 1844, Reguly writes:

'The first of these bear songs tells of the creation of the world and of man, and ends with the genesis of the bear. The subject of the second song is a woman who makes a bear oath which is later fulfilled. The third one deals with a famous bear hunter who was such a good bowman that he could hit an arrow shot into the air with three other before it fell to the ground. He killed ninety-nine bears but the hundredth tore him to death, etc. Then I recorded a song about the devil, telling of how God had exiled him in eternal swamps. Their customs and ways of living show such a striking similarity with those of the Lapps, Cheremiss and Votyaks that a description of the ancient culture and religion of the entire Finnic ethnic group can no longer be regarded as a task of great difficulty.'[13]

[11] Pápay (1905) p. 12.
[12] Pápay (1905) pp. 14–5.
[13] Pápay (1905) p. 12.

This remark indicates that, while collecting material, Reguly was busy with questions of local religion, too. His records supply scholars with valuable basic data for the study of the religious life of the Siberian peoples. Later research workers, particularly Bernát Munkácsi and Artturi Kannisto, while collecting material, paid considerable attention to bear songs and the bear cult. It is possible that it was Reguly's work that directed their attention to this important field of the Ob-Ugrian folk tradition.

Students of the religious life of the Ob-Ugrian peoples have collected relatively little material on their shamanistic beliefs. Reguly's notes throw some light on this problem as well. They show that the *ńajt* is a kind of a magic healer, corresponding to the shaman of the Altaic peoples, whereas the term itself is related to the Finnish *noita* and the Lappish *noajde*. If we recall the information provided by Pallas, according to whom the Mansi *ńajt* falls into a trance by means of beating a drum, there is no reason to doubt that the *ńajt* mentioned by Reguly is the shaman of the Mansi.

The interests of this pioneer concentrated not only on the Finno-Ugrian but also on the Turkic peoples. During his journeys to and from Siberia he did not neglect the opportunity to collect data on the Chuvash and Tatar peoples. We know from the plan he had worked out before starting on his travels that he had intended to visit Turkey. Evidently he wished to study both the Turkic and the Finno-Ugrian peoples. It is worth while emphasizing that Reguly was one of the first scientists to understand that data on the Finno-Ugrian and the Turkic peoples are of equal importance when studying the ethnogenesis of the Magyars. Although he himself was unable to carry out his plan, the mere fact that he drew the world's attention to this relationship was revolutionary.

1958 marked the one hundredth anniversary of Reguly's death. He died at the early age of 39, returning home from his travels, exhausted and diseased. This anniversary should not be regarded as an occasion for celebrations but rather as an appropriate time to recall his noble example and to re-examine the many problems he raised. His life which was entirely devoted to science and his great work have left their imprint on the works of Hungarian and foreign scholars alike and his collection constitutes an indispensable source for all those who study the European and Siberian peoples.

REFERENCES

Hunfalvy, P. (1864) *A vogul föld és nép* (The Vogul Territory and People), Pest.

Pápay, J. (1905) *Reguly Antal emlékezete* (In Memory of Antal Reguly), Budapest.

Toldy, F. (1850) Reguly s a finn–magyar kérdés (Reguly and the Finno-Hungarian Problem), in *Reguly Album*, Pest.

Zsirai, M. (1939) Reguly Antal emlékezete (In Memory of Antal Reguly), in *MNy*, Vol. XXXV.

Seite Cult and Drum Magic of the Lapps*

by

E. Manker, Stockholm

Fig. 1. The pre-Christian, animistic Lapps worshipped holy lakes and holy mountains; like the pyramid-shaped mountain which can be seen in the background. Photo by the Author

Nowadays when the church bells are rung in the scattered parishes of the Lapland, in Karesuando at Eastertide, in the peat-covered church-huts of Mellätno in the summer, or in the chapel of Fatmomakke in the autumn, the Lapps are seen to walk to the services with good Christian devotion. But in a show-case of the Lappish Department of the Nordic Museum, we can see a very different figure of a Lapp from the 18th century, dressed in leather, sitting and beating his drum. He is a *noid* engaged in the practice of his profession. Another show-case contains a variety of strange stones, which seem to be surrealistic idols, called *seite*. In a third show-case there sits a primitive wooden figure, strangely moulded by a joint effort of nature and human hands. In the remote Lapland, tradition has preserved places of ancient cult, and geographical names like *Passe*, *Ailes*, *Saiva*, etc. (Figs 1–3) which still indicate something holy or dangerous to the native and remind us of the faith that had existed before churches and chapels were built there.

From time immemorial the Lapps living in close contact with nature had worshipped nature, as shown by their religion which before their Christiani-

* The figures have been taken from the author's study published in *Acta Lapponica* Vols I, VI and XIII (resp. 1938, 1950 and 1957).

zation was shamanistic. Forces of nature were regarded by them as divinities, and all things—animals, plants or lifeless objects—had a soul. The dead continued to live in a world not much unlike that of the living. The Lapps lived in a spiritual world which was for them as real as the material world. This was pure animism.

Magic drums, with their multifarious pictures, have come down as documents of bygone mythical beliefs to the posterity. The sun, *Peive,* occupied a most important place in that mythology, not as a personified figure but as a cosmic, primaeval being whose emanations spread over the world as rays and carried other, more clearly shaped deities. One of these was *tiermes* or *hora-galles,* the god of thunderstorm (called also *ahttje,* i.e., 'father' or 'old man'), represented as a man and fervently worshipped. Another was *biegg-olmai,* who drove the winds out of, and back to, their caves. This central complex of figures, with its natural mythology, is in full harmony with the people's deeply felt dependence on Nature's power. The moon, called *aske* or *mano,* occupies a less prominent place on the drum, nevertheless it was a significant enough figure to become dangerous sometimes. Other natural divinities were *väralden-olmai,* 'the man of the world', a god of fertility and the protector of the reindeer and other animals; *rana-neida,* a female being, who was patronous of spring and vegetation; *leib-olmai,* 'the man of blood or of alder', who ruled over wild animals and venison and was thus the god of hunting; and finally, *tjas-olmai,* 'the man of waters', the ruler of lakes and fishing.

Several of these divinities can be clearly recognized in the nordic mythology. For instance, *peive* corresponds to the Sun-god of the Bronze Age,

Fig. 2. Impressive natural formations were often chosen sacrificial places like this cavity called *passeuksa,* i.e. 'the Holy Portal'. Photo by the Author

Fig. 3. Near Passekårtje, 'the Holy Fall', a rock with the vigorous profile of a man was worshipped as a *seite*. Photo by the Author

tiermes to Thor, *biegg-olmmaialden-olai* to to Freyr Njord, *vär* and *rana-neida* to Frigg. However, this should not lead us to believe that they have merely been borrowed. Among a primitive people, thunderstorm is a far too powerful phenomenon for its divine personification to be borrowed from a higher culture. With a fair amount of accuracy, the same may be said of the sun which in the arctic region is of vital importance, and this also applies to the wind, etc. Frequently two Lappish names are to be found for these and other deities and spirits, one showing connections with the nordic culture, the other being of native origin. Thus, *horagalles* is an exact translation of 'old man Thor', whereas the more ancient appellation *tiermes* is connected with the heavenly god *torym* of the Ostyaks (Khanty), and Voguls (Mansi). *Mano* is only a superficially assimilated form of the nordic word for 'moon' though the moon as a natural phenomenon and divine being was known much earlier under its Lappish name *aske*. These two examples reflect the following situation: the original Lappish ideas have melted with the conceptions of Scandinavian peoples and the former may have received new features, or even new names, from the neighbours.

Besides these natural divinities, the Lapps had a number of more abstract gods, like *jubmel* or *ibmel*, who is obviously identical with the Finnish *jumala* and is worshipped as the supreme god. Among the Swedish Lapps in particular the same position has been maintained by *radien-ahttje*, 'the father of the Ruler', the Ancestor. His image occupies a central place on the upper half of some drums. Usually he was accompanied by his son *radien-pardne*, through whom he acted among men, and by his wife, *radien-akka* (the old woman of the Ruler). This patriarchal trinity corresponds, in a striking manner, to the Christian Trinity, with the Holy Ghost rationally replaced by a female being. Sometimes this divine family was augmented by *rana-neida*, considered as *radien-ahttje*'s daughter: *radien-pardne* was also called *tjorve-radien*, i.e. 'horned-radien' and is often figured on the drums with an aureole of reindeer antlers around his head.

In this group of supreme divinities there is a large range of smaller, yet important gods, or rather spirits, which are sometimes led by *madder-akka*, 'the Ancestress'. She is followed by her three daughters: *sar-akka* 'the old woman of spinning', who dwelt under the fireplace, spinning threads of sinew and assisting births, as well as the fawning of reindeer; *uks-akka*, 'the old woman of the door', who watched the door, as her name indicates, and *juks-akka*, 'the old woman of the bow', who had a bow as her symbol and was the protectress of boys. The birth of a man occurred in the following stages: first, *madder-akka* received the soul of the child from *radien-pardne*, created a body in herself and passed it on to *sar-akka* who put the little man into his mother's womb and was responsible for his birth. Once the child was in the womb, *juks-akka* could interfere and change a girl into a boy.

A dreaded figure was *rota* or *ruto*, the demon of death and diseases. He was generally supposed to be sitting on a horse and therefore he is closely connected with Odin, the god of Death. The dead, called *jabmek*, lived in *jabmi-aimo*, 'the world of the dead' or *mubbe-aimo* 'the other world'. There was also another world called *saivo*, the Lappish paradise, where the dead—the *saivo-olmak*—lived like here on earth, but under happier circumstances, and they had their herds of reindeer, their huts, their game and fish, etc. Certain mountains and lakes were known as the dwellings of the *saivo* people, as a kind of Lapp villages or *sita* in the *saivo*, and were regarded as holy (*passe*). The *saivo* lakes were generally believed to have a double bottom. (A physical reason supporting this belief was found in springs which were seen as holes at the bottom of some shallow lakes.)

Originally the Lappish mythology does not seem to have included a devil. Strange as it may sound, the devil—*fuodno*—appeared on the scene with Christ. He is known under names which come from the Finnish *perkele*, such as *perkel* or *baergalak*. However, even without the devil which the theologians have rationalized, life was hard enough. The powers of nature, the divinities and spirits—even though not evil—could become rather dangerous if one was not on good terms with them.

Thus the world was full of dangers, including not only thunderstorms, bad weather, snow slides, treacherous ice, wild beasts and other lurking

Fig. 4. The shaman drum is the most important instrument by means of which the *noid*, or shaman, was able to get in contact with the gods. The figures on the drumhead symbolize both spiritual and worldly phenomena. (Nordic Museum, Stockholm)

foes, but also diseases and all sorts of evil that could not be understood. Everything was somehow linked up with the world of spirits, and one could not get along if one depended merely on one's own strength and common sense. One had to be able to emerge from oneself, to penetrate everything that was visible, in fact, to lift oneself still higher and enter the world of spirits and divinities. This was done in a transcendental state of ecstasy which was reached with the help of magic drums. Therefore, the father of every household had his own drum which he used to the best of his knowledge. But, as everybody did not possess the right 'power', difficult cases had to be transferred to those, who had greater talent. These 'well-informed men' came to be known as *noids*, or wise men—the magicians, the shamans of the Lapps. They do not seem to have formed a priestly caste, nor did they wear special shaman's clothes like their Siberian colleagues (only a sort of 'magic belt' is known) but lived and dressed like everybody else. Yet, they used their drums, offered sacrifices and performed other religious duties whenever the need arose.

Like other shamanistic peoples, the Lapps used the drum as a means of achieving exaltation (Figs 4, 5). While drumming, the *noid* fell into trance and often fell unconscious on the floor, where he could lie for a whole day. During that time, his spirit left him and wandered to the other world, where he performed great deeds, if he had luck. He learnt the will of the gods, saved the life of a sick man, found the trace of an escaped herd of reindeer, ascertained the whereabouts of a bear or a wolf, averted the plot of another *noid*, etc. During his journey he could use certain attendants—the 'noid escort'. In the *saivo* world, holy animals such as reindeer, birds and fish were at his disposal, and he succeeded in his battles with either their help, or appearing in their shape. A *saivo-sarva* (*sarva* = male reindeer) would occasionally be engaged in a desperate struggle with the 'magic reindeer' of another *noid*. If he felt exhausted, the *noid* strengthened himself with *saivo* water. Normally he returned victorious; sometimes, however, he was unfortunate and no more woke up from his state of unconsciousness (Figs 6, 7).

The Lapps also used their drums as means of divination and fortune telling. They seem to have developed this custom more than the Siberian peoples. This probably explains the preservation of a multiplicity of figures every one of which has its specific meaning: a god, a spirit or an ordinary object. Small objects, rectangular pieces of reindeer antlers, a ring, or bundles of rings were placed on the drumhead as a pointer. While the drum was being beaten, the vibration of the skin made that object jump about—over, among or right on to the figures. Sometimes the pointer was regarded as a jumping frog, an animal which is also used in other aspects of magic. Fortune was told or the will of higher powers explored by the movements of the pointer or by the place where it stopped on the drumskin (Figs 8, 9, 10).

When the Lapps were converted to Christianity, hundreds of drums were burnt in spite of the fact that, according to some records, people pretended that they used them as compasses were used by the mighty lords who had come to confiscate the drums. Collectors and those who were interested in matters of culture—including some of the zealous priests—managed however to save a number of drums, which gradually dispersed all over Europe. Later, when they were all gathered, it was found out that there are 71 drums, and a few fragments and imitations. Twenty-nine of these *tabulae sacrae* are preserved in the Nordic Museum. In caves and other hiding places in the distant corners of Lapland, there are probably still a few more odd drums concealed by their owners not so very long ago. As a matter of fact, some were used secretly as late as the end of the 19th century.

In some cases the *noid* received information through the drum concerning the kind of offering the gods wished to receive. Otherwise sacrifices were made in accordance with traditional rites. *Peive* was sometimes offered white reindeer, *tiermes* received occasionally a whole reindeer bull which was buried in the soil so that nothing but its upright antlers were visible, and its genitals were sacrificed to *väralden-olmai*, the god of fertility. *Tjas-olmai* was offered fish tallow which was smeared on the sacrificial stone or *seite,* whereas in emergencies, a horse was bought from a farm for *rota*, the demon of diseases so that he might ride away on it. The offering

Fig. 5. Various magic objects were hanging from the bottom of the drum. (National Museum, Copenhagen)

Fig. 6. This group of realistic symbols, representing wild reindeer, bear and hunter, decorates a magic drum kept in Rome. Tracings by the Author

Fig. 7. The pictures on the drum include sometimes the *noid* himself with his drum. Tracings by the Author

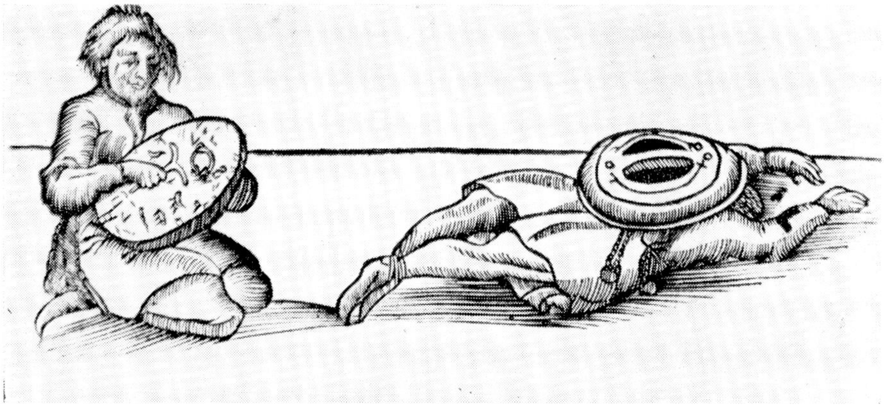

Fig. 8. A *noid* in trance, beating his drum. According to Schefferus, 1673

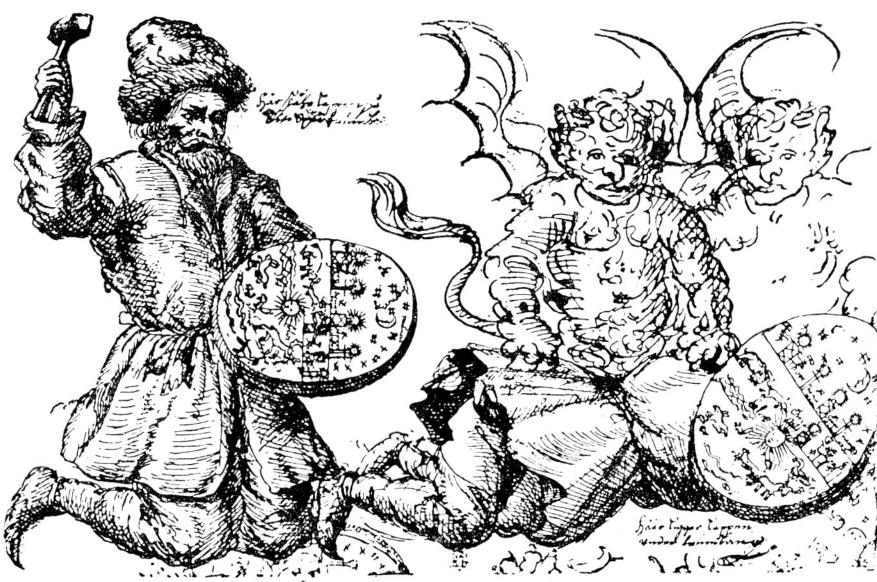

Fig. 9. The same theme interpreted by a missionary of the 17th century. The slumbering *noid* has fallen into the clutches of the devil

of a reindeer was frequently connected with a sacramental meal, in the course of which the nutritive parts of the beast were retained by the participants, and the antlers, bones, blood and tallow were sacrificed to the deity. In less critical situations, the sacrifice would be limited to small pieces of jewellery, coins, etc. There were also various other sacramental customs, which, more often than not, betrayed Nordic influence. At the sacrificial altar of *väralden-olmai*, for instance, blood was smeared on the pillar with which he was believed to support the world. Explorers drew a parallel between this and the German worship-column Irmensul. *Rana-neida* was offered a spindle which was sprinkled with blood. A brass or copper ring was hung in the chimney of the hut for protection against the Christmas moon. 'Sun-porridge' was eaten at Midsummer night and '*sarakka*-porridge' after a lucky delivery in honour of *sarakka*, the nurse of the divine world.

Major deities were offered sacrifices near or on a kind of a sacrificial altar which was built in the same manner as the provision-stands used by the Lapps up to the present day. Reproductions can often be seen on drums, and we possess the sketch of such a sacrificial stand, dating back to the 17th century. There are three 'hora-galles idols' on it with reindeer antlers between them and pieces of meat offered on sticks before them. Big stones and peculiar boulders were also used as sacrificial altars. Several of these are still known through tradition. Sacrifices could also be made in grottoes or rocky caves on cliffs, beneath or, on the summit of 'fjells' (mountains), in fens, beside waterfalls, in lakes and springs, or at a tree. These places of sacrifice belonged either to a single person, to a single family, but they might also be frequented by a clan community or the population of a whole district. In the course of an exploration of the Swedish Lapland which was conducted by the Nordic Museum, 507 such places were charted. They are generally well known for some conspicuous feature of the place itself or in its surroundings: extraordinary size, odd shape, etc. Quite often the sacrificial place is situated at the foot of a steep slope, or of a declivity of rubble where a thundering earth slide had once occurred. Considering the role which thunderstorms play in the great drama of nature, we may suppose that the place was sometimes pointed out by a flash of lightning.

In some of the places of worship roughly carved wooden idols were found, in others, *seites*, i.e. stone images of gods, which showed slight traces of human craftsmanship but were originally fashioned by nature herself: thus the deity must have shaped itself. A woodcut from the 17th century shows a Laplander worshipping a *seite*, which is erected on a bed of brushwood, in a semicircle of reindeer antlers: bare-headed and on bent knees the Lapp is approaching it (Figs 11, 12, 13). As the result of an excessive zeal and ill-advised collecting, most of the important sacrificial places have been despoiled of their *seites*. In other cases, they were destroyed or rolled away with Christian ardour. In some remote places, however, one may still hit upon a *seite* standing intact and majestic. The Lapps even today sometimes make a detour to visit a sacred locality.

In Lule and Pite-Lapland the *seites* are often called *storjunkare*—a late loanword. The notion of the *seite*, on the other hand, must have been an original element in Lappish religion. Explorers have given various interpretations to the conception of the *seite*. At times it was considered the *genius loci*, at others the worshipped god himself, his image, or his residence,

Fig. 10. A *noid* beating his drum. (From the Lappish Department of the Nordic Museum)

or again a powerful ancestor. To me, the *seite* seems a synthesis: an incarnation of the divine ruler of the place (region) and hence the centre of power of that place (region) as well. The idea of an ancestor is not farfetched either, since, in many cases, the *seite* is situated in the *saivo*-region, that is in the world of the dead.

All over the Swedish Lapland, the religion of the Lapps has developed along the general lines described in this study. Nevertheless, a considerable number of local variants have arisen within the area: a typical feature of the northern districts is the division of the divine and spiritual world into higher and lower spheres, as can be seen in the picture-fields of the drums. The Lappish world of deities in the Finnish and the Russian Laplands differs in many respects from that of the Scandinavian area. There are, however, many fundamental religious concepts and mythological figures

Fig. 11. The *seite*, a stone which is oddly shaped by nature, was worshipped as an idol, an incarnation of divinity, from where superhuman power emanated. Photo by the Author

which have accompanied the Lapps in their migration from East to West and from North to South. The Lapps have the basic elements of animism in common with all the other circumpolar tribes.

The Christianization of the Lapps began at a very early date but produced no results worth mentioning before the end of the 17th and the beginning of the 18th century. The most ancient document in this connection is a royal letter, dated 1389, by which Queen Margaret calls upon the Lapps to abandon their heathen faith and to integrate into the community of Christian nations. No records are available concerning the impact of this letter on the Lapps. Thirty years later, Erich of Pommern sent the first missionary, by name Toste, to the Lapland. The seed sown by him fell on stony ground—which is found in a great abundance in Lapland! By all means, the result was not to be boasted of. At the bottom of all these attempts at Christianization was the almost mythic 'lappkvinnan Margareta', a Lapp woman who, in her fervent ardour, travelled as far as Malmö to obtain an audience from her namesake the queen.

During the reign of Gustav Vasa and Charles IX, the endeavours were continued. The first churches of Lapland were built, among other towns, in Gråträsk, Arvidsjaur, Lycksele, Jokkmokk, Jukkasjärvi and Enontekiö, as a result of a decree issued in 1603. The Lapps presented themselves in full number at the churches, but at home, in the backwoods, they clung with even greater faith and confidence to their heathen *seite*s and drums. (The effect of baptism would sometimes be washed off with chewed alder-bark.) In some places, however, Christianity was victorious: tradition knows

Fig 12. Also primitive wood-blocks were worshipped as idols. (Nordic Museum)

of three Lapp brothers far up north who built the chapel of Rounala in 1606 at their own cost.

At the end of the 17th century, the conversion to Christianity took a more energetic turn. The Lapps were forced to attend church ceremonies and to surrender their drums, which were subsequently burnt. Those of the Lapps who showed resistance, stowed away their drums or made new ones, were sentenced to fines or flogging. In some rare cases, as in 1692 at Arjeplog (executed in 1693), resisting Lapps were sentenced to be burnt at the stake together with their drums, to serve as a warning. Thus, by the second half of the 18th century, drum magic was rooted out by administrative measures. But the old faith still survived and in the distant corners of Lapland well into the 19th century drums were secretly beaten and sacrifices made to *seites*.

Later on, Christianity became just as sincere and fervent among the Lapps as in the Swedish communities. Nevertheless, some reminiscences of the heathen form of worship have penetrated the new form, as can be established from the survival of certain ancient rites: voluntary donations are collected in the bell purse, but also coins are sometimes inserted into the wall of the church. Christianity assumed its strongest and the strictest form in Torne district, where Lars Levi Laestadius was active in the first

Fig. 13. Worship of *seites*, according to Schefferus, 1673. The *seite* is erected on a bed of brushwood inside a circle of reindeer antlers

half of the 19th century. Occasionally Laestadianism degenerates into fanaticism, and at the meetings of its adherents so-called 'motions' (Finnish *liikutuksia*) can be frequently observed, which in spite of all their Christian appearance remind us that something of the shamanistic ecstasy is still living.

Survivals of Totemism in the Hungarian *táltos* Tradition

by

B. Gunda*, Debrecen

The persons termed *táltos* are the most remarkable figures of Hungarian popular belief according to which these men (rarely women) possess superhuman strength or some extraordinary qualities. Although the figure and activities of the *táltos* have often been confused with those of the witch and the 'itinerant scholar' (*garabonciás diák*), the strata of popular tradition related to the problem have been fairly well sorted out by G. Róheim and V. Diószegi.[1] One characteristic fact—already noticed by Róheim—is that no one confesses that he is a witch, whereas one does admit that one is a *táltos*. In the days of witch trials, in 1725, one of the women in Debrecen publicly declared that she was a *táltos* and that she considered her art to be of divine origin.[2]

According to popular belief a child which is born with teeth becomes a *táltos*. From the first years on it is taciturn, melancholy, reserved, and keeps aloof from the others, but it is very strong for its age. As a grown-up, a *táltos* seldom enters a village, but prefers to roam about the fields and the pasture land, calling sometimes on people who live on lonely shepherds' farms. Often the man with the qualities of a *táltos* is himself a shepherd or a field warden. If he happens to drop in somewhere during his wanderings, he usually asks for milk, sour milk, cottage cheese and eggs. His predilection for milk is remarkable. He will even suck milk from the cow's udder; by looking at milk in a milk-jug, he can tell from which cow it has come. A *táltos* can not only prophesy and heal but he can also raise winds, storms, and cause fires, hail and floods. He can detect hidden treasures by looking at the sun, in a steel mirror, or at the shine of his own fingernail. At the end of the 18th century a *táltos* was accused of singing 'devilish songs' and of 'causing a fire and asking for help from the Prince of Darkness'.

The *táltos* is said to have obtained his art while in his mother's womb. A more widespread belief holds that a person who was born to be a *táltos* often falls into a trance-like state at a more advanced age and so acquires his art. On such occasions the would-be *táltos* has hysterical fits and is on the verge of fainting. He hears strange sounds and sees strange signs. His soul is carried away by supernatural beings. At the age of seven he has to go through an ordeal, he has to fight with a stronger *táltos*. During the time of the struggle, he usually appears in the shape of a bull or of a stallion. In the struggle the teeth of the weaker *táltos* are pulled out by the stronger one. The fight is continued in the sky to which they rise on wings. A great storm arises at the time of the fight. According to some informants, the

* Unless some other source is given, the information comes from the author's field of notes.

[1] Róheim (1925) pp. 3–35; Diószegi (1958).

[2] Róheim (1925) p. 8. Valesca Klein (1935) p. 374, is mistaken in considering the *táltos* a kind of a witch. Otherwise she gives a good compilation of information about the subject. Her treatise has hitherto escaped the attention of Hungarian scholars.

táltos candidates are trained by initiated persons of the opposite sex. It is also known that during the initiation ceremonies, the candidate has to climb up a tree, trunk or ladder in order to peep 'into the sky'.

The Hungarian *táltos* used a drum while exercising his magic. By beating it he was able to discern malicious persons. With the drum the *táltos* could conjure up supernatural beings, cure the sick and foretell the future. Some informants state that while performing his magic a *táltos* climbs up a tree. As headgear he wears an antler, a horn of an ox and feathers of various birds. The *táltos* can become invisible, he can change into a stallion, a red, dun or black bull, and—less frequently—into a pig, a goat, an owl, or a pigeon. Behind his shape as an animal there lurks generally his 'life spirit' or 'assisting spirit'. His assisting spirit is usually a horse which was identified with the magic drum. The identification of the magic drum with the horse has survived in a Hungarian riddle.[3]

Considering all these characteristic traits, first G. Róheim and recently V. Diószegi have proved that the Hungarian *táltos* corresponds in fact to the shaman of the Ural-Altaic peoples. We know that the *táltos* belief has an Ugrian substratum, which later on absorbed Turkic elements. Where the Ugric stratum begins and where the Turkic influence ends is, of course, somewhat problematic. G. Róheim's standpoint certainly deserves attention. He is of the opinion that the Hungarian *táltos* has taken over a form of the Asian shamanism at a state which is closest to the shamanism of the Turko-Mongolian peoples and hence is a Turkic rather than an Ugrian shaman. There is no doubt that shamanism was characteristic of the Magyars of the Great Conquest and that the function of the shaman was performed by the *táltos*.[4]

Róheim's standpoint is strengthened by the fact that the word *táltos* is a Turkic loan-word in the Hungarian language (Turkic *taltyš* $\sim$ *taltuš* 'beater, flogger', 'beating into a stupor, a swoon'), dating back to a period preceding the settlement of the Magyars in the Carpathian Basin, and it originally referred to some kind of magicians whose functions and characteristic features of behaviour, clothing, etc., corresponded to those of the shamans of the Turkic peoples.[5]

Pastoral life, with its closed and binding patriarchal order, has played an important part in keeping alive beliefs and ideas which are connected with the *táltos* and in preserving the eastern traditions of the Hungarians in general.[6] The shepherds have best preserved, in the most diversified forms, the *táltos* belief rooted in shamanism.[7]

As to the origin of the *táltos* it is of decisive importance that his ancestors were animals. Róheim mentions data from the middle of the 19th century from Csallóköz (a large Danubian island between Pozsony [Bratislava] and Győr), according to which the *táltos* is born with teeth, but if the fact is revealed by the midwife or someone else, the child cannot be brought up but dies. Horses especially can give birth to a *táltos*. The *táltos* possesses great knowledge and can tell where treasures are hidden in the

[3] For details, see the works of Róheim, Diószegi and Klein.
[4] Róheim (1925) p. 25 and (1954) p. 28; Diószegi (1958) p. 436.
[5] Pais (1958) pp. 261–95.
[6] Gunda (1940) pp. 26–49.
[7] Gunda (1949) p. 7 and (1943) p. 222.

earth. He can assume the shape of various animals, but he lives chiefly among people in the form of a shepherd or a farm hand. Every seven years he must go to an other country to fight his antagonist, an other *táltos*. The country of the *táltos* who gains the upper hand, will be very fertile.[8]

In Békés county a *táltos* infant was carried by a bull between its horns. Inquiries were made as to who his parents were but they were not discovered. According to one of the shepherds he was 'made by the bull', and that is why he was so strong.[9] His name was Bálint Bika (Valentine the Bull). In a village in Bihar county, a wolf jumped upon a girl and 'covered her'. The girl gave birth to a son who possessed the qualities of a *táltos*. The child was strong, swarthy, had a wolf's teeth in his mouth and at the age of seven had hair on his chest and loins. His strength was great so that the village lads were no match for him.[10]

During the period between 1920 and 1930 a strange fellow called János Balog was wandering around the villages of Gádoros and Szentetornya (Békés county). He earned his living by doing all kinds of odd jobs on the farms: he cleaned wells, trod dung bricks for fuel, set ovens, gelded animals, killed old useless cats and dogs, destroyed rats, buried corpses, etc. It was said that he was a *táltos* and 'did not do the work himself', but made someone else do it, although nobody could see who that somebody was. He only drank milk and could tell from which cow the milk came. He had calf teeth in the back of his mouth. Sometimes he disappeared for months and on returning he said he had gone to fight with a black bull. During the fight, he could be seen only in the shape of a red flame in the sky. 'Balog is fighting now' the old people used to say. He was the step-brother of the bull.

The old farmers also knew that János Balog had been born from a cow on a wet meadow near the village of Kisszénás. He had neither male nor female relatives. According to tradition, he was found by the herdsmen. One day he disappeared for good during a heavy wind and a thunderstorm.

The old farmers of Nagyszénás had also heard about this man called Balog. He himself is known to have said that he was a *táltos* but that his strength was waning. The inhabitants of the village believed that his mother was a cow which he had sucked until the age of seven. She had died from sheer exhaustion. Balog knew how to deal with animals. He could subdue a raging bull, and savage dogs ran away from him in fear.

The more sensible farmers only smiled at the stories that were told about Balog. Several of them knew him, but considered him a vagabond. The explanation they gave for his vanishing for years was that the authorities had arrested and imprisoned him for vagrancy and theft. In 1942 I asked the police station of Nagyszénás whether they knew a vagabond named Balog. They had heard from their former chief that a man called Balog had been arrested for vagrancy and theft and two farmers had come to the police asking that he should be at least temporarily released because he could cure their animals. The district court passed no sentence on him, as his mind was found to be 'disturbed'.

[8] Róheim (1925) p. 14.
[9] Szűcs (1943) p. 84.
[10] Szűcs (1943). According to Chuvash popular belief, if a woman is made pregnant by a bear, the body of her child is covered with thick hair. See Mészáros (1909) p. 93.

In the village of Öcsöd (Békés county) there was a *táltos* called Lajos Jónás whose father was a stallion. That is the reason why he could fight in the shape of a horse. Every seven years he had to go to a fight. When he returned, he was very tired, exhausted and covered with froth like a dray-horse dragging its load through the mud. Before he fought with the white stallion he would sleep for months in order to gather strength. But in his sleep he could see and talk, and even answer questions. That is how a relative of my informant knew that her husband would return safe and sound from World War I. This particular *táltos*, too, lived on nothing but milk. When he dropped in anywhere, a jug of sour milk was immediately placed before him. According to tradition he became quite prosperous in his old age; he owned a farm and fields. He was an excellent farmer and an expert in animal husbandry, because — according to his neighbours — his father was a horse.

Other informants hold that Lajos Jónás was an excellent farmer because he always read newspapers to which he regularly subscribed. He was also a regular fair-goer, thus he gathered experience and knew so much about animals that a veterinary surgeon could have learned from him. That is how Jónás, originally a poor man, became a well-to-do farmer as he advanced in years.

Between the villages of Békésszentandrás and Öcsöd (Békés county), on a pasture land called Horga, there lived a one-eyed shepherd who gelded animals. He was said to have been called even to the big manors to cure sick animals and he could tell from which egg would come a pullet and from which one a cock. In autumn when the cry of the rails could be heard or the wild-geese were passing overhead in their flight, he kept on gazing at the sky. He said that he had been brought up by cranes in a marshy meadow between Gödény hill and the river Körös. His mother was a crane who had a small red feather in one of her wings. That is why he was called János Daru (Crane).

According to another source from Békésszentandrás, the *táltos* men were brought up by cranes and ospreys and had neither father nor mother. Those who fought with stallions had stallions as fathers.

In Békésdoboz (Békés county), a fieldwarden named Károly Varga was supposed to be a *táltos*. In the summer he lived outside on the banks of the Körös and kept watch over the fields of corn and melons. He guarded the hay and the growing crops so that the shepherds should not drive their sheep into them. The wandering Rumanian artisans who used to come down from the Bihar mountains to practice their woodworking crafts were very much afraid of him. If he looked at their horses, these would throw themselves on the ground and ruin all the wooden vessels they were carrying on their backs. In the winter he sought shelter in a hut where he used to sit by a fire. He never went into the village, and so the farmers used to bring him food there. If one of them shared his meal, he was the first to touch the food. ('Neighbour, please, start eating so that we also may eat', they used to tell him.) In the winter Károly Varga often fished through a hole in the ice. The people of the village were sure that he was a *táltos*. If he was angry with a farmer, it never failed to happen that in the following summer the farmer's crops were destroyed by hail, or else his finest horses died. Károly Varga had a bright button (according to others, a spoon) in which he could see buried treasures. He never took off his embroidered peasant's cloak (*szűr*) from his shoulders, not even when it was shining with blackness. Tradition says that he wore this cloak because he had crane's or eagle's feathers on his shoulders, which nobody was allowed to see. There was a rumour that both his father and mother had been cranes (or, according to another variant, only his father). In the autumn when the cranes were flying, this *táltos* would say: 'One day I will also go with you!' Mihály S. Szabó's father once saw him rise into the air and his cloak drop from his shoulders. He ran to pick it up, but did not find anything on the spot, where it had fallen.

'*Táltos* people are not of human origin'—declared Zsuzsánna V. Szatmári of Békésdoboz, who believed that only God was stronger than a *táltos*. She had heard about a fieldwarden named Varga, a *táltos*, who kept a tame crane in his cottage. He often played a zither, but the zither always sounded like a drum. He lived on nothing but milk. Neither birds nor the beasts were afraid of him.

On the edge of the so-called Káka puszta (at Szarvas, Békés county) there stood an old solitary oak-tree. Once the shepherds noticed a raven—already a rare bird in this region at the turn of the century—sitting on it. The younger ones did not recognize the bird, only the old shepherds knew that the bird was a raven. For days it had been sitting on one of the branches of the tree. When they wanted to drive it away, it said: 'Come with me, Mihály, my son!' Then it flew away. That very same night an old herdsman named Mihály disappeared. At dawn when the cattle were driven out, his stick and coat were found in the enclosure and even a few bright feathers were discovered in his coat. Nobody knew his other name. He was a kind of *táltos*. He was taciturn and had great strength and could see treasures in a bright spoon. Even when he was asleep he could watch over the grazing cattle. When he drew a circle around a herd, there was no thief or highwayman who could drive the cattle away. Not even a rabid dog could cross the circle.

The old treasure diggers knew about this herdsman named Mihály and eagerly took him along with them at night when they wanted to do some digging in a hill. In Békésszentandrás he once confided to János Bagi that when he was flying he could see treasures even from above. There was a vervain soldered on his palm. When he put his hand on a padlock, a lock or a shackle, they opened at once.

In the village of Öcsöd (Békés county) the belief is still maintained that the seventh child in a family will become a *táltos*. He is born with teeth, and at the age of ten or twelve he must go to a *táltos* of another country to learn his art, which is transmitted to him in his sleep. A *táltos* can raise a storm, he knows how to cure sicknesses and he can see what is hidden in the sky and under the earth. He can foretell a war, a storm and an earthquake. The father of a *táltos* is either a wolf or a stallion. He is often born of a young hot-blooded woman who is unable to keep her passion under control and goes as far as the tenth village for gratification. 'One of them was Queen Maria Theresia!'—said my informant.[11]

[11] The informant lived earlier in the farmsteads of Öcsöd, where he heard about this *táltos*.

In the village of Panyola (Szatmár county) I have recorded that the father of a *táltos* is thought to be a stallion, a wolf or a bull. Even as late as forty or fifty years ago, to the east and south of Kalocsa (Pest county) there were extensive pastures where shepherds grazed their flocks, which were not attacked by wolves, and even the most cunning thief could not steal a single animal. All the shepherds had to do was to place their sticks next to the flock and the sticks watched over the sheep. They were *táltos* men whose fathers were wolves or dogs.

Before 1944 near Martonvásár (Fejér county), there was a big farm called Kismarton. In the summer hired German labourers who came from the villages near Budapest, used to work there. Once in the spring time when they went to the farm to work, a sickly man of about forty-five or fifty turned up some days later and cut wood, carried water and did other menial works for their kitchen. They all were sure that they had a *táltos*, who was born from some animal with teeth in his mouth, and that he had the power of healing. Not the hired German workers so much as the servant women had a superstitious fear of 'Ferkó' (that was his name) and avoided him whenever possible. Once when the children were jeering at Ferkó, a servant woman, called Mrs. Árki, scolded them, saying: 'Don't annoy Ferkó, he is a *táltos*. He will torment you and maim you . . . and then your mother can take you to the medicine man of Ősi!'[12]

I knew this *táltos* myself. I spoke to him several times, but could find out nothing about his life or his past. Usually he just mumbled unintelligible words. He was mentally unbalanced and suffered from haemorrhoids. His dull deep-set eyes filled one with terror. It is not surprising that the servant women were afraid of him. They probably linked up the *táltos* tradition with him because of his uncanny demeanour.

Certain other informants, though not explicitly saying that a *táltos* is born from an animal, yet stress those characteristics from which one may infer their bestial origin. In Transdanubia, in the region of Göcsej, a boy who is born with hair on his back, is supposed to become a *táltos*.[13] In Sárrét (Békés county) there was a *táltos* who had hair some ten inches long on the left side of his chest, ever since he was a child[14]. In Bihar county (village of Konyár) hair which is growing on the chest, back, or loins of a child is considered evidence of being a *táltos*. A story (from the Jászság) says that János Csorna was born with foal's teeth and was therefore a *táltos*.[15] In Püspökladány (Hajdú county) a *táltos* is generally believed to be born with teeth. Some *táltos* men have foal's teeth or wolf's teeth on either side in the back part of their mouths. Sándor Csompora, who was reputed to be a *táltos*, once looked in through the window of an enemy of his at the farm of Gáspártelek (belongs to Nagyszénás, Békés county), and was seen to

[12] Ősi is a small village in Veszprém county where a man lived who cured gullible people by saying prayers. He was frequently consulted by maid servants. One of them was his patient but she also went to a Budapest clinic for treatment. For details of his activities see Vajkai (1938). When the above data were recorded, he had been dead for some time, yet his marvellous successes in curing diseases were still a constant topic among the woman servants.

[13] Gönczi (1914) p. 175; cf. Diószegi (1958) p. 342.

[14] Szűcs (1936) p. 42.

[15] Kálmány (1914) p. 168.

have ox-horns on his forehead. He had to fight with a black bull.[16] This Sándor Csompora once fell asleep in a field and when he was found there, he had ox-horns on his forehead and nobody dared wake him. Later when he was told about this incident, he did not deny it and said only: 'It was lucky you did not wake me up for I would have trampled you to death!' He spoke a lot in his dreams and often mentioned a black bull. His fellow labourers said that on such occasions he was fighting with a bull.

In the Hungarian folklore literature there are only hints at the fact that a child born with teeth becomes a *táltos*.[17] The records do not describe the teeth. Evidently, the peasants who told the stories no longer remembered which animal's teeth were in the mouth of the *táltos* in question. But in earlier times the fact was probably mentioned since it helped them to infer his origin. For as we have seen above, a *táltos* who was born from a wolf had teeth of a wolf in his mouth.[18] But the field workers who recorded the cases did not ask what the teeth of the *táltos*es were like.

I am convinced that a further, thorough collection would reveal still many interesting facts about *táltos*es, the descendants of animals. However, it is evident from the part of the material which is surveyed in the present study that horses, bulls, wolves, cows, cranes, eagles or ravens figure among the ancestors, the parents or foster parents of *táltos*es. Elsewhere our material shows that their animal ancestors can only be inferred from physical traits with which they are born (they have the teeth of foals, calves or wolves; they have long hair on their chest or loins, they are seen with horns). We are also told that it is not advisable to marry a girl from a district which has been frequented by a *táltos* in bull form, because she will never desire her husband.[19] It is evident, though we have no direct references, that only a *táltos* child can be born from the love of a bull and a girl.

These ideas concerning the origin of the Hungarian *táltos* agree with the totemistic manifestations of the Turkic and Mongolian shamans.

In many cases we know that shamans also descend from animals. Among the Buryats, for example, it was a widespread belief that they originated from eagles.[20] Among the Yakuts the strongest shamans professed to have been sent by eagles.[21] The eagle, moreover, is a characteristic totem animal among the Yakuts. The clan which regards this bird as its totem is called *tojon törütäch*, 'descendant of the eagle'. If, after having prayed to the eagle for a child, a sterile woman bears a child, it is considered to have been conceived by the bird.[22] The Yakuts regard the eagle also as the ancestor and creator of the shamans. L. Šternberg writes: 'They believe that the eagle devours the soul (*gür*) of the child which has been chosen as a shaman, whereupon he (the eagle) flies *with sun and moon* into the field and, there, on a sacred birch planted for this very purpose, he hatches an egg which

[16] Data from Nagyszénás–Kiscsákó, Békés county.
[17] Diószegi (1958) pp. 123–5.
[18] Szűcs (1943) p. 84.
[19] Szűcs (1936) p. 43.
[20] Nioradze (1925) p. 2; Sandschejew (1928) p. 976; Harva (1938) pp. 465–6; cf. also Róheim (1917) p. 84.
[21] Data from Russian sources are quoted by Róheim (1917) p. 84; Nioradze (1925) p. 2.
[22] Šternberg (1930) p. 132.

he then breaks. The child which was concealed in it, is placed in an iron cradle at the foot of the tree. It is guarded and reared by the so-called "mother bird" and raised as an *Ajȳ Ojūna*, i.e. a shaman.'[23] The following information given by L. Šternberg is worthy of attention: 'Among the Kachs the eagle appears as the ancestor of the shamans, at the same time also as a totem animal and as the eponym and guardian of a certain clan.'[24]

It is remarkable that in the language of the Nivkhi (Gilyaks) the eagle has the same name as the shaman (*čam*), which indicates that a close relationship exists between shamanism and the eagle.[25]

From our point of view the observations made by L. P. Potapov among the Altaic Turks are particularly valuable. Various animals—the stag, the wild goat, the deer, but primarily the reindeer—were the totemic ancestors of certain tribes. Later, with the development and change of social ideas and organizations, these animals were regarded only as the ancestors of the shamans. Still later the reindeer became the chief spirit of the shamans. The shaman drum represents the totemic ancestor, which is at the same time the assisting animal of the shaman. The leather strings and bosses on the drum symbolize the horns, udders, or ears of the animal. Among the Abakan Turks (Khakas), the shaman drum has six dangling leather strings on the upper part. These are regarded as the animal's six nipples by which she feeds the shaman. The Selkups consider the drum a living animal—a stag. The name of the shaman drum or some part of it (*bur, bor*) also means a species of deer. Among the Altai Turks, the genealogical relationship with animals was remembered in regard to the shamans, whereas, the clan as a whole and its individual members had already forgotten the idea of its descent from an animal ancestor.[26]

It is also customary among other Turkic peoples to trace the origin of shamans to some animal. However there is no need here to go into further detail in this extremely intricate question. I have pointed out a few typical characteristics of Yakut and Turko-Altaic shamanism. In another connection V. Diószegi enumerates further important facts about the mother-animal of the shamans (which among the Yakuts can be an elk, a reindeer, a bear, a giant bull, a stallion, etc.).[27] A deeper analysis of this problem would lead us to animal ancestry, that is to say still farther afield, to totemism.

From what has been said, it is evident that the traditions which refer to the origin and the ancestry of Hungarian *táltos*es, undoubtedly preserve the memory of totemism. Their ancestors are always animals or birds and they often originate from the intercourse of a woman (generally a virgin) with a wolf or a bull. The story from Bihar county about a *táltos*, who was born from a wolf and a girl, resembles the Turkic tradition according to which this people traced its origin to a wolf. Kazakh and Bashkir genealogical legends consider the dog as their ancestor.[28] It is likely, that in the past,

[23] Šternberg (1930) p. 133. A good summary of the origin of the shamans is: Eliade (1951) pp. 76–9.

[24] Šternberg (1930) p. 144.

[25] Šternberg (1930) p. 144.

[26] Потапов (1935) pp. 134–52.

[27] Diószegi (1958) pp. 387 ff.

[28] Németh (1930) pp. 69–70 gives a detailed account of the problem of totemism, stressing totemistic names.

totemistic traits were more widespread in Hungarian popular belief than is generally assumed.[29]

In my opinion the reminiscences of totemism in the Hungarian people have their parallels in the totemism among the Altaic Turks. Among the latter it is the shaman, among the former it is the corresponding *táltos*, who preserves the memory of totemism and of the genealogical relationship with animal ancestors. Originally, larger social groups (families, clans, tribes) or their leaders kept alive the evidence of their descent from animals. The fact that the Magyars of the period of Conquest considered their sacred chieftain as the son of their totemic animal (the *turul*: a falcon-, eagle- or hawk-like bird) and that when the strength of their chieftain was waning, he was solemnly sacrificed, also supports this theory.[30] The descent of a social group or community from an animal was expressed by their belief regarding their chief. Evidence of the totemistic descent of the Hungarian clans, or possibly of other social units, may be found in the Hungarian proper names and clan names of the 12th and 13th centuries, which are of totemistic origin. (*Ákos* 'white bird, white falcon', *Karcsa* 'hawk, falcon', *Kartal* 'eagle, black eagle', *Tiván* 'falcon', *Bese* 'sparrow hawk', *Turul* 'falcon-, eagle- or hawk-like bird', *Kaplony* 'tiger, panther', *Thonus-Oba* 'pig-father', *Kurd* 'wolf', *Tege* 'ram', etc.)[31] Etymologically these are of Turkic origin and go back to a period preceding the settlement of the Magyars in the Carpathian Basin.

In the course of social development the relationship of the tribe, the clan and of their individual members with the totem ancestor has faded and only the *táltos*—like the shaman—has preserved the memory of the descent from an animal ancestor. The shaman, as well as the *táltos*, is thus an embodiment of older traditions.

Totemism raises also economic, social and psychological problems. It has survived in different peoples in various forms, depending on their stage of social advancement. I. Haeckel distinguishes between individual, tribe, clan (or group) and sex totemism.[32]

Individual totemism 'as an integral part of shamanism' can be found in Northern Asia, North America and Indonesia. Each shaman has a certain number of guardian and assisting spirits (Schutz- und Hilfsgeister). These, frequently animals, are also the individual totems of the shaman.[33] In the assisting spirit (Hilfsgeist) which appears in an animal form—as well as in the *táltos*—we recognize totem ancestors. The traditions about the origin of the Hungarian *táltos*, like the idea of the shaman's animal descent, actually point to a form of individual totemism. But it is probable that

[29] For a detailed account, see Róheim (1954) pp. 4 ff, 48 ff. For totemistic traces in folk tales and curses, see Gunda (1958) pp. 63–72.—The following works refer to the individual problems of Hungarian totemism: Molnár (1954) p. 67; Györffy (1958) pp. 12–87, 565–615; Pais (1959) pp. 76–80; Gyallay (1959). For the Ugric aspects of the question, see Steinitz (1938) and Haeckel (1946).

[30] Róheim (1917) pp. 58–99.

[31] Gombocz (1914); Pais (1922) pp. 31–3; (1949) pp. 37–43.

[32] Haeckel (1939) pp. 243–60.

[33] Haeckel (1939) p. 244. Falsirol (1941) p. 69 also points out the totemistic connection between the shamans and certain animals.

here this individual totemism is the result of regressive development. Just as the belief of animal descent was transferred from the tribe to the shaman, so also tribe, clan and family totemism changed to individual totemism.

REFERENCES

Diószegi, V. (1958) *A sámánhit emlékei a magyar népi műveltségben* (Survivals of Shamanism in Hungarian Popular Culture), Budapest.

Eliade, M. (1951) *Le chamanisme et les techniques archaiques de l'extase*, Paris.

Falsirol, O. (1941) *Il totemismo e l'animalismo dell'anima*, Naples.

Gombocz, Z. (1914) Árpád-kori török személyneveink (Turkish Personal Names from the Time of the Árpáds), in *MNy*, Vol. X, pp. 58–99, 241–9, 293–301, 337–42.

Gönczi, F. (1914) *Göcsej s kapcsolatosan Hetés vidékének és népének összevontabb ismertetése* (Concise Survey of the Region and People of Göcsej and Adjacent Hetés), Kaposvár.

Gunda, B. (1940) The Anthropogeography of Pasturing on the Great Hungarian Plain. Geographical Review, in *IRHGS*, Vol. LXVIII.

(1943) A néprajz és a magyar őstörténet (Ethnography and Hungarian Prehistory), in *A magyarság őstörténete* (ed. by L. Ligeti), Budapest.

(1949) A magyar népi műveltség keleteurópai helyzete (The Position of Hungarian Popular Culture in Eastern Europe), in *NT*, pp. 3–19.

(1958) A totemizmus maradványai a magyar táltoshagyományban (Traces of Totemism in the Hungarian *táltos* Tradition), in *ADDME* 1957, pp. 63–72.

Gyallay, D (1959) Lófi székely—lófő székely (Son-of-a-horse—head-of-a-horse *székely*), in *MNy*, Vol. IV, pp. 218–26.

Györffy, Gy. (1958) A nemzetségtől a vármegyéig, a törzstől az országig (From Clan to County, from Tribe to Country), in *Sz*, Vol. 92, pp. 12–87, 565–615.

Haeckel, J. (1939) Über Wesen und Ursprung des Totemismus, in *MAG*, Vol. LXIX, pp. 243–60.

(1946) Idolkult und Dualsystem bei den Ugriern. Zum Problem des Eurasiatischen Totemismus, in *AfV*, Vol. I, pp. 95–163.

Harva, U. (1938) Die religiösen Vorstellungen der altaischen Völker, in *FFC*, Vol. 125.

Kálmány, L. (1914) Ipolyi Arnold népmesegyűjteménye (Arnold Ipolyi's Collection of Folk Tales), in *MNGy*, Vol. XIII.

Klein, V. (1935) Der ungarische Hexenglaube, in *ZfE*, Vol. LXVI.

Mészáros, Gy. (1909) *A csuvas ősvallás emlékei* (Survivals of the Primitive Religion of the Chuvashes), Budapest.

Molnár, E. (1954) *A magyar nép őstörténete* (The Prehistory of the Hungarian People), Budapest, 2nd ed.

Németh, Gy. (1930) *A honfoglaló magyarság kialakulása* (The Ethnogeny of the Magyar Tribes of the Great Conquest), Budapest.

Nioradze, G. (1925) *Der Schamanismus bei den sibirischen Völkern*, Stuttgart.

Pais, D. (1922) Régi személyneveink jelentéstana (Semantics of the Old Hungarian Personal Names), in *MNy*, Vol. XVIII, pp. 26–34, 93–100.

(1949) Turul. Az Árpád-nemzetség totem-madara (Turul. The Totem Bird of the Árpád Clan), in *MNy*, Vol. XLV, pp. 37–43.

(1958) A táltos meg az orvos (The *táltos* and the Physician), in *MTANyIOk*, Vol. XII, pp. 261–95.

(1959) Totem-halmozás Álmos vezér származása körül (Accumulation of Totems around the Origin of Chieftain Álmos), in *MNy*, Vol. IV, pp. 76–80.

(Potapov, L. P.) Потапов Л. П. (1935) Следы тотемистических представлении у алтайцев (Traces of Totemistic Notions in the Belief of the Altaians), in *СЭ*, Nos 4–5, pp. 134–52.

Róheim, G. (1917) A kazár nagyfejedelem és a Turulmonda (The Great Kazar Chief and the Turul Legend), in *Ethn*, Vol. XXVIII, pp. 58–99.

(1925) *Magyar néphit és népszokások* (Hungarian Popular Beliefs and Customs), Budapest.

(1954) Hungarian and Vogul Mythology, in *MAES*, Vol. XXIII.

Sandschejew, G. (1928) Weltanschauung und Schamanismus der Alaren-Burjaten, in *A*, Vol. XXIII, pp. 538–60, 967–86.

Steinitz, W. (1938) Totemismus bei den Ostjaken in Sibirien, in *E*, Vol. III, pp. 125–40.
Šternberg, L. (1930) Der Adlerkult bei den Völkern Sibiriens, in *AfR*, Vol. XXVIII.
Szűcs, S. (1936) Táltosok és boszorkányok a Nagysárréten (*Táltos*es and Witches on the Nagysárrét), in *Ethn*, Vol. XLVII, pp. 39–48.
(1943) Adatok az alföldi táltosok ismeretéhez (Contributions to the Knowledge of the *táltos*es of the Great Hungarian Plain), in *Ethn*, Vol. LIV, pp. 84–5.
Vajkai, A. (1938) Az Ősi-i javasember (The Wiseman of Ősi), in *Ethn*, Vol. XLIX, pp. 346–73.

The Hungarian Shaman's Technique of Trance Induction

by

J. Balázs, Budapest

1. Many variants of the Hungarian verb *rejt* 'to hide, to conceal' can be found in Hungarian documents: Jókai C. 4. *elreyt-*, Vienna C. 6 *elreit-*, Munich C. 39 *reit-*, Virginia C.42 *reyt-*, Nádor C. 427 *reyt-*, Jordánszky C. 396 *el reyt-*, Born.: Ének. 493 *rőyt-*, Ver. *reüt, rüt*, Cal. *elritχ*-Zvon: Post. I, 20 *rőyt.*[1] There is, however, in Old Hungarian another verb which is strikingly similar: *rüt, rőt, röjt, réüt, rit, riüt* 'to enchant' and also the verb *el-rüttetik* 'to become enchanted'; Jókai C. 77 *el ryẘtetyk*, Nagyszombat C. 381 *el růtetheec*, Virginia C. 86 *rewitetik*, Jordánszky C. 781 *el-rewtettem*, Példk. 73 *elrytetet*, Debrecen C. 164 *el ritetik*, 207 *el rüttetik*, Érdy C. 502b *el reeẘteteek*, Teleki C. 63 *elrůtetneiec*, Born: Préd. 244 *rőtete.*[2]

Opinions still differ as to whether these two verbs are identical or resemble each other accidentally. Budenz[3] for one treats them separately saying that 'this *rüjt* or *rőjt* originally coincided with the verb *rejt* . . .'. Kálmán Szily expresses a similar opinion.[4] He says that the verb *rejt* which is identical in form with, but completely different in origin and meaning from, the verb *rejt* as used today, must be assumed to have existed in ancient times. Ödön Beke,[5] one of the linguists who have recently studied these verbs, has also maintained that they do not belong together.

On the other hand, Béla Vikár[6] thinks that old Hung. *réit* (with its variants *rit, rüt, rőt, riüt, röjt*) is an original form of the existing verb *rejt* 'to hide, to conceal' and that the two are really one and the same verb. This opinion is shared by Ernő R. Baboss[7] who says that the verbs *rejt* 'to entrance' and *rejt* 'to conceal' are probably identical. According to him, their connection could be better understood by assuming, with Vikár, an intermediate meaning, like 'to enchant, to practise magic'. According to Géza Bárczi's *Hungarian Etymological Dictionary*[8] the verb *rejt* is of uncertain origin. It is probably related to the verb *révül*, and 'its original meaning is "to entrance" → "to charm, to enchant", hence: "to make something disappear by enchantment" → "to hide": The meanings "to entrance", "to enchant", etc. are verified by ancient records'. Dezső Pais[9] who has recently subjected this question to the most thorough examination, and who has assumed a similar semantic evolution, considers the two verbs as belonging together.

[1] Szarvas–Simonyi (1890–1893) Vol. II, pp. 1389 ff.
[2] Szarvas–Simonyi (1890–1893) Vol. II, pp. 1389 ff.
[3] Budenz (1873–1881) p. 660.
[4] Szily (1921) pp. 45–6.
[5] Beke (1950) pp. 417–8.
[6] Vikár (1906) pp. 253–4.
[7] Baboss (1938) pp. 223–4.
[8] Bárczi (1941) p. 254.
[9] Pais (1949) pp. 6–7.

Which of these two opinions is correct? Are the two verbs entirely different, or do they have a common origin? As will be seen later, this question is not only of linguistic but also of ethnographic importance.

2. Pál Hunfalvy related the stem of the verb which means 'to enchant' to the Mansi (Vogul) substantive *réj* 'warmth'[10] According to him, a Mansi magician beats the magic drum until he falls into a trance; this is shown by the trembling of his body. Then it is said that *Târom réj täuen johts* 'the *réj* of Tarom (god) has descended on him', or *Târomnel réjl johtáues, pinves* 'God has afflicted, burdened him with *réj*', or *Târom jäni réj täuen johts, láting täuen ti pini* 'God's great *réj* has descended on him and has impelled him to speak . . .' From this we learn that the entrancement of the magician when divining or announcing visions is called *réj* by the Voguls; and we notice with astonishment that this word *réj* greatly resembles the Hungarian word *rév, réj*, which was found to be the stem of the derivatives referred to. The fact that the Hungarian and the Vogul languages use one and the same term to designate the same condition seems to prove that this word once belonged to the pagan Hungarian religion and that shamanism similar to what is still practised among the Voguls may have been practised among the ancient Hungarians as well. Adjectives which are used with the word *réj* in Vogul are: *jäni* 'great' or *pubi* 'marvellous, extraordinary'. The shamanistic trance is called 'great *réj*' or 'marvellous *réj*'.[11]

Hunfalvy points out that the Mansi *réj* is actually a verbal noun. József Budenz[12] agrees with him and quotes the following words from the cognate languages: Mansi Berezovo dial. *réj* (*réi*), *re* 'warmth', *reiñ* 'hot'; *rejtil* 'to become warm' (Russian согреться), *reil-* 'to get warm' (Hungarian melegedni), *reilt-* 'to heat, to make warm' (Russian согревать), Lozva dial. *rēn*, 'warm', *re* 'to burn (of the sun)', *reitil* 'to heat oneself', Konda dial. *rig* 'heat, sultriness' (Russian зной), Khanty (Ostyak) Berezovo dial. *rúu* 'heat, warmth, sultriness, fume'; *ruuñ* 'hot, sultry', Komi *ru* 'fume, vapour' (Perm dial. *šoni̊t ru* 'warm fume'), Udmurt *ǯual-* 'to burn', *ǯuat-* 'to set on fire'. Zoltán Gombocz[13] and Ernő R. Baboss[14] agree with this comparison. Géza Bárczi[15] admits the probability of a Finno-Ugric origin of this Hungarian verb, and of the aforesaid correspondence. Dezső Pais[16] thinks that the Mansi words which have been enumerated above 'can be accepted as cognates of the stem of the Hungarian word-family'. The vocabulary compiled by Móric Szilasi for Munkácsi's *Collection of Vogul Folk Poetry* contains, among others, the following words which may serve to complete the Finno-Ugrian data which we have above: Berezovo dial. *rēi̊'* ∼ *rei̊'* ∼ Middle-Lozva dial. *rī̊* ∼ Lower-Lozva dial. *rī* 'heat', Northern dial. *rēγiñ* ∼ Middle- and Lower-Lozva dial. *rēñ* 'warm'. According to Kannisto[17] the various Mansi dialects use the following words for 'warm, hot, heat': Tavda-Yanichkova dial. *rεi*, Lower-, Middle- and Upper-Konda dial. *rəγ*, Pelym

[10] Hunfalvy (1866) pp. 247–52.
[11] Hunfalvy (1866) pp. 249–50.
[12] Budenz (1873–1881) pp. 658–61.
[13] Gombocz (1913) p. 190.
[14] Baboss (1938) pp. 223–4.
[15] Bárczi (1941) p. 254.
[16] Pais (1949) p. 6.
[17] Kannisto (1919) p. 45.

dial. *riγ* N. and S. Vagilsk, Lower-Lozva dial. *rī*, Upper-Lozva dial. *rēγ*, Sosva dial. *rēγ*. He says that these words can be made to agree with the Finnish *räkki* 'heat' and the Hungarian *rejt* 'to entrance'.

Taking all this into consideration, Dezső Pais assumes a stem of Finno-Ugric origin—**rek*—from which he excludes the Finnish *räkki*.[18] In his opinion the Hungarian verb *rejt* is a derivative of **rek-*, formed by the factitive suffix *-t*. The same explanation is given by Mészöly[19] and Baboss[20]. Records like Vienna C. 114 *elrehec* (= elrejtsék), Döbrentei C. 55 *rehed* (= rejtsed), reflect the intermediary stages of the development *rëkt* > *rëχt* > *rëjt*. Ödön Beke is of a slightly different opinion.[21]

As we have seen, the above-mentioned scholars agree in saying that the verb *rejt* 'to entrance' is a derivative of a Finno-Ugrian word and can be collated with the listed Mansi words. Research and the frequently quoted study by Dezső Pais have so far shown that these verbs and nouns are, in one way or another, related to the shamanism of the Ugric period. Another problem, so far without any satisfactory answer, is the etymology of the Hungarian verbs which were mentioned at the beginning of this paper. Do both of them lead us back to the world of shamanism, or only one? Are they of completely different origin, or are they related to each other, or perhaps even identical?

3. A solution to the problem can be approached by examining the shamans' method of going into trance in the critical light of the most recent linguistic, ethnographic and archaeological results, and by going back to ancient records. The semantic interpretation of the Hungarian verb *rejt* 'to entrance' is supported to this day mainly by Hunfalvy's explanations and by Munkácsi's comments on the travelogue of the Russian Academician P. S. Pallas[22].

Concerning Mansi shamans and 'northern' shamans in general, Munkácsi writes: 'The visible result of the trance-inducing ceremony of the northern shaman is that he himself falls into a state of swoon-like unconsciousness, i.e. trance. Under the effect of fly-agaric or vodka which he consumed before the performance, he feels already fuddled at the beginning of the ceremony. The degree of his drunkenness and excitement is hightened by stamping, loud and excited singing, the strain which is caused by howling, the noise of beaten kettles and pots, and by the auto-suggestion that he is in contact or even struggles with the spirit.'[23] Here we learn that the shaman eats fly-agaric or becomes intoxicated with brandy and in this way he falls into a trance. When dealing with the shamanism of the Ugric peoples, K. F. Karjalainen says that the shamans of the Irtysh Khanty frequently use the fly-agaric as a narcotic.[24] Many travellers describe how the Siberian shamans eat fly-agaric and as a result fall into a trance.

[18] Pais (1949) pp. 6–7.
[19] Mészöly (1910) p. 442; (1928) p. 309.
[20] Baboss (1938) pp. 223–4.
[21] Beke (1950) pp. 417–8.
[22] Pallas (1773–1776).
[23] Munkácsi (1892–1921) Vol. II, p. 0380.
[24] Karjalainen (1921–1927) Vol. III, pp. 306–8.

It is to Åke Ohlmarks that we owe the most systematic compilation of such records.[25] He states that the fly-agaric (*Amanita muscaria* L. Pers.) is a most current drug among the shamans of the Koryaks, Itelmens (Kamchadals), Nenets (Yurak Samoyeds) Selkups (Ostyak Samoyeds) and the Khanty who are living near the Yenisei and Irtysh rivers. In a special chapter Ohlmarks deals with alcohol which is used as a drug by the shamans of many Siberian tribes.[26] He states that the shamans of the Irtysh Khanty, the Tungus, Saams (Lapps), Buryats and other peoples very often lose consciousness after drinking some intoxicating liquor. However, alcohol seems to be unable to compete with the fly-agaric as a trance-inducing drug. Words which mean fly-agaric constitute a widespread family in the Ob Ugrian languages and also in several other Finno-Ugrian languages. The corresponding derivatives are very informative from the point of view of semantics, since they point to the way in which the Finno-Ugrian shamans fell into trance—a way that must have been very ancient and very widespread.

Here is a list of the etymologically related names for the fly-agaric in different Finno-Ugrian languages:

Mansi *pańχ*, *pińka* 'fly-agaric', Northern dial. *pāŋχ*, Middle Lozva dial. *pθnk* 'fly-amanite', *amanita muscaria*, Khanty Northern dial. *poŋχ* 'fungus, fly-amanite', Irtysh dial. *pāŋχ*, *paŋχ*, Yugan dial. *paŋga* 'fly-amanite', *pā̰ŋχ*, Irtysh dial. *pŏnk* (*pṵ̄ŋ̰k̰ə̂m*) 'fly-agaric', Lower Demyanka, Fil dial. *paŋ̰χ* Verkhne-Kalymsk, Vartokovskoe dial. *pa̰ˋnŋˁk̰* Tremyugan dial. *pˁåŋ̰k̰*, Nizyam-Berezovo dial. *poŋ̰χ̰*, Kazym dial. *poŋˁk* 'fly-agaric', Mordvin *paŋga*, *paŋgo*, Mari *pongθ*, *pongo*, *poŋγθ* 'fungus'.[27]

The importance of this poisonous fungus in the ceremony of the Mansi shaman is demonstrated by the fact that in the Northern Mansi dialect (Szily 1921) *pāñχ* means 'drunkenness, delirium, intoxication'.[28]

The following words are derivatives which are used in Mansi to express drunkenness, trance: *pāŋχli pāŋχėtāli*, Lozva, Pelym dial. *pθŋkˁli* 'he is drunk, intoxicated'. Also in Khanty: Northern dial. *pà̰ŋ̰χtəm*, Irtysh dial. *pṵ̀ŋ̰k̰ʟ̂əm* 'to practise magic singing, after eating fly-amanite, to cure by magic' Upper Demyanka dial. *pàŋ̰γə̂t̰t̰à* 'to scream and bluster after eating fly-amanite', Tremyugan dial. *pˁåŋ̰k̰ə̂ʌ̂t̰āγз* 'to become intoxicated by fly-amanite, begin to sing under its influence'. The song of the shaman who has fallen into a trance caused by fly-agaric is designated in Khanty by the following words: Tremyugan dial. *pˁåŋ̰k̰ ā̰ɼ̰əχ*, *pˁåŋ̰k̰ə̂ʌ̂să̰uγзˁ* 'fly-agaric song, a song which a magician sings having eaten fly-agaric', Vasyugan dial. *paŋ̰ˁ lə̂mŋ̰ə̂ŋˁ* 'song which is sung after fly-agaric has been eaten'.[29]

According to T. E. Uotila[30] this family of words includes also the Komi (Wichm.) Sysola dial. *pagalni̮* (lose consciousness [e.g. by drinking alcohol]) (Wied.) *pagalny* (*pagavny*) 'lose consciousness, become disconcerted', (Sakhov) *pagavny* (Russian отравиться, убиться 'to be poisoned', 'to be killed') (Udorka-Vashka dial.), 'шататься' 'to stagger' (Sysola dial.). According

[25] Ohlmarks (1939) pp. 101–8.
[26] Ohlmarks (1939) pp. 109–13.
[27] All these data from Munkácsi (1907) pp. 343–4.
[28] Munkácsi (1892–1921) Vol. I, pp. 115–7.
[29] Munkácsi (1907) p. 343.
[30] Uotila (1930) p. 177.

to Uotila, the primary nominal of root these words is **pag*, which can be collated with the above names for the fly-agaric. He believes that this word existed also in Komi, a proof of which can be found in the following Komi adjectives: (Wichm.) Luza dial. *pagi̮r* 'sour, acrid taste of beer', (Wied.) *pagyr* 'sour, sharp, acute', (Wichm.) *pagi̮ra* 'stale beer', *pagi̮ra* 'of pungent taste'.

Uotila assumes that these adjectives can be derived from the noun **pag* 'fly-agaric' with the addition of the suffix *-r*, and that their original meaning is, then, 'tasting of fly-agaric'. This would prove that the Komi language formerly had a special term for the fly-agaric. Steinitz, however, doubts whether these Komi words are in their right place.[31] They may have received their meaning of 'to fall into trance, to faint' through the same semantic development as the Mansi and Khanty verbs of similar meaning which we have mentioned above.

According to Toivo Lehtisalo[32], the following words also belong to this category: Nganasan (Castr.) *fanká'am* 'to be drunk', moreover Selkup (Castr.), Kamas, Cheya, Cha, Upper-Ob dial. *pöŋer* 'drum',[33] Tym dial. *pə̑ŋgə̑r̥*, Narym, Lower-Vasyugan, Middle-Ket dial. *pə̑ŋgə̑r̥*, Upper-Ket dial. *pī̥ŋgā̄r̥*, Chaya dial. *pə̑ggə̑r̥* 'a special Samoyed music instrument'.

Lehtisalo[34] collates the Nenets verb *ṣ̌āmpā* 'to sing the shadow of the dead into the underworld by banging the magic drum', etc. with the Finnish noun *sampa: maansampa* 'lycoperdon', *puunsampa* 'a whitish fungus which is of the size of a pin's end on birches, alders and willows'. Now, if the root of this Samoyed verb is a noun meaning fungus, their semantic relation is the same as that of the following nouns and verbs: Khanty *pàŋ̣χ* ∼ *paŋ̀χ̬təm*, Nenets, Obdonsk dial. *jā̆ʙ'ɾē* 'to be drunk' Nenets (Reg.) *javebš* Lyamin dial. *βīp̆pī*, Sakhalinskaya dial. *β'īppī*, Nyalina dial. *β'īppī* Pur dial. *βīpī* 'fly-agaric' ∼ Enets, Khantaika dial. *jébi'ēr̄o*, Bayikha dial. *jebi'edo'* 'to be drunk'.

Relying upon this evidence for his argument, Lehtisalo assumes that the Finnish shamans also formerly used fly-agaric as a narcotic and falling into a daze under its influence may have been current already in the ancient Uralic period.[35]

In all these cases, the Finno-Ugric words meaning 'trance, daze, drunkenness', go back to a noun meaning 'fungus, fly-agaric', which, by metonymy, meant the dazed, entranced or drunken state itself (as in Mansi), whereas the verbal derivatives of the same noun stand in many Finno-Ugrian languages for 'falling into trance'. From the semantic point of view this requires no further explanation, since it is self-evident that the basic word of derivatives meaning 'to get into a state' is the word which stands for the substance inducing that state. There is, naturally, a cause-and-effect link between such basic words and their derivatives. The verbs which are formed from these nouns mean 'to provide somebody with something' or 'to be provided with something'. Hence it is not only from descriptions by travellers, but also from certain linguistic facts (derivatives of

[31] Steinitz (1944) p. 37.
[32] Lehtisalo (1928) p. 122; (1929) p. 131.
[33] Donner (1920) p. 111.
[34] Lehtisalo (1929) pp. 130–2.
[35] Lehtisalo (1928) p. 122.

Finno-Ugric words meaning 'fungus, fly-agaric') that we learn that the Finno-Ugric shamans used the fly-agaric as a narcotic from early times.

4. We may ask, however, whether in the Finno-Ugric languages all of the verbs meaning 'to fall into trance' go back to a noun meaning 'fungus, fly-agaric', and whether these Finno-Ugric words are loan-words or not.

Bernát Munkácsi considers the afore-mentioned words to be of Old Iranian origin.[36] He bases his opinion on the astonishing similarity between the semantic development of the Old Iranian word *baŋha-* and that of the Mansi word for 'fly-agaric'. According to Bartholomae this Old Iranian word has several meanings[37]: (1) The name of a plant (and its juice) which was also used for abortions; (2) The name of a drug made of that plant, at the same time the name of the narcotized state reached through eating the plant. Other occurrences of this word are: Old Ind. *bhaṅga-*, *bhaṅgā-* 'hemp; drug prepared of hemp-seeds', New Pers. *bag* 'henbane' (*bangī* 'senseless'), Armenian *bang* 'Hyosciamus', Afg. *bang* 'hemp'.[38] According to W. Geiger[39], in Persia hashish is made of hemp.

Whether Munkácsi's derivation is correct or not, can only be decided by scholars of Iranian. As far as semantic development is concerned, it is somewhat difficult to understand why the Finno-Ugrian peoples should call the fly-agaric by a word which originally meant 'hemp', since apart from their narcotic effect, these two plants have nothing in common. (Lehtisalo considers this adaptation of words improbable.[40]) Nevertheless, it should be noted that hemp-seeds may have played a part in the falling into trance of the shamans, as it still may do today. We will return to this point later.

Our task now will be to investigate the origin of the mysterious 'heat' or 'warmth', implied in the verbs *rejt: réül,* and in their Finno-Ugric equivalents.

Munkácsi writes: 'The falling into trance of the shaman is caused by the heat of the wind which arises from the sudden appearance of the spirit; it is by this hot wind that the spirit makes his presence known and it is this wind that renders his invisible being recognizable to mortals even if only in an unconscious state.'[41] In this work, and also in his comments to Volumes I and II of the *Collection of Vogul Folk Poetry,* Munkácsi quotes particular Mansi expressions which are related to the ceremony of the Mansi shamans. 'Terms for the shaman's falling into trance: *tḁ̄rėm—rē'i̊nė, pupi̊'-rē'i̊ni joχtawė* "he is seized by the heat of god, the heat of the idol" or *pupi̊ jäni̊, tḁ̄rėm jäni̊ rēγä tawän joχts* "the great heat of the idol, the great heat of god came upon him" or *pupi̊' rē'i̊l ti pinėstä* or *tḁ̄rėmnėl rē'i̊l pinwės* "the idol or god has sent heat on him". . . This notion can readily be understood if we bear in mind that when the shaman—according to his own belief—causes the divinity to appear, he is quite intoxicated and may, indeed, feel a strong sensation of heat in his head and body . . .' The expression *tḁrom rei täuän johts* was explained by Reguly's Mansi informant

[36] Munkácsi (1907) pp. 343–4.
[37] Bartholomae (1905) p. 925.
[38] Horn (1893) p. 53; Uhlenbeck (1898–1899) p. 194.
[39] Geiger (1882) p. 152.
[40] Lehtisalo (1928) p. 122.
[41] Munkácsi (1900) p. 168.

through the words: *lating täuän ti pini*. He translated them periphrastically as follows: God has sent heat on him (on the shaman), and has *placed his own word* into him.[42]

This shows that the shaman, when he falls into a trance, feels some kind of heat in his head and body. But what is the cause of this heat? According to the above explanation, the hot wind which touches the shaman in his trance is not really heat, but, as Munkácsi puts it, exists only in the imagination of the shaman who is in a daze, overpowered and entirely narcotized by the invocation of the divinity.

Again and again the word heat ('heat of the hot wind', 'heat of the idol', 'heat let down on the shaman by god') appears in the various expressions which are used to describe the Mansi shaman's state of trance. I am of the opinion, however, that its origin can be explained neither by the consumption of the fly-agaric nor by the drunkenness. Still less can we understand the nature of that *bluish smoke* which, according to Munkácsi[43], the Mansi 'imagine to arise around the shaman', as a distinctive mark of the divinity's arrival. In quoting this, Munkácsi relies on the Russian Academician P. S. Pallas, who says about the Khanty shamans[44]: 'The Ostyak sorcerers use a hand-drum like most of the Siberian shamans. During the trance they are said to get horrible convulsions in front of a great fire in the hut, until they are abandoned by the devils which, according to them, they have lured there, and able to give the required answer. During the convulsions, all those present make a continuous noise with kettles, bowls, etc., *until they imagine that they see blue smoke appearing above the sorcerer*[45], who, after the paroxysm he has gone through, looks for a long time as if he were unconscious and unnerved'. We can read about this mysterious blue smoke in a slightly different version in an earlier report. Munkácsi published several parts[46] of a work by the Swede, Johann Bernard Müller, printed for the first time in 1721 in which the author describes the invocation of divinities performed by Khanty shamans as follows: 'Bound, the shaman throws himself on the earth where he waits, with uneasy convulsions, to be possessed by the devil who will answer the questions put to him concerning future events, good hunting fields or will let him know about the means by which quarrelling people can be appeased. During this time, those who have asked for the prophecies, are standing around him, howling incessantly and making continuous noises with kettles and other resounding vessels, *until a blue vapour, which they consider to be the prophesying spirit, appears*[47] forcing those who are standing around to withdraw and seizing Satan's servant who is thrown to the ground . . .'[48]

Both reports agree in mentioning that blue smoke (Pallas) or blue vapour (Müller) appears during the conjuration of spirits and the ceremony of the shaman. According to Müller, the blue vapour 'which is considered to be the prophesying spirit' ('welcher für den wahrsagenden Geist von ihnen

[42] Munkácsi (1892–1921) Vol. II, p. 0381.
[43] Munkácsi (1892–1921) Vol. II, p. 0380.
[44] Pallas (1773–1776) Vol. III, p. 62.
[45] The italics are mine, J. B.
[46] Munkácsi (1899) pp. 285–92.
[47] The italics are mine, J. B.
[48] Munkácsi (1899) p. 287.

gehalten wird') appears before the eyes of the people who surround the shaman and make continuous rattling and howling noises. Then the spirit, in the shape of blue vapour, disperses those standing around, takes hold of the shaman and throws him to the ground (Müller). Pallas speaks of blue smoke instead of vapour. He does not maintain that it really appears before the shamans making a noise, but says cautiously that those present make a noise and shout 'until they imagine that they see blue smoke appear above the shaman', ('bis ihrer Einbildung nach ein blauer Rauch über dem Zauberer entsteht').

What can this mysterious blue vapour or smoke be? Munkácsi had nothing to go by except these reports, which were insufficient to solve the problem. He was also aware of the fact that not all of Pallas's work reflected the original observations of the author. 'The major part of the sections dealing with the Ob Khanty is probably based on the records of a student called Zuev who—following the instructions of Pallas who himself never was in that area—travelled in 1778 along the Ob up to the Arctic sea and explored this region very thoroughly in many respects.'[49] Hence, Munkácsi obviously knew that the data in this part of Pallas's book had been collected by Zuev, his young collaborator, although he could not have been acquainted with the original notes of the latter, since they were not printed until quite recently.[50]

G. D. Verbov who prepared this important work for the press, wrote in the introduction that the reports about the Khanty and Nenets in Volume III of Pallas's book were based on material collected by Zuev.[51] While he made use of the young student's material, Pallas at times either completely disregarded, or considerably shortened some very significant passages.[52] Among those which are treated particularly inadequately by Pallas, Verbov mentions Zuev's records on the Khanty shamans.[53] How true this is, can be seen if Pallas's description is compared with the following lines in Zuev's work about Khanty shamanism: 'They mostly act as shaman, if they repeatedly see something strange. When the shaman (ворожей) dreams of something appalling or wonderful, he assembles the inhabitants of the near-by yurtas in order to prophesy for them. A plain man who has had a dream may also consult the shaman about it. Divining is performed as follows: The fortune-teller or shaman (волхв) binds himself, lies on the earth, makes faces and then begins to rave (ломается). *Beside a big fire which is burning in the tent*[54] he utters unintelligible words while he is waiting for the devil to appear. It is believed that by answering the shaman's questions the devil foretells the future and shows him where the best hunting grounds are to be found. If a hunter gets into trouble, the devil will show him how to get out of it. The people who are eagerly waiting for the answer, shout incessantly, beat kettles and bowls and make a terrible noise around the shaman until a blue mist or smoke (синий туман или дым) appears above him. Then the shaman disperses those who are standing around him. *Anyhow,*

[49] Munkácsi (1899) p. 365.
[50] Зуев (1947).
[51] Зуев (1947) p. 15.
[52] Зуев (1947) p. 15.
[53] Зуев (1947) p. 15.
[54] The italics are mine, J. B.

he is always standing in smoke, because at the place where he mutters his words there is a fire smouldering all the time.[55] The shaman arises and jumps up and down like a maniac. Sometimes after this the shamans remain unconscious and insensible for more than an hour. When they regain their senses, they give cunning answers to the questions or say anything that occurs to their mind, or are simply raving. They relate in great detail how they saw the devil and what happened to them . . .'[56]

The original report differs from the description by Pallas not only because it is more detailed, but also because it solves the mystery of the blue mist or smoke, that surrounded the shaman. According to the shortened (and distorted) text of Pallas, the people who are standing round the shaman only imagine that they see blue smoke above him. Zuev, on the other hand, gives a simple and natural explanation: A dense cloud of smoke is rising from the constantly smouldering fire.

In completeness and explicitness, Zuev surpasses not only Pallas, but also Müller. His report is much more plausible: The shaman, hiding in blue vapour or a smoke-like substance, disperses the bystanders, faints and sinks to the ground. Zuev was a remarkable supporter of the enlightment who refused to believe in miracles and supernatural phenomena. Indeed, not only here, but in many parts of his work he proves to be a sensible observer with a materialistic outlook.[57]

If we accept the testimony of Zuev, who—unlike Pallas—collected his material concerning the Khanty shamans on the spot, we obtain reasonable explanations for the expressions which are connected with Mansi shamanism, cited by Munkácsi in several versions (*tḁrom rei, tḁ̄re'm rē'ine, pupi̊' rē'i̊nė* ['heat of god, heat of the idol'])[58]. The Khanty shaman described by Zuev most probably did not faint because he had eaten fly-agaric or drunk alcohol, but rather from the thick bluish fume or smoke to which Zuev made reference repeatedly.[59]

In addition to the consumption of fly-agaric, of intoxicating drinks and of tobacco smoke, another way of falling into a trance mentioned by the scholars is that of inhaling the smoke of various narcotics. Ohlmarks deals with this kind of trance induction in a separate article.[60] He says: 'Among many peoples a fourth way of arousing ecstasy by drugs is known: to fill the room . . . with the smoke of narcotics. In shamanism this custom seems to be of greater significance than the smoking of tobacco or the drinking of wine.'[61]

Speaking of the Nenets shaman, the *tadib*, Zuev says that when a shaman is invited for the trance ceremony he brings a drum and a dress with him.[62] He sits down in the middle of the tent, while those who are waiting for the prophecies take their seats around him. A big fire is made in the tent, and on it the drum is dried so as to produce a louder sound.

[55] The italics are mine, J. B.
[56] Зуев (1947) p. 45.
[57] Зуев (1947) pp. 43–7.
[58] Munkácsi (1900) p. 168, (1892–1921) Vol. II, p. 0381.
[59] Зуев (1947) pp. 43–7.
[60] Ohlmarks (1939) pp. 113–22.
[61] Ohlmarks (1939) pp. 113–4.
[62] Зуев (1947) pp. 46–7.

The shaman puts his dress on and fastens iron shackles on his hands and feet, which rattle whenever he moves. The attending Nenets shout out loud in a drawling voice: *hoi̯! hoi̯! hoi̯! hoi̯!*[63] Then the shaman, sitting in his central place, looks around with fierce eyes and stares up, and then down. Taking the drum in his hand, he begins to beat it, first gently, then louder and louder. Finally he jumps up, runs around in the tent and suddenly throws the drum away. 'For a long time he raves like this in the smoke, close to the big fire, until he loses consciousness and collapses. When he comes to, he begins to chant and sweats tremendously because of the great heat . . .'[64]

5. Why is the smoke or vapour which entrances the shaman blue? Both Zuev and Müller seem to stress the colour—and yet the smoke of burning logs could certainly not be called blue by anyone!

There is also an explanation for this mysterious blue. Karjalainen writes that the shaman of the Khanty living near the Demjanka river fills his hut with the smoke of a resinous bark.[65] The Tsingala Khanty burn the bark of a certain species of pine-tree (Pichta-Fichtenrinde) with which they fumigate the room.[66] Ohlmarks[67] quotes Shirokogorov as saying that the shamans of the southern Tungus inhale the smoke of *Laedum palustre* or of some other resinous plant, while the Manchurian Tungus use some kind of Chinese incense. The Nivkhi (Gilyak) shamans also use the smoke of *Laedum palustre* to daze themselves, whereas the Nanai (Golds) use some unknown plant for the same purpose. From Ohlmarks we learn that the shamans in the north-western regions of Mongolia and the neighbouring areas throw juniper twigs into the fire as incense. The Oroks use glowing aromatic *Taechom palustre*. The Tofa (Karagas) shamans throw the twigs of a thuya-like tree into the fire, which is smothered, and the glowing twigs send forth a dense, aromatic, w h i t e smoke which is then inhaled by the shaman. The Buryats burn the pine-tree bark, heather and other plants, while the female shamans in Eastern Turkestan throw anise grains on the glowing logs.[68] This kind of ceremony is summarized by G. Chubinov as follows: 'Among the majority of the more advanced Siberian peoples three kinds of plants are used for shamanistic performances: The bark of the silver fir (*Pinus picea* L.), thyme (*Thymus serpyllum* L.), and the juniper tree (*Juniperus sabina* L.). Among the Buryats, the Soyots and several other tribes, there are special stones and implements on which to place the glowing embers and the censing substances. The smoke and, particularly, the smell of these plants have a narcotic effect and are meant—like ecstatic states in general—to cause dimness of the consciousness . . .'[69] And again: 'In most cases, something is thrown into the fire: pine twigs, leaves of the marsh-rosemary, the bark of a silver fir, a piece of juniper tree or thyme, but also lumps of grease, butter and other food-

63 Erdődi (1937) pp. 228–33; Mészöly (1952) pp. 46–9.
64 Зуев (1947) pp. 46–7.
65 Karjalainen (1921–1927) Vol. III, p. 306
66 Karjalainen (1921–1927) Vol. III, p. 307.
67 Ohlmarks (1939) pp. 114–5.
68 Ohlmarks (1939) pp. 115–9.
69 Tschubinow (1914) p. 44.

stuff, so that the smoke should have a special smell.'[70] We see then, that the Siberian peoples frequently throw various plants, seeds and resinous bark into the fire before shamanistic performances. Therefore it is very likely that the smoke or vapour which was described by Müller and Zuev seemed blue, that is, different from normal smoke, because it came from resinous material, seeds or some plant which had been thrown into the fire. We have seen that the twigs of the thuya-like tree which are thrown into the fire by the Tufa shamans, produce an unusually thick white smoke. It may be safely assumed that the blue smoke comes from the burning of some other substance.

6. Returning now to the verb *rűt, rőt, röjt, réüt, rít, riüt,* which was used in Old Hungarian, we wish to remind the reader of those Mansi words meaning 'warmth, heat' which are said to be related to the root of this old Hungarian verb. We may suppose, with a fair degree of likelihood that its original meaning was 'to warm, to heat',—an opinion we share with Dezső Pais.[71] The shaman in a trance must have been heated by the big fire and by the stupefying, hot smoke rising from the vegetable matters which are thrown into the fire (the technical term used in Ugric shamanism for this status was *rűti, röjti, réüti*). However, the shaman is not only heated (*rűti*), (*rűtött*) but also concealed (*rejti*), (*rejtett*) from the eyes of the spectators by the thick smoke. Let us recall Zuev's testimony: when the Khanty shaman falls into a trance, 'he stands in smoke because the fire is burning all the time at the place where he is muttering his words'.[72] According to L. v. Schrenck[73], glowing embers are placed on an iron plate before the Nani (Ulchi) shaman, with the dried stalks of *Laedum palustre* scattered on it. *The shaman bends over the thick stupefying smoke* and, beating his drum which he holds up, sings a melancholy song in a low voice which gradually grows louder.[74] Chubinov says the same thing: 'The shaman of the Gilyaks and other aborigines of the Amur region *bends over a thick strong-smelling smoke*[75] which comes from the dry leaves of marsh-rosemary (*Laedum palustre*) placed on glowing embers'.[76] It is evident, that this dense smoke not only heats (*rűti, röjti, réüti, ríti*) but at the same time hides (*rejti*) the raving shaman.

What has been said so far can now, in my opinion, be placed alongside the semantic problems in our two Hungarian verbs. The striking resemblance between the Hungarian verb *rejt* 'hide, conceal', together with its 16th century forms *rőyt, reüt, rüt,* and the old verb meaning 'enchant' is not entirely accidental. Originally they may have designated two related stages of a shaman's falling into trance and it is probable that an essential semantic relationship existed between them. Long ago the two were one and the same verb. The first meaning may have been 'to warm, to heat' which later by metonymy or contamination changed into 'hide, conceal'.

[70] Tschubinow (1914) p. 42.
[71] Pais (1949) p. 7.
[72] Зуев (1947) p. 45.
[73] Schrenck (1891) Vol. III, p. 757.
[74] Ohlmarks (1939) pp. 116–7.
[75] The italics are mine. J. B.
[76] Tschubinow (1914) p. 44.

7. It still remains to be explained how the blue vapour or blue mist appeared above the entranced shaman. It is true that Zuev states this only as an alternative; but Müller, half a century before him unmistakably and exclusively mentions only blue vapour.

These two statements do not stand alone. T. Lehtisalo quotes an English traveller who describes a magic ceremony which he attended as an eye-witness in 1556 at the estuary of the Pechora. This report is interesting enough to be reproduced here in full:

'The sorcerer could be distinguished from the other men because he wore *a special hood which covered his face.*[77] He beat a big drum with a stick and screamed fiercely. The Samoyeds present took up the screaming until the sorcerer seemed to have lost his consciousness. At last he fell on his back and lay there as if he were dead. They explained to the Englishman that their spirit was now telling the sorcerer what the Samoyeds had to do and where they had to go (the sorcerer had been questioned on these points). Then all who were present shouted three times *ohu!* and the sorcerer rose and continued to sing. In the meantime, five reindeer were killed upon the sorcerer's orders and various acts of sorcery began. The sorcerer stabbed himself with the sword without being wounded; he made the sword red-hot and ran it through his body, the point coming out of his back. The Englishman made sure of this by touching it. *Then the Samoyeds boiled water in a kettle, brought a square seat on which the sorcerer squatted like a tailor, and the kettle with boiling water was placed next to him.*[78] After these preparations, the sorcerer tied a chamois strap four feet in length around his neck; a man was placed on each side of the sorcerer to hold the ends of the strap. *The sorcerer was covered with a long cloth and the two men pulled the strap.*[79] The Englishman heard a few objects fall into the boiling water. He was told that the head, the shoulders and the left hand of the sorcerer, torn off by the strap, had fallen into the kettle. The superstitious Samoyeds did not allow the Englishman to check what had happened, saying that anybody who saw things which a human eye was not supposed to see would die. Soon the songs and shouts were resumed and the Englishman saw a finger stick twice through the cloth which covered the sorcerer. The Samoyeds told him that it did not belong to the sorcerer, who was now dead, but to an animal which was unknown to them. In spite of a careful search, the Englishman could not find a hole in the cloth. The performance ended by the sorcerer appearing safe and sound beside the fire and explaining to the traveller that no one else would be able to solve the mysteries which the spirit had revealed to him in his trance.'[80]

When comparing this detailed description with the reports known up to now we find a new element in it: the sorcerer, completely wrapped up in a long cloth, is sitting near a kettle of boiling water, his face is covered with a kind of a hood. For us, this ceremony contains two noteworthy facts: first, steaming water is placed next to the sorcerer; second, the sorcerer is concealed from the bystanders, beside the kettle of boiling water.

[77] The italics are mine, J. B.
[78] The italics are mine, J. B.
[79] The italics are mine, J. B.
[80] Lehtisalo (1924) pp. 157–8.

According to Ohlmarks, steaming water figures also among the requisites of the shamanistic performances in Central Mongolia.[81] There are also other sources which state that the shaman becomes dazed by vapour. The oldest report is found in Book IV of Herodotus, recently considered by a certain K. Meuli in connection with this problem. Writing about the funeral ceremonies of the Scythians, Herodotus describes how they are purified after the burial: 'Three stakes are fixed in the ground so that they incline towards one another, and around them pieces of felt are stretched, as tight as possible. Inside this booth they put a dish, into which a number of red-hot stones are placed. Then the Scythians take some hemp-seeds, and creeping under the felt coverings, throw them down upon the red-hot stones; immediately the seeds begin to smoke, and give out such steam as no Grecian steam-bath can exceed. The Scythians enjoy it and make a tremendous noise shouting.'[82]

Commenting on this passage (Her. IV/73)[83] Meuli[84] says that this funeral purification rite must have been a form of shamanism. It is in this manner that the Scythians led the souls of the dead to their place in the other world, and, at the same time, purified their homes from the returning ghosts of the deceased. Many similar rites are to be found among the Mansi, the Samoyeds and the Buryats. The best known and most detailed description of such a shamanistic purification rite is to be found in the work of Radloff.[85] According to Meuli, the screaming of the Scythians during the ceremony as described by Herodotus is nothing but the usual loud shouting of the shamans in trance, while the hemp-seeds[86], which are heated on the glowing embers, cause their trance. Ohlmarks regards this explanation as a plausible one[87] adding that the shamans of the Eskimoes and other northern tribes are often half naked when they fall into trance, because—just like the Scythians—they get heated up by a sudorific.

The authenticity of Herodotus's report is supported by finds which have been brought to light by Soviet archaeologists from the Scythian tombs of the Gorno-Altai region. The excavations of the great grave hills (*kurgans*) which were covered with ice were started in 1865 by V. V. Radloff. Other expeditions reached the Altai mountains in 1927 and 1929. The latest excavations were started in 1947–1948 under the direction of the Soviet archaeologist S. I. Rudenko. The operations lasted for four years. In his recently published work about the results, Rudenko points out that the report of Herodotus about the life and the customs of the Eastern European and Asian tribes of the steppes are confirmed by the excavations.[88] According to Rudenko, the first two and perhaps the fifth of the recently discovered Pazyryk kurgans (*tumuli*) are from the 5th century B.C., whereas the others date from the end of the 5th or probably from the beginning of the 6th century B.C.—i.e. from the Scythian period instead of the Hunnish-

[81] Ohlmarks (1939) p. 115.
[82] Meuli (1935) pp. 121–76.
[83] Rawlinson (1859) pp. 65–6.
[84] Meuli (1935) pp. 122–4.
[85] Radloff (1884) Vol. II, p. 52.
[86] Meuli (1935) p. 135.
[87] Ohlmarks (1939) p. 120.
[88] Руденко (1953) p. 8.

Sarmatian period as was believed by others (S. V. Kiselev, K. F. Smirnov and M. I. Rostovtsev).[89] Two finds in the second Pazyryk kurgan are important for our present study.

In this second kurgan, in the north-western corner of the crypt, a bundle of six rods was found. An oblong four-legged vessel was lying underneath them filled with broken stones.[90] The rods were 122·5 cm long, 2 cm in diameter and 3 cm at their thicker end. Two centimetres from their upper end there was a hole with a strap through it. They were all covered with thin strips of birch rind.

Farther north, in the western part of the crypt, another bronze vessel of the 'Scythian kettle' type was found, also filled with stones. Above it there were six rods similar to those previously mentioned. Robbers had partly broken and knocked them over. The rods and the censer (курильница) were covered with a large piece of leather.[91]

Rudenko explains the purpose of these objects, emphasizing that the two vessels contained not only stones but also some hemp-seeds (*Cannabis sativa* L., *C. ruderalis* Janisch).[92] More hemp-seeds, some of which were burnt, were found in another leather-covered vessel which was tied to one of the rods above the 'Scythian kettle'. Also the handle of the censer was covered with birch rind. This was probably necessary because from the hot stones it contained the censer became too hot to be held by its handle.[93]

According to Rudenko, all these objects were used for performing purification ceremonies after the funeral, very similar to those described in great detail by Herodotus in his report on the Scythians who lived near the Black Sea. The censers found in the grave were used in the same manner as described by the Greek historian. Rudenko is probably right in concluding from this and from other evidence that, as a result of mutual influence, the nomadic pastoral peoples who were living on the vast steppes and mountains of Eastern Europe and Asia, stood, in spite of their different origin, close to each other in social structure and religious beliefs.[94]

It should be noted that Rudenko does not mention a connection between these censers and shamanism. This is natural, since the finds themselves do not prove its existence. Nevertheless, it must be borne in mind that among the peoples in whose life shamanism played a part, the purification ceremonies after the funeral were performed by the shamans, as can be seen from Radloff's famous description which we quoted above. It seems, therefore, in no way unreasonable to assume that the sudatories and censers which Herodotus and Rudenko described may have had a function in the shamans' falling into trance among the Eastern European and Asian nomadic peoples.

If this is so, we have a very ancient report of a narcotic vapour or smoke making the shaman fall into a trance. We may very well compare

[89] Руденко (1953) p. 360.
[90] Руденко (1953) Fig. 52a.
[91] Руденко (1953) p. 333.
[92] Hoops (1911–1919) Vol. II, pp. 445–7; Ebert (1924–1932) Vol. V, pp. 116–7.
[93] Руденко (1953) Vol. XXIV, Plate 2.
[94] Руденко (1953) p. 361.

Zuev's and Müller's descriptions of the mysterious blue vapour appearing above the shaman with Lehtisalo's travelogue and with passages in Herodotus.

Basing himself on Oldenberg's *Vorwissenschaftliche Wissenschaft,* Meuli describes a sacrificial ceremony of the ancient Hindus which also aims at achieving contact with gods or spirits through an ecstatic state. Seclusion (Sichabschließen), similar to what has been mentioned above, fasting until exhaustion and, originally, sweating were resorted to for this purpose. The priest who is to offer a sacrifice first takes a bath, then puts on clean garments and, with his head wrapped up, sits down on a piece of black antelope leather near the sacrificial fire in the sacrificial tent, '. . . the cover and the proximity of the fire must have caused a certain heat; there is a special expression for the sweat that appears on the man . . .'[95]

Referring to literary sources, Meuli says that sudatories were known from times immemorial to the Slavs, Germans, Greeks and Macedonians. Originally, like the Scythians described by Herodotus, the Greeks used glowing stones or cinders for this purpose. Their sudatories were situated in a hemispheric tent, or hut, and probably had about the same equipment as the Scythian ones which have been described above. The steam bath of the Finns, the *sauna,* is well known. Quoting passages from the Kalevala (e.g. Book XVII, 277 ff., Book XXVI, 31 ff.) some scholars maintain that they reflect ancient beliefs which are associated with the effect of magic. In this respect, Book XLV (pp. 197–204) is particularly instructive. Here old Väinämöinen, the 'chief shaman', prepares a vapour-bath of magical force, with the help of glowing stones, in order to expel the demons of disease.[96]

Accordingly, in the eastern part of Europe, sudatories represented a most ancient custom and—as may be concluded from what Herodotus has to say about the Scythians—probably played a certain role in exorcism among the ancient nomadic peoples of Southern Russia.[97]

The closed space increases the strength of the vapour and its stupefying effect. In the Scythian ceremonies as described by Herodotus, as well as in the ancient Hindu sacrificial rite which has been referred to, one point is essential. Those performing the ceremonies are isolated from the outer world so that they should be heated by the vapour as much as possible. This kind of seclusion is inseparable from magic sweating. An equally important feature of the shamanistic ceremonies in question is the fact that like a person inhaling, the shaman bends over the dazing smoke and sometimes covers his head with a towel to absorb as much as possible of its narcotic effect. It is typical of the trance-inducing ceremony as performed among the Nenets that the shaman hides in a kind of tent filled with hot vapour, as was evident in the travelogue we quoted above.

While investigating more ancient methods of shamanistic trance induction Mircea Eliade[98] refers to a frequently quoted but also much debated passage

[95] Meuli (1935) p. 134.
[96] '*Saattoi saunan lämpimäksi,* (He prepares a hot bath,
Kivet löylyn lyötäväksi' Heating stones for stroking with boughs)
[97] Meuli (1935) pp. 135–6.
[98] Eliade (1951) p. 351.

in Strabo's book about the Mysians: *καπνοβάται*[99] in which—some say—he was alluding to Getic shamanism. This report should be handled with caution since there is nothing to prove that *καπνοβάται* refers to shamanism.[100]

On the other hand, references to 'magic heat' (*chaleur magique*) in Mircea Eliade's work deserve attention. Descriptions by numerous travellers show that shamans—and in fact the sorcerers of primitive tribes in general—are in the habit of impressing spectators with performances near the fire, in smoke, with burning or glowing objects. Many wild tribes refer to magic power with words meaning 'burning, fiery', 'hot', 'heat', 'ignition', (e.g. the Dobu tribe, the inhabitants of the Solomon and the Rossel islands). We also know that the inauguration ceremonies of candidates for shamanhood frequently required performances involving the masterful handling of fire (*maîtrise du feu*).[101] We cannot agree, however, with the next step in Mircea Eliade's argumentation where he says rather arbitrarily that the methods of trance induction under discussion (heating, smoking, drunkenness, etc.) are of a later date and secondary in character, the use of such narcotics being typical of a degenerate technique of trance induction.[102] This is contradicted by common, and hence very ancient, elements in the vocabulary of Hungarian shamanism, common with Mansi and Khanty shamanism. Suffice it here to mention the noun which is implied in the stem of the Hungarian verb *réül*, the Khanty equivalents of which are e.g. Nizyam-Berezovo dial *rǒu̯' ṛèu̯'* 'warm ("neither hot nor cold")', Kazym dial. *rǒu̯' mǒu̯rǒu̯* 'mist which is rising from the ground', etc.[103] These variants from Khanty dialects clearly show that the noun meaning 'heat, warm' and forming the stem of the Hungarian verbs *réül* : *rejt* : *rejtezik* could have originally meant not heat which descends from heaven, but heat coming from some concrete earthly source such as fire, smoke or vapour.

Among the Khanty words in question, the following ones should be noted in particular: Kazym dial. *rǒu̯'* 'warmth, heat (in a house, in the bath-house, outdoors), sultriness (summer weather; indoors) vapour of a sauna; rising steam?; strength (of vodka); nutritiveness (of food)', Obdorsk-Berezovo dial. *ṛɒu̯* 'heat'. They refer clearly to a heat which is developed in an enclosed space, where the 'air is stuffy'. We do not know why the various authors have placed a question-mark after 'bathroom vapour', 'rising steam', but the meanings of the Khanty word under discussion fit well into the category which we have just analysed in connection with the shamas' ceremony.

An interesting parallel to the semantic development 'heat > daze' can be found in Russian. According to D. N. Ušakov's Etymological Dictionary the word угар means: 1. Suffocating, poisonous gas, produced by the incomplete combustion of coal (developed for instance in a stove, if the flue is closed too early, before the coal has burnt out). 2. A disease caused by the asphyxia. 3. Unconscious, ecstatic state, enrapture, etc.[104] An essentially identical semantic development can be observed in the case

[99] Strabo (1866) Vol. VII, 3.3, C, 296, 4–8.
[100] This question requires further studies which would exceed the scope of our paper.
[101] Eliade (1951) p. 414.
[102] Eliade (1951) p. 416.
[103] Karjalainen (1905) p. 208.
[104] Ушаков (1935–1940) Vol. IV, p. 876.

of the Russian noun чад: 'smoke, carbon monoxide, the foul smell of a burning greasy substance—that which dazes, renders unconscious.'[105]

Therefore we may say with a fair degree of certainty that if the old Hungarian verb *rít, rűt, rőt, röjt, réüt, riüt* meant 'to enchant' in terms of heating and dazing with vapour or smoke, then—by a transfer of meaning—the same verb could easily have meant also 'to hide, to conceal'. And have we not just seen that concealment is inseparable from getting stunned by heat?

8. It is clearly evident from other derivatives of the stem of the verb *rejt* that the original meaning of this stem was 'heat, warmth'. Examples of this are: *rekedő, rekkentenőnc, retkentenőnk, rekkenő, meg-reckent, rekkenés, reckentő.*[106] These and other examples show that the verbal derivatives of the root which means 'heat, warm' did not in general mean 'heating, warming', but the 'stagnation or warming of the air in a closed sultry space'.

Let us now examine other derivatives of the verb *rejt.* Once it had a reflexive derivative formed with the suffix *-z*: *rejtëz-ik* (the oldest record is Jókai C. 4 *reytezuen*) meaning Cal., MA. PPB. 'to hide oneself'. With a verbal prefix: *el-rejtëzik*; Cal, MA. (the most ancient record in the codices Vienna C. 169 *el reitezet,* ibid. 170 *el reitèzèt volt,* Munich C. 20 *reitezhetic,* Érsekújvár C. 303 *reytezeth*). From the end of the 16th century variants with *ö* (oe) of this verb are to be found: Kár. Bibl. I. 330 *rőytezhessel,* MA: Bibl. I. 256 *rőytőztec.*[107] The meaning of the old readings is always 'to hide, to conceal'.

It will be instructive to compare the above with the following words from the *Magyar Tájszótár* (Hungarian Dialect Dictionary): *elrejtëzkëdik* 'álhalott lesz = to fall into a state of apparent death, trance' (Tolna), *rejtözött, elrejtő(d)zött, elrejtëzëtt* 'álhalott = apparently dead' (Szeged, Tolna, Csongrád, Mindszent, Székely region). Ödön Beke completes these data with further material that is worthy of interest.[108] The first occurrence of this word is found in János Arany's epic poem *Toldi,* where *elrejtezik* means 'becomes apparently dead'[109]. E. Fél quotes: *ê vót rejtőzve* 'was apparently dead'[110]; *megrőjtődzött* 'was as if dead'[111]. P. Veres writes: *elrejtezett volna* '(as if he) were dead.'[112]

This is precious material indeed. Beke completes the picture with the following remark: 'The stem *rejt* of these words is not the verb *rejt* which means "to hide", but the ancient Hungarian verb *röjt, rűt, rőt, rit, réüt, riüt* "to enchant", and it is the factitive form of the verb *réül, révül, rivöl* "to fall into a trance".'[113] After the argument which has been presented in this paper, we feel that such a distinction is no longer acceptable. If what we have said so far is correct, the verbs *(el)rejtëzik* ~ *(el)röjtözik*

[105] Ушаков (1935–1940) Vol. IV, p. 1231.
[106] Balázs (1954) pp. 432–3.
[107] Szarvas–Simonyi (1890–1893) Vol. II, pp. 1391 ff.
[108] Beke (1950) pp. 417–8.
[109] Balázs (1954) p. 433.
[110] Fél (1941) p. 74.
[111] Fél (1941) p. 150.
[112] Veres (1950) Vol. I, p. 94. For further detailed data see Diószegi (1953) pp. 303–11.
[113] Beke (1950) pp. 417–8.

should not be considered as two separate verbs, but form the derivatives with the suffix *-z* of one and the same verb—*rejt* ~ *röjt*. The original meaning of the verb may have been 'to heat'—hence its subsequent meanings: 'to become dazed on account of heating → to be apparently dead'.

On the other hand, since we know the various methods of the trance induction, I feel I am on safe ground when I maintain that the meaning 'to conceal oneself, to hide from the bystanders' (traceable back to the time of the codices and, therefore, very old) developed by contamination. Actually the concealment of the shaman can be divided into two clearly separable stages: the shaman lost consciousness (*elrejtőzött*) by bending over the vapour or smoke, and by covering his face and body—or he concealed himself in a tent or hut which was entirely secluded from the outside world and filled with steam. In other words, whenever he lost consciousness in such a manner (*elrejtëzëtt*), he at the same time concealed himself, that is to say, disappeared from the view of the spectators (*elrejtőzött*).

9. Other derivatives of the verb *rejt* seem to prove this point too. First, the noun derivative *rejtëk* deserves our attention. The oldest records are: Vienna C. 171 *reitek*, Munich C. 23, 59 *reitec*, Nagyszombat C. 199 *reytőg*, Jordánszky C. 370 *reytek*, etc.[114] Its variants with *ö* (ö̊) are remarkable: Zvon: Post. II, 605 *rőytők*. Besides these forms with a short and closed final vowel, the following variants with a long and open final vowel are noteworthy: Mon: TME. V, 316 *rejték*, Nót PM. 53 *rejtéki*.[115] From this variant which is of much later date than the former we may conclude, together with linguists like Dezső Pais[116], that the form with the short final vowel is much older than the derivatives with *-ék*.[117]

This derivative appears also as a proper name. Very old documents mention it as the name of two villages in Sopron county, 1223, 1346: *Ruhtuhewr*, 1434: *Rettekewr*, actually: *Rejtëk-őr* (near the Lajta river);[118] 1265: *Ruhtuk*, 1326: *Rehtekh*, 1433: *Rwhtwk*, 1494: *Reithek*, actually: *Röjtök* (near Kapuvár).[119] The forms in which *e* alternates with *ü* and *ö* are very informative, because they show that the basic word may have been either *rüjt-*, *röjt-*, or *rejt-*. The semantic implications of the derivative *rejtëk*, *röjtök* are still more far-reaching. According to MA 1621, the word means 'refuge'. Other old dictionaries give the same meaning: PP 1708, PPB 1767: *rejtek* 'refuge, lair, cave', *rejtek hely*' hidden place', *rejtek kamara* 'secret room', etc.

As we see, the meaning of all these noun derivatives is 'a place of concealment'. This is natural if the word *rejtëk* is compared with similar derivatives of ancient origin, like *hajlok* ~ *hajlék*, *menedék* old Hung. and dialectal *nyomdok* ~ *nyomdék*, *szurdék* ~ dialectal *szurdik* ~ old Hung. and dialectal *szurduk* ~ *szurdok*, etc. all of which designate a place. The suffix *-k* is of Uralic origin (**-kk*); its Finno-Ugric equivalents are frequently used

[114] Szarvas–Simonyi (1890–1893) Vol. II, p. 1390.
[115] Szarvas–Simonyi (1890–1893) Vol. II, p. 1390.
[116] Pais (1918) p. 68, (1940) p. 301, (1941) pp. 346–7.
[117] For an up to date summary of the problem see Bárczi (1951) pp. 29–30, 167–8.
[118] Csánki (1890–1913) Vol. III, p. 627.
[119] Csánki (1890–1913) Vol. III, p. 627.

to designate the place of the action expressed by the basic word; cf. Mari (Wichm.) Kozmodemyansk hill dial. *βanžęm*, Yaransk dial. *βandzęm*, 'to cross, to go over something': Kozmodemyansk hills dial. *βanžak*, Yaransk dial. *βandzak*, Birsk dial. *βondžak* 'ford, swimming-place for horses (B), small bridge or crosspiece (over a stream)'; Mordvin, Erza (Reg.) *śolgams*, (Wied.) *śolgoms* 'zárni, bezárni (= to close)'; *śolguk* 'gát, itatóhely, víztorlasztás (= dam, horse-pond, barrage)', etc.[120] We may, therefore, conclude that the original meaning of the words *rejtëk* ~ *röjtök* was also 'place'. It was probably first used to designate the site where the action of the verb *rejt* ~ *röjt* took place.

There is, however, a recorded case in the Csallóköz dialect where *röjtök* means 'rekkenő idő (= oppressive heat, sultriness)'.[121] What is the origin of this meaning? What semantic evolution can be assumed to integrate the two different meanings of one and the same derivative, *rejtëk* ~ *rejtök*: 'hidden place, a place where something is concealed' and 'sultry weather'? Or should they be separated from each other like the verbs *rejt* 'hide' and *réüt* 'to enchant'? Budenz thought that they should.[122] But was he right? On the basis of our present argument I believe that he was mistaken. Bearing in mind the method of trance induction analysed above, we may accept it as a fact that the derivative *rejtëk* ~ *röjtök* of the verb *rejt* ~ *röjt* was not only 'a hidden place', but meant also 'sultry weather, heat'.[123] We have described in detail the sudatories of the Scythians and of the Nenets, the magic steam baths of the Finns and the sweating baths of the Hindus. It is possible that *rejtëk* ~ *röjtök* originally meant the sweating tent of the Ugric shaman. Analysing the meaning of the Moldavian Csángó word *regelye* 'hidden place in the forest, in a ravine, deep bed of a brook', Dezső Pais thinks that Hungarian or Székely magic and enchanting operations 'may have had a form or a part in which the sorcerer—like the shaman—really had to hide himself in the forest'. Explaining the original meaning of the word *rejtëk*, he says: 'It is quite possible that the word *rejtëk* was originally used for places that served as hiding place for shamans (sorcerers)—hence its meaning "refuge". '[124] This puts our hypothesis in its proper light. In fact, it may now be considered more than a mere hypothesis. If we analyse the shamanistic ceremonies of the related peoples and of the other Eurasian tribes, we must conclude that a frequent method used by the shaman for getting into a trance was to hide in a closed room, generally in a tent which was entirely secluded from the outside world, filled with vapour or smoke. In the oppressive heat of this closed space he lost consciousness and fell into a trance. Therefore, the word *rejtëk* ~ *röjtök* may have meant simultaneously such a place, e.g., a tent or a hut and, by extension, the oppressive heat which developed in that place.

10. In addition to this, we have another proof that the word *rejtëk* ~ *röjtök* meant not only 'a hidden place' in general, but also the sweating tent of the shaman. It is a well-known fact that different variants of the

[120] Budenz (1884–1887) p. 68; Györke (1935) p. 10; Lehtisalo (1936) p. 380; Bárczi (1951) p. 168.
[121] Kőrész (1872) p. 332.
[122] Budenz (1873–1881) p. 660.
[123] Pais (1949) p. 7.
[124] Pais (1949) p. 10.

word *rejtëk* ~ *röjtök* frequently appear in incantations (*regös* songs, sung by young masked peasant lads at their winter solstice performances, called *regölés*) in Transdanubia and in the Székely region. Sometimes the word is replaced by *rőt ökör, rétt ökör, rét ökör* and other distorted expressions, but if we accept Béla Vikár's[125] textual analyses we may trace one of the ancient original elements of our *regös* songs in the word *rejtëk* ~ *röjtök*.[126] According to this linguist the two first lines of the songs in their 'reconstructed original form' read as follows in modern pronounciation:

'*Rejtekem régi törvény,*	My refuge is an ancient law
Haj regül rejtem!'[127]	(Obscure, unsolved meaning).

In connection with the first line, Dezső Pais shares the opinion that 'originally it contained one of the variants of our modern word *rejtëk* (place of concealment)'[128] with the first person singular possessive suffix. Vikár, on the other hand, assumes, for prosodic reasons, that in the above text *rejtëkëm* may have been originally *réitököm, rëŋėχtėkėm* and mentions the antecedents **rëŋëjt* or **rëŋëχt* of the verb *rejt*.[129] I am of the opinion that Dezső Pais is right when he remarks that 'this is a superfluous complication of the problem'.[130] I believe that the origin of the verb *rejt* is clear enough and so is that of its derivative *rejtëk*. What is of importance for us is Vikár's argument for proving that the word *rejtëk* is an ancient element in our *regös* songs.

It has long been known that certain, perhaps the most ancient motifs of the *regös* songs are connected with Ugric shamanism. This point has been brought out in a somewhat confused manner by Gyula Sebestyén[131] and later by Béla Vikár[132] and others. Recently Dezső Pais demonstrated with exact and convincing reasoning that the *great road* 'along which the performers of the *regös* songs wandered or arrived, was somewhere near *the great road* of Bornemisza's magicians, i.e. in the realm of (shamanistic) magic'.[133] Let us compare this with a statement by K. Meuli, who says that the 'great journey' ('Die große Fahrt') which figures in shaman stories is not only an ancient and most frequent element in the magical shamanistic ceremonies, but one of the oldest and most remarkable motifs to be found in epic poems of shamanistic origin, or in fact, of epic poetry in general.[134] We can hardly doubt that there is a close link between this very old (if not the oldest) component of the Hungarian *regös* songs, and Ugric shamanism, shamanistic conjuration of spirits and trance induction. Therefore, the word *rejtëkëm* in our *regös* songs should simply and naturally be regarded as referring to the shaman's place of trance—the sudatory which we have mentioned so often, or some other place which is radiating

[125] Vikár (1907) pp. 21–39.
[126] For a different approach see Kerényi (1953) pp. 241 ff.
[127] Vikár (1907) p. 39.
[128] Pais (1949) p. 9.
[129] Vikár (1907) pp. 32–4.
[130] Pais (1949) p. 121.
[131] Sebestyén (1902) Vol. V, pp. 468 ff.
[132] Vikár (1899) pp. 97–9, (1906) pp. 101 ff.
[133] Pais (1949) pp. 9–10.
[134] Meuli (1935) pp. 164–76.

oppressive heat. Was it not at the very first stage, at the beginning of trance induction, that the shaman concealed himself in such a place?

We are unable to understand why and on what basis Vikár gave the following meaning to the verse in question[135]: '*My sorcery* is an ancient law[136], I change it (by magic) into a song'. This is a wrong interpretation of the word *rejtëkëm*. In our opinion, in the *regös* songs it must have no other meaning than elsewhere: 'my hiding (i.e. sweating, heating) place'. In the light of what has been said, the sense of the verse becomes quite clear. It is a reference to the place and the method of trance induction — to old customs and old 'laws'.[137]

ABBREVIATIONS

Born: Ének.	=	Bornemisza Péter, *Énekes Könyv* (Hymnal), Detrekő 1582
Born: Préd.	=	Bornemisza Péter, *Prédikációk egész esztendőre* (Sermons for the Whole Year), Detrekő 1584
Cal.	=	Calepinus szótára (Calepinus's Dictionary)
Gyöngy D: MV	=	Gyöngyösi István, *Murányi Vénus* (Venus of Murány)
Kár: Bibl.	=	Károlyi Gáspár, *Biblia* (Bible), Vizsoly 1590
KBártfa	=	*Bártfai kalendáriom* (Calendar of Bártfa), 1583
MA: Bibl.	=	Molnár Albert, *Szent Biblia* (The Holy Bible), Hanau 1608
MA	=	Molnár Albert, *Dictionarium Ungarico-Latinum*, Nürnberg 1604
MA 1621	=	Molnár Albert, *Dictionarium Hungarico-Latinum*, 3rd enlarged edition, Heidelberg 1621
MisocProgn	=	Misocacus Vilhelm *Prognosticon*, Kolozsvár 1578
Mon.Irok	=	*Monumenta Historica*, Vol. III
MonTMF	=	*Monumenta Historica*, *Török–magyarkori történelmi emlékek* (Historical Sources from the Turkish–Hungarian Period), Ed. by Hungarian Academy of Sciences
NotPM	=	*A peleskei nótárius pokolba-menetele* (The Descent to Hell of the Notary of Peleske), Basel 1792
PéldK	=	*Példák könyve* (Book of Parables), 1510
PP	=	Pápai Páriz Ferenc, *Dictionarium Latino–Hungaricum*, Lőcse 1708
PPB	=	Pápai Páriz Ferenc, *Dictionarium Latino–Hungaricum*, revised by Péter Bod, Szeben 1767
Szatm.Cent	=	Szathmár-Némethi Mihály, *Halotti Centuria:* Száz halotti prédikációk (Funeral Centuria: a Hundred Funeral Sermons)
Tyuk: József	=	Tyukodi Márton, *József pátriárka élete* (The Life of the Patriarch Joseph), Várad 1641
Ver.	=	Verantius Faustus, *Dictionarium quinque linguarum*, Venice 1595
Zvon: Post	=	Zvonarics Mihály, *Magyar postilla* (Hungaria Postilla), Csepreg 1626–1627

REFERENCES

Baboss, R., E. (1938) A gyújt-gyüjt-féle egytagú causativumok alaktanához (Contributions to the Morphology of Monosyllabic Causatives of the Type *gyújt- gyüjt-*), in *MNy*, Vol. XXXIV.

Balázs, J, (1954) A magyar sámán révülete (About the Trance of the Hungarian Shaman, in *Ethn*, Vol, LXV.

[135] Vikár (1907) p. 34.

[136] The italics are mine. J. B.

[137] After the publication (in 1954) of this paper in Hungarian, I found two important works which refer to the subject: Diószegi (1954) pp. 20–68 and (1958) especially pp. 59 ff and 295 ff.

Bárczi, G. (1941) *Magyar szófejtő szótár* (Hungarian Etymological Dictionary), Budapest.
(1951) *A tihanyi apátság alapítólevele mint nyelvi emlék* (The Foundation Charter of the Tihany Abbey as a Linguistic Document), Budapest.
Bartholomae, Chr. (1905) *Altiranisches Wörterbuch*, Strassburg.
Beke, Ö. (1950) Arany János nyelvéhez (On János Arany's Language), in *MNyr*, Vol. LXXIV.
Budenz, J. (1873–1881) *Magyar–ugor összehasonlító szótár* (Hungarian–Ugric Comparative Dictionary), Budapest.
(1884–1887) *Az ugor nyelvek összehasonlító alaktana* (Comparative Morphology of the Ugric Languages), Budapest.
Csánki, D. (1890–1913) *Magyarország történelmi földrajza a Hunyadiak korában* (The Historical Geography of Hungary in the Time of the Hunyadis), Budapest.
Diószegi, V. (1953) Adatok a táltos révülésére (Contributions to the Technique of Trance Induction of the Hungarian *táltos*), in *Ethn*, Vol. LXIV.
(1954) A honfoglaló magyar nép hitvilága ('ősvallásunk') kutatásának módszertani kérdései. (Methodological Problems concerning the Investigation of the 'Ancient Religion' of the Magyars of the Conquest Period, around the turn of the 9th cent. A. D.), in *Ethn*, Vol. LXV.
(1958) *A sámánhit emlékei a magyar népi műveltségben* (Survivals of Shamanism in Hungarian Popular Culture), Budapest.
Donner, K. (1920) Über die anlautenden labialen Spiranten und Verschlusslaute im Samojedischen und Uralischen, in *MSFOu*, Vol. XLIX.
Ebert, M. (1924–1932) *Reallexikon der Vorgeschichte*, Vol. V, Berlin.
Eliade, M. (1951) *Le chamanisme et les techniques archaiques de l'extase*, Paris.
Erdődi, J. (1937) Huj, huj, hajrá (*Huj, huj, hajrá*, a War-Cry of the Ancient Magyars), in *Szegedi Füzetek*, Vol. III.
Fél, E. (1941) *Kocs 1936-ban* (The Village of Kocs in 1936), Budapest.
Geiger, W. (1882) *Ostiranische Kultur im Altertum*, Erlangen.
Gombocz, Z. (1913) K. M. úrnak (To Mr. K. M.), in *MNy*, Vol. IX.
Györke, J. (1935) *Die Wortbildungslehre des Uralischen*, Tartu.
Hoops, J. (1911–1919) *Reallexikon der Germanischen Altertumskunde*, Vol. II, Strassburg.
Horn, P. (1893) *Grundriss der neupersischen Etymologie*, Strassburg.
Hunfalvy, PH. (1866) Régi magyar szók (Old Hungarian Words), in *NyK*, Vol. V.
Kannisto, A. (1919) *Zur Geschichte des Vokalismus der ersten Silbe im Wogulischen vom qualitativen Standpunkt*, in *MSFOu*, Vol. XLVI.
Karjalainen, K. F. (1905) *Zur ostjakischen Lautgeschichte*, in *MSFOu*, Vol. XXIII.
(1921-1927) Die Religion der Jugra-Völker, Vols I—III, in *FFC*, Nos 41,44 and 63.
Kerényi, Gy. (1953) A regös ének magja (The Core of *regös* Songs), in *Kodály-Emlékkönyv*, Budapest.
Kőrész, K. (1872) Csallóközi tájszók (Dialectal Words from Csallóköz), in *MNyr*, Vol. I.
Lehtisalo, T. (1924) Entwurf einer Mythologie der Jurak-Samojeden, in *MSFOu*, Vol. LIII.
(1928) Uralische Etymologien, in *MSFOu*, Vol. LVIII.
(1929) Sampa, Sammas (The Words *sampa* and *sammas)*, in *V*, Vol. XXXIII.
(1936) Über die primären ururalischen Ableitungssuffixe, in *MSFOu*, Vol. LXXII.
Mészöly, G. (1910) A rokon szó eredete (The Origin of the word *rokon*), in *MNy*, Vol. VI.
(1928) *Szoptani, *csusztani, *buktani (The Words **szoptani*, **csusztani*, **buktani*), in *MNy*, Vol. XXIV.
(1952) Az ugorkori sámánosság magyar szókincsbeli emlékei (Survivals of Shamanism of the Ugric Period in the Hungarian Vocabulary), in *MNy*, Vol. XLVIII.
Meuli, K. (1935) Skythische Schamanen bei Herodot, in *H*, Vol. LXX.
Munkácsi, B. (1892–1921) *Vogul Népköltési Gyűjtemény* (A Collection of Vogul Folk Poetry), Budapest.
(1899) A vogul nép ősi hitvilága (The Ancient Religion of the Vogul People), in *Nyk*, Vol. XXIX.
(1900) A vogul nép ősi hitvilága (The Ancient Religion of the Vogul People), in *NyK*, Vol. XXX.
(1907) 'Pilz' und 'Rausch', in *KSz*, Vol. VIII.

Ohlmarks, Å. (1939) *Studium zum Problem des Schamanismus*, Lund.
Pais, D. (1918) Sukoró és Homorog (The Names *Sukoró* and *Homorog*), in *MNy*, Vol. XIV.
(1940) Meleg és hideg (The Words *meleg* and *hideg*), in *MNy*, Vol. XXXVI.
(1941) Szarándok (The Word *szarándok*), in *MNy*, Vol. XXXVII.
(1949) Reg (The Word *reg*), in *MNyTK*, Vol. LXXV.
Pallas, P. S. (1773–1776) *Reise durch verschiedene Provinzen des Russischen Reiches*, St. Petersburg.
Radloff, W. (1884) *Aus Sibirien*, Vols I–II, Leipzig.
Rawlinson, G. (1859) *The History of Herodotus*, Vol. III, London.
(Rudenko, S. I.) Руденко, С. И. (1953) Культура населения Горного Алтая в скифское время (The Culture of the Population of the Gorny Altai in Scythian Times), Moscow–Leningrad.
Schrenck, L. (1891) *Reisen und Forschungen im Amur-Lande*, St. Petersburg.
Sebestyén, Gy. (1902) A regösök (The Hungarian Minstrels *regös*), in *MNGy*, Vol. V.
Steinitz, W. (1944) *Geschichte des finnisch-ugrischen Vokalismus*, Stockholm.
Szarvas, G. and Simonyi, Zs. (1890–1893) *Magyar Nyelvtörténeti szótár* (Historical Dictionary of the Hungarian Language), Vol. II, Budapest.
Szily, K. (1921) Elrejt (The Word *elrejt*), in *MNy*, Vol. XVII.
Tschubinow, G. (1914) *Beiträge zum psychologischen Verständnis des sibirischen Zauberers*, Halle.
Uhlenbeck, Chr. C. (1898–1899) *Kurzgefasstes etymologisches Wörterbuch der altindischen Sprache*, Amsterdam.
Uotila, T. E. (1930) Etymologioita, (Etimologies), in *V*, Vol. XXXIV.
(Ušakov, D. N.) Ушаков, Д. Н. (1935–1940) *Толковый словарь русского языка* (Explanatory Dictionary of the Russian Language), Moscow.
Veres, P. (1950) *Három nemzedék* (Three Generations), Vol. I, Budapest.
Vikár, B. (1899) Regös (The Word *regös*), in *MNyr*, Vol. XXVIII.
(1906) A regös ének (The *regös* Song), in *MNyr*, Vol. XXXV.
(1907) A regös ének (The *regös* Song), in *NyF*, Vol. XXXIX.
Viski, K. (1913) A szalontai nép nyelvéből (From the Language of the Inhabitants of Szalonta), in *NyF*, Vol. LXIX.
(Zuev, V. F.) Зуев, В. Ф. (1947) *Матерыалы по этнографии Сибирии* XVIII *века* (Ethnographical Material from 18th Century Siberia), Moscow–Leningrad.

Some Observances and Customs of the Mansi (Voguls) in Connection with Childbirth*

by

E.I. Rombandeeva, Leningrad

From the time of birth children of both sexes are considered by the Mansi to possess equal rights and to be equally 'pure'.[1] But this lasts only until sexual maturity, and so from the day when menstruation begins, a girl becomes *ikšam* 'impure'.[2] From now on until her death she must strictly adhere to the customs, the ways of living and the religious rites of her people lest she be punished by some *pupiγ*.[3] As the saying goes, 'she must remain unnoticed'. She must get used to the objects of her environment in order not to touch with her feet, or footwear, or tread on, or even step over anything that is forbidden. Nor may she touch or step over any object which could be lifted as high as her shoulders, or higher, excepting iron things.

The Mansi believed that the invisible 'impurity' of a girl or woman, which was the result of menstruation, materialized in her body in various ways at different times.

a) During menstruation a woman is said to be 'impure' up to her neck, even though she takes special care of her body: as soon as she reaches maturity she no longer removes her special hygienic girdle, the so-called *māńkaš*.[4] Since her whole body, with the exception of her head is 'impure', others who are living in the house, and even the house itself must be protected from her. For the menstrual period she retires to the *māńkol*[5] and lives there in seclusion. If she needs anything from the big house, she may only approach the gate. There, by knocking or shouting, she must ask one of the women to give her the required object or food.

During her stay in the *māńkol,* she is careful not to change her clothes or ornaments. As soon as she notices the menstrual flow, she undoes her

* The present paper was written on the basis of the authors' personal observations supplemented with material obtained during the expeditions to the Mansi settlements along the Sosva and Lyapin rivers in 1956 and 1958. Therefore, the ethnographic and linguistic terminology and facts given here should be considered as referring to the northern Mansi (Sosva and Lyapin dialect groups). The observances and customs which have been described in this paper were widespread with the afore-mentioned group of Mansi in the pre-Soviet era. Now, however, they occur as rare survivals of the past.

1 'pure'—not in the sense of 'physically clean' but clean in accordance with the religious customs of the people in question.

2 *ikšam*—spiritual impurity.

3 *pupiγ*—a spirit which was worshipped by certain families, a whole village, or the whole Mansi people.

4 *māńkaš*—a special hygienic girdle or belt, solid and complicated (sometimes also used as a *chastity belt* against impudent men); instead of cotton-wool this belt is stuffed with *ossi*, i. e., soft tender shavings from the wood of willow trees. In the past *ossi* were used also for drying and wiping the body and for washing dishes.

5 *māńkol*—a small house, unfurnished, sometimes almost unheated; built at a certain distance from the big house and from the roads.

ties (which are used as a weapon in social games against boys or men who try to be familiar) and lets them down to full length. Later on, in the big house, it will be a very complicated procedure to 'clean' the invisible 'impurity' from them. During her periods a woman must use a separate boat and if she travels in a common boat, she must sit down where she stepped in, because she must not step on, or over, pieces of boards which are lying on the bottom of a leaking boat.

b) After menstruation, the 'impurity' of a woman diminishes and is restricted to the front part of her foot, or only to her big toe. The footwear of a woman is always considered impure and, hence must not be lifted higher than the knee, or kept above any object other than iron. In the winter a woman's footwear is dried on the threshold, in the courtyard, and in the summer it is often suspended on a tree of straight growth, far away from the sight of children and men, in order not to transmit its 'impurity' to these 'pure' beings. If a man is contaminated by 'impurity' from a woman or a girl, he will be unsuccessful in hunting and fishing.

Once a girl is married she wears permanently a kerchief to hide (*kasstuŋkwe*) her face from men who are older than her husband. She is even forbidden to look at them and if she does cast an inadvertent glance at them, she will contract an eye disease, or even lose her eye-sight—so 'superior' are men to the women.

Before marriage a girl does not hide her face from men, but, like all women, she must do so before certain male *pupiγ*s and at holy places, viz. *ūs*es.[6]

If a woman approaches a man older than her husband, she calls him *àki*,[7] and she should address him through her kerchief indirectly: 'will he say . . .' or 'will he listen . . .' but never directly: 'say . . . listen . . .'. She must always use the personal pronoun in the third person singular but not in the second person singular.

c) During childbirth a woman is believed to be entirely 'impure' except for her hair.

As soon as a woman notices signs of pregnancy (*pōjtəs* in the Mansi language meaning literally 'something has discontinued'), she starts to observe the strict rules of conduct which govern that condition. For instance, in the 4th to 5th months of pregnancy she must possess a *sōs*[8] which she makes herself if she can, or which is made by some other woman with dexterous hands, who is considered an expert in making such doll. But the maker must not be a female shaman.

Preparing the *sōs* is a process which is concealed from strangers; it is connected with a feast. The person who prepares the *sōs* interrupts her work at various stages to pray and eat, and then resumes work, or turns to the religious consecration of the doll. When the doll is ready, a string of glass beads is hung around its neck and braids are fixed in the back. The waist is girded with a ribbon the ends of which hang loose at the back. The doll is held by means of this ribbon when necessary, so that it stays

[6] *ūs*—a place which was believed to be the abode of the *pupiγ*

[7] *àki*—a designation which was used by the daughter-in-law for the father of the husband and for all men elder than the husband.

[8] *sōs*—a female figurine as big as the thumb. It is made from the soft dried brown growth of the birch tree.

in a vertical position. The *sōs* is meant for the *šāń*[9] 'mother' and establishes a connection between the future mother and the *šāń*. If the woman has a *sōs*, the child will not be too big and the delivery will be easier.

The *sōs* is kept in the woman's *tuťšaŋ*[10] neatly wrapped up in a pretty shawl or in the skin of some small wild animal. Later on this is presented as a gift to the *pupiγšāή*.

Usually in the fifth or sixth month of her pregnancy the woman builds a very primitive cradle (*ampten sān*)[11] in which the infant sleeps for the first week of its life. A small warm coverlet (*lēŋkw*) of swan skin or the winter fur of a hare, as well as a small pillow, *sasnētmil*[12], made of the soft fur of deer, etc. are also provided for.

For herself the woman prepares a very shabby garment and a pair of old shoes which she will wear during the first week after the delivery (later on this dress will be hung on a tree in the woods, or will be exposed in a birch bark bucket, in a part of the forest which is never visited by people). She also weaves a *tōwar*[13], on which she will sit and sleep. All these things are kept ready for use and strictly hidden from the sight of men and strangers.

During the nine months of pregnancy the expectant mother herself, or some other woman who does not belong to the family, takes the *sōs* in her arms from time to time, kisses it and holds it in a perpendicular position by the ends of the ribbon with the tips of three fingers. The woman who is holding the *sōs* must divine whether the expectant mother will live through the birth, whether the child will be healthy, etc. This is revealed from the way the *sōs* moves. It is an auspicious sign if it swings smoothly and quietly; whereas if it hangs motionless or twitches, something unfavourable must be expected. The *sōs* must be taken more frequently in hand as the time of the birth approaches.

Just before confinement, the mother goes to live in the *māńkol*. She takes along the *tuťšāŋ* and the other objects which she has prepared for the event. In the small house she removes her good clothes, and puts on the old ones. Then she or one of the women makes a fire in the *šowal*[14], boils water and prepares flat stones with which to keep her body warm after the birth of the child. (During the delivery not only the close relatives who assist the woman but also girls over ten years of age are permitted to watch the patient and to share with her the labour pains.

[9] *šāń*—a *pupiγ* which is represented as a grey-haired woman who is believed to send the children who are supposed to be born. When she is not angry with the future mother, she sends a boy; otherwise, a girl.

[10] *tuťšaŋ*—a pretty little bag which is decorated with beads, ornaments, chamois coloured fringes, used for holding sewing accessories.

[11] *ampten sān*—(literally: a cup of birch-bark for the feeding of dogs) a primitive cradle, made of birch bark. It looks like a trough.

[12] *sasnētmil*—a sheet of birch wood, about twice the size of the palm of a hand. It is laid on the knees of the child together with a reindeer skin to prevent its feet from becoming crooked, and the blanket from getting wet.

[13] *tōwar*—a mattress made of bundles of long dry grass (with strings).

[14] *šowal*—chimney for the fireplace which is usually in the right corner of the house, facing the door. It is made in the following way: thin vertical poles are placed next to each other in order to form a large-diameter tube; their ends are bound together and they are smudged with a thick layer of clay mixed with a soft grass called *wānsi*.

The girls are told: 'Obey your mother! Look at the pain with which she brought you into this world!')

To facilitate the delivery a special apparatus is made. Two parallel vertical rods are set up about 1 m from each other; their ends are firmly attached to the floor and to the ceiling, and then a cross-bar is fixed to them at a height of 70 cm. At critical moments the woman leans with her breast against the rods and supports herself by her armpits.

In the case of a difficult or late delivery one of the women takes the *sōs* in her hands to find out what the trouble is and whether the child will soon appear. In this manner the woman may discover that the reason for the difficulties or for the delay might be improper language which the woman has sometimes used in connection with somebody, or a husband's love affair with another woman during the pregnancy of his wife, or previous infidelity of the woman who is in labour. If the husband is found guilty, one of the women goes immediately from the *māńkol* to the big house and tells him that his wife cannot recover and that *he* is to blame for it. If he does not wish his wife to suffer or to die, he must at once confess how many times, and with whom, he has betrayed his wife during her pregnancy. With this information the woman hastens back to the *māńkol* and makes one notch on one of the vertical rods for each time the woman has been deceived, according to the confession of the husband. If the patient herself is to blame, the woman turns to her and then makes one notch on the other vertical rod for each time when she has committed adultery. By this method the infidelity of both the husband and the wife is revealed.

During the confinement the husband must stay at home in the big house and wait for the child to be born.

The expectant mother customarily hangs her most beautiful kerchief on the front corner of the *māńkol*, for the *śāń*. The female shaman *ńājt-nē*[15] informs the *śāń* of its presence there. If the patient's condition is serious, the strongest male shaman is called for. He proceeds as follows: He takes an axe in his hands and binds a string to the axe and to its handle so that the axe is hanging with its edge upwards. Then he turns to some *pupiγ* in his thoughts and with the aid of the axe he discovers the cause of the woman's pains and states what kind of sacrifice is necessary. In extreme cases, the shaman prepares a *turmankol*[16]. Sometimes it is revealed to him that a certain *pupiγ*, or *ten-ajn-kuľ*[17] demands the patient herself as the sacrifice. It is believed that the shaman has the power to save her life by indicating somebody else, a complete stranger. This innocent person may die soon after the substitution. The shamans were greatly feared by the simple people because of this 'power' of theirs. They used to boast that

[15] *ńājt-nē*—a female shaman. She carries a thin and slender piece of board, the length of a finger, fixes the end of a knife to it, then lifts the board and sits waiting for the swaying of the suspended knife.

[16] *turmankol*—(literally: 'dark house') The shaman works at night in the house, which is completely dark, either alone or in the presence of many people. He finds out the cause of the illness in a conversation with the *pupiγ* by means of a musical instrument called *sāŋkwoltap*.

[17] *ten-ajn-kuľ*—something infernal, having invisible spiritual existence.

they would send to the other world any person who would refuse to pay due respect to them.

When the child is born there is general rejoicing. One of the women cuts the *pùkńi,* the placenta. When this is done, the child's face is smeared with blood so that the child will have red cheeks later on. The placenta is then tied down and the wound is sprinkled with dust from a corner of the hut. The woman who has cut the *pùkńi* will be addressed by the child later as *pùkńiśāń* (literally: 'navel mother'), and she, in turn, calls the child *pùkńipiγ,* or *pùkńiāγi* (literally: 'navel son' or 'navel daughter'). The child's mother must present this woman with a knife, and sometimes also with other objects. The *pùkńiśāń* washes the child and puts it into the cradle which has been supplied by the mother. Then a meal is prepared with which the birth of the child is celebrated for the first and last time.

The dress of the mother must be torn down in the front from top to bottom so that it becomes a gown. This happens because the Mansi say that the child dropped out through the belly and tore the dress. (After the birth of the child the mother wears a girdle which is made of hair to help her body regain its normal shape.)

Dressed in her worst clothes for one week, the mother must run out in the street, exposed to the cold in the winter, and to the mosquitoes in the summer. With this the first phase of purification is over,[18] and the woman may change her clothes for better ones. But she is not through with the ordeal yet: she has to undress in the frost, or expose her body to legions of mosquitoes, which finally purify her (the same method is resorted to during menstruation at any arbitrary place such as in a wood, on a road, etc.).

At the end of the first week the mother cleans the child of the invisible 'impurity' which is believed to have been transferred to it from her. For the purification of the child the *sōs* made for the *śāń,* as well as its *lālwa*[19], is used. Then the child is put to sleep in another cradle, or in another bed. The mother's old dress and the child's first cradle are carried to the woods and hidden in the most 'impure' spot, behind the *māńkol,* where all the first cradles of the children who are born in that *māńkol* are hung up. The *śāń-pājup*[20] (literally: 'the mother's basket') is also hung up on the branch of a tree, as high as possible. If the child is to live long, the basket must remain hanging there for a long time.

It is not until three months after the childbirth that the woman decides to make preparations for her return to the big house. For this purpose she must go through careful purification ceremonies to free herself from invisible 'impurity': she makes a *sāpjiv-χāp*[21] and adorns it with little rings which are made from very thin twigs. In the middle she makes a little hole, into which she lays the *sōs* (a piece of dry brown birch wood), lights it and places on it the *lālwa* which emits special 'purifying' odour as it burns. The water in which the child and the mother are washed is prepared in a special manner. While it is being warmed, the burnt *sōs* is

[18] *taktalaχtil*—the process of getting rid of the invisible 'impurity'.

[19] *lālwa*—a dried piece of castor i.e. a musk bag of the beaver.

[20] *śań-pājup*—an improvised cover of birch-bark in which the placenta is placed.

[21] *sāpjiv-χāp*—this looks like a toy ship made of rotted birch or willow which easily yields to the woman's knife.

thrown in, then it is grated into powder in the water and then dissolved. The water changes slightly its colour and becomes soft and velvety. A piece of chewed *lālwa* which mixes with it and dissolves is thrown into the water obviously for the purpose of rendering it more sacred. As soon as the water has settled, the woman washes herself first and then the child. When she is clean, she puts on her clothes. Then she lays on a stone a red-hot axe which steams and causes a hissing sound when it gets into contact with the moist *ossi* which have been thrown on it. With her child and with all things that were with her in the *māńkol*, the woman jumps over the steam shouting *kuχ-kuχ* thereby getting rid of her load of 'impurity', which is carried away by the steam. When this has been done, she takes the *sāpjiv-χāp* in which the smouldering *sōs* lies with the *lālwa*. The smell of the *lālwa* spreads through the whole house. The woman begins to carry the *sāpjiv-χāp* all over the *māńkol* holding it under objects and chairs. The smoke and smell of the *sōs* and *lālwa* clean away the last remains of 'impurity'. Later the woman holds also the child in the cradle over this little boat and steps over it several times herself. At last she is completely 'clean'. She can now go into the big house, yet not before one more ceremony has been performed: one of the women takes the child together with the cradle which is covered with a beautiful kerchief, while the mother grasps the *tuťšaŋ* in one hand, and the *sāpjiv-χāp* in the other. Then, having made one first step, they stop and put the *sāpjiv-χāp* on the floor over which the woman with the child jumps first, then the mother, saying: '*ś̌anttịγ-ponttịγ*'. (The meaning of these two words is not quite clear.) They now go out of the *māńkol*, but on the way to the big house the same ceremony is repeated several times.

When they reach the house the woman who is carrying the child knocks at the door and says: 'Hi! open the door, the woman (man) has arrived on four reindeer, on five reindeer', or 'the woman who owns horses, who owns cows has arrived'. Thereupon she opens the door a little with her elbow and shows the child, with its feet forward; then she closes the door again. This is repeated four times if the child is a girl, and five times in case it is a boy. Finally the woman flings the door open and jumps with the child into the house, saying: '*ś̌anttịγ-ponttịγ*'. Then the child is carried to the front corner where the cradle is set up. To celebrate the arrival of the child in the house, a feast is arranged.

The woman who carried the child to the house is from now on called *altumś̌āń* (literally: 'mother by carrying'). The kerchief with which the child was covered is given to the *altumś̌āń* as a present. The ribbon or string, which was placed on the knees of the child when it was brought to the big house, and pierced with four needles in the case of a girl and with five in the case of a boy, is also presented to this woman. The needles represent skunks. Later also the *altumś̌āń* has to present her *altumāγi* (literally: 'daughter by carrying') with a gift.

Besides the *pùkńiš̌āń* and *altumś̌āń* the child must have also a *pernaŋś̌āń* ('godfather') and a *pernaŋāś̌* ('godmother') who usually give crosses to the child and expect to receive a present. No christening ceremony is performed.

It is believed that some time after the birth of the child the soul of a dead person whose will to live has been strong moves into it, i.e. the child 'clashes with it' (in Mansi *l'aχtχati*)[22]. The souls of one or several dead persons may move into the body of a newborn child. A dead man may transmit his soul five times to children who are born at different times, a woman may transmit hers four times, because a woman has four souls whereas a man has five.

The soul of a dead person is believed to enter the child when it begins to cry at night. It is supposed to be disturbed by something or somebody. At daytime when it is sleeping, it is carefully moved with the cradle; then somebody—usually one of the women—lifts it carefully while thinking of one of the dead relatives. If the child becomes suddenly very heavy they have found out who troubled it and it will stop crying.

If a child is still-born or dies during the first week of its life, it is not buried in the common cemetery, but in the dry soil of a completely isolated pine forest. When a commemoration service is held for the child, no bonfire is lit at the grave but warm food is brought from the house. This divergence from the normal funeral repasts, where everything is prepared on bonfires in the cemetery, is due to the belief that the child, being so small, can neither light a bonfire nor eat anything warm.

LITERATURE

(Šavrov, V.) Шавров, В. (1871) Краткие заметки о жителях Березовского края (Short Notes on the Inhabitants of the Berezovsk District), in *ЧОИДРМУ*, April–June, Book Two.

[22] *l'aχtχatuŋkwe* (possibly from the word *l'eŋntχatunkwe*) 'clash against one another' as the soul of the dead person against the child.

Two Purification Rites in the Bear Cult of the Ob-Ugrians

by

B. Kálmán, Debrecen

The bear cult of the Ob-Ugric peoples, probably the best developed and most interesting of all cults connected with this animal among the Finno-Ugric peoples, has given rise to a considerable amount of studies which are available in the literature.[1]

In this paper I wish to deal with two ceremonies in some detail: sprinkling with water (or pelting with snow) and fumigation.

1. THE SPRINKLING WITH WATER

A great many religions (Hinduism, Mohammedanism, etc.) prescribe ritual washing or baths, for it is natural that the ways of keeping the body clean should be adapted, in a symbolic way, to the 'purification of the soul'. Thus in Christianity the rite of baptism, or the use of holy water 'cleanses one from sin' and renders the object or person who has been sprinkled with holy water, sacred or pure.

With this in mind we would like to examine similar rites which have been performed in the course of bear festivities. Here too sprinkling with water implies spiritual purification and deliverance from sin.

Water-sprinkling rites are often mentioned by research workers and occur also in the bear songs themselves.[2]

The lucky bear hunter has snow thrown on him in winter and is sprinkled with water in summer. This act features among the ceremonies of the bear cult on three different occasions.

a) According to Gondatti[3], when the bear is killed, the hunters throw snow at each other in winter, and moss and earth in summer for the purpose of purification. Patkanov also describes this custom: 'When hunting has come to a successful end, the hunters pelt each other with snow, and in summer with earth, or else they sprinkle each other with water, a custom which I could never understand'.[4] Pavlovskij[5] also mentions it, after Gondatti and Kharuzin.

But we have no other reports about this rite. Neither the Mansi (Vogul) nor the Khanty (Ostyak) bear songs refer to it. Apparently it was restricted to a small area which was inhabited by the Southern Khanty. Pavlovskij[6]

[1] Munkácsi and Kálmán (1952) pp. 15–164; Kannisto and Liimola (1958) p. 114; Kannisto, Virtanen and Liimola (1958) pp. 113, 339–83 and the bibliography given therein.

[2] Cf. e.g. Munkácsi and Kálmán (1952) pp. 74–8; Kannisto, Virtanen and Liimola (1958) pp. 360–1; Kannisto and Liimola (1958) p. 404.

[3] Karjalainen (1921–1927) Vol. III, p. 197.

[4] Patkanov (1900) Vol. II, p. 127.

[5] Павловский (1907) p. 178.

[6] Павловский (1907) p. 178.

and Karjalainen[7] are of the opinion that it is probably a religious custom, a symbolic purification of the hunter from the sin of having killed the bear. Of course, it may also be interpreted as an attempt to mislead the bear, by pretending that they did not recognize it because of the storm or snow-fall which is faked deliberately in order to bring about a phenomenon which could be blamed for their 'mistake'.

b) A more general custom, more characteristic of all the Ob-Ugrian peoples, is for the people of the village, and particularly the women, to pelt the home-coming bear hunter with snow in winter, or to sprinkle him with water in summer. As the hunter is carrying the skin of the killed animal, much of the water falls on the bear, too. Since most bear songs refer to it, this custom must be old, yet nobody before Munkácsi had mentioned it. His Konda Mansi informant describes this manner of welcoming the hunter home in the following sentence: 'The women jump at him and begin to pelt and sprinkle him with snow and water'.[8]

Pápay writes about the rite as follows: 'As they approach the village, they utter loud shouts. At this signal all the village folk go out to meet the hunter, and in winter they sprinkle each other with snow and in summer with water, and then they carry it (the bear) into the village. When they reach the house of the hunter, the scene with the snow and water is repeated'.[9]

A similar description is given by Kannisto: 'The hunters and the villagers start to shout loudly and begin to pelt each other with snow, or sprinkle each other with water in summer'.[10] After the carcass of the bear has been carried into the house of the man who killed it, and the village people have swarmed in to take a look at it, the hunter picks up a birch-bark box which has been filled with snow and placed before the bear. The hunter 'turns it toward the people and strikes the bottom with his fist so that the snow falls all over the people and the box lands on the floor. In this manner the bear is said to take part in the merry snowball fight of the villagers'.[11]

A similar rite which was practised by the Khanty who are living along the lower course of the Ob river is described by Karjalainen. Along the upper course of the Ob, in the region of the mouth of the Vakh, this rite is connected with washing in a dish.[12] The religious character of the rite is still more evident in the case of the Konda Khanty. When the festive march reaches the village: 'One of the women holds a water jug in her hand, the other a censer. The skin is smoked, then sprinkled with water three times. The man who sprinkles, "frightens" the bear with the words *χors, χors, χors*! Also the hunters are sprinkled. In Tsingala every one who passes by sprinkles water on the bear; the killer of the bear is thrown into the water, if he is caught, or at least his head is made wet. Every newcomer is sprinkled with water.

[7] Karjalainen (1921–1927) p. 200.
[8] Munkácsi (1892–1921) Vol. IV, p. 416.
[9] Pápay and Fazekas (1934) pp. 13–4.
[10] Kannisto and Liimola (1958) p. 404. For slight local variants of the custom, see Kannisto, Virtanen and Liimola (1958) pp. 348–9.
[11] Kannisto and Liimola (1958) p. 404.
[12] Karjalainen (1921–1927) Vol. III, p. 201.

This is a game, but it has probably a serious background upon which the purificatory, i.e. exorcising ceremonies were based, which were supposed to prevent anything improper from happening'.[13]

Many references are found in the Mansi bear songs to this snow fight and sprinkling which accompany the home-coming of the bear hunters:

> 'A snow game is played around me'.[14]
> 'Many girls with scarts on
> Welcome us, playing snow games,
> Welcome us, playing water games.
> Many young men with arrows in hand
> Receive us, playing snow games.
> Many young men with hows in hand
> Receive us, playing water games'.[15]

That this 'game' is not simply a display of joy but a rite of religious significance is proved by the occurrence of a bear song with a tragic ending. There an old man from the Windy Lake (*votiŋ tūr ōjkä*) brings home his wife who has been killed by the bear, together with the slaughtered beast. His younger brothers come out and receive him with great lamentation, yet 'they sprinkle water on their elder brother'.[16]

This motif frequently turns up in Kannisto's northern bear songs too:

> 'The snow game is being played,
> The water game is being played'.[17]

The bear songs which were cited above are from the northern regions (the Sosva and Upper Lozva rivers). In the bear songs from the southern dialect area (middle section of the Lozva) as they have been recorded by Reguly (some of them were put into the northern dialect by Munkácsi), this phase in the ceremony is always linked up with the same simile: The sprinkled hunter and the sprinkled bear skin are compared to a male squirrel which is swimming in the water. In the songs from the middle course of the Lozva we hear only about sprinkling with water and no mention is made of pelting with snow. It should, moreover, be observed that, whereas in the north the pelting with snow is done by men and women together, in the Mansi language area farther south, and among the southern Khanty, the sprinkling is done only by women. In the Mansi bear songs on which Munkácsi has commented this rite occurs in four places. One of them is: 'Like male squirrels swimming in water—damn it!—so the little women soaked us through and through'.[18]

[13] Karjalainen (1921–1927) Vol. III, p. 200.

[14] Munkácsi (1892–1921) Vol. III, p. 101.

[15] Munkácsi (1892–1921) Vol. III, p. 317; similar references are found in several other bear songs. Munkácsi (1892–1921) Vol. III, pp. 146, 247, 365, 476.

[16] Munkácsi (1892–1921) Vol. III, p. 380.

[17] Kannisto and Liimola (1958) p. 130. See also pp. 49, 93, 115, 148, 158, 191, 206, 228, 239, 337.

[18] Munkácsi (1892–1921) Vol. III, pp. 342–3. See also pp. 203, 295, 354–5.

In the western bear songs which Kannisto cited this motif does not occur. In an eastern prose fragment from the Konda region we find the following sentence:

'Six wet dances, seven wet dances he [the hunter] dances with him [the bear]'.[19]

A song from the Tavda region (south) concludes as follows:

'Three women poured water on it [the bearskin]'.[20]

These two short references reflect the same kind of peculiarities as those which were mentioned in the songs from the west in the recordings of Reguly.

Yet among the Khanty bear songs reference is made to this practice only in one single longer bear song which is given by Patkanov:

'The many women of the town
The many men of the town
From a mass of water
Have lifted a nice cloud
To the sky.'[21]

Patkanov adds a comment here: 'During the bear festivities it is customary to sprinkle each other with water. In the winter people pelt one another with snow'.[22]

As a summary of the foregoing we may say that the custom of pelting the home-coming hunter with snow or sprinkling him with water is, in all probability, a purification ceremony of religious origin, by means of which the bear hunter and the participants in the celebration were believed to cleanse each other from sin. Purification is important for two reasons: (a) the bear must be made to believe that they are not to be blamed for killing it and eating its flesh and, (b) only decent, sinless people may remain in the proximity of the bear. Infantyev comments also on the fact that the guests who arrive at the feast must be sprinkled with water before they approach the bear and pay honours to it.[23] According to Kannisto[24], on the second day of the bear feast, after the meat has been consumed, the host sprinkles his guests with water. This act, of course, has also as its purpose the purifying of the guests from the sin which they have committed (the eating of the bear meat).

c) The third occasion for the snowball battle (although not mentioned in the bear songs) is at the end of the bear feast. Gondatti tells us that while men are carrying the bear out, women pelt each other with snow, sprinkle each other with water and dance in the house where the bear feast is held.[25] Kannisto watched the following scene among the Sosva Mansi: '. . . the moment the bear and the men who are carrying him enter the courtyard, a violent snowball fight begins where nobody is spared.

[19] Kannisto and Liimola (1958) p. 252.
[20] Kannisto and Liimola (1958) p. 397.
[21] Patkanov (1900) Vol. II, pp. 202–3.
[22] Patkanov (1900) Vol. II, p. 236.
[23] Инфантьев (1910) p. 79.
[24] Kannisto and Liimola (1958) p. 411.
[25] Karjalainen (1921–1927) Vol. III, p. 224.

Even the bear gets his face full of snow. In the general confusion, two men take hold of the bear's table and carry it behind the house, to the sacrificial field.'[26] This game, an artificial storm, is probably staged for the purpose of diverting the bear's attention from the fact that the feast has come to an end and it must be carried out of the house, which might offend the bear.

If after the festivity the bearskin is sold to a Russian merchant, a few shovelfuls of snow are cast after him on his departure, according to Infantyev.[27] This again is an act to soothe their conscience: the 'snow storm' prevents them from seeing the merchant and they are thus unaware of the removal of the bearskin.

The 'snow and water game', then, has two components: The sprinkling with water is originally a purification rite which cleanses the bear hunter and the participants in the feast from the sin of killing the bear and of eating its flesh. The pelting with snow is partly a development of this symbolic rite. The fact that this water, like the parts of the bear's body and the implements used at the bear feast has a special name different form the ordinary one, argues for the cult origin of the water-sprinkling.[28] The snow fight, on the other hand, is one of the various types of deception which belong to the bear feast and 'mislead' the bear. Their aim is to make the bear believe, through the evocation of the snow storm, that it was not recognized or passed unnoticed when it was carried out from the house where the bear feast had been held. For a similar reason the eyes of the animal are covered when the bear meat is eaten, the crowing of a raven is imitated, the men perform scenes of the bear's life, for which they put on masks.

2. FUMIGATION

The fumigation rite is a ceremony common to several religions. The original purpose of appyling smoke was, as in the case of water, probably also physical—for cleansing, or getting rid of skin diseases and parasites. When fumigation became a symbolic religious rite, the pleasant odour of the smoke (among the healing incenses there were some with a pleasant smell) in addition to its cleansing and healing effect enhanced the solemnity of the occasion. The rite of fumigation for cleansing and exorcising evil spirits was widespread among Jews, Christians and in several oriental religions.

The use of smoke for purification purposes is frequent with women among the northern peoples. For their 'impure periods' (menstruation, labour and delivery) women withdraw to a separate 'little house' which they are allowed to leave only after fumigation. Moreover they must fumigate every object which they have touched during those periods.[29] If they happen to step on, or even over, a piece of man's clothing, it has to be smoked before it can be used again.[30]

[26] Kannisto and Liimola (1958) p. 417.
[27] Инфантьев (1910) p. 80.
[28] Cf. Kannisto, Virtanen and Liimola (1958) p. 352.
[29] Зуев (1947) pp. 63–4.
[30] Kannisto, Virtanen and Liimola (1958) p. 395.

Fumigation is part of the rites which are associated with the killing of the bear and the feast of the bear. Mostly beaverskin, birch tinder, and—when no tinder is available—juniper or pine branches are used to produce smoke.[31] Several types of occasions when purification by smoking is resorted to are mentioned in our sources.

a) The Khanty alone are reported to embrace and kiss the bear[32] after killing and before flaying it. They eat of it, too, but the meat is smoked first.[33] Similar customs occur among other northern peoples.[34]

b) Only the Khanty bear songs refer to smoking before carrying home and 'cradling' the bear.[35] On the other hand, in the case of the northern Mansi, when the bear is carried home 'a censer, a bowl where burnt tinder is smouldering, is put into the sledge with the bear'.[36]

c) When the hunters arrive, the bear is smoked simultaneously with, or previous to, the sprinkling. Evidence of this custom is gained chiefly from the northern bear songs recorded by Kannisto:

'By hands smoky with tinder fire
 I was attacked.'[37]
'By a hand which offers the morning (smoke of)
 beech tinder, I am surrounded
By a hand which offers the evening (smoke of)
 beech tinder I am surrounded.'[38]
'An old woman with arms grown old,
An old woman with legs grown old,
Walks up to me with (sweet-)smelling
 tinder-fire in her hand.'[39]

Numerous traces of this rite are found among the Khanty. With reference to the southern Khanty we have Karjalainen's reports.[40] Pápay, on the other hand, collected songs with this motif from the northern Khanty.[41] Steinitz recorded a bear song with the following lines from the Khanty:

'To the small hunting lodge
They brought (me).
By two brave heroes, grown up together,
With tinder fire and beaver musk in their hands
We were received . . .'[42]

[31] Cf. Kannisto, Virtanen and Liimola (1958) pp. 360–1.
[32] Cf. Munkácsi and Kálmán (1952) p. 67.
[33] Karjalainen (1921–1927) Vol. III, p. 197.
[34] Holmberg (1926) pp. 21–3.
[35] E. g. Pápay and Fazekas (1934) pp. 57, 81–2; Steinitz (1939) Vol. I, p. 318.
[36] Kannisto and Liimola (1958) pp. 403–4.
[37] Kannisto and Liimola (1958) p. 93.
[38] Kannisto and Liimola (1958) p. 108.
[39] Kannisto and Liimola (1958) p. 130. See also pp. 148, 191, 216, 239, 337.
[40] Karjalainen (1921–1927) Vol. III, p. 200.
[41] Pápay and Reguly (1905) p. 256; Pápay and Fazekas (1934) pp. 60, 83, 86, 91, 96.
[42] Steinitz (1939) Vol. I, p. 318.

d) Kannisto mentions in his ethnographic studies[43] that a censer lies on the threshold of the house where the bear feast is held. It is over this that the hunters step into the house and carry the bear inside. The bear songs also mention this custom.[44]

e) On the occasion of the bear festivities tinder glows on the table of the bear during the entire course of the ceremony.[45] And in the morning the housewife smokes the bear. This is also mentioned in the bear songs:

'There stands the woman, with her hair up,
Wife to the very good son of your hostess,
Her hand fragrant with the smoke of evening birch-tinder
Has been standing before you for a long time,
Her hand fragrant with the smoke of morning birch-tinder
Has been standing before you for a long time . . .'[46]

It is noteworthy that the smoking, after the bear has been brought home, is done exclusively y women both in the Mansi and in the Khanty songs (those of the berkal region excluded). According to Pápay the explanation is as follows: 'The women ... holding smoking birch-tinder in their hands, surround the bear. In this way they are, in fact, cleansing themselves, for according to Ostyak belief, women are impure and may approach holy objects only after such purification rites'.[47] There is no doubt about the correctness of this statement. At the beginning of the present paper I have given quotations which confirm this theory. Yet, however correct it may be in itself, every rite which is connected with the smoking of the bear cannot be accounted for by this single explanation.

This smoking of the bear during the home-coming and during its reception, as well as the steadily glowing tinder on the bear's table, are partly symbols of respect and partly meant to drive away the evil and harmful thoughts of the bear. That is to say, the aim of smoking is partly tribute and partly exorcism.

Self-purification underlies most of the fumigation rites of women. The aim of cleansing is evident also in the custom which demands that the bear hunter step over the purifying smoke before entering the house where the feast is held.

The constant or frequent burning of incense in the house where the bear feast is held ensures the solemnity of the occasion and serves the cleansing and exorcising purpose of keeping away the evil spirits.

However much the sprinkling with water and the use of incense recall similar Christian rites, primarily in the Orthodox Church, it is unlikely that they have anything to do with Christian influence. These symbols are very widespread, and in many places their origin goes back to a period much earlier than Christianity. They have only a single primitive element in common which is purification.

[43] Kannisto, Virtanen and Liimola (1958) pp. 348–9.
[44] E.g.: Kannisto and Liimola (1958) pp. 206, 229.
[45] Kannisto and Liimola (1958) pp. 356, 361.
[46] Munkácsi (1892–1921) Vol. III, p. 195, also p. 211; Kannisto and Liimola (1958) p. 366; Steinitz (1939) Vol. I, p. 330.
[47] Pápay and Fazekas (1934) p. 14.

REFERENCES

Holmberg, U. (1926) Über die Jagdriten der nördlichen Völker Asiens und Europas, in *JSFOu*, Vol. XLI/1.
(Infantyev, P. P.) Инфантьев, П. П. (1910) *Путешествие в страну вогулов* (Journey to the Land of the Voguls), Saint Petersburg.
Kannisto, A. and Liimola, M. (1958) Wogulische Volksdichtung IV, in *MSFOu*, Vol. 114.
Kannisto, A., Virtanen, E.A. and Liimola, M. (1958) Materialien zur Mythologie der Wogulen, in *MSFOu*, Vol. 113.
Karjalainen, K. F. (1921–1927) Die Religion der Jugra-Völker, Vols I–III, in *FFC*, Nos 41, 44 and 63.
Munkácsi, B. (1892–1921) *Vogul népköltési gyűjtemény* (Collection of Vogul Folk Poetry), Budapest.
Munkácsi, B. and Kálmán, B. (1952) *Manysi (vogul) népköltési gyűjtemény* (Collection of Mansi [Vogul] Folk Poetry), III/2, Budapest.
Patkanov, S. (1900) *Die Irtysch-Ostjaken und ihre Volkspoesie*, St. Petersburg.
(Pavlovskij, V.) Павловский, В. (1907) *Вогулы* (The Voguls), Kazan.
Pápay, J. and Fazekas, J. (1934) *Északi osztják medveénekek* (Northern-Ostyak Bear Songs), Budapest.
Pápay, J. and Reguly, A. (1905) *Osztják népköltési gyűjtemény* (Collection of Ostyak Folklore), Budapest.
Steinitz, W. (1939) *Ostjakische Volksdichtung und Erzählungen I*, Tartu.
(Zuev, V. O.) Зуев, В. О. (1947) Материалы по этнографии Сибири XVIII века (Ethnographical Material from 18th Century Siberia), in *Тр ИЭ*, Vol. V.

On Some Ancient Anthropomorphic Images from West Siberia

by

V. Moszyńska, Moscow

The majority of the sacred figures (*toŋχ, pupiγ*) in the ethnography of the Ob-Ugrians can be divided into: (1) zoomorphic and (2) anthropomorphic groups. The former, which probably preserves totemistic notions of the remote past will not be dealt with in the present paper. Of the anthropomorphic figures, which are rather variegated, only a few have been selected for analysis, and one of the possible ways of their development will be followed up.

In his study on the religious conceptions of the Ob-Ugrians, Karjalainen maintains that, in folklore, the word *toŋχ* is often used instead of the word *χatai*; the former may also have the meaning 'the ghost of the dead', of those dead, however, who have long been confined in a new frame, a new body. Only such a person can become, through transfiguration, a *toŋχ* who having been in this life in command of faculties greater than those of ordinary mortals, possesses a higher and more powerful spiritual personality than his fellow-creatures, and whose actions are said to exercise a fair amount of good or bad influence on the living. Analysing the word *toŋχ*, Karjalanien comes to the conclusion that it is synonymous with the word *is* 'soul', with the sole difference that *is* refers to the soul of all deceased persons whereas only a person of distinction can become a *toŋχ*.[1] Therefore *toŋχ* may be assumed to be the name for the representation of some very powerful member of the clan, usually of some ancestor.

As seen from the available material, such figures could be made of various substances.[2] Of special interest are the wooden ones with masks of a different substance. For instance, B. Munkácsi refers to J. B. Müller's information about a wooden idol that has a facial mask made of white sheet iron, in which the eyes, nose and mouth have been cut.[3] V. N. Chernetsov, on the basis of information collected among the Mansi (Vogul) people, maintains that it was not the wooden but the metal figure which was kept inside the wooden case that was worshipped as a sacred idol.[4] Gr. Novitskii makes repeated mention of figures which consist of a metal and a wooden part. When describing the 'idol' he found in the Shorkorovskii *yurtas* (tribal village) he points out that it consists of a log wrapped in cloth, and only the upper part of it is a metal mask. ('The idol which was found in the Shorkorovskii *yurtas* has in the middle a small log, fifty years old, wrapped in cloth; the upper part is a mask in the likeness of a man, carved out of iron plate.'[5])

[1] Karjalainen (1921–1927) Vol. I, p. 49.
[2] Новицкий (1884) pp. 50, 56–7, 82, 100; Pallas (1786) p. 332; Munkácsi (1904) pp. 67, 69; Karjalainen (1921–1927) Vol II, pp. 41–2; Гондатти (1888) pp. 7, 16–7.
[3] Munkácsi (1904) p. 69.
[4] Чернецов (1927) p. 23.
[5] Новицкий (1884) p. 50.

Karjalainen reports that among the Vasjugan Khanty (Ostyaks) the heads of such figures are made of small pieces of stone which have been wrapped in hides and cloth. The same author refers to Witsen, according to whom the Khanty people had figures of the clan's ancestors with faces roughly sculptured out of stone.[6]

The wooden parts of these figures were, of course, liable to delay and had to be replaced at regular intervals of, say, three, five or seven years. This is in itself an evidence of the main and really sacred part having been the one which was made of metal or stone.

Together with the archaeological finds from Western Siberia there have come to light figures representing human heads or faces which are made of various material and with varied techniques. Some of these are dealt with below.

1. A mask from the Makar-Visyng-Tur settlement (on the Lyapin, a left tributary of the Northern Sosva river)[7], cast of white bronze (size 7·1 cm,) represents an oblong human face of Europeoid type (Fig. 1). The nose, which is straight and remarkably long, courses down half the length of the face in the form of a somewhat protruding edge. The eyes, the brows,

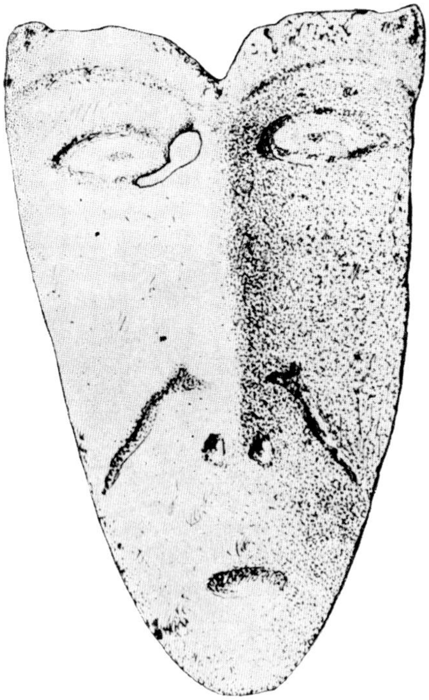

Fig. 1. A mask from the Makar-Visyng-Tur settlement

Fig. 2. A mask from Leushi

[6] Karjalainen (1921–1927) Vol. II, p. 42.
[7] Чернецов (1953a).

the mouth and the furrows starting from the base of the nose are represented by hollows and deep lines which gives evidence of the figure's having been cast in a mould. On the reverse side there is a small ring by which the object can be hung up.

The mask was discovered in an ancient settlement of the Ust-Polui culture, a complex that may be dated back to the last century B. C.

2. A mask from Leushi[8] (basin of the Konda river), cast of white bronze (size 11·7 cm) represents a human head, the pointed helmet of which is decorated with peculiar ornaments (Fig. 2). The oblong face has a long straight nose. A short handle protrudes from the chin obviously for the purpose of fastening the mask to an object. This mask which was found at the site of a burned-down sacred building near the village of Leushi is an object of archaeological value and probably originates from the neighbouring old settlement of Leushi which, by analogy of the ceramics found there, dates back to the first century A.D.

3. A mask from Kintusov[9] (upper reaches of the Salym river, a left tributary of the Ob), cast of white bronze (size 9 cm), represents a human head, the pointed helmet of which is decorated with a row of circular ornaments (Fig. 3). The features of the face, the long nose, the eyes and the mouth are represented schematically. A short handle protrudes from the chin.

According to V. N. Černecov, the mask was found in an archaeological complex dating somewhere between the 10th and 12th centuries.[10]

4. A mask from Cape Pochevash (on the right bank of the Irtysh river near Tobolsk)[11], carved from a panel with traces of silver covering still visible (height 9·1 cm), represents the head of a man with a helmet (Fig. 4). The features of the face are rendered schematically, a handle protrudes from the chin.

This mask is an isolated find and can hardly be connected with the settlement at the Cape Pochevash. Nor has it yet been possible to figure out the epoch to which it belongs.

5. Among the archaeological finds of anthropomorphic character from Western Siberia there are also sculptures.

Sculptured man's head[12] (Tui river, a right tributary of the Irtysh) in dark-green serpentine (height 10·7 cm), carefully cut and polished (Figs 5a, b, c, d). The realistic representation of the head permits us to describe the person anthropologically. He must have been a man of the Europeoid type with a low skull, a thin and long face and a narrow forehead. The nose is strongly protruding, with a well-formed high bridge, broad and arch-like at the base. The superciliary ridges are distinct though only slightly developed. He had a broad mouth with thin lips. The square protruding chin and the powerful lower jaw are emphasized. The mouth is rendered by a straight deep-cut line, the superciliary ridges are semioval, the eyes deep, round holes leaving room for inlaying some other material. The ears which are pierced probably for the fixing of earrings, are represented rather

[8] Чернецов (1957).
[9] In the Tobolsk Museum, No. 5200.
[10] Чернецов (1957) Plate XLVII–10, p. 225.
[11] In the Tobolsk Museum, No. 5581.
[12] Ibid. No. 5840.

Fig. 3. A mask from Kintusov

Fig. 4. A mask from Pochevash promontory

schematically. The head is covered with a closely fitting helmet which is adorned along the rim with an indented design. The neck turns into a shaft which is round in the front and flattened at the nape. The upper part of the nape is bordered by a rectangular protrusion which can be seen very well in profile. The lower end of the shaft, except for the flattened part, is edged by a little round roll.

Another piece of sculpture (from the shore of the Ir river, a right tributary of the Ishim river) represents a two-headed or rather two-faced creature.[13] It is made of dark-green serpentine. The faces resemble each other and, therefore, may be described together. They show a narrow-faced Europeoid type, like the one which has been described above. The massive lower jaw is made markedly conspicuous. The eyes are represented by round hollows, the mouth by a straight carved line. The ears are carved rather minutely; in one of them even the spiral of the auricle can be seen. This two-faced head is completed by a long cylindrical shaft, the lower end of which is bordered by a roll.

Unfortunately, no data are available concerning the circumstances under which these two pieces of sculpture described under 4 and 5 were found,

[13] In the Omsk Museum.

nor is it known which archaeological complex they are connected with. Therefore it is rather difficult to date them. However, by considering the style in which they are featured and by comparing them with other sculptures from West Siberia they may be assumed to date from the last century B.C. or the first century A.D.[14]

These stone figures from the Tui river and from the Ir river are comparable to with each other not only because they have been finished with the same technique and of the same material but also because both have a shaft which serves, in all probability, the purpose of attaching wooden parts as an 'extension' to them. This is particularly visible on the piece from the Tui river: the flat back of this one's shaft must have been closely adaptable to a similar semioval shaft. The fastening was secured by the roll on the lower end which prevented the frame from slipping off.

One more detail is worth mentioning. It has been said that the eye sockets of the Tui sculpture are deep holes in which inlays of some other material could have been placed. Some sacred Ob-Ugric representations of tribal ancestors are known to have such inlaid eyes. P. S. Pallas[15] for instance mentions wooden representations of 'idols' whose eyes are set in. The figure of *sūjur ēkwa*, a creature which was worshipped on the Lyapin river, had similarly inlaid eyes. According to a legend, the *sūjur ēkwa* sometimes closes its eyes—a bad omen which foretells disaster.

Fig. 5a. A sculptural representation of a man's head, Tui river. (Front view)

The *šoχriŋ-ōjka* also had inlaid eyes. K. D. Nosilov describes this figure as follows: '. . . resembling a man, in the darkness of the barn there sits the idol wrapped in furs, shawls and girdles, with three-pointed caps on its head, made of black, blue and red cloth. It is hard to see anything of the idol's face for the furs, shawls and laces and because the caps are drawn over its eyes. In this dark corner of the barn, with its monstrous nose, distorted and protruding, it seemed to me so hideous, its tin eyes stared at me so blindly that I felt a strong impulse to turn away'.[16]

When a person dies among the Ob-Ugrians, an effigy is made of him.[17] The method used varies from district to district. These effigies are described

[14] For particulars see Мошинская (1952).
[15] Pallas (1786) p. 332.
[16] Носилов (1904) p. 88.
[17] Новицкий (1884) p. 46; Шавров (1871); Белявский (1833) pp. 98–9; Гондатти (1888) p. 43; Старцев (1928) p. 124.

Fig. 5b. A sculptural representation of a man's head, Tui river. (Right-side view)

in detail and discussed by V. N. Černecov.[18] Here it should only be noted that they were not all made of wood. Sometimes, for instance, even the hair of the deceased was included in the representation. As he points out, the terms which are used to designate the figures are varied. Besides the names *moχar* and *akań* the word *soŋət* is widespread, particularly among the Khanty. The words mean deceased, dead, image of the dead, and skull. Analysing the conceptions of the soul among the Ob-Ugrians, V. N. Černecov comes to the conclusion that the effigy of the deceased, his *soŋət*, is the receptacle of the fourth reincarnation of his soul. This explains why the hair of the deceased is used for the preparation of his image. Connecting the idea of the fourth soul which continues to live in the body after the death with the meanings (skull, deceased, image of the fourth soul) of the word *soŋət*, he infers that *soŋət* is the surrogate for the body of the deceased, the surrogate for the deceased himself.[19] Karjalainen comes to a partly similar conclusion when he says that the figure which has the hair and the clothes of the deceased, and in which his soul continues to live according to the conception of the Khanty, is the deceased himself.

As has been mentioned, a figure is made after the death of each individual and it is kept in the family of the deceased for a certain period of time. The figure is particularly significant if the deceased used to be a prominent member of the tribe. This was noticed by scholars in the 19th century in connection with the effigies of shamans. 'When a shaman dies', says V. Šavrov, 'a puppet is made in honour of his memory, and this is worshipped as a deity not only by the women of his clan, but also by the men of other clans who knew him'.[20] Paasonen remarks in his dictionary that the word *toŋχ* also has the meaning of 'dead hero', and Karjalainen, analysing the notion *toŋχ*, comes to the conclusion that it stands for a deified member

[18] Чернецов (1959).
[19] Чернецов (1959).
[20] Шавров (1871).

of the clan. According to him, not every dead person can become a *toηχ*, but only one who possessed extraordinary faculties.[21]

According to the Ob-Ugrians, the revived soul inhabits the skull. Owing to this belief, a special relationship has been surmised between the two, and the custom of scalping which prevailed among these peoples as recorded in history and folklore, is connected with this belief. S. Patkanov remarks that, according to the Khanty, the human soul, which has been deprived of its scalp, is no longer able to regenerate, and finally dies. Hence heroes endeavour to retain their scalps even after death.[22]

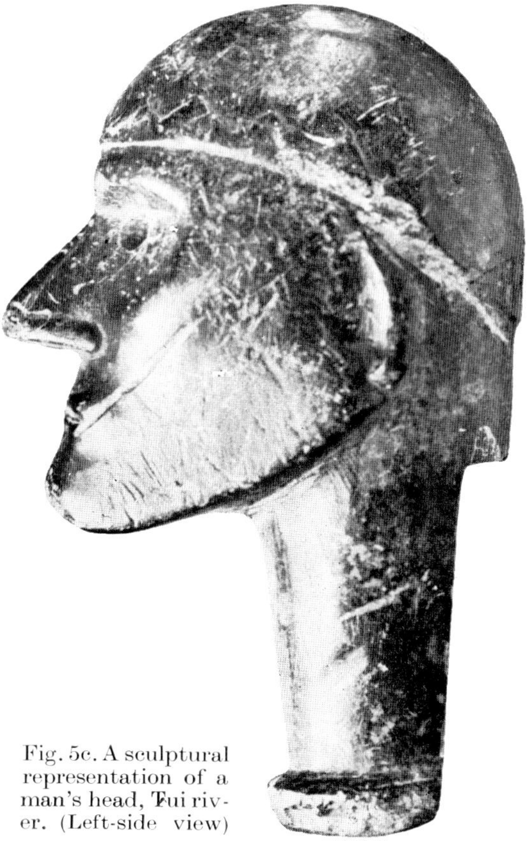

Fig. 5c. A sculptural representation of a man's head, Tui river. (Left-side view)

The custom of preserving parts of the skeletons or the skulls of some prominent and venerable members of the clan (particularly of the forefathers) is a well-known and widespread phenomenon. It has been recently observed and described in detail by B. Söderberg among the Bakuta[23], and it is also well known among the peoples of Siberia. We may refer in this connection to the ethnography of the Yakuts and Yukagirs.[24] This custom may be related to that of preservation of the bear skulls by the Ob-Ugrians. The repeated burial of skulls known from archaeological data is also presumably linked up with it. Some tombs in the burial places of the Ananino culture are of special interest in this respect, because they differ from others by their specially rich equipment, and by the fact that the only human remains interred in them are skulls.[25]

As a summary we may say that the hypothesis, according to which stone effigies and masks were meant to replace the skulls of the ancestors, seems to be justified. They formed a part of anthropomorphic representations, and are very similar to those known to us from the ethnography of the Ob-Ugrians.

[21] Paasonen (1926) p. 259; Karjalainen (1921) p. 49.
[22] Patkanov (1900) p. 30.
[23] Söderberg (1956) p. 109.
[24] Окладников (1949) pp. 222–3.
[25] Пономарев (1892) pp. 424–8.

Fig. 5d. A sculptural representation of a man's head. Tui river. (Half-right view)

One question still remains to be answered. In what historical period and under what circumstances could the veneration of the anthropomorphic representation of male ancestors attain such a wide expansion?

A careful analysis of the historical development of the Ob-Ugrian clans shows that their composition changed from time to time. This may be accounted for by the periodic disintegrations, due to various reasons, and the subsequent branching off of the settlements. Consequently, a chronology may be set up as regards the date of origin of the various clans. It is remarkable that the effigies of the ancestors of the most ancient clans, and of the collateral ancestors almost invariably display zoomorphic characteristics.[26] Among the ancestors of the more recent clans, however, zoomorphic and anthropomorphic figures can be found side by side. Sometimes certain indirect factors help us establish the age of the clans with anthropomorphic ancestors. For instance, the seven *š̊oχriŋ-ɔ̄jka* brothers with anthropomorphic features are represented as blacksmiths. And we have sufficient reason to surmise that clans whose ancestors had been characterized as 'old smiths' must have originated in the early Iron Age, when blacksmiths were particularly esteemed members of a clan.[27]

A certain anthropomorphic ancestor of the Khanty from the *Š̊ham Ās* creek appears as a two-faced creature whose features show a close likeness to the stone effigies from the Ir river. Characteristically, an anthropomorphic ancestor is invariably referred to as *tal' aγən kentup χum* 'man in a pointed helmet'.

From this fragmentary evidence material alone we may conclude that it was in the early Iron Age that the veneration of the anthropomorphic figures of the male ancestors became widespread. The end of the first millenium B. C. was characterized, in the Ob-Irtysh region, by large-scale migrations which brought along the wide diffusion of special tools and weapons, and the growth of settlements.[28] This was probably a period when the power of the war leaders increased, the migrations of

[26] Чернецов (1939) pp. 24–7.
[27] Чернецов (1953b) pp. 230–1.
[28] Чернецов (1953b) pp. 231–4.

some groups of population became more intensified, and the filial clans came into existence. It is probably with this same period that the wide distribution of the representations of the clan ancestors as human figures with metal or stone faces should be associated. We are, however, justified for more than one reason to believe that, in some cases, anthropomorphic representations of the ancestors existed already at a much earlier date. This is proved anyway by the stone sculpture of a man's head brought to light in 1957 at the prehistoric settlement Samuski IV on the Tom river, in a complex which belongs to the Bronze Age.

REFERENCES

(Beljavskij F.) Белявский, Ф. (1833) *Поездка к Ледовитому морю* (A Journey to the Arctic Ocean), Moscow.

(Černecov, V. N.) Чернецов, В. Н. (1927) Жертвоприношение у вогулов (Sacrificial Rites among the Voguls), in *ЭИ*, No. 7.

(1939) Фратриальное устройство обско-угорского общества (The Fraternal Structure in Ob-Ugric Society), in *СЭ*, Vol. II.

(1953a) Бронза Усть–Полуйского времени (Bronzes of the Ust-Polui Period), in *МИА*, No. 35.

(1953b) Усть–Полуйское время в Приобье (The Ust-Polui Period in the Ob Region), *МИА*, No. 35.

(1957) Нижнее Приобье в 1 тысячилетии нашей эры (The Lower Ob Region in the First Thousand Years of our Era), in *МИА*, No. 58.

(1959) Представления о душе у обских угров (Concept of the Soul among the Ob-Ugrians), in *Тр ИЭ*, No. 51. (Transl. in *Studies in Siberian Shamanism*, ed. H. M. Michael, University of Toronto Press, 1964.)

(Gondatti, N. L.) Гондатти, Н. Л. (1888) *Следы язычества у инородцев Северо-Западной Сибири* (Traces of Paganism among the Nationalities of North-West Siberia), Moscow.

Karjalainen, K. (1921–1927) Die Religion der Jugra-Völker, Vols I–III, in *FFC*, Nos 41, 44 and 63.

(Mošinskaja, V. I.) Мошинская, В. И. (1952) О некоторых каменных скульптурах Прииртышья (On Some Stone Sculptures of the Irtysh Region), in *КСИИМК*, Vol. XLIII.

Munkácsi, B. (1904) Ältere Berichte über das Heidenthum der Wogulen und Ostjaken, in *KSz*, Vol. V.

(Nosilov, K. D.) Носилов, К. Д. (1904) *У вогулов* (Among the Voguls), Saint Petersburg.

(Novickij, Gr.) Новицкий, Гр. (1884) *Краткое описание о народе Остяцком* (A Short Description of the Ostyak People), Saint Petersburg.

(Okladnikov, A. P.) Окладников, А. П. (1949) *История Якутии* (The History of Yakutia), Vol. I, Yakutsk.

Paasonen, H. (1926) *Ostjakisches Wörterbuch nach den Dialekten an der Konda und am Jugan*, Helsingfors.

(Pallas, P. S.) Паллас, П. С. (1786) *Путешествие по разным местам Российского государства* (A Journey to Various Places in the Russian Empire), Part Two, Book One, Saint Petersburg.

Patkanow, S. (1900) *Die Irtysch-Ostjaken und ihre Volkspoesie* , Part Two (with Russian Translation of the Irtysh Ostyak Texts), Saint Petersburg.

(Ponomarev, P. A.) Пономарев, П. А. (1892) Ананьинский могильник (The Ananino Burial Ground), in *ИОАИЭКУ*, Vol. X, No. 4.

Söderberg, B. (1956) Ancestor Guardian Figures among the Bakuta, in *E*, Vols 1–2.

(Starcev, G.) Старцев, Г. (1928) *Остяки* (The Ostyaks). (A Socio-Ethnographical Treatise), Leningrad.

(Šavrov, V.) Шавров, В. (1871) Краткие заметки о жителях Березовского уезда (Short Notes on the Inhabitants of the Berezovsk District), in *ЧОИДРМУ*, April–June 1871, Book Two.

Khanty (Ostyak) Sheds for Sacrificial Objects

by

J. Kodolányi Jr., Budapest

The sheds for food storage of the Ob-Ugrian peoples are erected at some distance from the main buildings in which they live. These structures are little wooden buildings standing on posts. Similar constructions can be found throughout Northern Europe, from Finland to Western Siberia. Buildings of the same kind have been used in Khanty areas for storing sacrificial objects. The term 'shed' was obviously applied to these because of their resemblance to the houses in which victuals are stored.

K. F. Karjalainen dealt with sacred sheds in his monograph on the religious life of the Ob-Ugrian peoples[1] and also makes mention of such buildings in his correspondence.[2] He distinguished sheds for spirits and sheds for sacrificial objects. His studies of the subject led him to the conclusion that sacred sheds probably originated at a much earlier period, and sheds serving as dwelling places for spirits were of a later development. The few buildings found in Tsingala furnished Karjalainen with important data. From a letter to his fiancée we know that it was only by great efforts that he could manage to take a look at the inside of these buildings, and he obtained practically nothing from informants as to their purpose.

In the summer of 1898, János Jankó, a member of the Zichy expedition, visited all the Khanty settlements in the valleys of the Irtysh, Ob and Yugan rivers. His aim was to collect material (objects, photographs, and notes) relating to the culture and anthropology of this people. During his journey, in the course of which he happened to reach Tsingala almost at the same time as Karjalainen, he studied the sacrificial sheds of that region. The notes he made on this occasion are still available in his diary. He examined the structures in the company of a certain Tsingala merchant called Baturin who helped him collect data (Fig. 1). In his diary Jankó described his experiences as follows:

'First of all he escorted me to the little sheds. By a branch of the Irtysh, south of the village, at a slightly raised place the field is uneven and betrays by unmistakable signs that once a village had stood there. This is corroborated by the Ostyaks themselves. We found three small buildings in different conditions, all three surrounded by elder-bushes. Supposedly they were the property of different clans at different times. The Ostyaks will not admit that they are sacred sheds, although it is a known fact that to this day sacrifices are still offered in them. As Baturin told me, Ss. Peter and Paul's Day is one of the principal days for sacrifice.

[1] Karjalainen (1918) pp. 200–4.
[2] Karjalainen (1921) pp. 40–2.

Fig. 1. The sacred shed at Tsingala (Photo by János Jankó)

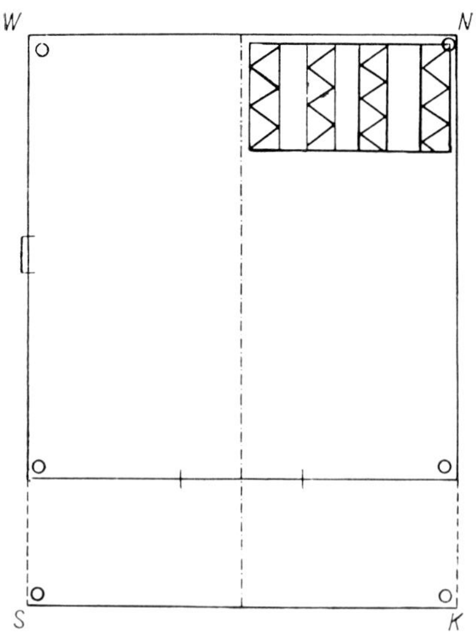

Fig. 2. Horizontal cross section of the sacred shed (after Jankó's sketch)

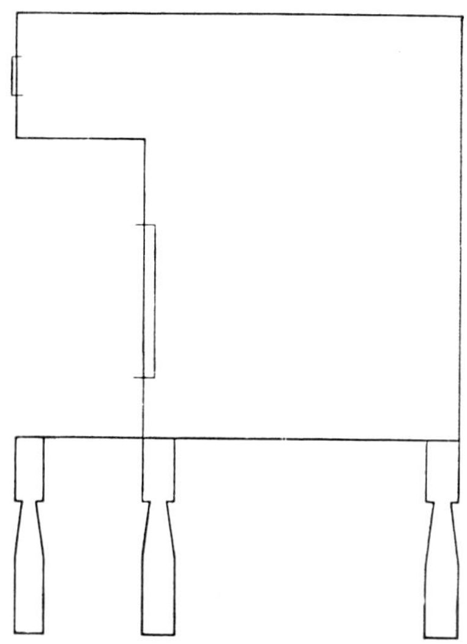

Fig. 3. Vertical cross section of the sacred shed (after Jankó's sketch)

Fig. 4. The drawing on the table in the sacred shed (after Jankó's sketch)

'According to Baturin there was an Ostyak, who was not very prosperous in spite of his efforts to improve his lot. So he sacrificed his horse, then his cow and finally his sheep to the devil, but to no avail. Then the man turned to Baturin and asked him what god to adore now that he had ruined himself for the old one. Baturin explained to him that it was too late, since the horse, the cow and the sheep were gone, and sent him to the pope of the Orthodox church.

The first sacrificial house is exactly like a shed for food storage in the purest style, only it is much smaller. It has no roof, its six supporting posts have collapsed, and one corner is overgrown with elder. Among the luxuriant elder bushes, on a little four-legged bench, we found a little quadrangular, iron-strappedbox with a padlock, which contained shreds of an old embroidered shirt, women's caps (*kalpak*) and shoes. The second shed was similar to the first in every respect except for the door which, unlike that of the first, was double, and for the four sides not higher than three or four beams. The third was completely intact, with elderbushes right in front and behind and with nettles almost as tall as a man growing nearby. Nettles always are the landmarks of places inhabited by Ostyaks. The groundplan of the shed and its cross section along the line —.—.—.— are shown in Fig. 2.

'The building is made of strong timber and stands on six posts, each about 0·85 m high. Its ground area is 2 square metres with a small balcony in front, about 0·58 metre wide, covered by a projecting part of the building. The door is 0·56 × 0·66 m and opens at about 0·10 m from the ground. The only window is 0·16 square metre. Without the posts the entire building is about 2 m in height, 0·58 m of which is taken up by the roof-structure, the front of which is developed into an attic covering the balcony. The rest is one with the only room of the little house (Fig. 3).

'The building is called *tabās*—a name which usually meant a shed for storing food in the Ostyak language. The back right corner of the room contained a low four-legged table (height: 0·50 m) with a drawing on it (Fig. 4). On the ground, around the table, traces of blood, hair and feathers could be seen. I asked the Ostyaks what kind of hair it was, and, after a lengthy debate, they said that it was from a bear. It was quite light, so light indeed that it looked like the hair of a calf, though they would not admit this. There was a cross-beam on the left side of the shed from which hung two red felt ribbons with one knot tied in the first and four, in the other. When I began asking questions, the Ostyaks expressed astonishment on their faces and replied that some child must have played and hung them up there. It was, however, again quite sure that they were not telling the truth.

'We went out of the narrow shed in which there was hardly room for all of us and examined the copse of elders. Soon I came across the jawbone of a horse. It was obvious that a horse had been sacrificed here, and it was likewise obvious that my Ostyaks had been camouflaging when telling me that the bones were of fallow-deer. But I pretended to put up with the wrong scent, for I knew that I would need these people the next day again. All three sheds were surrounded by elders. (In Ostyak, elder is *amp-loχon-juχ*, *amp* meaning "dog", *loχon* "currents, strawberries, berries", and *juχ* "tree".) The elder has a religious significance and that is the reason why it was found around this sacrificial shed.'[3]

Jankó took a photograph of the shed (Fig. 1) and, as we have said before, measured the building and made a rough sketch of its ground plan (with the table) and a sectional view. The design on the top of the table is reproduced in a separate sketch.

The reports of both Karjalainen and Jankó bear witness to the circumstance that these religious buildings were found in a state of decay not long before their final extinction. They came across but a few of them during their journey. By offering this contribution of János Jankó's observations and drawings as supplements to Karjalainen's reports, we wish to pay a tribute, however modest, to the memory of the scholar who followed in the footsteps of Antal Reguly.

REFERENCES

Karjalainen, K. F. (1918) *Jugralaisten uskonto* (The Religion of the Ugrians), Porvoo.
(1921) *Siperian matkoilta* (Journeys in Siberia), Porvoo.

[3] Hungarian National Museum. Ethnological Archives of the Ethnographic Museum, Budapest.

An Ostyak Tale from M.A. Castrén's Manuscripts

by

W. Steinitz Berlin

In 1843–1845, the very years when A. Reguly was collecting Vogul and Ostyak epic folk poetry, M. A. Castrén also stayed with the Ostyaks. In the preface to his *Versuch einer ostjakischen Sprachlehre* published in 1849, Castrén explicitly refers to A. Reguly: 'More precise information on the relationships of the mentioned (Ugric) languages is awaited for by the learned circles of the world from a born Hungarian who has recently returned from his philological-ethnographic journey he made to the Ostyaks and Voguls.'

Castrén's linguistic and ethnographic notes on his expeditions lasting for years in the north of Eurasia met, after his premature death (1852), a happier fate than A. Reguly's collections. In that same decade A. Schiefner published on behalf of the Petersburg Academy of Sciences *M.A. Castréns Nordische Reisen und Forschungen* in twelve volumes. Yet Schiefner did not publish Castrén's entire material, and so his manuscripts, which are now kept in the Helsinki University Library, still contain interesting data which may be of importance to linguistics, ethnography and folklore. While in Helsinki on the occasion of the celebration of the 75th anniversary of the Finno-Ugrian Society in December 1959, I found in Castrén's Ostyak notes, among other things, an unpublished Ostyak tale he had taken down. In view of the afore-mentioned relationship between A. Reguly and M. A. Castrén it seems to me appropriate to publish, in this memorial volume dedicated to A. Reguly, who was the first collector of Ostyak epic poetry, this original sketch of an Ostyak tale.[1]

The tale belongs to the Aarne–Thompson 1535 type of funny story, 'The Rich and the Poor Peasant', which—evidently taken over from the Russians—is very popular with the Ostyaks and Voguls too; see W. Steinitz, *Ostjakische Volksdichtung* I, p. 101 ff., II, p. 155 ff. While, however, the stories cited there display a genuine Ostyak-Vogul background and the redundant style of Ostyak-Vogul fairy-tales, our story is told in a primitive style and has an essentially Russian milieu. There are in it a number of Russian loan-words, and in two cases the story-teller used Russian words first, which he then interpreted in Ostyak (see the annotations to the text). The story at issue is obviously a Russian fairy-tale which was adopted relatively recently, but which has been slightly 'Ostyakized' in style and milieu.

[1] I thank the Deputy Director of the University Library, Dr. S. Haltsonen, for the photostat of Castrén's note and his courtesy in permitting me to publish the tale.

OSTYAK TEXT[2]

1. *Oluga udōt,*[3] *i tau tajōt. Tawet Ruďetna wetta pardai. Oluga tajōt i sagar; wetta pardai. Oluga i puraš tajōt; wetta pardai. Oluga i amp tajōt,*[4] *wetta pardai. Oluga tau sox*[5] *jaga werōt,*[6] *sagar sox ńīr werōt,*[7] *puraš*[8] *sox mil werōt, amp*[9] *sox pōs werōt.*

2. *Oluga menōt woǯa. Oluga woš xowatta menōt, xōda tangōt. Oluga xōt tibena perdagat*[10] *tēwōt. Oluga kiimedje eikjiídet. Oluga angermetna tjaras neng jet. Togos xujdat*[11] *jit. Oluga kolbets tibeja tangot. Tjaras neng ängermetna xùit jit tjarasedat. Togos xuj kolbets tibeja pogotmet.*

3. *Tjaras neng togos xuit menot kolbets sunga Oluga pungeda. Njut ingestegen: 'Ma nemem Oluga.' 'Ma nemem Tjaras.' Tjaras nerget: 'Vina tua. Min jantjemen.' Oluga nerget: 'Tjag*[12] *armedem.'*[13] *Tjaras nerget: 'Ad arga. Wičemna pelestemen.' Wičena pelesegen.*

4. *Tjaras nerget: 'Šam omde njat-sem-tagat*[14] *oxteja. Menaba üt-njereme.*[15] *Min kim lattatmen.' Oluga sir nox-pidot. Olugan tjaras it pogotmaj kolbets tibeja; kolbets tibeja it kerkot.*

5. *Oluga kinlattot. Tjaras-neng xuitna tegettjäj; xuit nerge(t)*[16]*: 'Kolbets tibeja medóje kerkot.' Tjaras-nenget nerget: 'Kobašnja it kerkot.'*

6. *Oluga kew*[17]*-xooda joxtangot; padja werot.*[18] *Oluga pesan kutteba oomastaj. Oluga teedä jäntjä jiwot. Oluga padarta jiwot: 'Kolbets tibena mettä uut.' Tjaras nerget: 'Atatta wegenj ent termet li? weet xaera wax*[19] *medem.' Oluga nerget: 'Menem njur xoomdje meja. Nèng wijada jux dčubat!'*

7. *Oluga tangot kolbets tibeja. Oluga wardjémet*[20] *xoomdiat. Tjaras àit (ajit) xoongat. Ket pùneŋ boosēt sekremet;*[21] *tomet perdajit kerget. Teu kim xoogatmót. Oluga kolbets tibiwet nox kidot. 'Ujedená?' — 'Perdäjit kerguu. Xatj teewajuu.' Olugaja wèt xaera wax mejaj.*

[2] Castrén's text in general is easily legible and rendered verbatim here (except that Castrén's ŝ is replaced by š, etc.). Separation of words and sentences, as well as punctuation, are mostly my work. The notes that follow here include variants and marginal notes by Castrén (translation and notes in [ ] are mine.—W. St.).

[3] *ut* [lives].

[4] O. *tait i amp* [O. has . . .].

[5] [First:] *sogaiwet.*

[6] *O. tau sogewet werda pardet jaga* [O. causes a rug to be made of the horsehide].

[7] *Oluga werot sagar sox njiiret.*

[8] *Oluga puraš.*

[9] *Oluga amp.*

[10] åt ett och annat, всякое.

[11] det är en kamrat till köpmans hustru [that is a friend of the merchant's wife].

[12] *tjeg.*

[13] *tjag armedem* запою (armedem пою).

[14] *njat-sem-tagat* скатерть ['nose-eye-cloth', originally most likely 'napkin'; cf. KT 983b].

[15] иди, подстерни. [*üt* is vaguely written and uncertain.]

[16] [First:] *neng.*

[17] [First:] *kamenien.*

[18] [First:] *sdorówoittot.*

[19] 500.

[20] 'förföljer', *wardjem* 'jaga, drifva, förfölja [to chase, drive, pursue]'.

[21] *puneŋ bos* 'hund-handske'; *sekremem* 'ударить'; NB. köpmannen är subj. [merchant is subj.]

8. *Oluga jox menot. Rugetna ingistaj: 'Xottja wax wijėn?' — 'Wošna köndćaj: taw sox jaga köndčaj, sagar sox njir köndčaj, purš sox mil köndčaj, amp sox pos*[22] *köndčaj.*[23] *Tünenget. Wèt xaera wax wijem.'*

9. *Rudjet*[24] *tawdat wedet, taw sox jaga jondet, sagar sox njir jondet, purš sox mil jondet. Rudjet tudajit Olugana wodša. Oluga nerget: 'At jiwada. Ma mendam tinésta.' Oluga menot wodša, teg*[25] *xaedjet.*

10. *Ugum guja*[26] *jastet: 'Mettä wojet jidet.*[27] *Sora tuada puškajet!' Saldatna estemdajit puškajat. Inot*[28] *perda atmajit.*

11. *Olugaja wet xaera wagat mejïj. Oluga jox menot, jastot, xoodang adada*[29] *jastet: 'Nèngatten*[30] *teruk*[31] *menet, tin portitet.' Kuš tagestet, endamet.*[32]

12. *Rudjetna xandep čundet*[33] *nox wejáj; tau sox xaera axtaj.*[34] *Oluga padtja tiwedot. Tau sox xaer tibe gerkot. Oluga tuwaj jinga esteda. Oluga jinga esteda juwaj. Jug endam. Adamet jux köndča menet.*

13. *Tjaras jit, ogodat jit. Tjarasna sekremaj taw sox xaera. Oluga kin divdot. Tjaras nerget: 'Mand esta!' Oluga nerget: 'Tanga!' Olugana wès tox jondaj. Oluga ogda dedôt, jox menoot.*

14. *Xajadetna joxtaj; tjaras it pogotmaj jenga. In xajat jox menet. Oluga metten ōgot jogat; totta ar wax tait.*[35] *'Xottja wax wéjen,*[36] *Oluga?' Oluga nerget: 'Ma jangam tagemna wax argem.' — 'Mungat taua!'*[37] *Oluga nerget: 'Taw sox xaera jondada!' Sox xaera jondadet. Olugan xaereja estajit, jinga estajit. Oluga jox jiwot.*

TRANSLATION[38]

1. There lived Oluga; he[39] had a horse. The Russians ordered his horse to be slain. Oluga had a cow; it was ordered to be slain. Oluga had a pig; it was ordered to be slain. Oluga had a dog; it was ordered to be slain.

[22] [Not clear.]
[23] Allt detta är efterlängtad (*köndčaj*) i staden.
[24] [First:] *ruget*.
[25] [Superscribed:] *teua*.
[26] (*ugum gui*) 'embetsmännen [Beamter]' ['leader, chief'].
[27] нечистый дух идет 'an unclean spirit is coming' [literally: 'some beasts are coming'].
[28] Тех унесло.
[29] *xocdaŋat* 'домашний' [*χctəŋ ăt* 'house companion'].
[30] ваши.
[31] [The word is not clearly legible; cf. Kr. *truk* 'suddenly' < Russ. вдруг.]
[32] ждут, der finnes ingen (köpare) [there is none [buyer]]. [Castrén's insertion 'buyer' has no sense and is even grammatically wrong.]
[33] Folket borttog trappstegen (*čunt* 'steg' [for Irtysh *tšumət* id.]).
[34] man hängde [they hung]. [*axtaj* for Irtysh *eχət* 'to hang' (*eχtaj*)].
[35] [The sentence from ';' onwards has been added subsequently].
[36] [Both above *ė* and above *j* stands a clear dot.]
[37] [The first *a* is illegible.]
[38] In addition to the notes, Castrén wrote in places between the lines translations in Russian or Swedish; they are worked into the present translation made by me.
[39] Castrén uses the feminine personal pronoun for Oluga several times in his Swedish annotation; in the Russian annotations, which come direct from the story-teller, however, only masculine forms appear (Олуга ушел, вышел, зашел, etc.). If we were really dealing with a woman, this would certainly have been explicitly stressed in Ostyak, since the name Oluga was not customary in Ostyak and could not indicate the gender in itself.

Oluga made a rug of the horsehide, leather boots of the cowhide, a cap of the pigskin, gloves of the dogskin.

2. Oluga went to the town. He went through the town and entered into a house. Oluga ate up everything in the house. Oluga looks out. Then he sees: a merchant woman comes, she comes with her friend. Oluga went into the cellar. The merchant's wife sees: her husband comes with [other] merchants. [The merchant's wife] shoves her friend into the cellar.

3. The merchant woman's friend went to the corner of the cellar up to Oluga. They ask each other: 'My name is Oluga.' 'My name is Merchant.' The merchant says: 'Get schnapps, we want drink!' Oluga says: 'I begin to sing aloud.' The merchant says: 'Don't sing! Let's change our clothes [?].'[40] They changed their clothes [?].

4. The merchant says: 'Put a candle on the table-cloth. Go, spread it out. We go out of here.' Oluga went on up [the stairs]. Oluga pushed the merchant down into the cellar; he tumbled down into the cellar.

5. Oluga went out. The merchant woman was scolded[41] by her husband; the man says: 'Something has fallen in the cellar.' The merchant woman says: 'The kneading trough has fallen down.'

6. Oluga went into the stone house and saluted [those present]. Oluga is offered the middle seat at the table. Oluga started eating and drinking. Oluga began to talk: 'A ghost is in the cellar.'[42] The merchant says: 'Are you not strong enough to chase him away? I give you five bags of money [kopecks].' Oluga says: 'Give me a leather whip! You take clubs!'

7. Oluga went into the cellar. Oluga chases him with the whip. The merchant climbs slowly up. He strikes both dogskin gloves [together?]; they fell down. He ran out. Oluga came up from the cellar. 'Have you seen him?' [They replied:] 'We tumbled down head first. He has all but devoured us.' Oluga receives five bags of money.

8. Oluga went home. The Russians ask him: 'Where did you get the money from?' 'In town they look for a horsehide rug, look for cowhide boots, look for a pigskin cap, look for dogskin gloves. These are costly. Five bags of money I got [for them].'

9. The Russians slaughtered their horses [cows, pigs] and sewed horsehide rugs, cowhide boots and pigskin caps. Oluga leads the Russians as far as the town. Oluga says: 'Don't come [further on]! I go to trade.' Oluga entered the town, they remained behind.

10. [Oluga] says to the [town] chief: 'An evil spirit is coming! Bring on here guns quickly!' The soldiers discharged the guns. They [the Russians] were swept away head over heels.

11. Oluga was given five bags of money. Oluga went home and said, to the house companions he said: 'You people went away suddenly [?]. They have wasted the price.' However [much] they waited [for the men], they are not there [any longer].[43]

[40] *wičemna, wičena*: the translation is uncertain. Irtysh *witš* (*wiš*) is given as 'boundary' by Castrén, Patkanov, Paasonen, KT 260. Only Paasonen has recorded Konda *witš* 'osa/Teil'. 'Clothes' is Irtysh *ütš, üš*.—Also the suffixes are not entirely clear.

[41] Irtysh *təγəťťə* 'to vex, scold' (KT 983).

[42] Irtysh *məttə* 'something; ghost'.

[43] Cf. the annotation to the Ostyak text.

12. The Russians took the stairs high; [instead] a horsehide sack was hung up there. Oluga went out to shit and fell into the horsehide sack. They carried Oluga away to throw him into water. There is no pole there [to push him under the water]. The men go to look for a pole.

13. The merchant comes with his sledge. The merchant strikes at the horsehide sack. Oluga got out. The merchant says: 'let me get in!' Oluga says: 'Get in!' Oluga sewed up the mouth [of the sack]. He got on the sledge and drove home.

14. The people came [back] to him; the merchant was pushed into the water. Then the people went home. And lo! Oluga comes with his sledge,[44] he has much money there. 'Where did you get the money, Oluga?' Oluga says: 'Where I was[45] there is much money.' [The people say:] 'Lead the way [there]!' Oluga says: 'Sew a sack of horsehide.' They sewed a sack of [horse-] hide. Oluga put them into the sack and threw them into the water. [After that] Oluga went home.

[44] Composition and translation are not wholly clear. Castrén renders the sentence in the annotation as follows: 'Олуга санки собирает'. Where he gets the meaning 'собирает' I cannot tell. But his rendering does not fit in well with its very meaning either. One might as well expect *ogdedat jogat* 'comes with his sledge'. It might be rendered more accurately this way: 'Look, that's Oluga! There comes a sledge.' As to *metten*, see Paasonen 1225.

[45] 'In the place I traversed.'

On the Trail of Ostyak (Khanty) Mythical Songs

by

E. Vértes, Budapest

The most complete description of the religious beliefs of the Ob-Ugrians has so far been given by Karjalainen in his three-volume work[1] (published in Finnish and German) based on his own collections on the spot and on the early literature. He could not have been familiar with Paasonen's Ostyak collection (which probably contains relevant material) because it is, unfortunately, still in manuscript and thus inaccessible for scientific research. Nevertheless, the sample sentences in Paasonen's posthumous dictionary[2] are useful contributions to the study of the religious beliefs[3] of the Southern Ostyaks, more closely the Konda Ostyaks. These sentences in the Konda dialect can be found under various entries not necessarily of mythical content. It is rather difficult to handle them because of the frequent lack of Finnish and German translations which, if they exist, usually give only the contents. In most cases, the interpretation or explanatory comments are given only in Finnish, whereas about one third of the examples are not translated at all. Hence, we wish to complete Karjalainen's monograph with a few data on the extinct Southern Ostyak group by arranging the sample sentences of Paasonen's dictionary in semantic groups with a word-for-word translation of the Ostyak text. The exactness of the translation may sometimes involve a somewhat unnatural word order or style in English.

1. 'According to Karjalainen, the Ostyaks call the celestial god *Num-tūrəm* "*tūrəm* residing or living above".[4] Paasonen quotes a single pertinent expression from the Konda region: *num-turə̑m-χui̯-kǻk̓з* 'свѣтлый богъ | heller gott'[5] or, more correctly, in the interpretation of Paasonen's relevant entries, 'upper-heaven man-father'.

According to Karjalainen's records, the celestial god has several other names in the folklore, such a *num-säŋkə* 'oberes Licht'[6] which occurs also among Paasonen's sentences:

1. *i̯əγpăχsȧŋən num sȧŋkə i̯əγetȧ*	The two brethren to their 'upper-light' father
kitməštȧ pȧi̯kəštȧ pitγən[7]	began to pray, to implore.

[1] Karjalainen (1918, 1921–1927).

[2] Paasonen (1926).

[3] There are no useful data in this respect among the Eastern (Yugan) Ostyak sample sentences in Paasonen (1926).

[4] 'Die gewöhnlichste Benennung des *Himmelsgottes* ist bei den Ostjaken *Num-tūrəm* der oben weilende oder wohnende *tūrəm*', Karjalainen (1921–1927) Vol. II, p. 250.

[5] Paasonen (1926) 1388. His material is quoted here, as far as possible, in a simplified form, according to Steinitz (1950) p. 22.

[6] Karjalainen (1921–1927) Vol. II, p. 251; cf. also pp. 260 ff and 271 ff.

[7] Paasonen (1926) 773. The word-for-word translation of the Ostyak texts is given in the parallel columns. For the Finnish or German translation as given, if any, in the dictionary, see the foot-notes.

2. *nŏŋ ăḱḱȧ χui̯,*	Thou *ăḱḱȧ* man,
ḱürtȧ χui̯,	*ḱürtȧ* man,
ot mənȧ ket,	so that thou shouldst not depart [= perish]
num sȧŋkə iəγəmnз	by my upper-light father
estitȧi̯əm	I have been lowered [to the earth]
kărt-seγər,	on an iron chain's,
wăχ-seγər	metal chain's
təi̯nз[8]	end.

The attribute 'upper' can be omitted, since *sȧŋkə* 'kirkas, selvä (esim. vesi, taivas, teekeittiö), klar, hell, heiter; himmel, gott',[9] as an epithet to the words 'man' or 'father', can only refer to some celestial being.

3. *sȧŋkə i̯əγəmnз pȧrtəm*	destined by my celestial father
i̯əm weŋəm[10]	My good son-in-law,

4. *sȧŋkə χui̯ ḱȧḱз,*	Celestial man father,
kăt kŏren estəm	thy two legs lowered
χui̯ ḱȧḱз usən[11]	thou art a man father

In his work quoted above Karjalainen writes about the celestial deity, his helpers and other celestial spirits.[12] According to him there is *one* predominating celestial deity, and there are *several* subordinate spirits. Among the Konda Ostyaks, who were exposed to strong foreign influence, Paasonen found three prominent and distinctly classified members in the celestial hierarchy. The possible relationship[13] between these three deities, and the idea of the Holy Trinity as taught by the missionaries, or rather as thought of by the people, seems to require and deserve a more thorough examination. Next to *sȧŋkз* stands the deity called *pȧi̯rə̑χḱз,*[14] whose name could probably be discovered in the Ostyak *pairəχšə*, recorded by Karjalainen, and in the last member of the Vogul (Mansi) *nom-tōrəm-pairəχś* 'der obere Gott-Herr'.[15]

[8] 'Jottet sinä joutuisi hukkaan, laski ylhäinen kirkas isäni minut maan päälle rautaisten vitjojen nenässä | damit du nicht verloren gehen solltest, senkte mein hoher klarer Vater mich auf die Erde an eisernen Ketten', Paasonen (1926) 712; cf. No. 14.

[9] Paasonen (1926) 2117.

[10] 'Ylijumalan (minulle) määräämä hyvä vävyni', Paasonen (1926) 1686.

[11] Paasonen (1926) 2117.

[12] Karjalainen (1921–1927), Vol. II: Die Geister des Himmels pp. 250–312. Der Himmelsgott pp. 250–95, Die Gehilfen des Himmelsgottes und andere himmlische Geister, pp. 295–312.

[13] Karjalainen (1921-1927) Vol. II, too, takes into account the influence of the Holy Trinity: pp. 252, 279, and mainly p. 305.

[14] '. . . eräs "taivaallinen" jumala, jota rukoillaan lähinnä *sȧŋkз* jumalan, ylijumalan jälkeen | einer von den "himmlischen Göttern", dem Obergott *sȧŋkз* zunächst an Rang', Paasonen (1926) 1595; cf. Paasonen (1903) p. 16 and (1902) p. 127.

[15] Karjalainen (1921–1927), Vol. II, pp. 251, 296; cf. also *Pairəχšə*, p. 296.

In the celestial hierarchy the third god, according to Paasonen, is called *kȧrt-wărəš, wăχ-wărə̑š təi̯nз omə̑ste χui̯.*[16] This deity does not appear in Karjalainen's work.

The following two sentences may, to a certain extent, reveal the hierarchy of the three celestial deities, though not as clearly as in Paasonen's description.

5. *sȧŋkə i̯əγetȧt*	with their celestial father
əi̯ təŋkər pitȧ,	one shoulder [?] [extending] to his mate
əi̯ wăn pitȧ	one axis [?] [extending] to his mate,
omə̑ttəŋən păi̯rə̑χḱз,	both are sitting *păi̯rə̑χḱз,*
păi̯rə̑χḱəi̯ȧt	with *păi̯rə̑χḱз*
kȧrt wărəš—təi̯ətnз	on the iron chain's end
omə̑stə χui̯[17]	sitting man

The following comparison should be understood as *sȧŋkз*'s words:

6. *nŏŋ, păi̯rə̑χḱз,*	Thou, *păi̯rə̑χḱз*
mă̑nnə kəntšȧi̯em	over me
ərtȧ nŏŋ ot utȧ,	thou must not prevail,
nŏŋ kȧrt-wărə̑š təi̯ətnə	thou, on the iron chain's end
omə̑stə χui̯	sitting man
păi̯rə̑χkənə (ɔ: *-ḱ-*) *kəntšȧ*	over *păi̯rə̑χḱз*
ərtȧ nŏŋ ot utȧ[18]	thou, must not prevail

sos-turə̑m might possibly represent a deity not recorded by Karjalainen, though under the entry *sos* 'kärppä hermelin' Paasonen considers it to be a name:

7. *sos-turə̑mnз wet-tüi̯*	by *sos-turə̑m* in [his] five-fingered
tŏŋtȧŋ untȧŋ ketȧ wəi̯ȧi̯[19]	[?] good [?] coniferous hand it is taken.
8. *kŏš sem păγətmot,*	as the eye glanced there
sos-turə̑m i̯ŏχtot[20]	*sos-turə̑m* came.

[16] 'Rautaisten vitjain? päässä istuva mies (eräs jumala) | eine Gottheit', Paasonen (1926) 665, and also 'eräs korkeampi jumal'olento, joka ylhäällä ilmassa tarkastelee ihmisten elantoa ja ilmoittaa näkemänsä ensin *păi̯rə̑χḱз* jumalalle, joka sitten vuorostaan ilmoittaa asiat *sȧŋkз* jumalalle. *enə-ikə-i̯ir* ja *χutəm-puχə̑t-i̯ir* nimisissä suuremmissa kolmen kylän yhteisesti viettämissä uhrijuhlissa uhrataan hänelle samoin kuin kahdelle viimemainitulle valkoinen tai vaaleakavioinen hevonen ja häntä rukoillaan kolmannessa sijassa *sȧŋkз*:n ja *păi̯rə̑χḱз*:n jälkeen | höheres Götterwesen', Paasonen (1926) 2901. In several of his entries (66, 340, 596, 1201) Paasonen describes how the sacrifices are offered to the above-mentioned three deities and to the local spirits.

[17] '—istuvat (istuivat) vierekkäin | —sitzen (sassen) nebeneinander', Paasonen (1926) 1809. The dash probably replaces the missing part of the full name of the deity, cf. above.

[18] 'Sinä, p., älä ole suurempi minua [minuun | verraten]', Paasonen (1926) 731.

[19] '*sos-turəm* ottaa (liinan) viisisormiseen hyvään (?) käteensä', Paasonen (1926) 2645.

[20] 'Kun vilkaisi, niin (jo) tuli *sos-turəm*', Paasonen (1926) 1654.

9. *kărəpăst ikənз* by old man *kărəpăst*
ketȧ pȧγtȧi̯ is seized in hand
sos-turə̂m[21] *sos-turə̂m*.

As the word *turə̂m* means 'sky, air, world, god, devotional picture',[22] so *sos-turə̂m*, too, might indicate a celestial deity or one of the forms in which he appears.

The following fragmentary sentence, taken from an unknown context, refers to the notion of God in general, and not to a specific spirit:

10. *ȧr turəmȧt ńuttət*[23] many gods they swore to.

2. Many sample sentences contain the name of a spirit called *toŋχ*:

11. *mətoi̯ənnз mońk̀ȧi̯ən* what is being narrated by thee,
tăm toŋχ this spirit,
χoχəttз i̯eməŋ məγȧ![24] is it running to the holy land!

12. *χŏta uttз toŋχət sot kȧštȧi̯*[25] hundreds of spirits living [any]where are conjured up [by witchcraft]

13. *toŋχ enə i̯es i̯estəsəi̯*[26] Spirit's great wailing is wailed [?]

Sentences referring to spirits more definitely identified:

14. *i̯uχ-toŋχ χŏr-ikə,* forest-ghost-shaped old man,
ikit pek̀ȧ at the old man
ettitt(ə)‿omatət n̥tȧ[27] to look he has not the ability.

Karjalainen refers frequently to 'hairy-eyed' spirits.[28] This epithet is also to be found in Paasonen's text:

15. *pünəŋ sempe kŏrəŋ unt-toŋχ*[29] hairy-eyed, footed forest ghost.

Karjalainen translates the word *toŋχ* as 'spirit, hero'[30] and regards the *meŋks* as demons, inferior and usually evil beings[31]. However, he quotes examples in which both words are parallel expressions.[32] According to an

[21] '*k*. ukko tarttui *sos-turə̂m*:iin', Paasonen (1926) 1657.
[22] 'Taivas, ilma, maailma; jumala, pyhäinkuva | Himmel, Luft, Welt; Gott, Heiligenbild', Paasonen (1926) 2678.
[23] 'Vannoivat monen jumalan kauttaa', Paasonen (1926) 1509.
[24] Paasonen (1926) 1767.
[25] 'Kaikki *toŋχ*-haltijat mainitaan, missä asunevatkaan', Paasonen (1926) 671.
[26] Paasonen (1926) 303.
[27] 'Ukko on metsä-*toŋχ*:in näköinen, hän ei voi katsoa ukkoon (se on niin hirveännäköinen)', Paasonen (1926) 1541.
[28] Karjalainen (1921–1927), Vol. II, pp. 229, 234, 372.
[29] 'Karvasilmäinen jalkasin kulkeva metsä-*toŋχ* | haaräugiger, zu Fuß gehender wald-*toŋχ* (= -geist)', Paasonen (1926) 2600; cf. Steinitz (1941) p. 47.
[30] 'Geist', 'Held', Karjalainen (1921–1927), Vol. II, p. 212.
[31] Karjalainen (1921–1927), Vol. II, pp. 371 ff.
[32] Karjalainen (1921–1927), Vol. II, p. 213, see note, p. 374, and also *Unt-toŋχ*, *Unt-meŋk* and *Ur-meŋk*, p. 229.

interpretation in Paasonen's dictionary, *toŋχ* is a 'spirit'[33], whereas *meŋk* is a malevolent forest goblin[34]. In the sample sentences only the latter occurs:

16. *ur-meŋk,*	Forest goblin,
mȧnt tšək3 wətȧ	to take me, very much
wərȧtȧŋən,	if thou wishest, [under thy power],
eŋkrəstəmən[35]	let us curse each other [= let us match witchcraft].

Two parallel designations[36] occur in one sentence for an evil spirit possibly representing the devil:

17. *kŏľȧ werəm*	for the evil spirit made,
tu̯ȧ werəm	for the devil made
i̯əγnüret kənttšȧi̯[37]	father vengeance is sought for

The *ȧi̯-urt*, alias *enə-ik3* is the local spirit of the settlement of Nakhrach.[38] There is a sentence hinting at it:

18. *ȧi̯ urt tăpasȧ*	into *ȧi̯-urt*'s cabin
tŏwn3 tə părəi̯ȧi̯[39]	by him [a hole] is drilled.

3. Examples referring to sacrifice:

19. *pəsȧn saχə̑t χăttətȧ pitət.*	The top of the table to push further on they began.
wərəŋ woi̯3 i̯iret kořkḱȧi̯, [sic! E. V.]	A gory animal-sacrifice was offered,
χăttəŋ woi̯3 i̯iret kořkḱȧi[40]	A light-coloured-animal-sacrifice was offered

20. *χutə̑m ükəs pȧγtot.*	He slaughtered three oxen.
ik3 i̯iresot	The old man offered,
poresot[41]	[food] sacrifice he offered.

Concerning sacrifice of dogs, seldom recorded:

21. *səγəŋ ȧmp i̯irət tŏttə waχə̑t,*	tailed-dog sacrifices were required there
təγəŋ ȧmp i̯irət tŏttə waχə̑t[42]	braided-hound sacrifices were required there.

[33] 'Haltija | Geist', Paasonen (1926) 2600.
[34] 'Eräs paha, metsässä asuva haltija; . . . | ein böser, im Walde wohnender Geist', Paasonen (1926) 1201.
[35] Paasonen (1926) 144. Cf. 'jos kovin haluat minut ottaa, niin kirotkaamme toinen toisemme!' Paasonen (1926) 2751, and Karjalainen (1921–1927), Vol. II, p. 229.
[36] Cf. Karjalainen (1921–1927), Vol. II, p. 383.
[37] Paasonen (1926) 2660.
[38] Paasonen (1926) 66, 141; cf. Karjalainen (1921–1927), Vol. II, p. 188.
[39] 'Hän kavertaa (veistä vääntämällä) reikiä *toŋχ*-aitan jalkoihin', Paasonen (1926) 1623.
[40] 'Heille (k)annetaan verisiä uhreja', Paasonen (1926) 449.
[41] Paasonen (1926) 340.
[42] Paasonen (1926) 2158.

So long as the offered sacrifice has not been performed, the image of the sacrificial animal, cut out of birch bark, will hang in the house of the devotee.

22. *măn us ota χŏr wăńḱəm*[43]	I, for [one] more year [?], have cut out the image.

4. We know of two fragments of prayers. One is that of the hunters:

23. *i̯àγət-neŋ,*	Between a hostile woman,
iàγət-χui̯ kŏtn3	a hostile man
mànt ot χăi̯à[44]	do not leave me.

And this is how the sacrificing shaman prays:

24. *àt χeḱḱ3*	if [there] remains
toŋχət sot wànt3,	a spirit [from] hundred promontories,
àt χeḱḱ3	if [there] remains
meŋkət sot wànt3	a forest goblin [from] hundred promontories,
ot utaŋ[45]	it should not stay [away].

5. The number of sentences referring to sorcery is comparatively large. The following may be the words of a shaman in a trance:

25. *àrγəm ăte̯mnə*	with a chant, by myself [= by my own power]
àrγài̯əm,	I have been chanted [here],
mońḱəm ăte̯mn3	with a tale, by myself,
mońḱài̯əm[46]	I have been told [here] [= I have conjured up myself here]

26. *àr piḱ ot werà*[47]	Many magic tricks do not perform

27. *nŏŋ tšătšə wotšəŋ χui̯,*	thou [art] man of thy own castle,
tšătšə puχtəŋ χui̯,	of thy own hamlet,
nŏŋə mütà[48]	do magic thyself!

[43] Paasonen (1926) 2955.
[44] 'Sanoo metsämies rukouksessan: älä jätä minua (*sàŋk3*), vihamiesten ja -naisten keskelle', Paasonen (1926) 261.
[45] 'Älköön jääkö pois (? mainitsematta t. tulematta) yhtään *toŋχ*- ja *meŋk*-haltijaa (oik. *toŋχ* :ien, *meŋk* :ien sadoista niemistä?)', Paasonen (1926) 596.
[46] 'Laululla minut tänne laulettiin | durch Gesang wurde ich hierher gesungen', Paasonen (1926) 38. As regards the travel of the shaman or of his soul cf. Karjalainen (1921–1927), Vol. I, p. 83, (1921–1927), Vol. II, pp. 334–5, (1921–1927), Vol. III, pp. 317, 328, etc.
[47] 'Älä tee monia konsteja (sanoo appi vävylleen, joka muuttaa itsensä milloin miksikin)!' Paasonen (1926) 1787.
[48] 'Sinä oman linnan mies, sinä oman kylän mies, itse loihdi!' Paasonen (1926) 2727.

28. *măn toŋχət sot,*	I, spirits, a hundred,
meŋkət sot	forest goblin, a hundred,
sărnáγem[49]	have conjured up.

The magician might be the grammatical subject of the following parallelism:

29. *piḱəŋ χui̯ piḱət mătəmtət,*	breathing man's breath they touch,
olχᵊŋ χui̯ olχᵊt mătəmtət[50]	living man's life they touch.

Words, still uninterpreted, can be found in the spell, such as the question-marked entry *păḱḱən*:

30. *imз ḱərttá pitot,*	the woman began to charm
imз usᵊstá pitot	the woman began to be [?],
imə nerγət:	the woman is shouting:
păḱḱen wokət,	*păḱḱen wokət*
păḱḱen wokət[51]	*păḱḱen wokət*

6. Parallelisms frequently occur also in fragments on the dead:

31. *măn ḱaŋχə̑ otməi̯əm,*	I dreamt of the dead,
maŋχə̑ otməi̯əm[52]	I dreamt of the deceased.

32. *ătə məssir ḱaŋχənз*	by some, dead,
ătə məssir maŋχᵊnз	by some deceased
pətem us tə wəγətái̯[53]	my ears are tinkling

tárən and *χut* are uninterpreted entries in Paasonen's dictionary but Karjalainen's above-quoted work helps us explain the following phrases:

33. *tárənát mənəm χui̯,*	man, departed with *tárən* [to the other world]
χuttát mənəm χui̯[54]	man, departed with *χut* [to the other world]

[49] 'Loihdin esille kaikki *toŋχ-* ja *meŋk*-haltijat, "поминалъ и угощалъ" ich zauberte alle *toŋχ-* und *meŋχ*-Geister herbei', Paasonen (1926) 2090.

[50] Paasonen (1926) 1535.

[51] '(on olevinaan loitsua) | (soll eine zauberformel vorstellen)', Paasonen (1926) 1603. It is with a certain reserve that the infinitive *usə̂stá* can be translated by 'to be' on the basis of the entry *ūsə̂-* (2833) and by assuming an *-s-* formant; cf. Ganschow (1956–1957) p. 241, (1965) p. 31. As far as the meaning is concerned, the verb with a full vowel in the second syllable *ùsẹsəm*, *ùsẹ̄·stȧm* 'haukotella | gähnen', as in 2832 and Karjalainen (1948) p. 86b, or *ūsə̂ttá* 'heristää, uhata lyödä | eine Faust machen, zu schlagen drohen' recorded by Karjalainen (1948) p. 90a from the southern (Krasnoyarsk) dialect would fit much better in the text.

[52] 'Näin unissani vainajan', Paasonen (1926) 898.

[53] 'Mikä vainaja pannee korvani soimaan', Paasonen (1926) 898.

[54] 'Sanotaan muistojuhlassa toisen tai oman käden kautta surmansa saaneesta vainajasta | wird bei der Gedächtnisfeier für den durch fremde oder eigene Hand gestorbenen gesagt', Paasonen (1926) 2434.

34. *tȧrən wek̓*	*tȧrən*-looking
χut wek̓	*χut*-looking
mɔnɔm χui̯[55]	man, departed [to the other world]
35. *unt wȧn nămə̂s tȧtot,*	the body breathed a short remembrance
porə̂χ wȧn nămə̂s tȧtot	the stomach breathed a short remambrance
tȧrən χăwə̂ttə̂ χui̯,	ascending man [to become] *tȧrən*,
χut χăwə̂ttə̂ χui̯[56]	ascending man [to become] *χut*.

These three phrases apparently allude to a dead person transformed into a spirit and held in mythical veneration, like the *soi̯əm toηχ*[57] arisen from the drowned man. Karjalainen, too, explains more than once that, after their deaths, persons of authority are eventually honoured with a veneration like that due to spirits.[58] He mentions two spirits of the lower regions, causing diseases: *tā̆rən, tērən* and *kul, χul̓*.[59] The former was originally not an infernal spirit, as he himself says, since sacrificial offerings were hung on trees for him, which would not normally be the custom in connection with infernal spirits.[60] Since, according to the Ostyaks, the dead can send diseases, they may easily be imagined to turn into spirits causing maladies. And since this spirit was originally not an infernal one, the same verb ('ascend') might have been employed to refer to the dead turning (i.e. 'ascending') into a spirit, as the one used to describe the ascending moon or star. Note that Karjalainen, too, cites several phrases using *tā̆rən* and *χut* in parallel expressions, like *tärən-pənt, χut-pənt* 'Kriegspfad, Unfriedenspfad', or in the Vasyugan epithet of the fire: *tā̆rən-tuγət, kol-tuγət*.[61]

The translation of the following sentence is not quite clear:

36. *tăm ătȧmз*	this man,
i̯əm ătȧmз	a good man,
n̥tȧmȧ i̯əwot,	turned into nothing;
tŏw titət mənot	his soul went
tor wȧntšet mȧη məγȧ,	lake . . . to land,
tŏw wȧntšet mȧη məγȧ[62]	sea . . . to land [= he died]

55 'Suku *toηχ*:iksi muuttunut toisen käden kautta surmansa saanut taikka itsemurhan (muula tavalla kuin hirttäytymällä) tehynt sankari, tai muu mies jossakin suvussa | zum Stamm-*t*. gewordener, durch eigene od. fremde Hand (nicht durch Erhängen) gestorbener Held oder auch ein anderer Mann einer Familie', Paasonen (1926) 2600, 2434, cf. 1201.

56 Paasonen (1926) 2434.

57 'Veteen hukkunut *toηχ* haltijaksi muutunut ihminen (ei sankari) | im Wasser umgekommener, in einen *toηχ*-Geist verwandelter Mensch (nicht ein held)', Paasonen (1926) 2600, cf. 2209; Karjalainen (1921–1927), Vol. II, pp. 246–8.

58 Karjalainen (1921–1927), Vol. II, pp. 212, 375 and cf. previous note.

59 Karjalainen (1921–1927), Vol. II, pp. 355–62, 343–50.

60 Karjalainen (1921–1927), Vol. II, p. 359.

61 Karjalainen (1921–1927), Vol. II, p. 348.

62 'Tämä ihminen, hyvä ihminen kuoli; sielunsa meni "paratiisiin" | dieser Mann-ein guter Mann, starb; seine Seele ging ins "Paradies" 'Paasonen (1926) 2944.

7. The Ostyaks have different ideas about the soul.[63] Phrases can be quoted showing that certain mental parts of a human being may leave the body behind:

37. *i̯emәŋ urt, pәsәŋ urt,*	holy hero, mighty hero,
mәtȧ mәγȧ nӑmәs pӑnә̂n[64]	what land thou put thy soul on.

Other expressions for the soul:

38. *titәŋ χui̯ titәt ottәt,*	the spirited man's spirit he prolongs,
olχә̂ŋ χui̯ olχә̂t ottәt[65]	the living man's life he prolongs.

Among the above fragments containing mythical material, there is only one (No. 1) marked by a vertical line, meaning verse, in Paasonen's dictionary. These short fragments themselves reflect the characteristic features of Ob-Ugrian poetry.

Most phrases exhibit parallelism[66] (see Nos 1, 2, 5, 6, 17, 19, 21, 23, 24, 25, 27–38), some represent *figurae etymologicae*[67] (Nos 13, 25, 29, 38). Steinitz and also Fokos suppose the lack of inflecting elements (case-endings and postpositions)[68] to be characteristic of Ob-Ugric songs, and particularly of archaic hymns. The texts quoted also illustrate this phenomenon (Nos 24, 31, 35). A certain intrinsic pulsation serves as a basis for breaking into lines the sentences not displaying any of the said characteristics of poetic language (see Nos 4, 7, 8, 9, 11, 14, 16, 18, 23). What is more, even our one-line examples exhibit that verse-like pulsation (Nos 10, 12, 15, 22, 26, and by *figura etymologica*, also No. 13). It is noteworthy that the names of spirits are either short, that is, easily adapted to verse, or verse-like. For instance:

koi̯-ȧtәmpә sewәŋ χui̯,	braided man [from] yonder the moor,
koi̯-ȧtәmpә weḱәŋ χui̯[69]	handsome man [from] yonder the moor.

Without knowing the context, it is, of course, difficult to state for certain that these fragmentary sentences are parts of mythical hymns, yet such hymns do seem to be, and apparently come from a stage of civilization when they were sung. We feel entitled to accept this conclusion on the strength of the resemblance between the shamanistic chants of the Samoyeds and the above quotations.[70] Although Steinitz's observations on the difference in the text of Ostyak songs when sung and when dictated, are undoubtedly

[63] Karjalainen (1921–1927), Vol. I, pp. 23–49.
[64] 'Mihin maahan hankit lähtöä?', Paasonen (1926) 1767.
[65] 'Lisää hengelliselle (elävälle) miehelle henkeä | verlängert dem lebendigen Mann das Leben', Paasonen (1926) 1535.
[66] Steinitz (1941) pp. 29–41; cf. Gáldi (1960) pp. 149–75, esp. pp. 160–8.
[67] Steinitz (1941) pp. 41–6.
[68] Steinitz (1941) p. 10, note 23, p. 18; Fokos (1956) pp. 61–95.
[69] 'Suon takainen palmikkoinen mies, suon takainen kaunis mies', Paasonen (1926) 596, 800.
[70] Cf. Gáldi (1963).

reliable (the former may include elements increasing the number of syllables[71]), Paasonen's sample sentences may equally well be fragments of songs, without any modification. If this is so, Steinitz's theory about Ostyak prosody and that of its critics[72] should be reconsidered thoroughly.

REFERENCES

Austerlitz, R. (1958) Ob-Ugric Metrics. The Metrical Structure of Ostyak and Vogul Folk Poetry, in *FFC*, No. 174.

Fokos, D. (1956) A névragozás történetéből (From the History of the Declension of Nouns), in *NyK*, Vol. LVIII, pp. 59–95.

Gáldi, L. (1960) A finnugor népi verselés tipológiai áttekintése (A Typological Survey of Finno-Ugric Popular Versification), in *I*, Vol. XLVIII, pp. 149–75 (to be continued as the publication of author's lecture delivered at the Finno-Ugric Section of the Linguistic Society, Budapest, 25th November, 1958).

(1961) Robert Austerlitz, Ob-Ugric Metrics, reviewed, in *NyK*, Vol. LXI, pp. 1931–5.

(1963) Zu einigen Problemen des Versbaus in den samojedischen Schamanengesängen, in *Glaubenswelt und Folklore der sibirischen Völker*, Budapest; (and in the present volume).

Ganschow, G. (1956–1957) Die Verbalbildung in der ostjakischen Sprache, in *Wissenschaftliche Zeitschrift der Humboldt Universität zu Berlin*, Vol. VI, pp. 241–6.

(1965) Die Verbalbildung im Ostjakischen, in *Ural-Altaische Bibliothek*, Wiesbaden.

Karjalainen, K. F. (1918) *Jugralaisten uskonto* (The Religion of the Yugrians), Porvoo.

(1921–1927) Die Religion der Jugra-Völker. Vols I–III, in *FFC*, Nos 41, 44 and 63.

(1948) *Ostjakisches Wörterbuch*, Ed. by Y. H. Toivonen, Helsinki.

Paasonen, H. (1902) Über die türkischen Lehnwörter im Ostjakischen, in *FUF*, Vol. II, pp. 81–137.

(1903) Matkakertomuksia vuosilta 1900–1902 (Description of a Journey from the Years 1900–1902), in *JSFOu*, Vol. XXI/5.

(1926) *Ostjakisches Wörterbuch nach den Dialekten an der Konda und am Jugan*, zusammengestellt, neu transskribiert und herausgegeben von Kai Donner, Helsingfors.

Steinitz, W. (1941) *Ostjakische Volksdichtung und Erzählungen*, Vol. II, Stockholm.

(1950) *Geschichte des Ostjakischen Vokalismus*, Berlin.

[71] 'Füllelemente', 'Füllvokale, Füll- oder Dehnsilben, Füllsuffixe oder Füllworte', Steinitz (1941) p. 4.

[72] Steinitz (1941) pp. 1–29; Austerlitz (1958); Gáldi (1961) pp. 131–5.

Funeral Rites among the Enets (Yenisei Samoyeds)

by

G.D. Verbov* Leningrad

On September 10, 1938, I arrived at the nomadic settlement of Galchikha and learnt that the wife (from the Solda clan) of the Enets Sado Osaku had died. She had already been dead for two days. Her body lay to the left of the entrance in the tent. It had been wrapped, I was told, in hides. The lower edge of the tent was slightly raised at the feet of the deceased. Within the tent (shared by another family) life went on as usual: the fire was burning, tea was boiling in the kettle, the children were playing. The Enets, having procured some planks, were nailing together a *pəmb'* (a box-like coffin) with iron nails. They had previously measured the body of the deceased. Once the coffin, which was assembled outside, near the tent, was ready, people started to dismount the tent, while the widower in the tent was gathering the dead person's belongings to be placed with her in the coffin. He cut some slices of bread and put them into a small pan with a lid, then he produced a knife, a cup and a saucer, etc. A woman living next door, entered the tent, scraped some brick tea into a cloth and added it to the rest. All this was then put into a tin box with a lid.

Later they took down the tent leaving only six poles standing. The widower put the spare summer sandals of the deceased in the bottom of the *pəmb'* and with the help of the neighbour spread two pieces of old reindeer hide over them. The body was taken out of the tent, the feet forward, covered with old sacking and laid in the coffin, the lid of which was nailed down.

Since the graveyard was at a distance of one kilometre and a half or so, the Enets asked the schoolmaster to lend them a horse. The *pəmb'* was put on a cart, and the little procession started. Beside myself, it consisted of the widower, his mother-in-law and neighbours such as an old man called Monkansi with his son and an 'Enetsized' Nenets called Yar.

The graveyard was situated in the north-eastern corner of the little island on which the Enets settlement was located. The burial place rose on one or one and a half metre above the flat and marshy surface of the island. The coffin was taken off the cart and lowered to the ground. The Enets then started to work out the four cardinal points, and after some reflection old Monkansi unerringly pointed to the north, in which direction the feet of the deceased had to be placed. The coffin was measured and they started to dig the grave with the two spades they had brought with them. They had measured the coffin for a second time very accurately to avoid superfluous work and probably so as to make the coffin strictly fit into the grave with the sides touching the walls as required by ritual. Osaku was digging with the greatest zeal. The grave was made 5–10 cm deeper than the height of the coffin, which was then let down into it.

* The material here published is taken from the archives of the ethnographer G. D. Verbov, killed in action in 1941. His legacy is now kept in the Archives of the Ethnographical Institute, Academy of Sciences of the USSR, stak No. 2.

To the right of the head of the coffin Osaku put the pan with the slices of bread, three enamelled bowls and an axe, the edge of which he had previously sharpened by rubbing it against an iron spade. On top of these objects they placed a *nadorč*[1] and the handle of a scraper used for preparing hides, after having removed the iron part. Again near the head were placed the tin box with the cup, the saucer, the knife of the deceased and the tea-kettle wrapped in a rug, and also an (empty) tin snuff-box with a hole in it. A tea-kettle and a pail each with a hole previously pierced in it, were put near the grave. The coffin and the objects in the grave were then covered with folded reindeer hides that had been used for lining the tent. It was from these that the woman in our company had cut the strings, and it was also she who had cut into pieces the rope by which the coffin had been attached to the cart, throwing the pieces away at a distance of about ten steps from the grave. For this purpose she had used the knife of the deceased, which was then laid into the tin box. To the right of the head a vertical board was forced into the ground, and the grave was strewn with the sand dug from it and covered with turf to prevent the sand from being blown away in dry weather.

When everything was ready, Osaku who, the whole day and during the funeral, had been quite calm and willing to chat with us turned to the grave and, though he was not crying, began to lament covering his face with his hands. This lasted for about one minute. Then the procession started for home. Suddenly, however, Osaku turned again toward the grave, drew a line across his footprints with the spade and staring straight in front of him uttered a few words in a rather loud voice. Since I do not understand the Enets language, they failed to convey anything to me. When an opportunity presents itself I shall find out what the words meant.[2] They were probably a farewell formula. On the way home nothing more was said and no one looked back.

According to old Monkansi, there exists also another custom when leaving the grave : at a respectful distance a stick is placed across the footprints and several others are stuck into the ground at its two ends. Then, turning towards the grave and pointing to the north they shout: 'Do not come our way ! your way is yonder above !' Then a big camp fire is made and they purify themselves. This custom is also observed by the Nenets.

When I was there, however, non of this was done. The old man told me that it was usually done at a later date. The custom seems to disappearing or, perhaps, they were just shy because of my presence. Incidentally, everything used for attaching the body (the shroud, fastening the coffin to the cart), is called *haľmer ju*[3], which is the same as the Russian бабий узел (literally: 'woman's knot'). The old man also told me that according to their custom a shred of red cloth with two beads sewn on it, had been placed on the deceased. The beads are meant to be eyes with which to find the right way to the other world. The same custom was observed among the Nenets. When we arrived at the settlement, Osaku's tent was already pitched at another place, since life cannot go on where somebody had died.

[1] A board used for the processing of hides.
[2] We did not find these words in G. D. Verbov's manuscript.
[3] Literally: 'the knot of the deceased'.

On Some Problems of Versification in Samoyed Shamanistic Songs

by

L. Gáldi, Budapest

1. As regards rhythm, the shamanistic songs are not entirely independent of the other genres of Samoyed folk poetry. Thus, when examining the material at hand concerning the rhythm of Samoyed shamanistic songs, it might be instructive for us to begin with the Kamas Samoyed lament which has been dealt with by the Hungarian-born Professor J. Lotz in two special studies, namely: *Structural Presentation of the Kamassian Lament*[1] and *Kamassian Verse*[2]. The lament in question is the only existing Kamas Samoyed folk-song recorded phonographically.[3] This was taken down in 1914 by Kai Donner from his chief informant, Avdakeja Andžigatova. Later it was published by Aulis Joki in an Appendix to K. Donner's Kamas dictionary.[4] In his analysis Lotz comes to the conclusion—a rather striking one it seems—that the Kamas verse belongs to the 'pure-syllabic' type, which includes also Mordvin and Hungarian versification.[5] Within this type Lotz endeavours to trace the prosodic features of this text. Making a phonetic transcription, and splitting it up into semantic units, he gives the following arrangement:

6 6	6	*kijen maˀlεpijem*	*sākər məjasaŋpə*	*iˀpəlε kojōpi;*
5 3	5	*(maˀlεpinε t́ym*	*kyk noˀt́ə*	*ösērlε̄ paˀpi!)*
6	6		*sākər məjasaŋpə*	*iˀpəlε kojōpi,*
6	6		*səri t́əlamsaŋpə*	*māla kojōpijəˀ;*
4	6		*kys alωət́ə*	*iˀpəlε kojōpi;*
6	3/6		*ikaˀ örenkət́ə*	*iˀpəlε māla kojōpijam,*
5	7		*tykanpəkat́ə*	*t́yršylε kojōpijam;*
7/3	5	*kijen koˀlajlapijem*	*t'usanpə*	*māla kojōpi;*
	6			*tyj təsεŋpə em ku!*
7	4		*iˀ pəlε jātosaŋpə*	*teˀ lāmpijəˀ,*
6	6/3		*šytēne t'ysεŋpə*	*pārt'emaˀ tōlāmpi, kampijəˀ*

[1] Lotz (1953).

[2] Lotz (1954) pp. 369–77.

[3] Lotz could not find the original cylinder in Helsinki and used therefore the text in Donner's manuscript which he published in facsimile: Lotz (1954) p. 372.

[4] Donner (1944) p. 87.

[5] In a recent paper Lotz (1954, p. 376, note 22) sums up his typology which he had set up with little regard to historical development: 'The basic regularity refers generally to the number of syllabic pulses in such frames (i.e. in a semantically well-distinguished segment or "colon"). Further metric differentiation depends on the use of the prosodic features: stress, pitch and duration. Thus there are the following types of verse: (1) Pure-Syllabic (e.g. Kamassian, Mordvin, Hungarian "national" versification, etc.): (2) Syllabic-Prosodic: (a) Dynamic, based on differentiation of stress levels (English, Russian, etc.); (b) Melodic, based on differentiation of pitch classes (Chinese); and (c) Quantitative, based on duration (Classical Greek, Hungarian "quantitative" versification).' We shall have the opportunity to come back (in section *3*) to a certain use of the vocalic quantity, at least in Yurak Samoyed.

Relying on this arrangement, and on the earlier translation made by K. Donner, Lotz gave his own English version and a prosodic analysis of the text. Since at this stage of versification the correct interpretation of the text is indispensable (and alone can account for rhythmic fluctuations), let us first examine Lotz's translation[6]:

'My black mountains where I used to wander have been left behind;
(My land where I wandered grew my golden grass!)
My black mountains have been left behind;
My white summits have been left behind;
[All] of my strength has been left behind;
Of my large clan, I alone am left behind;
Of my family, I, straying alone, am left behind;
My lakes where I used to fish have been left behind;
Now I do not see them!
My hut-poles are rotten,
My sewn birchbark cloths are all shrivelled; they are gone'.

Lotz sums up the results of his prosodic analysis as follows[7]:

'The results of our analysis show that this Kamassian song-text *has a metric structure of hexasyllabic phrase cola*[8], *with slight variations.*[9] The song begins with a three-colon verse presenting the theme of the black mountains, the place of nomadic wandering left behind. This line is matched by another three-colon line containing an association with the lost pastures of golden grass. Then there is a repetition of the last two cola of the first line, matched by an identically constructed line; these refer to the objective circumstances of the black and white mountains and are parallelled further by a third line containing a subjective reference to the diminished vital strength of the tribe. This is a transition to the two following subjective lines, matched parallelly, about the dying out of the kin. There then follows a three-colon parallel verse matching the first line as if it were the beginning of a free strophic construction, but this is broken up immediately and this part is concluded by the subjective one-colon exclamation: "Now I do not see them!" The final couplet is constructed of two-colon verses, a parallel objective picture of the decaying nomadic hut symbolizing the end of the nomadic way of life and of the Kamassian tribe with a word of repetition, indicating finality, as the end of the song.'

Lotz is right when he speaks of a certain 'inner consistency' in the construction of the text, which, in his opinion, is naturally supported by several additional factors, such as the recurring parallelism and the 'syllabically determined phrase cola grouped into sentence-verses'. Of course, his most important statement refers to the 'hexasyllabic phrase cola'. The question to decide now is, whether anything similar occurs in other Samoyed verses, especially in the shamanistic songs with which we are now concerned.

[6] Lotz (1954) p. 373.
[7] Lotz (1954) p. 377.
[8] According to Lotz's own definition (1954, p. 376), the '*colon*' is 'a numerically characterized phrase in a poetic text'; his second definition is closely connected with this: 'A metrically defined sentence is called a verse (line).'
[9] The italics are mine, L. G.

2. Indeed, verse-like structure is evident, in the first place, in two sacred texts recorded again by K. Donner[10] from the Kamas. These are two short prayers. The rolling incantation-like rhythm of one of them even breaks through the German translation, being, of course, more pronounced in the original, where the parallel groups of words, as well as the 'flexional rhymes' and the 'word-repeating rhymes' (formed by the repetition of the last word of a line), make this rhythm more pregnant:

mɯ̄dɯ̄źĕʼ bŏďəlăm.	6 (3, 3)	Mit dem Opfer speise ich.
i̯ăzăŋbɒʼ,	3	Unsere Mutter,
abazaŋbaʼ,	4	Unser Vater,
ďɯ̆n ei̯et,	3	Geist der Erde,
ďɯbɯn, ďɯ̆ʼn, bɯn ei̯et	6 (2, 1, 3)	Geist der Welt, der Erde, des Wassers,
sānəlaʼ ig bŏi̯ārγɯ̆t.	7 (3, 1, 3)	verachtet eure Nüsse nicht!

Naturally, it would be useless to look for hexasyllabic lines formed 'with slight variations' in this text as in the lament analysed by Lotz. Here we have no more than two lines that can be considered hexasyllabic. Based on this text alone, it would be possible for us to presume quite hesitantly that a kind of short hexasyllabic line was the basic pattern of the Kamas Samoyed verse. But there is another, very important principle to remember: we have to go beyond the pattern based on pure 'phrase cola' if we want to touch upon the vital element which is common in every verse, i.e., the dynamically conceived *rhythm*! A partition of text into lines with the necessary rests at the end of each line, will in any case yield a certain rhythm. At the same time, however, we must find out whether the number of syllables contained in the individual phrases of the verse ('cola') are all rhythmically equivalent. A perfect equivalence is *a priori* improbable. From Finno-Ugric analogies it would seem more plausible that certain syllables bearing main stress or secondary stress would dominate the surrounding ones to a greater or lesser extent and preserve this property of theirs even in a song-text, i.e. in a 'musical' performance. T. Lehtisalo has already pointed out in connection with the Yurak Samoyed language that word accent and sentence stress cannot be disregarded. In his opinion, the stressed syllables in a sung text 'carry more dynamic force than in ordinary speech'[11] and therefore the true character of the verses is determined not only by the number of syllables, which is rarely fixed but—and these are again his words—by 'eine bestimmte Iktuszahl'.[12]

M. A. Castrén obviously had this distribution of ictuses in mind, when writing in his travel notes as early as 1840: 'A Samoyed, when not singing but only reciting a song, insists on a certain rhythm which became familiar also to my ears. This rhythm cannot be reduced to any definite metre, *yet shows a certain inclination to the trochee*'.[13] Hence, we may distinguish within the lines (or possibly even within certain sections thereof) minor units which extend from one stressed segment to the next; these can be

[10] Donner (1944) p. 100.
[11] Lehtisalo (1922) p. 84.
[12] Lehtisalo (1936–1937) p. 3.
[13] Castrén (1853) p. 193. The italics are mine, L. G.

termed — with due precaution — 'cola', or *beats* if the musical segmentation is known. Scanning our texts from this point of view, we shall soon reach a more general conclusion: it is not the hexasyllabic lines which seem to occur most frequently but lines in which, in conformity with the distribution of stresses, the presence of two segments may be presumed. In our first song almost every line belongs to this type, and so does the initial line of our prayer: *muduźẽ boďᵊlăm*. Even two-beat lines can be found in the following part of the second prayer (a characteristic 'hunting prayer'), published by K. Donner:

<table>
<tr><td>mănă | ďuksu | sɪrẹ̄rla'</td><td>Mir entgegen jage er</td></tr>
<tr><td>amnᵊy̭uba sī̯n | k̇urābə</td><td>einen männlichen sibirischen Hirsch,</td></tr>
<tr><td>bəʌān | kurābə</td><td>einen männlichen Elch,</td></tr>
<tr><td>ṣ̌ɪʙᴅə̂n | kurābə</td><td>einen männlichen Bären.</td></tr>
</table>

Prayers and shamanistic charms often completely merge in Samoyed folklore. In a typical shamanistic verse 'asking for a rich guarry', like the one below recorded by T. Lehtisalo, there are most likely two-beat parts only. A blind forest Ẏurak shaman *kq̇ľľeât ŋ̡āxannī̯* living in Nailina[14] performed for Lehtisalo the following prayer (No. 107):

<table>
<tr><td>β'it kāŋ βäεkku!</td><td>Wasserfürst-Alter!</td></tr>
<tr><td>ŋäεiββokkujjēj | kā'mrām'</td><td>Meinen armen Kopf beuge ich.</td></tr>
<tr><td>kāʌ' ʌ'ettāj | kāttaχe̮t!</td><td>Fische für mich, wenn ich doch fangen würde!</td></tr>
<tr><td>jieďześi | mi̯ẹ̄χẹ̄t!</td><td>Ohne Krankheit, wenn ich doch wäre!</td></tr>
</table>

The tendency toward a two-beat segmentation is clearly discernible in another shamanistic prayer, or rather in a kind of invocation; we find it in the introductory lines of Lehtisalo's text No. 106:

<table>
<tr><td>χᵉ χᵉ χēī!</td><td>He he hei!</td></tr>
<tr><td>śeȧn | nūm</td><td>Wieviele Num's,</td></tr>
<tr><td>śeȧn | käεχäε</td><td>wieviele Kaehe's</td></tr>
<tr><td>ťajjeä | mi̯e̮,</td><td>vorhanden sind,</td></tr>
<tr><td>ťśukku | δeȧn</td><td>sie mögen sich an dieser Stelle</td></tr>
<tr><td>kā'amjeämť!</td><td>niederlassen!</td></tr>
</table>

Or in text No. 104 (from the same informant):

<table>
<tr><td>nūm | βäεkku!</td><td>Num-Alter!</td></tr>
<tr><td>ńīḿijjōm | χäεmae̮r</td><td>Lasse dein eines Auge</td></tr>
<tr><td>tar'ńeȧŋ·jōm | χi̯ŋŋī̯rae̮ptāt!</td><td>nach hier schauen!</td></tr>
<tr><td>ŋ̡aśkə | jĭetsəä | jȧγguŋŋäεjjī̯ä!</td><td>Mögen die Kinder keine Krankheit haben!</td></tr>
</table>

All these shamanistic prayers, especially No. 106, are in prose intended for recitation, but they are interspersed with more lofty passages, with two-beat segments recalling by their structure Samoyed heroic songs or some-

[14] Lehtisalo (1947).

times even the rhythm of heroic songs about the shaman himself. In most cases, two-beat lines can be heard distinctly for instance in Lehtisalo's text No. 54 from the Arkhangelsk region, but recorded at Obdorsk:

tāďib'e | *tḙ̄tta* | *ńɯ*	Dem Sohne des Zauberer-Wirtes
śiβ-χaβ | *jǟľľe*	doch sieben Tage
jı̇ľľeδa | *χāјјɯ̄ββḙ,*	blieben übrig
jə̄śśe | *ťɯ̄ḿɯmta*	die Eisenbleche der Feuerstätte
tāraχal | *lāββi̮*	stiess er mit dem Fusse.
tāďib'ēm | *puχū*	Das Weib des Zauberers
nōi tōјјē | *ťśāmta*	an seiner Tuchdecke
ni̮χi̮·p | *tāmpiδȧ*	zupft, sagte:
mā: 'jēββants, jɯrkkaď!'	'Mein kleiner Liebling, stehe auf!'
mā:	sagte:
'tāďib'e | *tḙ̄ttānt*	'Dein Wirt der Zauberer
pı̇ľľi' | *βāllakkaδa,*	liegt im Sterben,
ŋāјјēnta' | *χānnaδa*	für seinen Körper ein Opfer
χādāś | *βoḙββāβ'?'*	zu töten, ist es schlimm?'

3. There is still one more conclusion to be drawn from the texts recorded by Lehtisalo: many of the presumable ictuses (to be found in Yurak Samoyed generally on the first syllable of the words)[15] are frequently accompanied and strengthened by the long duration of the stressed syllable or rather by its prolongation. All this goes to show that not only the dynamic stress holding the segment together but also the longer duration of the vowel, by which the ictus is strengthened, has its importance, even though as a secondary factor, in the segmentation of the Samoyed verse.[16]

Quantity seems to support the secondary stress on the third syllable of some quadrisyllabic words. If this is the case, two-beat segmentation may be surmised also in such lines of the late Yurak Samoyed shaman *ńāśśī jāmmāl* (Lehtisalo No. 83) in which only one long word fills the whole line. This 'murmured' text containing a great number of phonetically expressive word forms begins as follows:

śēńńen | *nāńńi'*	Früher
χunnan | *nāńńi'*	irgendwann,
sōmta | *βānni'*	zur Zeit
mɛ̄ľľeŋ | *kanna*	meines Gutwerdens (d.h. als ich Zauberer wurde)
xḙdḙmm: | *:ī̮*	in meinem
ŋȧεl | *nańńi'*	Traume
βī̮βŕŕɯδù'	ihr Rauschen
tōββa | *nōntu'.*	hörte man kommen.

[15] Collinder (1957) p. 426, referring to Lehtisalo, says: 'the first syllable can always carry the stress if the word is spoken isolated.'

[16] On the relationship between stress and quantity Collinder comments (1957, p. 426): 'The accentuation of disyllables depends on quantity. If one syllable is short and the other is long, the stress is on the long syllable.' Hence, there is a certain relationship between stress and quality also according to Collinder.

In the song of the shaman *t́iʙit́śē ńēml'i* conjuring up the spirits of the dead, recorded by Lehtisalo (No. 82) there seems to be a division into two or, not infrequently, into four segments. If so, Lehtisalo's text given below should be divided into rhythmic units consisting of two lines each. The melody, if it were known to us, would, we feel, support this assumption:

jāβmāl	Mein Grossvater
jirīkkoβ!	vom Oberlauf des Ob!
ŋār'i·ttaľe	Noch näher herbei
ḿēŕitsun!	eile!
śiďe ńūn	Zweier Knaben
ńēbe!	Mutter!
χunnānānt	Wohin
χäeβnān?	bist du gegangen?
śiďeńuβ!	Meine zwei Söhne!
χunnānānďi'	Wohin
χäεβnādi'?	seid ihr gegangen?
pannōŕē	Spitzkopf
jirīkkōβ!	mein Grossvater
χunnānānt	Wohin
χäeβnān?	bist du gegangen?

4. The examples hitherto analysed permit us to draw some general conclusions concerning certain properties of the two-member verses of the Samoyed shamans. To examples from these two dialects let us now add further proofs, as can be obtained from the analysis of the oldest Samoyed shamanistic texts ever recorded and of the texts that have recently become accessible. In our historical survey we must, naturally, rely on Castrén whose Samoyed texts were recorded in the 1840s but are obviously much older. For reliable renditions of them we are indebted to T. Lehtisalo.[17] In Castrén's collection we find two important texts: A heroic song of shamanistic character (No. 5), rhythmically very close to Lehtisalo's text (No. 54), and a true shamanistic song (No. 19) probably dictated late in 1842 to Castrén by a shaman.[18] Since these two texts are rhythmically rather different one from the other, it will prove worth while quoting a few lines here. Compared to other Samoyed songs, No. 19 contains uncommonly long lines. It is even probable that segmentation prevails in them, each line containing four segments instead of two. Here, too, as in Lehtisalo's text (No. 83), the expressive use of a great number of nasals is conspicuous.

tāvida njammaguem'ah	Dann aber ergriff ich,
nuum beau njammaguem'ah	meinen Himmelsbaum ergrift ich.
seämeän njāsanjin	Wieviele sind meine Kameraden,
suum buda kāmarŋaje'eh!	mögen sie sich vor mir beugen!

[17] Castrén (1940).

[18] According to Lehtisalo's introduction: 'The shamanistic song No. 19, in which the division of the text into words is incorrect in several instances, was probably dictated by the shaman who, in December 1842, acquainted Castrén with the problems of shamanism in the village of Syomza.' Castrén (1940) p. XV.

āta njim barttamuada'ah,	Auf ihre Füsse plötzlich sprangen sie,
tāveda lōrvadā'ah.	dann aber nahmen sie in ihren Schoß.
'seävin anji njōtsko'oh	'Das kleinste meiner sieben,
nū(v)un dain pōdertsun,	mein Himmelsrentier schirre an,
teän jīnjen moaptsun'.	am Lenkriemen meines Rentiers halte!'

In Castrén's song No. 5 (from which lines 209–24 are quoted below) the well-known rhythm of the heroic songs can readily be recognized. This example shows, how variegated the texts can be made if one-member lines (as No. 210) and three- and four-member lines (213, 217–8) are inserted among the two-member ones:

tjikī juoltsendo'	Die Zeit
taevioih,	kam,
waelje | *waäsakoh*	der Wahelje-Alte
ānin | *doh*	kam wieder
opoj | *ŋäe* | *ninje.*	mit einem Beine.
peänsertah | *idā,*	Seine Zaubertrommel hingen sie auf,
peänserta | *peänā,*	seine Zaubertrommel zog sich fest an,
ānin | *sāmbadā,*	wieder schamanierte er,
uäsako | *jānamāda* | *jānguh,*	der Alte lässt nicht nach,
opoi | *ŋäe* | *ninje* | *tarā.*	mit einem Beine hüpft er.
uäsako | *mah:*	Der alte sagte:
'tiēi | *hāsauva*	'Der gestrige Mann
pījekos | *tōvi̥ih,*	kam als Hermelin,
mōnsang | *hānawaeda,*	brachte den Schenkel fort,
häememda	wohin er ging,
jiherau.'	weiss ich nicht'.

5. In the next text, again recorded by Castrén, the shaman appears as an ermine.[19] By the motifs it contains, this text is closely related to the shamanistic song (*tȧδiβe śō*) recorded by Péter Hajdú in Leningrad in 1957 from V. Ledkov, a Yurak Samoyed of Bolshaya Zemlya (see Hajdú 1960):

χōra ǰīδm‿ʙiśȧm'	Ich gebrauchte ein Wassermausmännchen
ānart́ś me̮̥aβ'	um auf dem Wasser zu gehen,
χōra‿ṕījȧm'	ich gebrauche ein Hermelinmännchen
jōndert́ś mee̥aβ'	um zu Fuβ zu gehen.

[19] The names, of expressive value, of small fur animals also occur in Saam (Lapp) incantations, as quoted by P. Ravila (1934) p. 112.

[:saôppənȧ̯ᴅ̂ᴅžαm :]	Dzo',	Dzo',	Dzo',	Dzo'
[:pŭŏìᴅaγa̯ᴅ̂ᴅžαm :]	,,	,,	,,	,,
[ŏăŕrä̀ᴅža̯mα :]	,,	,,	,,	,,
[:néäδäᴅžȧ̯mα :]	,,	,,	,,	,,
[Mein Mäuschen :]	tso,	tso,	tso,	tso
[:Mein Hermelinchen :]	,,	,,	,,	,,
[:Mein Eichhörnchen :]	,,	,,	,,	,,
[:Mein Marderchen :]	,,	,,	,,	,,

Here, at last, we have a sung melody, and we can study the interrelationship between the text and the tune. The melody of Ledkov's shamanistic song, as transcribed by the Hungarian composer Erzsébet Szőnyi is as follows:

6. The Yurak Samoyed shaman song recorded on a wax cylinder in the Pushkin House, Leningrad, (No. 1402 B) and transferred in 1957 to tape by Vilmos Diószegi, is of great interest. Diószegi has also brought back to Budapest the text of the song written in Russian characters and a Russian translation by V. Ledkov. Relying on these sources Péter Hajdú has put the song down in phonetic transcription and has given a more exact Hungarian translation. As will be seen, this is not the song of the shaman himself, for it is not he who conjures up the dead, but it represents a phase of the shamanistic performance in which an important part is played by his assistant and also by those present. The chorus-like participation of the onlookers is not quite unknown: in describing the trance-induction of the Samoyed shamans K. Donner made mention of it:[20] 'The beating of the drum (by which the shaman conjures up the spirits) becomes gradually louder and rhythmically more pregnant, the shaman begins to sing. Having sung a few strophes, he waits until those present repeat his song. This chorus-like echo is led by his assistant who acts during the ceremony almost like a acolyte or an altar-boy. The song and the thumping of the drum suddenly becomes still louder, more passionate, and this is due not so much to the shaman's words, but rather to the *fervent emotion with which they participate in the act.*' Yet neither K. Donner, nor V. Diószegi has quoted any chorus sung in a collective ecstasy. This is why the text of wax cylinder No. 1402 B is worthy of attention also because of its content:

χāεχε ḿàď'taďi-tāďib'ôβ!	Shaman of the holy tent!
sattâ, ŋuľi', satta, satta tāďib'ôβ!	Strong, very strong, strong shaman!
ŋarka ŋarka tāďib'ôβ!	Great, great shaman!
ŋarka marra ťund taďib'ôβ!	Great shaman of the sand tongue!
taďi-taďi, tāďib'ôβ!	Shaman, shaman!
χunnorkād tāďib'ôβ!	Runaway [to the spirits] shaman!
χa'mandāda' tāďib'ôβ?	What does he say, shaman?
χa'mandāda', tāďib'ôβ?	What does he say, shaman?
χa'mandānanna, tāďib'ôβ?	What is he saying, shaman?
χa'mandānanna, tāďib'ôβ?	What is he saying, shaman?
marra ťund tāďib'ôβ!	Shaman of the sand tongue!

[20] Cf. Donner (1926) p. 111. His observations were given a wider ethnographic interpretation by Diószegi (1958) pp. 337 ff. The italics in the text are mine, L. G.

śunę t'ud tāďib'ôβ!	Shaman of the bluish sand tongue!
χi̮nnokkād tāďib'ôβ!	Singing shaman!
marra t'und, marra tund tāďib'ôβ!	Shaman of the sand tongue!
χi̮nnokkād tāďib'ôβ!	Singing shaman!
śunę t'und tāďib'ôβ!	Shaman of the bluish sand tongue!
ŋangęm χaŋa, tar̄t'śēm χaŋa tāďib'ôβ?	What is he asking for, the shaman?
χaďi'lata tāďib'ôβ!	Crooked shaman.[21]
ŋamgęm χāŋa tāďib'ôβ,	What does the shaman desire,
ŋarka χartta tāďib'ôβ?	The great shaman himself?
ŋarka marra t'und tāďib'ôβ!	Great shaman of the sand tongue!
χumna tūrta tāďib'ôβ?	Where is he going, the shaman?
tupka s'umb taďibôβ!	Shaman of the axe head!
marra t'und tāďib'ôβ!	Shaman of the sand tongue!
marra t'und tāďib'ôβ!	Shaman of the sand tongue!
χańńä' ṕāda tāďib'ôβ?	Where does the shaman want to go?
jēnnar' ŋamgę t'ikki̮ tāďib'ôβ?	What is he like in reality, the shaman?
χa'mandāda' tāďib'ôβ?	What does he say, the shaman?
jennar' ŋamgem puna (?) tāďib'ôβ	What is he like in reality, the shaman?
pudaŕena tāďib'ôβ!	Searching shaman!
χādi'lata taďib'ôβ!	Crooked shaman!
χańńā ṕāda tāďib'ôβ?	Where does the shaman want to go?
χūrkka mārkad tāďib'ôβ?	Which town is the shaman from?
χańńā ṕāda tāďib'ôβ?	Where does the shaman want to go?
χurkka marki̮ tāďib'ôβ?	The shaman of which town?
ŋamgęm puna (?) tāďib'ôβ?	What is he looking for, the shaman?
ŋamgęm purta t'ikki̮ t'āďib'ôβ?	What is this shaman looking for?

Of all the shamanistic songs mentioned so far, this text with its 'litany-like rhythm' is obviously the closest to a kind of isometric pattern for which, in all probability, its social function is responsible. The monotonous rhythm of the short parallel sentences contributes considerably to the rise of the ecstasy of those present. The grouping of the ictuses in the lines is already indicated by some of the five lines of the introductory part. Thus line 3: *ŋarka ŋarka tāďib'ôβ* clearly suggests the 'trochaic' rhythm mentioned by Castrén: ⏊ ∪ ⏊ ∪ ⏊ ∪.

In spite of certain fluctuation in the number of syllables, beginning from line 6, the melody, probably accompanied by some tom-tomming, comes to light with full clarity. The melody of lines 6 and 7, for instance, clearly intones the melodic line of the shamanistic song transcribed by E. Szőnyi:

[21] In the Russian translation: шаман с якорями 'a shaman with anchors', but this seems to be only an approximate interpretation.

In lines 6 and 7 we have it in the following musical form:

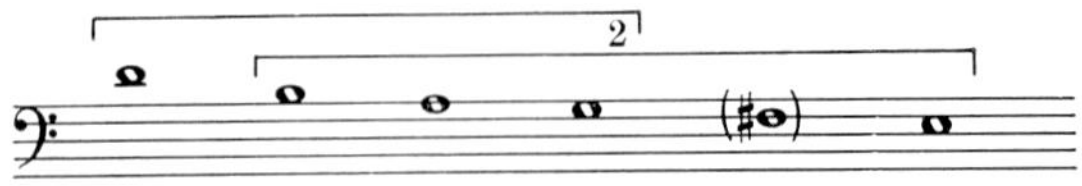

7. Great as the variety of the quoted texts is, we shall try to sum up our observations and return to our point of departure, since J. Lotz's conception of the Samoyed 'hexasyllabic' line can now be placed in a better perspective. Apparently, he examined only one song and thus failed to notice that he was actually dealing with a form that is known to have come up in the folk poetry of nearly every extant Samoyed language—a form, some variations of which are admittedly rather often hexasyllabic.[22] Samoyed versification may be called 'pure syllabic', provided the principle of syllabism is not interpreted as isosyllabism or a tendency toward it. Where isometry can be found in texts with two ictuses (based on syntactical and word stresses), it is due not to the identical number of syllables but—as is best seen from the above example—to the principle of an equal number of measures. It is under the prevalence of these principles only that in the domain of heterometric rhythm a tendency toward isometry has to be reckoned with.

Thus, the Samoyed shaman songs in general evince heterometry, but a kind of heterometry whose prosody and rhythmic structure have their own rules of thumb, i.e. are governed by specific correlations. Very seldom do less than two segments occur in one line, and never more than four. The number of syllables may, in principle, vary from one to four, yet this applies chiefly to two-segment lines or to lines without segmentation. Nevertheless, a line of 4,4,4,4, i.e. 16 syllables seems inconceivable and even lines of 12 syllables are very rare. An analysis of the shamanistic songs alone does not, however, enable us to go deeper into this question.

8. Finally it seems appropriate to comment upon the opinion voiced in 1911 by Aladár Bán on Samoyed versification: 'As far as versification is concerned, Samoyed folk poetry is on the same level as that of the Voguls and Ostyaks, and the external form should rather be called rhythmical prose than regular verse. Lyrical songs alone display some regularity. Even alliteration is incidental and rare'.[23]

This opinion cannot be accepted without certain reservation, because the association of elementary semantic and rhythmic units based on parallelism—a phenomenon I called 'heterometric concatenation' in another paper of mine on the Ob-Ugrian peoples[24]—is far from asserting itself at that advanced stage of development in Samoyed songs. This principle, on the other hand, is clearly recognizable, for instance, in the 'Magic Word of the Water Saint' recorded by Antal Reguly in the Upper-Lozva dialect of the Mansi language[25], at about the same time as Castrén was making

[22] No Selkup examples have been quoted, because there are not any at our disposal. Cf., however, the Selkup texts published by Castrén (1940), especially the short songs, Nos 5–8.

[23] Bán (1911) Vol. IV, p. 234.

[24] Gáldi (1960); a recent summary of Ob-Ugric versification: Austerlitz (1958).

[25] Ed. by Munkácsi (1892–1896) Vol. II, pp. 390–1.

his Samoyed collection. Suffice it here to quote a relevant passage (which also contains the characteristic parallelisms resulting in concatenation):

Rḁ̄śä sāt rḁ̄ś,	Whose beach is the seven beaches,
vōľä sāt vōľ:	Whose river bend is the seven river bends:
Numi-Tḁ̄rėm jäγännė	Your father Numi-Tarem, —
ḁ̄ltėn sät surmėl,	With seven silver roof-holes
jäγännẹ sātwėsän;	Has enchanted you, he, your father;
Numi-Tḁ̄rẹm āsännė	Your father Numi-Tarem, —
sårńi sāt sūrẹm-sūntėl,	With the orifice of seven golden roof-holes
jäγännė sātwẹsän.	Has enchanted you, he, your father.
χuľltγ́intep sāt a'i̊rėl	With seven undulating whirlpools
sātwėsän.	Has summoned you.

Obviously, parallelism is used in this Mansi text in a more consistent way than in any Samoyed text. It is none the less obvious, however, that the rhythmic forms occurring here are not far removed from the rhythm of the above-quoted Samoyed chorus-like song.

Samoyed folk poetry and its shamanistic manifestation have not reached the degree of symmetry prevailing in the Mansi songs. Since, however, Samoyed heterometry contains, potentially at least, almost every essential element of Finno-Ugric folk poetry, versification in Samoyed folk poetry may be assumed to represent an earlier stage of evolution than that represented by the Finno-Ugric forms known to us. That state can be presumed to have existed in the earlier development of every Uralian people; late traces thereof survive in the majority of the Finno-Ugric examples with free number of syllables, first of all in the well-known Mari heterometry.[26]

REFERENCES

Austerlitz, R. (1958) The Metrical Structures of Ostyak and Vogul Folk-Poetry, in *FFC*, Vol. 174.

Bán, A. (1911) *Egyetemes Irodalomtörténet* (General History of Literature), Ed. by Heinrich, G., Budapest.

Castrén, M. A. (1853) *Reiseerinnerungen aus den Jahren 1838–1844*, St. Petersburg.

(1940) Samojedische Volksdichtung, a collection, edited by T. Lehtisalo, in *MSFOu*, Vol. LXXXIII.

Collinder, B. (1957) *Survey of the Uralic Languages*, Stockholm.

Diószegi, V. (1958) *A sámánhit emlékei a magyar népi műveltségben* (Survivals of Shamanism in Hungarian Popular Culture), Budapest.

Donner, Kai (1926) *Bei den Samojeden in Sibirien*, Stuttgart.

(1944) Kai Donners Kamassisches Wörterbuch nebst Sprachproben und Hauptzügen der Grammatik, in *LSFU*, Vol. 8.

Gáldi, L. (1960) A finnugor népi verselés tipológiai áttekintése (A Typological Survey of Finno-Ugric Popular Versification), in *Irodalomtörténet*, Vol. XLVIII, pp. 149–75. (On the basis of a report to the Finno-Ugric Department of the Hungarian Linguistic Society on 25th November, 1958.)

[26] For the rhythm in the national versification of the Mari see Sebeok (1956).

Hajdú, P. (1960) Szöveg és dallam viszonya a szamojéd énekekben (The Relation of Text to Melody in the Samoyed Songs), in *Néprajz és Nyelvtudomány*, Vols III–IV.
Lehtisalo, T. (1922) Yurakkisamojedien lauluista (On Yurak Samoyed Songs), in *KV*, Vol. II.
(1936–1937) Beobachtungen über die Jodler, in *JSFOu*, Vol. XLVIII/2.
(1947) Juraksamojedische Volksdichtung, in *MSFOu*, Vol. XC.
Lotz, J. (1953) Structural Presentation of the Kamassian Lament, in *IHUHT*, Vol. VI.
(1954) Kamassian Verse, in *JAF*, Vol. 67.
Munkácsi, B. (1892–1896) *Vogul népköltési gyűjtemény* (Collection of Vogul Folk Poetry), Vols I–IV, Budapest.
Ravila, P. (1934) Reste lappischen Volksglaubens, in *MSFOu*, Vol. LXVIII.
Sebeok, T. A. (1956) *Sound and Meaning in a Cheremis Folksong Text*, For R. Jakobson, The Hague.

How Sereptie Djaruoskin* of the Nganasans (Tavgi Samoyeds) Became a Shaman

by

A. A. Popov Leningrad

Explorers who assiduously studied Siberian shamanism on the spot, revealed a long time ago that, prior to becoming shamans, the selected persons suffered from physical and psychic diseases for years.

In every case, these diseases were accompanied by peculiar hallucinations and visions of great importance. It was said that during the illness both the constitution and the mental frame of the selected persons underwent a thorough change, while special assistant spirits escorted them on their way to the different deities and spirits, and made them acquainted with each other. The spirits taught the future shaman which deity or spirit to address in this or that case during his trance.

The leading of the selected person by the spirit is described in detail in the story of Djaruoskin. This story, however, is somewhat different from similar narratives collected from shamans of other Siberian peoples. In these, it is the escorting spirit that shows the way and tells the selected man, what god or spirit he will encounter. In our story the escorting spirit says nothing and the future Nganasan shaman has to find out everything himself.

Before turning to the story of Djaruoskin, let us say a few words about Nganasan shamanism.

Just as among the other Siberian peoples, a *nga* or shaman must be a man whose ancestors were shamans and who has inherited the gift of the shamans. According to an old author, a shaman's son, who was desirous of acquiring his dead father's power over the spirits, made a replica in wood of his father's hand and it was this symbol of power that was passed on to the shamans as a heritage.[1]

It is worth mentioning here that not only the spirits of the shaman's ancestors, but many other deities and spirits took part in the selection of the future shaman, as for instance, the master spirits of water and of earth and even *Ospa* 'pox' who was considered a Russian female spirit. If among the ancestors of the candidate there were Evenki (Tungus) or Nenets (Yurak Samoyed) shamans, their spirits also participated in the selection, which meant that, although knowing neither of these languages, the novice used them in his trance when the assistant spirits moved into him.

*

Now, let us look at Djaruoskin's story.

When I was a young man I used to dream of all sorts of insignificant things just like any other man. But once, I saw myself going down a road

* *Djaruoskin*, a distorted form of *Iarockii*, the surname given to the shaman by the Russians.

[1] Третьяков (1869) p. 425.

until I reached a tree. With an axe in my hand, I went round the tree and wanted to fell it. Then I heard a voice saying: '(Fell it) later!' and I woke up.

Next day the neighbours said to me: 'Go and fell a tree for the *kuojka*[2] sledge!' I set out, found a suitable tree and started to cut it down. When the tree fell, a man sprang out of its roots with a loud shout. I was petrified with fear (from this unexpected event). The man asked: 'Where are you going?' 'What could I do—I am going to my tent.' 'Why, of course, since you have a tent, you must go there. Well, my friend, I am a man, who came out of the roots of the tree. The root is thick, it looks thin in your eyes only. Therefore I tell you that you must come down through the root if you wish to see me.' — 'What sort of a tree is that?' I asked. 'I never could find it out.' The man answered: 'From times of old, it is of this tree that the *kuojka* sledges have been made and the shamans have been growing from. Rocked in the cradle, they become shamans—well that's what this tree is for.'—'All right, I shall go with you.'

Whilst preparing to descend, the man turned to me and said: 'Just take a look at me and find out who I am?' His clothes at the sides reminded me of the wild reindeer's hide during moulting time. I did not ask him why he was wearing such clothes. Then again he said: 'Don't be afraid of me, but find out why there are such patterns on my *parka* (winter garment), both black and white patches?' I replied: 'On the left side you have the white spots because you dress in the attire of the (spirit of the) first snow; the black spots on the right side resemble the spots of the earth appearing in spring from under the snow—because you put on the attire (of the spirit) of the melting snow.' My companion turned his back and, taking my hand, said: 'Now let us go to our hosts!' I was afraid and thought that I was lost.

As I looked round, I noticed a hole in the earth. My companion asked: 'What hole is this? If your destiny is to make a drum of this tree, find it out!' I replied: 'It is through this hole that the shaman receives the spirit of his voice'. The hole became larger and larger. We descended through it and arrived at a river with two streams flowing in opposite directions. 'Well, find out this one too!' said my companion, 'one stream goes from the centre to the north, the other to the south—the sunny side. If you are destined to fall into a trance, find it out!' I replied: 'The northern stream originates from the water for bathing the dead, and the southern from that for the infants'. — 'Yes, indeed, you have guessed right', said he.

Then we set out on the shore of the northern stream. My companion always led me by the hand. We saw nine tents before us, the nearest one being tied round with a rope. A tree stood on each side of its entrance—one on the northern, the other on the southern side. 'What do you think these trees grew for?' asked the man. I replied again: 'One of the trees is bright as if the sun were shining on it. Because the parents bring up their children loving and fostering them, this must be the (protecting) spirit of the children.' Here my companion (as if to confirm my words) clapped his hands and smacked his knees with the palm of one hands. 'The dark

[2] *kuojka*—holy family relics such as stones, anthropomorphous and zoomorphous figures made of wood or metal. They were transported in special sledges.

tree is the tree of the moon, the tree of birth, enabling the women to fix the date of the birth according to the moon.' The man again clapped his hands and slapped his knees. Then he asked: 'What is the meaning of the bars that hang horizontally above the fireplace of the tent?'[3] Suddenly I found one of the bars in my hand and struck my companion with it. 'These bars are the borderline between two daybreaks, the backbone of the firmament. The northern bar is the beginning of the polar light, the southern one—the beginning of the cycle of dawns.' When I said this, my companion praised me. I got frightened. 'And what is the tent tied round with a rope?' asked my companion. And I said: 'When men go mad and become shamans, they are tied with this rope'. (I was quite unconscious and was tied up too.)

Then it seemed to me that we were in the street. We entered the first tent where we found seven naked men and women who were singing all the time while tearing their bodies with their teeth. I became very frightened. 'Now I shall explain this to you, myself, because you will not guess it anyway', said my companion: 'Originally, seven earths were created and it is through the spirits of these seven earths that (men) lose their minds. Some just start singing, others losing their minds, go away and die; others again become shamans. Our earth has seven promontories with a madman living on each of them. When you become a shaman, you will find them yourself.' — 'Where can I find them—you have led me to the wrong place'—thought I. 'If I do not lead you to see (the spirits), how could you make magic for the insane? If you find the spirit of madness, you will begin to shamanize, initiating (new) shamans. You must be shown all the ways of diseases.'

We came out of the first tent and went to the second that was placed on the northern promontory. The whole tent was covered with hoar frost and tied round in the middle with a black rope. Around the smoke hole, the tent was covered with something red. 'Down there (to the south) is Asondu—they are Tungus', said my companion. 'This is their tent. The black rope will serve you to cure stomach diseases, while the red stuff will help you heal madness that comes from headaches. The middle rope will serve you to cure epidemics. You will find out the meaning of this rope later on. When you will enchant, Tungus spirits will come to you and I, myself, do not understand them. When you enchant a Dolgan or a Tungus, come here and you will learn it.' The frost-covered tent had two smoke holes, one of copper and the other of iron. We entered the second tent, but found nobody there. 'Let us go back', said my companion.

We came out of the tent through another door and went into another tent which seemed to be covered with fishing nets. Inside, the fire was scarcely flickering. We found a disfigured old woman there in worn boots and otherwise naked, except for her upper clothes; she warmed herself by the fire. There was a dim light in the tent and shadows were flickering everywhere. The old woman asked: 'Do you know me?'—'Find the answer yourself!' said my companion to me. I replied: 'When a child is born, there is the afterbirth too, you are its spirit, are you not?' My companion and the old woman slapped their knees in astonishment. 'You are a good

[3] The bars that serve to hang the tea pot or the kettle over the fire.

guesser!' Then they asked again: 'Why is the fire dark?' And I answered: 'When a child is born, a new fire is kindled. You too, you are sitting here like a housewife who lights a new fire.' Again they slapped their knees in astonishment. 'In the southern part of the tent, the fire is very weak. Human beings are purified after birth by being fumigated with fire. This is the origin of the purifying fire.' 'That's right', said both of them. The tent had two kinds of *ńuks*[4]—some were frost-covered, some were white. I guessed the meaning of these *ńuks* too. 'We wrap up the dead in *ńuks* made of wild reindeer hide. Here, these rimy *ńuks* are the *nil'tis*[5] of the said *ńuks*. The white *ńuks* are the *nil'tis* of our leather garment.' Both confirmed the correctness of my words.

We went over to the fourth tent that stood in the middle of the water, behind three freezing waters. One part of it was covered with seven reindeer hides, instead of a *ńuk*, while the other was spread over with rippling water. I said: 'Doesn't the shaman make his clothing out of seven reindeer hides? And since the other part of the tent is covered with waves, it must belong to the spirit of the water—that is why it is standing in the middle of the water.' 'You will shamanize, you go downwards', they said. We entered the tent where we found an old woman sitting among heaps of children's clothes and killed dogs. On each side of the tent there were two white salmons. 'Do you recognize me?' asked the old woman. 'Find it out', said my companion. And I replied: 'This is the earth where we shall have to come in lean years. Here, she (the old woman) will show us where to find game and fish.' 'Find it out!'—they repeated. 'Seven hides—these I shall find when I shamanize, I shall find and recognize them. Every man, when he becomes a shaman, makes himself a seat. I shall shamanize for seven days, sitting on these seven hides. Two fishes: one of them means that we blow on the fire and breath goes out of us. The breath has a soul. What is it when the fish (opening its mouth) is panting? When we light the fire, it flares up on its back. Seeing the fire on the back of one fish, the other blows it out. Therefore, in imitation of this fish, a wooden fish was made. When a child is born, men act in a similar fashion. One of the fishes is full of roe—the roe floats above. The child is to be rubbed with the roe. When a child is born, we smear it with grease. This is what it means. Then throwing the grease into the fire we feed the house *kuojka* and also smear its face with grease. When people become hungry, it is this that gives enough to eat. It is this (the fish) you have to ask.' The southern side was closed with hides. Looking closer, I found that they looked like seven apertures instead of hides. From outside, they looked like hides, from inside, like apertures—through which one could look out. 'Why should I find this one out?' 'We shamans have seven resting places, henceforth you will find them. The seven apertures mean that when a man sinks under the water but has still some air left and you happen to be there, then you come and save him.'

We came out of the tent—the northern side was covered all over with ice. 'You will find this out yourself', said my companion. 'Don't come here, this is the way of another shaman.' When I submerged, I arrived

[4] *ńuk*—tent, cover of conical hut.
[5] *nil'ti*—one of the life substances ('soul'? A.P.).

at these places, and it seemed as if I were swimming in the water. We went to another tent. On each side there stood an iron trunk. A one-horned reindeer doe was tied to the trunk on the right side, while a stag, with bruised antlers, was tied to the left trunk. One tine was broken, the other twisted. I tried to guess what the two animals were tied with and found that they were tied with the rays of the sun. 'Well, that's strange'—I thought—'how is that these (ropes) do not break?' Although they were twisted, some of the threads were broken and hanging straight, like sticks. 'Do you know this rein-hind?' they asked me. I answered: 'No, I do not.' 'When you will be a shaman (surely you think that this is a real trunk—this is the spirit of everything humans do) and (during the ceremony) the men will beat your drum with the drumstick,[6] this trunk will split up. The rein-hind is the origin of the *kuojka* (made of stone or wood).' 'And what is this rope for?' 'It serves to brand the reindeer's fawn that is presented to the moon.' 'Why is the reindeer one-horned?' 'Every man who becomes a shaman, makes divinations about the reindeer whose hide will serve him as a dress. When you become a shaman, (don't) ask them (the wild reindeer for the clothes) and don't make yourself a suit of the simple wild reindeer's hide. Provide yourself with clothes, but, first, ask for permission from the mother of the wild reindeer, she will give you instructions about the wild reindeer whose hide is to be used for your clothes. One of the wild reindeer is a stag, he is the master spirit of the wild reindeer stags. It is this spirit whom you will have to ask to know which tree you are to make your drum of, or else your life will not be long.'

Walking round the tent, I saw that all the *ńuk*s were bright with decorations and fringes. We make fringes for ourselves, similar to the antlers of the reindeer. For we take hair from every animal to sew it on to the fringes of the attire similar to the cover of this tent. My companion said: 'Then, when you come here, ask questions, and if you are given hair of an animal, make yourself fringes of it'. Then I looked up to the smoke-hole and saw nine human figures made of iron. I don't remember any more, nor how I reached them, but I began to hit them with a stick, saying: 'When shamanizing in the clean tent,[7] I shall ascend by them'. 'You will be a great shaman indeed, you find out everything', said (my companion) clapping his hands. 'If this is so, I shall surely be a shaman. But I don't want to be a shaman', said I to myself. 'No, you will be a shaman, since you have seen all these things'—they said.

We entered the tent and found there seven moon-figures made of copper, similar to those that are on the shamans' clothes. 'Behold, they are yours', said (my companion) and began to give me the figures. I did not take them. 'No, take them', said he and brought forth seven suns which he showed me. 'What is this?' I thought, 'probably I shall have to enchant for seven days'. Then he gave me three times seven figures of the sun and said: 'Find it out!'—'I don't know.' My companion said: 'You, being a new shaman, stand up (i.e. recover) and cure thrice seven men from their illness'. When I shamanize, I walk round all these tents.

[6] The Nganasans, when appealing to the shaman during the ceremony, hit the drum with a stick.

[7] The feast of the clean tent, a spring festivity celebrated together with the 7 or 9 day ceremony of the shaman.

I came out of this tent and reached another. I think it was mine. To me, it seemed a strange one, not my own. People were sitting around the fire, men on one side and women on the other. I went in, not as a man but as a skeleton; I don't know who gnawed me off, I don't know how it happened. As I took a close look at them, they did not look like real human beings but like skeletons which had been dressed. At the bottom of the tent, there was a seven-bladed anvil. I saw a woman who looked as if she were made of fire. I saw a man holding (a pair of) pliers. The woman had seven apertures on her body. From these, the man pulled out iron pieces as from the fire, placed them on the anvil and struck them with the hammer. When the iron cooled down, the man replaced it in the aperture of the woman's body as if it were fire. Although there was fire in the fireplace, the man did not make use of it. I looked round. Near the fireplace there was a woman, stirring up the fire with animal hides, while the sparks were flying on every side. The man took a piece of iron, placed it on the anvil and hit it with the hammer which consisted in reality of seven small hammers on a single helve. At every blow on the iron, sparks rose into the air and flew out through the smoke hole. My companion asked me: 'What do you think, what tent have we entered?' 'I don't know', said I. 'However, it must be here that the pendants of the shaman's clothes are forged and it is probably these people I have to ask (for pendants), for my clothes. The man (= shaman) descends from many places, this is surely one of them.' 'This is not all of them, anyway', they said. I tossed my head back and began to look at the smoke hole. 'What are those seven figures in the upper part of the tent?' I asked. 'They are the spirits of your future saw-toothed pendants', he said. 'Do you give me these pieces of iron?' 'No, the time has not yet come for it', they said. I began to feel uneasy. 'Then why do you make me guess?' 'Who are these two beings, the man and the woman; are they humans?' 'So that is the origin of the shamans', I said. 'Indeed.' This is my fate—to lose my mind. 'Whenever you become a shaman, ask them for permission to make yourself clothes and a drum. Ask them also to give you reindeer for your clothes; if you come to this tent they will provide you with the necessary things. The sparks are birds, catch them, imitate them, we have birds, geese made from them, on the back of the dress.' When I entered as a skeleton and they forged, it meant that they forged me. The master of the earth, the spirit of the shamans, has become my origin. When a shin-bone or something else is hit and the sparks fly, there will be a shaman in your generation.

Then we came out of this tent and I began to look round. In front, there was a great river with sandy shores, and a hill with two tents on it. I began to guess what kind of a hill this was. It seemed as if the tent were standing on an iceberg, behind them were black spots of earth appearing from under the melting snow. The nearer tent was covered all over with white *ńuk*s, whereas the more remote one had *ńuk*s with checkerboard pattern.

When we went closer, we saw that tents were standing on both shores of the river. The checkered tent stood on the black spots of earth (on account of the melting snow), while the white tent was behind the river. It seemed as if I had come back to the river I had encountered at the beginning of the journey. One stream of the river continued southwards, the other northwards. 'Find it out', said my companion. 'How long will you make me guess things? Anyway, when I become a shaman, the origin of my shamanship will be here. Whenever I submerge, I shall descend this water.' 'You will implore the place of confluence of these streams. When you submerge you will return swimming in the southern flow. Your throat, similar to the stream, will begin to talk, to conjure up this spirit. At the checkered tent, the upper parts of the smoke hole, the *ńuk*s are made of poor, small-haired hides. What does it mean?' I said: 'The diseases devour everybody, yet they spare the half of mankind so that it may reproduce itself. Behold, these (the black squares of the checkerboard) cover the surviving men with the blackness of diseases.' 'Now that we have arrived here, I will leave you alone', said my companion. 'If you return, you will be a man, if not—you will die. Henceforward I cannot lead you anymore. I have led you to all the origins and ways of diseases. Shamanizing, you will find your way, by yourself.'

Standing near the tent, I looked at the river and saw a woman passing by. She was quite red, her face and hair included, and her dress was checkered. Some of the squares were red, the others blue. 'Here', said my companion, 'I have brought you here. It was only the way of the big disease that I have not showed you yet. That's why I have brought you here. You will shamanize with the big disease, take care not to curse but to implore when you shamanize. Can you guess who this woman is?' I: 'Her body is quite red and her dress is strange. The shaman gets up and sits down when shamanizing. Why, he sits on different litters, obviously, the woman carries these litters of mine with her. The red disease (measles) sometimes occurs, she seems to be its mother. There were checkered *ńuk*s on the tent. This means that the tent was covered with the clothes of the woman. She took off the *ńuk*s and put them on.' I looked at the tent—it was quite red. 'If in case of such disease (measles) I shall come here and appeal to her I shall surely heal (the sick)', said I. 'Yes', said my companion, 'I have told you already that I shall leave you. Well, my friend, you must not even look at this tent, whoever may be there. You must not go there. I have brought you to this tent, where your forefather, the famous shaman died of pox. I leave you here to make friends and not to die of this plague.'

I became terribly frightened. The woman said: 'You have come from far away. Yet, I am obliged to send you back.' And she breathed on me three times. As she was breathing I began to recognize the place. Yet she did not let me go immediately, but said: 'My friend, there is a tent there, but you must not cross the river, and, from now on, you should only come up to me. If you want to obtain advice from that tent (from its female occupant), I shall give you advice. By now, she will have surely come to know that you are here. I shall notify her. When she opens her mouth and blows, the fog sets in. It is from this fog that men get sick.'

Meanwhile, my companion stood sideways and listened. Then the fog came. 'Find it out!' said the woman. I began to get angry, but still said: 'Half of the fog is the breath of men; when I am a shaman, I shall be able to rescue the *nil'ti* of the dead from this fog.' 'Yes, indeed', said the woman. 'Here is a reef[8] with a red tent on it. If the sick is cured, the woman will come from the opposite direction. If the *nil'ti* falls beyond the reef, beyond

[8] A rock rising above the water surface in the river.

the limit of the fog—that is the limit of life—he cannot be saved.' Then I said to myself: 'I am sure that I have reached the place whence every man descends' and, turning towards the woman, I said aloud: 'You are surely the mistress of the earth who has created all life'. 'Yes, that is so', said she. 'Well, my friend, we had three children, the second lives in the red tent, the eldest is beyond the river. You return from these places. If you want to enter their tents, you shall die. They are half *barusis*, half *nguos*.[9] Take care to look in the direction in which you have to return.'

Then I made a few steps forward and, looking round, saw seven stone peaks. On one of them, there were seven willows, on the other seven thin trees, and so on—seven plants were growing on each peak. 'What's that?' thought I. Everywhere on the plants, there were nests of all kinds of birds. On the highest peak, in the middle, a bumble-bee was hatching her young. Her wings were of iron. She looked as if she were hatching while lying in the fire. The woman pointed at these nests and said: 'You are surely tired, come here, lie down, and find out what that is!' 'This is the bumble-bee, who creates (plant) for the tinder, so that none of the men should be in want of it. These seven peaks are the origin of every plant: the future shamans go around them. In these nests there are spirits—the master spirits of all the running and flying birds and game.' 'What can you see on the shore?' the woman asked. 'I can see two stone peaks.' 'Find it out!' I replied: 'When we reached at one of the peaks—this is the mistress of water, we can ask her for fish. The other peak is black. When a child is born, it is placed on a layer made of punk wood. Half of the peak is covered with such punk wood, the other half, with moss. When he gets here, the shaman can cure the child, in case of disease. The spirit of this lives here.'

Then I left them and saw nothing but the earth. It seemed as if I were going along a river. One shore was pebbly, the other covered with coloured stones—yellow ochre and black earth (graphite). A woman began to talk: 'The coloured stones are iron, copper and different metals. When you become a shaman, you make yourself pendants of them, therefore you go to them.' Going on along the shore I saw two peaks; one of them was covered with bright-coloured vegetation, the other was black earth all over. Between them, there appeared to be an islet with some very nice red plants in blossom on it. They resembled the flowers of the cloudberry. 'What is this', I thought. There was nobody near me, but I found it out myself. When a man dies, his face becomes blue and changes: then the shaman has nothing more to do. As I noticed the red grass grew upwards, the black downwards. Suddenly I heard a cry: 'Take a stone from here!' The stones were reddish. Since I was marked out to survive, I snatched up a red stone. What I thought to be flowers were stones. Someone said to me: 'When you have a clean tent made, have a fire lit with flint. When you get to your own tent, speak about this with the men, not sitting like this, but shamanizing, because you are a shaman singing with the throat of nine diseases.'

Suddenly I recovered my senses: I must have been lying for a considerable time, near the root of the tree. So I felled that tree and made a *kuojka* sledge of it. This was our ancestral *kuojka*. Wherever I go shamanizing, I always hear the terrible songs on the peaks. Before coming to myself,

[9] *barusi*—malevolent spirits, *nguo*—benevolent divinities and spirits.

I heard a voice saying: 'If you become a shaman, you will live a long time'. This was twenty years ago and I was not married. When I am shamanizing, I see a road to the north. When I am looking for a sick man, the road is narrow like a thread. I do not know who is leading me, in front I see the sun and the moon. On (the lower) part of the narrow road there are conical ramshackle tents; on this (road) you go for the breath of the man. The other part of the road (leading upwards) is quite entangled—I do not understand why. The man who is to recover has a breath like a white thread, while he who dies has one like a black thread. Going along the road, you look sideways and you proceed. Then you find the man's *nil'ti* and take it.

Half of my spirits comes from the men of the forest (the Tungus).

It was new moon, a bright weather in March when we three set off. I had been blindfolded with the hide of the wild reindeer and sent afoot to find a tree for the drum. The companions followed me on reindeer. The tree suitable for the drum, makes sounds like a drum. I ran forward, with my eyes covered, listening. The spirits do not give immediately the opportunity to find the tree, they mislead us. There are as many as three sounding trees instead of one—the third has to be chopped down.

Now, you are going to one of these trees—it is also coming towards you. 'Am I to take this tree?' I asked myself and began to sneak up to it as if it were a wild reindeer. But if you try to take it, somebody pushes you away so that you leap aside. 'Don't do it, or you shall die!' This is the tree where the spirits of your family live—the tree that protects your family against death and diseases. Realizing your mistake, you walk on before being pushed away.

Then you hear another tree sounding (like a drum) and coming towards you and you prepare to catch it. But you must not do so. This is the tree where the breath of your entire flock and the breath of the flocks of your home folk get, 'mixed'. If you touch this tree, you die.

Finally, I see a third tree, it stands and does not move. I begin to steal up to it. The tree says: 'Come, come, I am for you!' Then I fell it. My companions did the work with the axe, not I. They must by no means be relatives. If, for all that, the tree is a false one and not the real one, you implore in vain when bending the hoop—that would break anyway, and then somebody belonging to the shaman's kin would die. But if you suspect this, do make a drum of this hoop, even if it will be a bad one, but previously perform a ceremony, so that nobody should die.

When, for a man's illness, I make ceremonies to the evil spirits, the latter would say: 'Here, I have surrendered to you, what is he going to give me?' I ask: 'What you require for him I shall settle'. 'The ill man has to kill a certain wild reindeer', says the disease. The man indeed kills such a wild reindeer, gives me its hide and I make a new dress of it for myself. It may happen that the spirit does not speak sincerely and says: 'He should kill a wolf, a fox or some other game'. But in reality, the ill man kills a reindeer.

REFERENCE

(Tretjakov, P. I.) Третьяков, П. И. (1869) Туруханский край (The Turukhansky Region), in *ЗРГООГ*, Vol. II.

The Classification of Samoyed Shamans

by

P. Hajdú, Szeged

The shamanist peoples of Siberia used to classify shamans in different categories according to their capabilities and powers. In general there were three such categories. In the first category belonged the most powerful and learned shamans, who excelled in communicating with spirits and superior beings, and who practised magic and possessed wonder-working powers; the second category consisted of shamans who did not perform wonders or did so in a limited scope only, and who were reputed diviners and healers; and the third category included people with sensitive nervous systems who distinguished themselves by their skill in dream-reading and in healing minor diseases. Other peoples, however, made a distinction between good and evil (white and black) shamans, according as they were usually in contact with good or evil spirits.[1]

This classification of shamans, of course, may have implied many other elements too (the nature or power of their helping spirits, etc.), so that it is impossible to set up a complete system of classification. In many places the shamans' names varied by category, and also many outward marks distinguished them from one another, e.g. differences in the equipment, unequal social standing and the kind of authority they enjoyed.[2]

In this paper we are going to examine, from the point of view of linguistics and ethnography, the range within which the shamans were classified among the Samoyeds, and we shall analyse the idiomatic expressions of this classification. We shall try to achieve this aim by examining the Samoyedic words for 'shaman' and the related terms.[3]

[1] Nioradze (1925) pp. 85–7; Harva (1933) pp. 320–2; Ohlmarks (1939) pp. 52 ff.

[2] The classification of shamans existed, in addition to the Siberian peoples, also among the Lapps who live far to the west of the Samoyeds. Holmberg writes that division among the Lapp *noita*s (shamans) was notable. The really powerful shaman was capable of doing amazing things. He could hold a glowing stone in his hand, throw himself into fire without being burnt. See Holmberg (1915) pp. 96–7. T. Itkonen mentions three categories of Lapp shamans: 1. shamans who are able to fly; 2. shamans who, though unable to fly, can leave behind their body and transfer their soul to an animal for a certain time; 3. shamans who work magic by means of their voice. There are among them good and bad shamans. See Itkonen (1948) pp. 331–2. The classification of shamans is recognizable among the American Indians too. The Siberian shaman categories have up to now been dealt with far less thoroughly than those of the Indians. See Park (1938) pp. 92–3, 100 ff., with bibliography.

[3] In this place we shall not examine the religious expressions which may have served as a basis for the formation of words for 'shaman, sorcerer, prophet', etc. (e. g. Nenets Obdorsk dial. *jīlćetā* 'losen'), but which have no such derivatives we know of. Neither shall we deal with those words which, by their derivation and meaning, could have denoted the shaman, but which were used rather to distinguish dignitaries of the Christian church: e.g. Nenets *hahedaŋuda* 'Priester', noted down by Castrén (1855) ~ Obdorsk dial. *χäᵋχεdaŋkuda* 'Täufer' noted down by Lehtisalo (1956) p. 180a; Ustye dial. *χäεχe ḿādan‿todalāmbada* 'Kirchendiener' (Heizer des Zelts des Geistes)', noted down by Lehtisalo (1956) 494b; Nganasan *koikuˀsie*, Enets Khantaika dial. *kahafonde* Bayikha dial. *kihofonde*, Kamasian *ābes* 'Priester', etc.

1.1. The classification of shamans by their capabilities can be traced back more or less to the sources concerning the shamanism of the Samoyed peoples. V. Vereščagin[4] wrote in 1848 that the Nenets (Yurak Samoyed) shamans (*tāďiƀe, тадибэ*) and their helping spirits (*tāďēрси, тадебцio*) were divided into good and evil shamans and helping spirits.

A paper published by Islavin in 1847, on the other hand, says that not everybody can be a shaman, only one before whom a *tāďēрси* (helping spirit — P. H.) has appeared; whereupon the appointed person, who had the feeling for the vocation in him (even though a female), turned to an experienced *tāďiƀe*[5], under whose guidance he appropriated the shamanistic art.[6]

Starcev also says that in the belief of the Nenets there are good and evil, strong and weak, shamans according to their capability of performing the duties entrusted to them. Good shamans do not die, their souls continue to live in the moon, but the evil ones perish while they are still on the earth.[7]

Interestingly enough, the literature on the Samoyeds supplies relatively little information about good and evil shamans. Besides generalities, Lehtisalo states the important fact that the distinction between good and evil shamans can be traced in the beliefs connected with the dead. Namely, according to the Nenets, the dead dwelt in three different parts of the underworld. When introducing the souls of the dead into the nether world, the good shaman led them to a lucky country, the bad shaman directed them to an intermediate region, and the evil shaman left them in the worst marginal sections of the underworld.[8]

Lehtisalo in his work on Nenets mythology also refers to the classification of Nenets shamans by capabilities, but his description is confined to few categories of classification: 'Der unbedeutendste Zauberer ist der "Traumseher" und der "Seher", die die Waldjuraken "die aus dem Wurzelloch Schauenden" nennen. Der erstere sieht im Schlaf vergangene und zukünftige Ereignisse, der letztere sieht z. B., wenn ein Gast kommt, die Spuren vorher, beim Gelingen des Fanges "die Augen des Glücks", d.i. die des Fisches, Renntieres u. ä. Sie losen und gewinnen Klarheit über gute und schlimme Dinge sowie darüber, wann ein jeder Geist Opfer wünscht, und auf Grund davon wirken sie als Opferpriester. Sie schlagen die Trommel nicht und sind zu so schweren Aufgaben wie die Zurückbringung des gestohlenen Schattens eines Kranken nicht geeignet.'[9]

This picture is completed by data from G. N. Prokofjev, which E. D. Prokofjeva later inserted in one of her works. Accordingly, the Nenets shaman of Bolshaya Zemlya and the Yamal peninsula, when making the drum, offered the flesh of the slain reindeer to the divinity with whom the shaman believed he was in contact: The *выдутана* (Enets *будтодэ*, see below) sacrificed to the supreme godhead, *нум*, while the *дьяно* and the *самбана* (Enets *саводэ*) sacrificed to the evil spirit.[10] To this we have to

[4] Верещагин (1848) p. 153.
[5] Emphasis by the author.
[6] Иславин (1847) pp. 109–10.
[7] Старцев (1930) pp. 123, 126–7.
[8] Lehtisalo (1922) p. 99.
[9] Lehtisalo (1924a) p. 145; cf. also Прокофьева (1956) p. 627.
[10] Прокофьева (1951) p. 145.

add only that the above term *дьяно* (*d'ano*) for shaman figures obviously by mistake as the name for the Nenets shaman, since this word, for phonetic reasons, cannot possibly be a Tundra Nenets form. In fact, the initial *d'* is unknown in this Nenets dialect, and the published form (*d'ano*) can be only an Enets word.

1.2. Similar differences are found also among the Selkups (Ostyak Samoyeds): the real shaman is the one who, by divine predestination, stands as an intercessor between man and the spirit world, and that is always a male individual. Of course, there are both men and women who divine the future or practise healing, but they are unable to reach the rank and title of a real shaman.[11] The same author remarks in another place that soothsaying and similar functions are reserved to the so-called little shamans.[12]

A more detailed account of the classes of Selkup shamans is rendered by Prokofjeva. In her description the would-be shaman, having received the drumstick, withdrew to an empty tent or took to the deserted tundra, that is to say, he began to practise the art of drumming and shamanistic chanting in some desolate, unfrequented place. It was in this period that he contacted the spirits and that from a 'mortal being' *kuli̯l' söl'qup* (*кулыль сӧлькуп*) he was transformed into a shaman *te̯ti̯pi̯* (*тэтыпы*).[13] Afterwards the young shaman obtained his equipment piece by piece; after the drumstick he got the drum, then the apron, the shaman's mantle and boots, and finally the shaman's cap and the staff.[14] Before obtaining the cap, he might practise his art in either a light or a dark tent (that is either with or without fire). But after obtaining the symbol of the highest grade of shamanhood, the cap, he never shamanized without fire.[15] This description thus reveals that the Selkup shamans also had different powers. There were old, famous and learned shamans who possessed the shaman's complete equipment, and there were younger ones, 'novices', whose outfit was far from complete.[16]

1.3. The threefold classification of shamans can be made out most clearly among the Enets, where the highest category was represented by the so-called *budtode* (*будтодэ*), who communicated with spirits dwelling in heaven (he had a special shaman's costume, drum and staff). In the language of the Enets the shaman of the middle grade was called *d'ano* (*дьяно*); he protected men from evil spirits (he had a drum only). In the last category belonged the *sawode* (*савodэ*); who had neither a shaman's costume nor a drum, and who had the duty of communicating with the dead and officiating at funeral ceremonies.[17]

2.1. In Nenets the generic word for shaman was *tād'ib̕e*. The word was known as early as the 18th century: *tadi̯b* (*тадыб*)[18], *tadyb* 'Zauberer'[19]. Descriptions of a more recent date dealing with the Samoyeds, and Nenets

[11] Donner (1923) p. 143.
[12] Donner (1933b) p. 242.
[13] Прокофьева (1949) p. 338.
[14] Прокофьева (1949) p. 341.
[15] Прокофьева (1949) p. 338.
[16] Прокофьева (1949) p. 339.
[17] Прокофьева (1951) p. 126.
[18] Зуев (1947) pp. 43, 46.
[19] Pallas (1776) p. 75.

glossaries, also often mention the word: *tadibei* 'Zauberpriester'[20], *tabedi* [!] (*табеди* [!])[21]; *tadibe, tádibu* (*тадибэ, màдибу*)[22]; Arkh. *tatibė, tádibė* 'Schamane, Zauberer'[23]; *màдибе* 'shaman'[24]; *tádabe*[25]; *tadibe* (*тадибе*) 'shaman'[26]; *tāďb́ä-hiberï* 'Zauberer'[27] (the second part means 'man'), Kanin dial. *tādibe* 'jós, varázsló' (soothsayer, sorcerer)[28]; *taďibe* 'Zauberer'[29], *tādibea* 'Zauberer, Schamane'.[30] Noted down by Lehtisalo: Obdorsk dial. *tāďeb́ē*, Lower Ob, Sjoida, Ust-Tsilma dial. *tāďib́e*, Nes dial. *tāďiβ'eä*, Nyalina, Lyamin dial. *tāťśiṕeä*, Kiselevsk dial. *tāťiṕeä* 'Zauberer, Schamane', Nyalina dial. *ńie tāťśiṕeä* 'Schamanin'.[31] In today's Nenets literary language the word *tadeb́a* (*тадебя*) is 'shaman'.[32]

Equivalents of this word for 'Schamane, Zauberer' exist in Enets, Selkup and Kamas, too: Enets Khantaika dial. *tārebe*, Bayikha dial. *tādebe;* Selkup Karasino dial. *tytebe, tytebel-gum*, Bayikha, Tas dial. *čueʒebel-gum*, Narim dial. (?) *tutebe, tuetebel-gum*[33]; *тэтыпы*[34]; Kamassian *thārbu, ťårbə*.[35]

It should be noted in this connection that the Nenets *-d-* ~ Selkup *-t-* ~ Enets Bayikha dial. *-d-*, Khantaika dial. *-r-* equivalence is common and regular. In Kamas the medial *-d-* sound would be expected regularly, but here we have to do with a sound equivalent different from the normal, as in the case of Kamas *bera* ~ Nenets *pād*, Nganasan (Tavgi Samoyed) *foadai* 'Sack'. The initial sound *č-* in Selkup is not surprising; the terminal *-l* occurring in some forms is a frequent adjectival suffix in Selkup. Apart from this suffix, the words for 'shaman' in all Samoyedic languages seem to be derivatives. Namely, the *-ṕeä, -b́e*, etc. terminal morphemes can be identified with the same deverbal (and denominal) suffixes that occur in the following examples: Nenets *śūdab'e* 'riesenhafter Geist' (<: *śūda-rťś* 'pfeifen'), Enets *kābe* 'tot' (< : *kā-ro* 'sterben'), Selkup *kūbie* 'tot' (< : *ku-ak* 'sterben'), *ile-be-l* 'ausgelebt' (< : *ila-η* 'leben'), Kamas *ku-be* 'tot' (< : *khü-läm* 'sterben'), etc.[36]

If in the word *tāďib́e* and its family we rightly recognize a deverbal nominal suffix *-b́e* (or, in a certain sense, a formant of perfect participle), then we may come closer to ascertaining the etymology of the word. We cannot identify the root word that gave birth to *tāďib́e* for 'shaman', but we can see one of its derivatives in a Nenets word belonging in the domain of reindeer breeding: Obdorsk dial. *tādarā* 'brünstig, geil sein (Tier, Vogel);

[20] Klaproth (1823) p. 165.
[21] Степанов (1835) p. 59.
[22] Верещагин (1848) p. 153.
[23] Schrenck (1854) p. 272.
[24] Дунин-Горкавич (1910) p. 1.
[25] Рычков (1916) p. 189.
[26] Кушелевский (1868) p. 147.
[27] Sprogis (1952) p. 145.
[28] Budenz (1890–1892) p. 351.
[29] Reguly (MS) cf. Budenz (1890–1892) p. 351.
[30] Castrén (1855).
[31] Lehtisalo (1956) p. 447b. His notes are given here in a simplified transcription.
[32] Терещенко (1955).
[33] Castrén (1855).
[34] Прокофьева (1949) p. 438.
[35] Donner (1944). As will be seen later, we may possibly include here the Nganasan word *tuoptatā* 'shaman's helper', too. See section 4.2.
[36] Lehtisalo (1936) p. 257–8.

h a l b v e r r ü c k t s e i n',[37] Lower Ob dial. 'brünstig sein, sich begatten (von Renntieren; auch von dem Koitus der Menschen . . .)', Obdorsk dial. *tādarāmbada tī* 'brünstiges Renntier', Tas dial. *tādāwī* 'h a l b v e r r ü c k t',[38] Kiselevsk dial. *tātmaj* 'von der Begattung erschöpft'.[39]

The verb *tādarā* shows the deverbal augmentative suffix *-rā*: cf. Obdorsk dial. *sedorā* augm. < : *sedā* 'nähen', *numrā* 'hochmütig sein' < cf. Lyamin dial. *numaś* id., etc.[40] The unidentifiable root **tādā-* must sometimes have thus denoted in general the state of irritation, trance, ecstasy, and this must have given, by means of a suffix with a **-pe* element, the word for 'shaman' meaning originally 'in a trance', 'ecstatic'. The root word died out, or at least there is no trace of it, but its derivative with the suffix *-rā* came into use to denote half-wittedness, foolishness and, in the terminology of reindeer breeding, the state of the sexually excited animal. From the semantic point of view there is no difficulty in this interpretation—trance, the ecstatic state, this most important sign of the shaman's activity, could easily become a basis for naming the shaman, and the referring of trance to another kind of ecstasy—in this case sexual excitement—is equally understandable and not surprising.[41] The meaning 'halb verrückt sein' of the verb also follows readily from the meaning 'to be entranced'.

Neither does the phonetic form present any problem in the above interpretation. Against the medial *-d-* of the verb, anyway, there is in the Nenets word for 'shaman' a medial *-d'-* (*-δ'-*) sound (its regular equivalent in the Forest dialects is *-ťś-*). The Nenets *-d'-* (~ Forest dial. *-ťś-*)—like lots of similar phenomena—can be explained with the position before front vowels (cf. Obdorsk dial. *χilttarā* 'vornübergebeugt sein' ~ *χilťerťś* 'sich herabbeugen., *ṕidā* 'scheren' ~ *ṕit ťeḿe* 'Eigentumsmarke', *ŋidā* 'aufhängen' ~ *ŋiďoś* 'Henkel'.[42]

2.1.1. The word *tāďib'e* in Nenets is not only the generic term for 'shaman', it also gave birth to several items of the vocabulary of shamanism. The verb for 'to practise magic' is one of its derivative forms: *tadébtengu* (*тадéбтенгу*) 'shamanize'[43], *tadebtenggo(s')* [*тадебтенгго(сь)*] 'shamanize'[44], *tādieibťeŋudm*, *tādieibtieŋudm* 'ich zaubere'[45], Obdorsk dial. *tāďepťśeŋkō*, Syoida dial. *tāďepťeŋko* 'schamanisieren', Nyalina dial. *tāťśipeat!* 'schamanisiere!'; Kiselevsk dial. *tāťipeätŋa* 'er schamanisiert'[46]; the same holds of the word for 'helping spirit of the sorcerer': *tadebciju* (*тадебцiю*) 'spirits subjected to *Num*'[47], *tádébzo* 'Geist'[48], *tādiebťu*, *tādieibťu*, *tādiebću*, *tādiebćo* 'dienender Geist des Schamans'[49], Obdorsk dial. *tāďeptsū*, Syoida dial. *tāďeīpťśò*, Ust-Tsilma

[37] Emphasis by the author.
[38] Emphasis by the author.
[39] Lehtisalo (1956) p. 477.
[40] Lehtisalo (1936) p. 189.
[41] By way of analogy we may refer to Dezső Pais (1949) p. 13. He thinks that the Hungarian words *réj*, *rév*, *reg* may have denoted, besides magic ecstasy, also other kinds of dizziness, for example, inebriation, and it is on the basis of this supposition that he connects the Hungarian word *részeg* 'drunken' with this family of words.
[42] Lehtisalo (1956) pp. 185b, 387b, 34.
[43] Дунин–Горкавич (1910) p. 2.
[44] Терещенко (1955).
[45] Castrén (1855).
[46] Lehtisalo (1956) p. 478a
[47] Верещагин (1848) p. 159.
[48] Schrenck (1854) p. 273.
[49] Castrén (1855).

dial. *tāďeptso*, Semzha dial. *tād'epťšo*, Nyalina dial. *tāťšipeapťšu* 'helfender Geist des Zauberers'[50]. A name of the shaman's song is also related with that word: Nyalina dial. *tāťšipātmae* (= nom. actionis) *kinnapts* 'Zauberlied'[51] (literally: 'charm song'). I also took note of a compound word from the Bolshaya Zemlya dialect: *taδ'iбe͜źō* 'shamanistic song'. With the denominal verbal suffix *-mma-* which means 'to become something' the Sakhalin dialect has *tāťšipiemmae* '(er) wurde ein Schaman'.[52]

2.1.2. Combinations formed with the Nenets word *tāďiбe* as second member of the compound serve to denote shamans of different categories.

The following expressions denote the strong, powerful shaman: Nyalina dial. *χataε tāťšipeä* kraftvoller Zauberer (cf. *χataε* ~ Obdorsk dial. *sata* 'kräftig')[53]; Syoida dial. *βedūtana tāďiбe* 'Zauberer, der in anderen Welten umherwandert'[54]. The attribute *βedūtana* comes from Obdorsk, Syoida dial. *βī?* 'das Wandern des Zauberers in anderen Welten' (which has a *-d-* root, cf. gen. sing *βīdan*) or from one of its verbal derivatives. The adjective *βedūtana* is namely the present participle of the hypothetical verb **βedūta* and is formed with the suffix *-na* which is very frequent in Nenets.[55] The term means also 'shaman wandering about the other world' (or, in the light of what is said in footnote 55, 'shaman cutting up his body'). On the other hand, the Nenets verb *βedūtā* has a regular equivalent in the Enets root *budto-*, and its present participle with the suffix *-de*, *budtode* (*бydmodэ*), refers to the highest-ranking shaman among the Enets.

It is probable that Pur dial. *ŋitteлma* 'Zauberergreis' and also its dialectal variants referred to the shaman of the highest category. Today, however, the word is used mainly in a different sense, so we shall later deal with it separately (see section *4.2*).

The strong, powerful shaman is likewise denoted by the following two terms: Lower Ob dial. *śiďeńā? tārāna tāďiбe* 'Zauberer, der sowohl im Himmel als in der Erde wohnende helfende Geister hat' and *nūwūńān͜tārāna tāďiбe* 'Zauberer, dessen helfende Geister im Himmel wohnen'.[56] Literal translations of the two expressions are: 'shaman jumping in two directions' (that is, towards the heavenly and the earthly spirits simultaneously) and 'shaman jumping to heaven'.[57] To explain these terms, we have to make a little detour.

[50] Lehtisalo (1956) p. 478a.
[51] Lehtisalo (1956) p. 187a.
[52] Lehtisalo (1956) p. 478a.
[53] I have found the corresponding Tundra Nenets expression in the following line of a Nenets shamanistic song (No. 1402 B), recorded on a wax cylinder before the war and preserved now in Leningrad (it was re-recorded on tape by Vilmos Diószegi): *sata, ŋuľiᵒ sata tāďib'ōβ* 'strong, very strong shaman'. During the blockade of Leningrad the wax cylinders in question were very badly damaged, and the transliteration of the Nenets texts is by the Nenets poet V. Ledkov, student of the Herzen Institute. I have noted down the above line of the shamanistic song by comparing the Nenets text taken down by him with the song recorded on tape.
[54] Lehtisalo (1956) p. 70b.
[55] The word *βīᵒ* must have been used not only to denote wandering in the other world. As appears from its Forest Nenets equivalent (Pur dial. *βuj*) which means 'das Schneiden des eigenen Leibes in schadloser Weise (vom Zauberer)', we have to do here with a word which denotes the wonderful, incomprehensible powers of the shaman of the highest category.
[56] Lehtisalo (1956) p. 477b.
[57] Cf. also Lehtisalo (1930) pp. 21–2.

Prokofjeva says that the highest-ranking Enets shamans *budtode* (*будтоде*) had two drums: one to evoke the spirits of the 'upper world' (that was called *nano peddi* [*нано педди*] = 'heavenly drum'), the other to contact the spirits of the 'nether world' *ďano peddi* (*дьяно педди* = 'earthly drum'). The latter was at the same time a tool of the shaman of the middle category *ďano* (*дьяно*), who communicated with the spirits of the 'nether world', but neither he nor the lowest-ranking *sawode* (*саводэ*) could have the *nano peddi* (*нано педди*), the drum to summon the 'spirits of the upper world'.[58]

This description explains why the Nenets called their most eminent shamans 'jumpers to heaven' and 'jumpers in two directions'. It is evident that these two terms referred to the shamans of the highest category, who had the power to communicate with the heavenly spirits or with both the heavenly and the earthly (underground) spirits.

We may presume on the Enets analogy that corresponding in the Nenets language to the middle category of the Enets *ďano* (*дьяно*) is the Obdorks dial. *jāńān‿tārāna tāďib'e* 'Zauberer, dessen helfende Geister in der Erde, unter der Erde wohnen', literally: 'shaman jumping in the earth (underground)'.[59]

It is likely that Nyalina dial. *kompurtae tāťśipeä* 'Wahrsager, der alles weiss', literally: 'shaman who finds out (everything)'[60] (cf. Nyalina dial. *koś*, Obdorsk dial. *χoś* 'finden'), signifies the lowest category.

2.2. According to Prokofjeva the word *sambana* (*самбана*)[61] is also used to denote the shaman of the lowest category. She mentions this with reference to Prokofjev, from the Bolshaya Zemlya and Yamal peninsula dialects as an equivalent of *sawode* (*саводэ*), the lowest-ranking shaman among the Enets. She gives the following definition of his functions: '... a shaman partaking in the burial ceremony, performing the burial ceremony, guiding the souls of the deceased to the world beyond...'[62] According to Lehtisalo, the Obdorsk dial. word is *sāmpana* 'Zauberer, der die Schattenseele des Verstorbenen mit Liedern unter Begleitung der Zaubertrommel in die Unterwelt führt'; Lower Ob dial. *sāmpana tāďiƀe* 'Zauberer, der ein Jahr nach dem Tode die Angehörigen des Verstorbenen in der Weise reinigt, dass diese wieder auf der Jagd und beim Fischfang Glück haben'.[63] In fact, it is a derivative of a widely used technical term taken from religious life. *Sāmpana* is the present participle of a Nenets verb: Obdorsk dial. *sāmpā*, Kiselevsk dial. *χampaš*, which means 'die Schattenseele des Verstorbenen mit Liedern unter Begleitung der Zaubertrommel in die Unterwelt führen'.

Mention should be made here of Lower Ob, Ust-Tsilma dial. *sāmpadana* 'wer unter Begleitung der Zaubertrommel Zauberlieder singt', and Semzha dial. *sāmbadana χüẃeŕi* id., which are equally present participles from Tas dial. *sāmbadā* 'Zauberlieder unter Begleitung der Zaubertrommel singen'.[64]

[58] Прокофьева (1951) p. 147.
[59] Lehtisalo (1956) p. 477b.
[60] Lehtisalo (1956) p. 196a.
[61] Прокофьева (1951) p. 145.
[62] Прокофьева (1951) p. 338, note 3.
[63] Lehtisalo (1956) p. 406a.
[64] Cf. also Reguly *sambedaŋu* 'hexen', see Budenz (1890–1892).

Prokofjeva's statement that this word denotes the shaman of the lowest category is based partly on Prokofjev's account and partly on the fact that its etymological equivalent, Enets *sawode* (*саводэ*), also refers to the lowest category. Like the Enets *sawode* (*саводэ*), the Nenets *sāmpana* communicates with the world of the dead.[65] The difference between the two is that the former possessed no drum, while the latter did, and it was to the beating of the drum that he guided the soul of the deceased to the nether world.[66] Even though the Nenets *sāmpana* and the Enets *sawode* (*саводэ*) were essentially of the same category, they must have been distinguished from the lowest-ranking kind of sorcerers, the group of Nenets dream-readers.

2.2.1. Derivatives of the verb *sāmpā* are also the following expressions: Obdorsk dial. *sāmpadapts*, Tas dial. *sāmbadaps* 'Zauberlied', *samburc* (*самбурц*) 'shaman's *parka* (winter clothes made of fur)'[67]; *samburtsja* id.[68]; *sambutōršē* 'shamanize'[69]; *sambadawa* (*самбадава*) 'the shaman's secret' or 'shamanistic ceremonies'.[70]

The word is known in all Samoyedic languages, except Nganasan: Enets *sawode* (*саводэ*) 'shaman of the lowest category'[71]; Khantaika dial. *saburiŋaro*, Bayikha dial. *sabu-deŋado* 'zaubern' ~ Selkup Karasino, Bayikha, Tas dial. *sumba-ŋ* 'singen, zaubern', Tas dial. *sumbe-del-gum* 'Zauberer', Karasino, Chaya dial. *sombe-rnaŋ* 'zaubern', *somberi-gum* 'Zauberer'[72]; Tas dial *zūmвı̑-tı̑l' k̬up* 'der Schamane'[73]; *sumpı̑tı̑l' qup* (*сумпытыль қуп*) 'a shaman man'[74], etc. ~ Kamas *sāma-l'am* 'zaubern'[75]; *sāmo-l'àm, sāmə-l'àm* 'schamanieren'[76].

2.2.2. Uotila[77] compared these Samoyed words with the Finnish word *sampo* and its variant forms (e.g. *sampi, sammas*). In agreement with Friis, he holds that the mysterious *sampo* found in Finnish mythology may have been a magic drum, which might be the original meaning of the word. Leaving aside the question of what *sampo* might have been, we wish to make a few remarks on Uotila's etymologizing.

Uotila professes in accord with Setälä[78] and Karjalainen that we have no definite knowledge of the use of a magic drum in the Proto-Finno-Ugrian or Proto-Uralic period. A comparison between Finnish and Samoyedic would, in any case, presume the existence of the thing in Proto-Uralic. Uotila tries to overcome this difficulty by taking the meaning 'zaubern' of the Samoyedic verbs for the original meaning of the word, and *sampo* for a deverbal noun formed from the verb **sampaa*. He also thinks, however, that the Finns learned the use of the magic drum probably from the Lapps, and in such circumstances it would be strange for the name of an unused

[65] Прокофьева (1956) p. 628.
[66] Lehtisalo (1924a) pp. 131–5.
[67] Прокофьева (1951) p. 127.
[68] Castrén (1853) p. 192.
[69] Рычков (1916) p. 189.
[70] Немирович—Данченко (1881) 103, p. (1884) p. 77.
[71] Прокофьева (1951) p. 126.
[72] Castrén (1855).
[73] Prokofjew (1931) p. 477.
[74] Прокофьева (1949) p. 338.
[75] Castrén (1855).
[76] Donner (1944).
[77] Uotila (1929) pp. 126–30.
[78] Setälä (1902) pp. 142 ff., (1932) pp. 477 ff, 600.

object to emerge and survive, although the verb **sampaa* for 'zaubern' cannot be identified in the Baltic Finnish languages.

Setälä compared the word *sampo* to the Lapp *cuobo* 'Frosch'. This comparison makes us suppose an initial affricate, while the Samoyedic forms point to an initial *ś-*.

As concerns the family of Nenets *sāmpā*, Lehtisalo is of the opinion[79] that it is presumably related with the Finnish word *sampa* for 'sieni, rahka' (fungus, foam). This Finnish word has survived in names of diseases such as *sampa*, *sammas*, *sampaat* 'lapsilla tavattava suun limakalvon tauti; leuassa olevat rauhaset niiden turvoksissa ollessa: eräs hevostauti' (disease of the mucous membranes of the mouth of a child; glands of chin in swollen state; a disease of horses). This word is met also in the compounds *maansampa* 'maanmuna' (puffball), *puunsampa* 'valkea, neulannaupin kokoinen rahka koivussa, lepissä, pajuissa' (white fungus the size of a pinpoint on birch, on alder, on willow). According to Lehtisalo, the meaning 'to practise sorcery' first referred to magic in connection with ecstasy caused by eating fungi. Semantically he refers to the Khanty verb *paŋχtəm* 'laulaen loitsia kärpässientä syötyään, parantaa loitsulla' (practise magic by chanting after eating fly-agaric) (<: *paŋχ* 'kärpässieni' = fly-agaric). For phonetic reasons, Lehtisalo naturally separates the comparison of Finnish *sampa* ~ Nenets *sāmpā* from Finnish *sampo* and its derivatives. There are many reasons for us to disagree with this undoubtedly ingenious interpretation. The comparison is made dubious by the very fact that the accord appears only in the two remotest members of the Uralic language family. Also, it conflicts with the fact that the meaning 'fungus' cannot be identified in Samoyedic and even the meaning 'to practise magic in a fungus trance' is unknown there. Of course, it could be expected that some suffixed form of the word for 'fungus' should have the verbal meaning 'to practise magic'; but Samoyed *sāmpā* is a form without suffix—if we suppose it to be the equivalent of Finnish *sampa*.

In the light of these problems I have attempted a new interpretation of the Samoyed word *sāmpā*[80].

I think it is worth considering that the Samoyed words which mean 'to practise magic' may be related to the Manchu-Tungus verb *ša-*, *sa-*, *ha-* 'to know' from which, according to Diószegi[81], the Manchu *saman* and the Tungus *sama*, *śama*, *saman*, *šaman*, *χaman* 'shaman' are derived.

We may take the medial sounds *-mp-*, *-mb-* and *-b-*, *-m-* of Nenets *sāmpā*, Enets *sabu-*, Selkup *sumba-* and Kamas *sāmo-* to be not only equivalents of the Uralic medial consonant group **-mp-* but also verbal formants. The affixal form **-mp-* in the Samoyedic languages occurs as a durative-continuative formant. If we separate the hypothetical derivative morpheme **-mp-*, the initial sound of the root *sā-*, *su-* presents itself like that of the Uralic words beginning with an **ś-*. The Uralic **ś-*, however, lost its palatalization already in Proto-Samoyedic, so we have to reckon with a Proto-Samoyed root **sa-*. Such a form, however, may not only have survived from Proto-Uralic, but can also be regarded as the result of an

[79] Lehtisalo (1929) pp. 130–1.
[80] Hajdú (1954) pp. 53–6.
[81] Diószegi (1947) pp. 211–2.

old borrowing. The source, direct or indirect, may have been the Tunguso-Manchurian verb *sa-* 'to know'. Though this meaning cannot be identified in Samoyed, this perhaps does not challenge to our comparison. Namely, the word may have come into Samoyed already as a term pertaining to shamanism.[82]

2.3. The lowest place among the Nenets sorcerers and shamans was occupied by the dream-readers, in whose compound names we find as second member, not *tāďiḃe*, but a word denoting 'man', and so it may be inferred that they belonged actually only in the peripheral sphere of the shaman categories: Nyalina dial. *jutīsāmī ńieśeä* 'Zauberer, der nur aus dem Traum wahrsagt', and *jutirtaε ńieśeä* 'Zauberer, der aus dem Traum wahrsagt'.[83] The adjective *jutīsāmī* is a form of Nyalina dial. *jutī* 'Schlaf, Traum' with the possessive suffix *-sāmī* (~ Tundra dial. *-sawäεj*; Forest dial. *-samej*: e.g. *tī* 'reindeer': *tēsamej* 'one possessing reindeer'[84]). The first expression thus means literally 'dreaming man' (i.e. visionary; one who sees dreams). The second expression is a verbal derivative. The augmentative infinitive Nyalina dial. **jutirś*, corresponding to the augmentative Obdorsk dial. *jūderťś* 'träumen' (<: Obdorsk dial. *jude* id.), cannot be found in dictionaries, but in the word *jutirtaε* we have to see the present participle of this verb 'to dream' with an augmentative affix ('dreaming man').

Synonymous with these latter two names is the expression noted down by Lehtisalo as the Forest Nenets name of the lowest-ranking shaman[85]: Nyalina dial. *wenuj śī ńieśeä* 'Mensch, der durch ein Wurzelloch sieht'[86] (literally: 'root-hole man'); cf. also: *wānu śimnaε χinirŋaε* 'er sieht durch ein Wurzelloch (sagt man von einem Menschen, der im Traum od. dank seinem Scharfblick wahrzusagen versteht od. von übernatürlichen Dingen weiss)' (literally: 'he looks through a root hole'). The explanation of the expression lies in the Nenets fancy that the evil, underground spirits dwell in the earth, some of them beneath the roots of trees.[87]

2.4. The following two words, denoting the helper of the shaman, do not belong among the names for shamans in the strict sense of the word. Yet we have to deal with them because the shaman's assistant had an important part to play in the ceremonies and might be a small shaman himself,[88] or an unqualified person for that matter. It was his duty, before starting the ceremony, to hold the shaman's drum over the fire (thus the skin tightens, and the sound of the instrument becomes more powerful). He had to repeat to the audience the sometimes unintelligible words chanted by the shaman, as if to interpret them, and when the shaman fell into

[82] To derive the Samoyed forms from the Tunguso-Manchurian tongues is admissible even if we do not rely on the root *sa-* 'to know' for an explanation of Tunguso-Manchurian *saman*, etc., but we adopt the view of Németh (1913–1914) pp. 240–9 and of Laufer (1917) pp. 361–71 according to which these words have the root *sam*, which in turn is related with the word *kam* (*qam*) for 'shaman' used in certain Turkic languages. See also Lehtisalo (1959) p. 171.

[83] Lehtisalo (1956) pp. 145–6.

[84] Веробов (1935) p. 57.

[85] Lehtisalo (1924a) p. 145.

[86] Lehtisalo (1956) p. 54b.

[87] E.g. *pārne*: Lehtisalo (1924) pp. 41–2.

[88] Castrén (1853) p. 193 and (1953) p. 138.

ecstasy, he continued beating the drum, because failing that the shaman's soul could not get back to his body.[89]

The two words in use to denote the shaman's assistant are: Lower Ob dial. *ṕel'ej* 'Gehilfe des Zauberers, der den Gesang des letzteren wiederholt',[90] and *tel'tana* (*тельтана*) 'a shaman's assistant'[91] or Obdorsk dial. *teltaŋkoda* 'wer die letzten Worte des Sängers oder Sagenerzählers wiederholt, wenn dieser von Zeit zu Zeit innehält'.[92]

The first word is a derivative with the suffix *-j* from the Uralic word Obdorsk, Syoida, Lower Ob dial. *ṕel'e* 'Hälfte, Stück'. A derivative equally with the suffix *-j* from the word for 'Hälfte' denotes, in addition to the shaman's helper, the reindeer which is harnessed, in a pair, to the right of the leader, or, in a team of three, four, five or six, farthest to the right from the leader reindeer put on the left side, or which in general runs beside the leader, that is on its right: Lower Ob dial. *ṕɛl'ej*, etc. 'das Renntier ganz rechts im Gespann; auch überhaupt ein Renntier, das neben das Lenkrenntier angespannt wird'. 'Dieses Wort bedeutet im Jurakischen auch überhaupt ein Gespannrenntier, das kein Lenkrenntier ist.'[93] Cf. also Tas dial. *χaborta ŋaεrχam ṕel'εjŋāβ* 'das Elchbeinartige (Renntiername) spannte ich als Seitenrenntier an'.[94] In addition, this same word with the suffix *-j* and its further derivative with the diminutive suffix *-ko* may also denote the empty, left-side part of the tent: Obdorsk dial. *ńīd'é ṕɛl'ej ńāna parkaśiʔ ńāmtīʔ* 'seine Söhne sitzen nackt auf der linken Seite des Zeltes'; Obdorsk dial. *ṕɛl'ejku*, Lower Ob dial. *ṕiɛl'ejku*, Syoida dial. *ṕɛl'ejko*, etc. 'leere (nicht mit einem Bett versehene) linke Seite des Zeltes'.[95] Namely, we have to know that the fireplace, or the two iron rods which are placed over the fire and from which the handle of the kettle is hung, divide the Samoyed tent into two halves, of which the part to the right from the entrance is the more important. The couches are usually on this side, the other being left empty. If, however, two families happen to live in the tent, the left side is arranged in the same manner as the right half. The more distinguished family, that of the host, occupies the right side, and the less honours left half goes to the 'co-tenant'.[96]

The way the name developed from the word denoting 'Hälfte' in these cases was that the notion to be named was taken to be a component of the superior and closely related category. A component but inferior part of the right-hand section of the tent divided by the fireplace is the other *half*, the left-hand section of the tent. The reindeer put to a sledge farthest to the right forms a unit with the leader, but is also subordinated to it. In fact, this appears also from the way of harnessing the reindeer: namely, a

[89] Lehtisalo (1924a) pp. 153–5 and (1937) p. 12; Прокофьева (1951) pp. 148–9; Castrén (1953) p. 138; as for the helper of the Nganasan shaman, see Попов (1936) p. 100. The Lapp shaman's helper had a similar function. According to Olsen the shaman's helper rouses the sorcerer from his trance: he looks for the sorcerer's soul in the mountains, under the lakes, and carries it back from there. Holmberg (1915) pp. 105–6.

[90] Lehtisalo (1956) p. 376.

[91] Прокофьева (1951) p. 148.

[92] Lehtisalo (1956) p. 482.

[93] Lehtisalo (1932) p. 23.

[94] Lehtisalo (1956) p. 376b.

[95] Lehtisalo (1956) p. 376b.

[96] Попов (1948) pp. 84–7; Прокофьева (1956) pp. 618–9.

slab is strapped, through the upbends of the sledge-runners, to each of the two outside boards of the seat, and it is through the holes of that slab that the loose lead-rein is threaded in, one end of it being fastened to the hame's belt of the leader reindeer and the other to that of the outside animal (that is, that standing on the outside in a team of three, four or five), so that the two animals have to pull with the same force for the lead-rein to remain in its place. Should more than two animals be put to the sledge, the reindeer in between the two outside ones are tied to slabs fixed, especially for this purpose, on the length of tug stretching between the two outside slabs.[97] The team of reindeer thus has two fixed points—the two outside animals. Of these the left-side reindeer, the leader, is the more essential (it is by this animal that the team is guided), and the other is rather its helper, its complement, or, figuratively speaking, one *half* of it.

In the light of such names it is not surprising that the shaman's assistant is named by a derivative of the word for 'half', for the two persons form so much a unit in the shaman's trade that the shaman's job would not be perfect without this helper. A similar phenomenon may be seen in the Middle Hungarian *fél*, today's *feleség* 'wife', the basis of which is that married couple is viewed as a unit consisting of two halves.

The other name of the shaman's assistant, *telt'ana* (*тельтана*), is the present participle of Obdorsk dial. *teltā* 'die letzten Worte des Liedersängers, des singenden Schamanen oder des Sagenerzählers wiederholen (singend, wenn es sich um ein Lied handelt), der Wiederholer des Sängers oder Sagenerzählers sein'.[98] Another participial derivative of this verb is *teltaŋkoda* with its above-mentioned variant. This name, as appears from its interpretations does not refer solely to the shaman's helper, or belong solely in the shamanistic vocabulary.[99]

3.1. The Enets shamans, as we have already said, can be classed in three categories.[100] The *budtode* (*будтодэ*) communicates with the heavenly spirits, he has a complete shaman outfit (clothes, drum, drumstick, staff). The name corresponds etymologically to Nenets *wedūtana*: it is a present participle with the suffix *-de*.

The middle category is represented by the *d'ano* (*дьяно*), who protects people against evil spirits and whose instrument is the drum. A type identical with the *d'ano* (*дьяно*) may be the Nenets *jāńān tārāna tāďiƃe* 'the shaman jumping into the depth of the earth'. Enents *d'ano* (*дьяно*) and the first part of the Nenets expression spring etymologically from the same root: Nenets *jā* 'Land, Erde', Enets Khantaika dial. *jā*, Bayikha dial. *ďa* 'Erde, Stelle' (cf. also Selkup Narim dial. *ču*, Bayikha, Yelogui dial. *ťū*, etc. 'Ton, Erde', Kamas *ďu* 'Land, Erde', etc.). While, however, Nenets *jāńaʔ* (∼*jāńān‿* before a word beginning with a *t-*) is a compound of *jā* 'Erde' and the

[97] Lehtisalo (1932) p. 120 and Figs 14, 15, 16; Василевич–Левин (1951) pp. 68–9.

[98] Lehtisalo (1956) pp. 481–2.

[99] Sakhalin dial. *täeptakotae*, synonymous with this word, refers to the person repeating the chanted verses: 'wer die letzten Worte des Sängers oder Sagenerzählers wiederholt, wenn dieser nach einiger Zeit innehält'. Lehtisalo (1956) p. 480b. The participle may possibly be applied to the shaman's helper too, at least that is what can be inferred from the meaning of the verb underlying the formation: Pur dial. *täejptaś* 'die letzten Worte des Liedersängers, des singenden Schamanen . . . wiederholen' (ibid.).

[100] Прокофьева (1951) p. 126.

postposition *ńāʔ* 'towards', Enets *ďano* is formed from the word *ďa* with the adjectival suffix *-nō* (cf. Khantaika dial. *fionō* 'der äussere' < : *fio* 'das äussere'; *teinō* 'das oben befindliche' < : *teʔi* id.).[101]

We have already spoken of the present participle *sawode* (*саводэ*) referring to the lowest category of shamans. Here we point out only that from Castrén's notes we can infer a verb-root Khantaika, Bayikha dial. *sabu-* 'zaubern', while the shaman's name quoted by Prokofjeva points to a verb-root *sawo-* (*саво-*). This discrepancy becomes understandable if we have in mind that the equivalent of Castrén's intervocalic *-b-* in the Khantaika and Bayikha dialects is, as evidenced by Donner's examples *-w-* in the Karasino dialect of Enets:

Castrén		Donner	
ába	∼ Karasino dial.	*āba* ∼ *āwa*	'Schneehuhn'
éba	∼ Karasino dial.	*ewwa*	'Kopf'
tábu 'Haarwurzel'	∼ Karasino dial.	*tāwu*	kurzes Renntierhaar'
ťubae	∼ Karasino dial.	*ťuwwaj*	'Finger'[102]

The exact source or dialectal origin of the Enets linguistic data published in Prokofjeva's treatise is not known to us, but some of the recorded forms otherwise agree with a few words noted down by Donner[103] from the Karasino dialect (e.g. *pagge* (*паггэ*) 'shaman's *parka* [winter clothes made of fur]' ∼ Castrén *fagge,* cf. also *fē* 'Baum' ∼ Donner Karasino dial. *pɛ̄*; *peddi* [*пeддu*] 'drum' ∼ Castrén *féďiʔ* ∼ Donner Karasino dial. *pφeďi,* etc.).

3.2. An examination of the three names for shamans in Enets shows that the terminology here points to categories which are more settled and more differentiated than in Nenets. The corresponding Nenets names are attributive compounds whose second part is usually the word *tāďiľe* 'shaman'. Though Enets has an etymological equivalent for the latter (Khantaika dial. *tārebe,* Bayikha dial. *tādebe* and *тарыбы,* respectively[104]), and presumably it serves also here as a name for shaman in general, regardless of category, yet more recent sources do not quote it in this sense. It is obvious, however, that originally this word was used in attributive construction with *budtode ďano* and *sawode.* The constructions **budtode tādebe* 'shaman wandering in the other world', 'shaman cutting himself', **ďanō tādebe* 'earthly shaman', **sawode tādebe* 'shaman who guides the dead into the nether world' were responsible for the shortening by which the meaning of the qualified word was taken over by the attribute, thus giving birth to three new words for 'shaman'.

4.1. According to Castrén, 'shaman' in Nganasan is *ŋaʔ* (gen. *ŋadan*) with a radical *d-* (derivatives of this: *ŋatadandem, ŋatadaseam* 'zaubern', *ŋatadandatem* 'ich bin im Begriff zu zaubern', *ŋatadafeatem* 'ich gehe zaubern', *ŋatadańantendem* 'ich will zaubern', *ŋatadakiʔem* 'ich fing an zu zaubern'). Popov, on the other hand, mentions the shaman's name in the (presumably

[101] Lehtisalo (1936) 123.
[102] Joki (1955–1956) pp. 12–5.
[103] Cf. Joki (1955–1956) p. 13.
[104] Степанов (1835) p. 59.

plural) form *ŋida*.[105] ('In the notions of the Nganasans, the shamans–*ŋida* are representatives of all *ńā*, to whom they are also patronous in respect of evil deities and spirits. The fortune and well-being of all Nganasans are depending on them').[106] The words *ŋaˀ* and *ŋida* are undoubtedly related to each other, and, from the point of view of vocalism, they compare as Castrén's *samu* 'Mütze', *anai* 'aufs neue noch'[107] with Popov's *si̮mu* (*сьму*) 'cap', *ani̮ja* (*аньја*) 'first'.[108]

The origin of this word has not yet been investigated. For phonetic reasons it cannot be identical with Nenets *ŋā* 'unterirdischer Totengeist', Enets *ŋā* 'Gott' (~ Tavgi *ŋua* 'Gott'), but such an explanation would not be convincing semantically either.

About the origin of the word we also receive no information from the neighbouring, other than Samoyedic, languages, but for its phonetic form it seems to be an original Samoyed word. For this very reason we may broach the idea that Nganasan *ŋaˀ* (*ŋida*) is related with Nenets *ŋiterma*, which denotes the shaman who died and became a helping spirit. Cf. Nenets *ŋytarma, itarma* 'Geist; Schamane nach dem Tode'[109]; Obdorsk, Lower Ob and Tas dial. *ŋiterma* 'Totenpuppe, die man nach dem Tode eines Greises anfertigt (Obdorsk dial.), Totenpuppe, die der Zauberer aus einem Stück Holz von dem Sarg eines gestorbenen Zauberers anfertigt und dessen Angehörigen gibt, damit sie 5–6 Jahre aufbewahren (Lower Ob and Tas dial.)'; Oksino, Ust-Tsilma dial. *iterma* 'Kopf des Zauberers, der früher nach dessen Tod genommen, angekleidet und aufbewahrt wurde (Oksino dial.), ein in aller Zeit aufbewahrter "*šaïtan*" (Ust-Tsilma dial.)'; Semzha dial. *iterma βesko* 'aus einem Stück Holz vom Sarg eines gestorbenen Zauberergreises in der Gestalt eines Menschen geschnitzte und bekleidete Puppe, die als *tāďebtso* aufbewahrt wurde'; Sakhalin dial. *ŋitermae* 'ergrauter alter Mann'; Pur dial. *ŋittełма* 'Z a u b e r e r g r e i s'[110]; *ŋiteлman‿ŋäejβan‿лi* 'aus Holz gefertigte und in dem unzertrennt abgezogenen Fell eines kleinen Renntierkalbs[111] aufbewahrte Figur eines Renntierbullen, die der Sohn des Zauberers nach dem Tod seines Vaters herstellt ("wörtl. Schädel des Zaubererg reises")'.[112]

The linking of this word to Nganasan *ŋaˀ* (*ŋida*) is justified because the meaning 'Zauberergreis' is also known in the Pur dialect of Forest Nenets, and thus it is probable that the word was originally known also in Nenets in the sense 'old shaman' (= great, learned shaman).

[105] In the plural this word ought actually to end in a glottal stop. In this case, however, we can take it that the transcription of the plural mark (glottal stop) has been omitted. Our knowledge of the Nganasan dialects being very sketchy, we must even accept the possibility that this form shows a dialectal variant in the s i n g u l a r which has preserved the original form of the word. The Russian context in which it occurs still refers rather to the plural form.

[106] Попов (1936) p. 83.

[107] Castrén (1855).

[108] Попов (1948) pp. 5, 115.

[109] Castrén (1855).

[110] Emphasis by the author.

[111] This is evidently a reference to the shaman's animal shape.

[112] Lehtisalo (1956) p. 35a. With regard to the beliefs related to *ŋiterma*, see Castrén (1853) p. 264 and (1953) p. 184; Lehtisalo (1924) pp. 141–2. The linking of *ŋiterma* to Nganasan *ŋaˀ* is to be found in a paper of Sauvageot (1927) p. 468.

As a result of linking the Nenets with the Nganasan word we have to regard the terminal element *-rma* of *ŋiterma* as an assimilated suffix, just as in Obdorsk dial. *l'inťśerḿe* 'Rauschbeere',[113] ~ *lynďermea*[114] or Obdorsk and Syoida dial. *ńedarma* 'grosser Weg' <: *ńeda*.[115]

Nganasan *ŋaʔ* (~*ŋada-* ~ *ŋida*) and Nenets *ŋite(rma)* may have presumably denoted a highly respected 'chief shaman'. This is borne out by Popov's quoted description of the Nganasan shaman, and notably by the meaning of the Nenets words which denote not only 'Zauberergreis', but also the likeness of a shaman who after his death was revered and even became an assistant spirit. And this is possible only in the case of famous shamans held in high esteem.[116]

The scarce material available from Nganasan does not enable us to elaborate on other shaman categories. The classification of shamans in the neighbouring peoples and the high development of shamanism among the Nganasan, however, point to the existence of such a classification of shamans here, too.

This assumption seems to be supported by B. O. Dolgikh, who writes about a certain Djufage Kosterkina who was 'the greatest Nganasan shaman' of the 'twenties'.[117]

On the other hand, we do not know what to think of what Dolgikh writes in another place, namely that Nganasan has the word *ŋo* (*нго*) which means 'shaman'.[118] Is it possible that we have to do here with a word independent of *ŋaʔ*? Hardly believable. Rather we can suppose that what we have here is a distorted, phonetically inaccurate notation of this word.

4.2. The Nganasans call the shaman's assistant *tuoptatā*. A. A. Popov writes: 'At the shamanistic ceremony, the presence of a co-singer assistant *tuoptatā* is required. In his absence, no shaman will shamanize; he is the immediate guiding assistant of the shaman.'[119]

An explanation for the word *tuoptatā* can be attempted in either of two ways. It may have derived from a verb which means 'to repeat the last lines of the (shaman's) song', but which, according to our sources, is unknown in Nganasan, and whose Forest Nenets equivalent can be quoted from the Pur dialect[120]: *täεiptāś* 'die letzten Worte des Liedersängers, des

[113] Lehtisalo (1956) p. 235a.

[114] Castrén (1855).

[115] Lehtisalo (1956) p. 322.

[116] It should be noted in this connection that quite a number of good spirits are to be found in the mythology of the Ket (Yenisei Ostyaks). They include the heroes *bataj* ~ *batar*, *bogotpyl* (Donner, 1955), all of whom were believed to have been living men at one time. One group of them comprised the great, wonder-working shamans, who after death became good spirits, the most revered among them being one called *doh*: see Anučin (1914) p. 7. Cf. also Donner (1933a) p. 93: '*dōγ* is the name of an old hero, who is also called the first YO shaman'. See also Donner (1955) pp. 36–7. One of the principal helping spirits of the Tungus, Yakut, Buryat and Altai-Turkic shamans, too, is the soul of the dead shaman predecessor, and in some places it is even carved in wood and dressed like the *ŋiterma* among the Nenets. See Harva (1933) pp. 305 ff., and Ohlmarks (1939) p. 175. The dead shaman's likeness was constantly taken along also by the Yukaghirs. See Долгих (1952) p. 78.

[117] Долгих (1951) p. 9.

[118] Dolgikh derives from this the clan name *ŋomde* (*нгомде*), but his hypothesis is not convincing. See Долгих (1952) p. 44.

[119] Попов (1936) p. 100.

[120] Lehtisalo (1956) p. 480b.

singenden Schamanen oder des Sagenerzählers wiederholen (diese pflegen stets nach einiger Zeit innezuhalten, wobei der neben ihnen sitzende Gehilfe, der "Dolmetscher", die letzten Worte wiederholt, und zwar singend, wenn es sich um ein Lied handelt)'; cf. also Lyamin, Sakhalin dial. *täεptāt*! 'sei der Wiederholer', Kiselevsk dial. *täεptaχartˀ* id. The present participle of the Forest Nenets verb denotes the repeating singer: Sakhalin dial. *täεptakōta* 'wer die letzten Worte des Sängers oder Sagenerzählers wiederholt'. This word stands well the semantic comparison with Nganasan *tuoptatā*: there are also other examples in which the shaman's assistant is denoted by words of similar meaning (cf. Nenets *teľtana* [*тельтана*] in section 2.4).

Comparison between Nenets and Nganasan words, anyway, comes up against a phonological difficulty. For the diphthong *äε* of the Nenets first syllable the equivalent in Nganasan would be *oi*, *oai*, *aᵉ*, *ai*, or sometimes *a*:

	Nenets	Nganasan
ŋäεw	'Freiwerber'	~ *ŋoibi*
χäεχe	'Geist'	~ *koika*
ŋäε	'Fuss'	~ *ŋoai*
χäεwa	'Knochenmark'	~ *koaimu*
päεľi	'dunkel'	~ *faemei*
ŋäεwa	'Kopf'	~ *ŋaewúa*, *ŋaiwúa*
ŋäept	'Kopfhaar'	~ *ŋābta*, etc.

In our example, however, we have the relation Nenets *äe* ~ Nganasan *uo*, which is unusual and to which we find something similar, at most, in the case of Nenets *täεwā* 'ankommen, erreichen' ~ Nganasan *tuaro* (first person singular: *tuˀam*) 'kommen' (in this latter, however, the formation of the *u* is evidently due to the internal bilabial consonantal sound which has since disappeared).

On the other hand, the Nganasan word can be brought into relation also with the family of the Nenets word *tāďiƀe* (~ Enets *tārebe*, Selkup *tytebe*, Kamas *thārbu*). If we start from the hypothetical verb root **tādā*-, it strikes us that the element *-t-* (*-d-*) is missing in the Nganasan word, although it actually ought to be there too. There are examples, however, which show that the intervocalic *-d-* in Nganasan drops out in some cases. For example, Nenets *χādī* 'Fichte' (~ Enets Khantaika dial. *kari*, Bayikha dial. *kadi* ~ Selkup *kāde* ~ Kamas *koˀd*) ~ Nganasan *kua* 'Tanne'; Selkup Karasino dial. *sātaŋ*, Middle Ob, Ket, Natspumpokolsk dial. *sādaruŋ* 'waten' ~ Nganasan *suárum*; Enets Khantaika dial. *kūri*, Bayikha dial. *kude* 'Löffel' ~ Nganasan *kui*; Nenets *wāda* 'Wort' ~ Nganasan *buadu* id., but: > *buaˀfandum*, *buaˀkiem* 'sprechen'.

As for the vocalism of the first syllable, it is found that for the Nenets *ā* the Nganasan equivalent, in addition to *a* and *o*, is often *u*, *ua*:

Nenets		Nganasan
χāpt	'Ochs'	*kuabtáˀa*
mādarťś	'bellen'	*mutárum*
χāsawa	'Mann'	*kuajúmu*
χānā	'tragen'	*kuandaˀama*
maraŋka	'Moorbrombeere'	*muraˀka*
χādī	'Fichte'	*kuˀa*, etc.

The word root corresponding to Nenets *tādibe* (Enets *tārebe*, Selkup *tytebe*, etc.) thus may be **tuop-* in Nganasan, and so we have to reckon with a noun derived from a hypothetical verbalized root **tuopta-*. The element *-t-* of the suffix *-tā* may be identical with the element *-t-* of the adjectival and participial suffix *-tua*.[121]

We can rightly ask the question whether *tuoptatā* formerly denoted the shaman himself and, if so, how it came to denote the shaman's assistant. In my opinion the shaman in Nganasan may have been denoted, at best, by the portion **tuop(ɘ)* of this word, or by an earlier corresponding form (**tudop*[ɘ]). For some reason (it might have been taboo) this word for shaman was replaced by a different name and survived only in its verbal derivative (**tuopta*) whose form with the suffix *-tā* may formerly have been, with the word *kuajumu* 'Mensch', part of a compound word for 'sorcerer man'. This compound may have been used to denote a sorcerer, soothsayer, etc. of minor capabilities, and through this stage it was abridged to an independent word *tuoptatā* and later degraded to denote the shaman's assistant.

Of the above explanations, the first seems to be more obvious, but the second one is also defendable, so that I cannot decide in favour of either of them.

5. As has been said in connection with Nenets *tādibe*, the Selkup word for shaman is *tetịpị* (*тэтыпы*)[122], *təᴅьʙь*[123] (ɔ : *tə̂ᴅịʙị*) ∼Karasino dial. *tytebe*, *tytebel-gum*[124] (see section *2.1*).

Another Selkup name for shaman occurs in the following examples: *sumpịtil' qup* (*сумпытыль ӄуп*) 'shaman man'[125], Tas dial. *zūmʙịtịl' kup* 'der Schamane',[126] *zumʙьtьḷ* (ɔ: *zumʙịtịl'*) *qup* 'shaman man'[127], *sumbedel-gum* 'Zauberer'.[128] These are present participles of the verb *sumpịqo, sompịqo* (*сумпыӄо, сомпыӄо*), which means 'to work magic in light tent (that is, by fire)'.[129] As to the etymology of the word, see section *2.2.1.*

Prokofjeva[130] mentions a third name for shaman: the expression *ӄамытырыль ӄуп* 'shamanizing man', which is again a present participle from the verb *ӄамытырӄо*, which means 'to work magic in dark tent (that is, without fire)'.

To outline the background of these expressions, we have to recall that the Selkup shaman obtained his equipment piece by piece, and that until

[121] Lehtisalo (1936) pp. 281–2, 320.
[122] Прокофьева (1949) p. 338.
[123] Прокофьев (1935) p. 128.
[124] Castrén (1855).
[125] Прокофьева (1949) p. 338.
[126] Prokofjew (1931) p. 477.
[127] Прокофьев (1935) pp. 54, 109.
[128] Castrén (1855).
[129] Прокофьева (1949) p. 338; Прокофьев (1935) p. 54. For the various derivatives of the verb, cf. *zumʙtae* (ɔ: *zumʙtä*) 'песнь'; *lōzị-irat zumʙtäsä nil'čịk zumʙäptịmʙa* '...чорта старика напевом так распевая...', *zumʙäptịmʙa* 'распевает' < : *zumʙäptịqo* 'напевать, распевать', etc. See Прокофьев (1935) pp. 126, 106, and 99, 107. The adjective сомпыль occurs often in words denoting the shaman's outfit; с. ӱкы 'шаманская шапка', с. поркы 'шаманская парка', с. кита 'шаманская сабля': Прокофьева (1949) pp. 359, 370, 338. Here and below, I have changed certain signs of Prokofjev's phonetic transcription to the symbols more usual in this literature, e.g. ь, *i̊* = *ị*, *ae* = *ä*, *ə* = *ə̂*, *ḷ* = *l'*.
[130] Прокофьева (1949) p. 338.

he had obtained the shaman's cap he could equally work by fire or without fire (in light or in dark tent). However, the shaman who already had the shaman's cap, and thus the highest shaman grade, could practise only by fire.[131] The above two expressions denote the shaman practising by fire and the one performing without fire, but they imply classification only so far as the learned, old shaman in possession of the shaman's cap could only be a *sumвit̮ḭl' qup* while his younger mates who had not attained this grade were called either, *sumвit̮ḭl' qup* or *qamit̮ir̮ḭl' qup*. Namely, they were allowed to officiate either in light or in dark tent.

To all this we could add that, according to Prokofjev,[132] the difference between practising in a dark tent and in a light tent was that the shaman in the first case officiated w i t h o u t t h e d r u m. To illustrate the two methods we may quote from Prokofjev the following sentence: *jomвa čēl̮it zumвa, pit qamit̮irni̮, qamit̮irmi̮ aj zumвa* '*jomвa* shamanizes in the day time (with a drum), at night he shamanizes (without a drum); shamanizes (without a drum), and shamanizes (with a drum)'.[133]

The origin of the verb *qamit̮irqo* is unknown; but the sillable *qam-* strikingly reminds us of the word *kam* 'shaman' existing in the Siberian Turkic languages and is actually the origin of the Russian word *камлание* for 'shaman's ritual', too.[134] The Selkup root *qam-*, *qap-* can be recognized in the word *kapšit, kapšin* (*капшит, капшин*) 'drumstick'.[135]

In addition to the above three names, a fourth word for 'Zauberer, Priester' is mentioned by Castrén and Károly Pápai: Narim dial. *kèdehul-gup, kēdehul-gup*[136] and *khödzsüi-χupp* 'shaman',[137] the first part of which is related to Castrén's *kèdeǯak* (first person singular) 'zaubern' (as to the perfective affix *-ča* ~ *ǯa*, cf. *qonɒalɛl'čak*, first person singular from *qonɒalqo* 'to be going to bed')[138] and *kèd* 'Zauberweisheit' (the second part of it is a word for 'man').[139] The origin of the Selkup word *kèd*, to my knowledge, has not so far been investigated. Prokofjev mentions the word in the form *kətь* (ɔ: *kəti̮*) for 'strangeness, jugglery'.[140] We can suppose that the same root gave rise to the verb *kətьqo* (ɔ: *kəti̮qo*) 'to say', which Castrén noted down as follows: Narim dial. *kadap*, Middle Ob, Ket dial. *kānnau* (*-nn-* < *-tn-*), Upper Ob dial. *kannam*, Karasino dial. *ketam*, Natspumpokolsk dial. *kätnam*, Bayikha, Tas dial. *kètam, kètejam*, Tas dial. *ketendam, ketteľam* (freq.), etc.,[141] Tym dial. *kɛt-* 'sagen, sprechen',[142] and which is corresponding to Nenets Obdorsk dial. *χetā* 'erzählen, sagen',[143] and possibly to Lappish Lule dial. *keɔttu-* 'erzählen'.[144]

[131] Прокофьева (1949) p. 338.
[132] Прокофьев (1935) p. 123.
[133] Прокофьев (1935) p. 108.
[134] Harva (1933) p. 299; Vasmer (1953).
[135] Прокофьева (1949) p. 337.
[136] Castrén (1855).
[137] Pápai, see Hajdú (1953) p. 162.
[138] Прокофьев (1937) p. 113.
[139] Castrén (1855).
[140] Прокофьев (1935) p. 114.
[141] Castrén (1855).
[142] Donner (MS), cf. Lehtisalo (1924b).
[143] Lehtisalo (1956) p. 185a.
[144] Lehtisalo (1928) p. 133 and (1924b) p. 156; cf. also Hajdú (1955) p. 259.

This name for shaman refers mostly to a person whose profession is soothsaying, that is, to the 'sorcerer' of the lower categories.

Investigation into the Selkup shaman names has not made it possible to establish conclusively the correspondence between the particular names and categories, and we cannot know if there were sharp boundaries between the categories. The denominations anyhow lead to the conclusion that they were used with some differentiation, and thus it may be supposed that they are linguistic remnants of those categories. The incomplete picture is due also to the scantiness of the material.

6. Of the Kamassian names for shaman we also know only little. The old Samoyedic name, as we have already seen, is in use here too: *thàrbu*[145], *tʿārbə, tʿå̄rbu* 'Zauberer, Schamane', *tʿå̄rbə̄n* or *tʿå̄rbun pʿərɑa* 'Pelz des Zauberers'[146].

An explanation for our scanty knowledge about the Kamassian shamans is that at the beginning of this century the Kamas no longer had shamans and sorcerers. The latest students of them, Kai Donner, found not a single shaman among the remaining Kamas numbering about 50 people at that time.[147]

7. It can be seen from the foregoing that the classification of shamans according to capabilities existed among the Samoyeds. The categories and the relevant names are summed up in the following table:

Categories	Nenets	Enets	Nganasan	Selkup	Kamas
General names	*tāďiƀe*	*tyrebe*	(?**tuop: tuoptatā*)	*t̯el̯ip̯i̯*	*thārbu*
I. Strong (wonder-working) shaman	*wēdutana tāďiƀe* *śiďeńā̄? tārāna t.* *nūwūńā̄? tārāna t.* *χataε t.* (*ηiterma*)	*budtode*	*ηa?*	↑ *sump̯it̯il̓ qup* ↓	
II. Medium shaman	*jāńā? tārāna t.* ↑ *sāmpana (t.)* ↓	*ďano*		*qam̯it̯ir̯il̓ qup*	
III. Small shaman	*kompurtae t.* ↓	*sawode*		*k̯edehul-qup* ↓	
IV. Dream-reader, soothsayer	*jutissāmi ńieśeä* *jut̯irtaε ńieseä* *wenuj śī ńieseä*				

The table gives us a summary of what we have set forth in detail to show that the boundaries between the different shaman categories are not

[145] Castrén (1855).

[146] Donner (1944). The other word for 'Schamane' (*khöwü?*) indicated in Castrén's glossary (1855), as is pointed out by Joki (1952) p. 200, got that meaning only through a printer's error. The correct meaning of the word is 'Schaum'.

[147] Donner (1923) p. 250.

always sharply and rigidly drawn, and that in certain tongues (Nganasan, Kamassian) the categories cannot even be distinguished. This is due partly to the extinction of shamanism (Kamassian), partly to the scantiness of the available material (Nganasan). The abundance of the Nenets names for shaman is promising further additions to our stock of shaman names.

The distinction between shamans is most visible in Enets: here the shamans of the various categories are distinguished from one another not only by their functions, but in general also by their costumes and equipment.

Despite the fact that Nenets names for shaman referring to the particular categories are rather numerous, the differences between categories are nevertheless not so manifest as in Enets. This is accounted for by the fact that the different grades of Nenets shamans cannot be distinguished outwardly by their costumes and equipment. Namely, in the course of this century, the shaman costume among the Nenets gradually ceased to be the symbol of the shaman's office, and the shamans' influence also declined to a great extent.

Authentic descriptions of the Nenets shaman's costume are available already from the 18th century.[148] As Vereščagin[149] wrote in 1848, the shaman could be distinguished by his costume from the ordinary Samoyed. In 1855 the Archimandrite Venyamin wrote in detail about the Nenets shaman's costume of the Mezen region,[150] and Castrén[151] also remarked that the Nenets shaman wore a special costume. Lehtisalo, however, had the following to say: 'Bei den Juraksamojeden habe ich nicht so vollständige Zauberanzüge gesehen wie sie sich bei vielen östlicheren sibirischen Völkern finden.'[152] According to him it is mostly the shaman's cap (and, of course, the drum) which still belongs to the shaman costume and equipment. L. Heidenreich also wrote about the Kanin Nenets that their shamans had no special costume.[153] At the same time, however, Starcev noted that the shaman's costume was still to be found on Bolshaya Zemlya and on the Yamal peninsula.[154] To be sure, he also gave us to understand that the special shaman's costume was rarely worn.

With the slow disappearance of the shaman's costume[155] the shamans' influence also declined. We know not only from Gejdenrejkh's description that the shamans had no particular influence on the Kanin Nenets.[156] Also

[148] Зуев (1947) p. 46; Pallas (1776) pp. 75–8.
[149] Верещагин (1848) p. 153.
[150] Венямин (1855) p. 118.
[151] Castrén (1853) pp. 192–3 and (1953) p. 138.
[152] Lehtisalo (1924a) p. 147.
[153] Гейденрейх (1930) p. 26–7.
[154] Старцев (1930) p. 129–30.
[155] The disappearance of the shaman's costume is observed not only among the Nenets. It is no longer to be seen among the Ob-Ugrians or the Lapps either: Karjalainen (1918) pp. 553–4, and Holmberg (1915) p. 111. The latter writes in another place: 'Generally, shaman's costumes are beginning to decline everywhere, although the belief in shamanism still prevails.' (*AUFA*, I/2, 9.) Then he remarks that the first to disappear from the shaman's equipment are the gloves and boots, and what survives longest are the mantle and the headgear (op. cit., 10). 'The other magic instruments, such as the drum, would seem to have been more essential to the shaman, and their use has therefore been able to survive that of the costumes.' Holmberg (1922a) p. 9, and Harva (1933) p. 332; see also *idem*, *KSVK* II, Holmberg (1922b) p. 109.
[156] Гейденрейх (1930) p. 26–7.

Starcev noticed[157] that the Nenets felt differently about shamans than did their Selkup neighbours: they did not call them up so often, and if one of their prophecies did not come true, their authority fell considerably. It is this attitude of the Nenets that appears in their saying that the big shamans, if there are any, are very rare.

Lehtisalo writes: 'Bei einigen Samojeden ist der Glaube an die Geister und die Furcht vor den Zauberern sehr gering; wir haben es hier also mit einer ähnlichen Erscheinung wie der verschiedenen Einstellung zur Religion bei entwickelteren Völkern zu tun. Im allgemeinen herrschte unter den Samojeden allgemeines Bedauern darüber, dass es nicht mehr so gute Zauberer gebe als die Vorväter waren.' Moreover, according to Lehtisalo, even the Forest Nenets were so suspicious that the shaman received his pay only if the sick patient had recovered through his cure.[158]

The decline of the authority of Nenets shamans may account for the fact that the Nenets trusted the Enets shamans more than their own. On the other hand, the Enets in special cases, although their shamans yielded great authority over them and had preserved their system of classification from ancient times,[159] called upon the Ket and Selkup shamans whom they thought to be even more powerful and stronger than their own.[160]

Among the Enets and the Nganasans, as well as among the Selkups, the shamans enjoyed far greater authority than among the Nenets, and also the shaman's costume had not gone out of use to the same extent. It should not be overlooked either that they obtained the shaman's outfit piece by piece, according as their capabilities grew. This was definitely so among the Selkups, where the shaman obtained first the drumstick, then the drum, later the apron, the shaman's costume and boots, and finally the shaman's cap and the staff.[161] Indicative of this is the fact that the three categories of shamans among the Enets were outwardly distinguishable: only the *budtode* had the complete shaman's equipment.[162] Of the Nenets shamans and shaman costumes we have, unfortunately, no such detailed description as of other Samoyed shamans.

The linguistic material reveals that a common name for shaman (the *tāďiḃe* type) can be recognized in all five Samoyedic languages, attesting the ancient origin of shamanism. Notwithstanding this, we have the impression that this word is used rarely in most Samoyedic languages or is used only to denote 'shaman' in general, and is even replaced mostly by designations of more recent origin. In the face of this phenomenon we have two things to bear in mind. On the one hand, new names were needed to denote the shamans of different categories. New words were coined for this purpose, or the word for 'shaman' was supplied with attributes (which later, being abridged, naturally denoted by themselves the relevant shaman).

157 Старцев (1930) p. 129–30.
158 Lehtisalo (1924a) p. 166.
159 How things stand now we do not know.
160 Прокофьева (1951) p. 126, cf. also note 2.
161 Прокофьева (1949) p. 341. The Ket shamans were also of three categories, and they too obtained their outfit in a definite order (drumstick, frontlet, apron, boots, gloves, drum with a new drumstick, staff, mantle and crown, a second staff); see Анучин (1914) p. 33.
162 Прокофьева (1951) p. 126.

On the other hand, in certain cases the taboo may have played a role in the decline of the ancient Samoyedic word for 'shaman'. Such phenomena are observed not only in connection with animal names but also in the denomination of the instruments of the shaman.

We can mention for example the names of the shaman's drum. The Samoyedic languages have a common word for this most important shaman instrument: Obdorsk dial. *ṕenťśer*, Lyamin dial. *ṕienśeär* 'Zaubertrommel'[163], etc.; Enets *feďi* (gen. Khantaika dial. *feďiđo?*, Bayikha dial. *féďiro?*)[164], *neддu*[165], Karasino, Khantaika dial. *φeďi, pφeďi*[166]; Nganasan *feanďir* id.; Selkup Ket, Chaya, Upper Ob dial. *pöŋer* id. and 'Violine'[167]; Tym dial. *pə̂ŋgə̂r*, Chaya dial. *pə̂ggə̂r*, etc. 'besonderes samO. Musikinstrument'[168]; Kamassian *phīri*[169], *pʿēri* 'Zaubertrommel'[170]. This word, however, is comparatively rarely used. In addition, especially in the shaman's songs, quite different names are also given to the shaman's drum. For example, Nenets *χarw ni̮n* [ɔ: *ŋi̮n*] (*харв нын*) (correctly: *нгын*—P. H.) 'larch-tree bow' ~ Enets *kamo iddo* (*камо иддо*) id.; Nenets *nuw? nin* [c: *ŋi̮n*] (*нув' нын* [ɔ: *нгын*]) ~ Enets *na iddo* (*на иддо*) 'heavenly bow'; Nenets *minutja ṕa* (*минутя пя*) 'bent wood'; Nenets *χarw pud majona? aw* (*харв пуд майона'ав*) 'covered larch-tree hoop', etc.[171]

Lehtisalo's Nenets texts fully support this observation. In the shaman songs he noted down there often appear for drum the expressions 'bow-wood', 'singing larch-tree', 'that which has twice seven rim pegs'.[172] For example, Lower Ob dial. *ŋin-ṕāḿi? χinukād* 'Sänger meines Bogenholzes' (i.e. the spirits), *ŋin-ṕāḿi? jalibā ṕāχawāś* 'Mein Bogenholz zu schlagen begann ich', *ŋin-ṕāmta ńamŋāda* 'Sein Bogenholz nahm er in seine Hand', *χinukād ŋin-ṕāḿi'* 'Mein singendes Bogenholz', *śiďe śīw χäεŋorota ťiχinta ṕenawī?* 'Die mit zwei(mal) sieben Rahmenzapfen Versehene ist ganz glatt gespannt', *śiďe śīw χäεŋorotam ńāmpaś* 'die mit zwei(mal) sieben Zapfen Versehene (d.h. die Zaubertrommel) in der Hand haltend', etc.[173] Cf. also Tas dial. *χaŕeta χinota χārw ṕāḿi?* 'mein selbst singender Lärchenbaum'.[174]

Lehtisalo draws far-reaching conclusions from the fact that the word denoting 'bow', 'bow-wood', often occurs instead of the regular word for drum (*ṕenťśer*). He thinks that the use of the shaman's drum among the Samoyeds (as well as among those Finno-Ugrian peoples whose shamans used it at all) was a secondary development, and the shaman's magic instrument to chase away the spirits was originally the bow. He mentions

[163] Lehtisalo (1956) pp. 378–9.
[164] Castrén (1855).
[165] Прокофьева (1951) p. 146.
[166] Joki (1955–1956) p. 13.
[167] Castrén (1855).
[168] Donner (1920) p. 111.
[169] Castrén (1855).
[170] Donner (1944).
[171] Прокофьева (1951) p. 146.
[172] To stretch the drumhead properly, protruding pegs (= Obdorsk dial. *χäεŋor?*, Wb. 180b) are fixed upon the drum's rim. See Lehtisalo (1924) p. 149; Прокофьева (1951) p. 146.
[173] Lehtisalo (1947) pp. 477, 484, 485, 496, 503, 500.
[174] Lehtisalo (1956) p. 166b.

in this connection that the dead person's relations, after being 'cleaned', were given a small bow and arrow to ensure success in hunting. And then, during the ceremony, the shaman unleashed the bow in several directions around the *śīmsi*[175] post to frighten away the spirits.[176]

The bow and arrow certainly played a part in Finnish mythology: if someone shot an arrow at the effigy of a man, this latter would be bewitched.[177] Also the Skolt-Lapps held that the sorcerer could shoot out an arrow causing anybody coming near to fall sick. In olden times, moreover, in some places those who set out on a journey shot a magic arrow in front of them lest some harm should come to them on their way.[178] Harva in another place also refers to an older source, according to which the Lapps who wanted to bewitch an enemy, either near-by or far away, made a small bow and arrows of reindeer's horn. At the same time they made an effigy of the enemy and shot the arrow at the part of his body they intended to hurt. But Harva remarks also that, judging by the terminology, this custom must have been of Scandinavian origin. Namely the Lapps call this magic arrow *gand* or *dirri*, and these words come from Scandinavian *gandr* 'magic wand' and *dirri* 'magic arrow', to which similar superstitious procedures are attached also among the Scandinavians.[179]

Even if this Lappish fancy were not of Scandinavian origin, it would not follow from it, nor from similar Samoyed superstitions, that the bow as a magic instrument should have preceded the drum.

In beliefs of this kind we have only to see one of the many ways of bewitching or of self-protection against bewitching, but neither these instances nor the cover-names 'bow-wood' for shaman's drum as illustrated by Nenets examples make it necessary to consider the bow a primitive magic tool preceding the drum.

Harva also points out that by Lehtisalo's assumption the drum was used originally only to exorcize, to frighten away the spirits.[180] This, however, is unlikely, because the drum's principal function was to enable the shaman to contact his helping spirits.

The above-mentioned cover-names denoting the drum ('larch-wood bow', 'heavenly bow', 'bow-wood', 'singing bow-wood', 'bent wood', 'singing wood', 'covered larch-wood hoop', 'that which has twice seven rim pegs'), if examined within their system, bear evidence that they were based on certain p r o p e r t i e s, c h a r a c t e r i s t i c s of the tabooed object to be named: the shaman's drum was made of larch-wood, it has rim pegs, gave out sound, was bent and hoop-shaped, and we can take for such a characteristic also its resemblance to the bow, for both things are of b e n t w o o d[181] and of identical material (larch-wood). Accordingly, the Nenets

[175] The *śīmsi* is a vertical post set up in front of the entrance of the tent and serves to fix one end of the iron rod from which the kettle's handle is hung. The *śīmsi* is at the same time a sacred object in Samoyed mythology.

[176] Lehtisalo (1930) pp. 23–4.

[177] Harva (1948) p. 251.

[178] Itkonen (1948) pp. 344–5.

[179] Holmberg (1915) p. 109.

[180] Harva (1933) p. 356.

[181] The sketch of the Lapp 'magic bow' described by Harva is also reminiscent of an almost regular semicircle. Holmberg (1915) p. 109.

names of 'bow-wood' for the shaman's drum can be considered such cover-names characteristic of the object denoted as are, for example, the expressions 'four-legged beast', 'reindeer-killer', 'long-tail', 'the tailed one', for wolf.

In the light of the examples quoted in connection with the shaman's drum, it must not be surprising that, for reasons of taboo, more than one name should have come to denote 'shaman' among the Samoyeds.

To all this we may add that Karjalainen also was puzzled by the variety of Khanty and Mansi names for shaman. He found it strange that this person invested with such important functions should not have a single name on the whole Khanty territory. The Old Finno-Ugric name (∼ Finnish *noita*) occurs only among the Mansi (*ńait*), but here also we meet a wide range of other names, probably of a later and local origin (Tavda dial. *sātəp-kum* 'medicine man', North and Losva dial. *pupiγ kāstənə χum* 'man invoking an idol or god', etc.). The name which is most widely spread and known practically everywhere among the Khanty is a word combination beginning with East Khanty *tšēpän* (*tšēpänəŋ-χoi* 'sorcerer', Vakh dial. *tšīpānəŋ-kū* 'medicine man', etc.).[182] Widely used expressions are also *t́ərtəŋ-χoi* 'prophet, the learned one', Vakh, Vasyugan dial. *jol* '(great and little) magician', Tremyugan dial. *mūltə-kō* 'praying man', North Khanty *semvōjaŋ-χo* 'seeing man', etc.[183] The Ob-Ugrian names for shamans cannot at first sight be classed in clearly distinguishable categories, and that is certainly due to the decline of shamanism or to the fact that the place of the genuine shamans was taken over by different kinds of soothsayers, dream-readers and medicine men. A similar picture is shown also by Lapp shamanism, in which—although three categories of shamans can still be distinguished—there are in addition several names for the various kinds of seers and medicine men.[184]

We have cited these examples in support of our findings concerning the Samoyed names for shamans. Accordingly, we can really assume that the original names for 'shaman' were tabooed words, and this is the reason for their less general use or their far too general meaning. On the other hand, it appears also from our material that the various shaman names also denote different categories of shamans, and this fact too had a part in their formation.

Part of them appeared simultaneously with the decline of shamanism to distinguish the dream-readers, medicine men, and soothsayers having no supernatural powers who replaced the gradually disappearing 'genuine' shamans.[185]

[182] This Khanty word (taken over also by the Mansi), as I have demonstrated recently (*NyK*, LVII, 250 ff.) is of Zyrian origin.

[183] Karjalainen (1918) p. 551.

[184] Itkonen (1948) p. 345.

[185] I owe thanks to my friend Vilmos Diószegi, who made available to me the material of the 'Archives of Shamanism' arranged by him in the Budapest Ethnographic Museum.

(Anučin, V. I.) Анучин, В. И. (1914) Очерки шаманства у енисейских остяков (Essays on Shamanism among the Yenisei Ostyaks), in *СбМАЭ*, Vol. II, No. 2.
Budenz (1890–1892) Jurák-szamojéd szójegyzék (Yurak-Samoyed Glossary), in *NyK*, Vol. XXII.
Castrén, M. A. (1853) *Reiseneri nnerungen aus den Jahren 1838–1844*, ed. by A. Schiefner, St. Petersburg.
(1855) *Wörterverzeichnisse aus den samojedischen Sprachen*, St. Petersburg.
(1953) *Tutkimusmatkoilla Pohjolassa* (Study Tours in the North), translated by A. J. Joki, Helsinki.
Diószegi, V. (1947) Samán (The Etymology of the Word *samán*), in *MNy*, Vol. XLIII.
(Dolgikh, B. O.) Долгих, Б. О. (1951) Обрядовые сооружения нганасанов и энцев (The Ceremonial Equipment of the Nganasan and Enets), in *КрС*, Vol. XIII.
(1952) Происхождение нганасанов (Origin of the Nganasan), in *ТрИЭ*, Vol. XVIII.
Donner, K. (1920) Über die anlautenden labialen Spiranten und Verschlusslaute im Samojedischen und Uralischen, in *MSFOu*, Vol. XLIX.
(1923) *Siperian samojedien keskuudessa* (Among the Samoyeds of Siberia), Helsinki.
(1933a) Ethnological Notes about the Yenisei-Ostyak, in *MSFOu*, Vol. LXVI.
(1933b) *Siperia. Elämä ja entisyys* (Siberia: Its Life and Past), Helsinki.
(1944) *Kamassisches Wörterbuch*. Ed. by A. J. Joki, Helsinki.
(1955) Ketica. Materialien aus dem Ketischen oder Jenissei-Ostjakischen. Notes by K. Donner. Rev. and ed. by A. J. Joki, in *MSFOu*, Vol. CVIII.
(Dunin-Gorkavič, A. A.) Дунин-Горкавич, А. А. (1910) *Русско-остяцко-самоедскiй практическiй словарь* (Practical Russian–Ostyak–Samoyed Dictionary), Tobolsk.
(Gejdenrejkh, L.) Гейденрейх, Л. (1930) Канинские самоеды (The Kanin Samoyeds), in *СС*.
Hajdú, P. (1953) Pápai Károly szelkup szójegyzéke (Károly Pápai's Selkup Glossary), in *NyK*, Vol. LIV.
(1954) Etimológiai megjegyzések (Etymological Notes), in *NyK*, Vol. LVI.
(1955) Etimológiai megjegyzések (Etymological Notes), in *NyK*, Vol. LVII.
Harva, U. (1933) *Altain suvun uskonto* (The Religion of the Altaic Peoples), Helsinki.
(1948) *Suomalaisten muinaisusko* (The Ancient Religion of the Finns), Porvoo–Helsinki.
Holmberg, U. (1915) *Lappalaisten uskonto* (The Religion of the Lapps), Porvoo–Helsinki.
(1922a) The Shaman Costume and Its Significance, in *AUFA*, Ser. B, Vol. I, Fasc. 2.
(1922b) Noidanpuku ja sen aate (The Shaman Costume and Its Symbolic Significance), in *KSVK*, Vol. II.
(Islavin, V.) Иславин, В. (1847) *Самоѣды въ домашнемъ и общественномъ быту* (The Samoyeds in Their Home and Social Life), Saint Petersburg.
Itkonen, T. I. (1948) *Suomen lappalaiset* (The Lapps in Finland), Vols I–II, Porvoo–Helsinki.
Joki, A. J. (1952) Die Lehnwörter des Sajansamojedischen, in *MSFOu*, Vol. CIII.
(1955–1956) Kai Donners kleinere Wörterverzeichnisse aus dem Jurak-, Jenissei- und Tawgisamojedischen, Katschatatatarischen und Tungusischen, in *JSFOu*, Vol. LVIII, No. 1.
Karjalainen, K. F. (1918) *Jugralaisten uskonto* (The Religion of the Ugrian Peoples), Porvoo–Helsinki.
Klaproth, J. (1823) *Asia Polyglotta*, Paris.
(Kuševlevskij, Ju. I.) Кушелевский, Ю. И. (1868) *Сѣверный полюсъ и зѣмля Ямалъ* (North Pole and Yamal Land), Saint Petersburg.
Laufer, B. (1917) Origin of the Word Shaman, in *AA*, Vol. XIX, No. 3.
Lehtisalo, T. (1922) Jurakkisamojedien lauluista (On the Songs of the Yurak Samoyeds) in *KSVK*, Vol. II.
(1924a) Entwurf einer Mythologie der Juraksamojeden, in *MSFOu*, Vol. LIII.

Lehtisalo, I. (1924b) Zu den samojedisch–arischen Beziehungen, in *MSFOu*, Vol. LII.
(1928) Uralische Etymologien, in *MSFOu*, Vol. LVIII.
(1929) Sampa, sammas (The Words *sampa* and *sammas*), in *V*.
(1930) Kertomus työskentelystäni jurakkisamojedi Matvei Ivanovič Jaadnjen kanssa (Account of My Work with the Nenets M. I. Jaadnje), in *JSFOu*, Vol. XLIV, No. 4.
(1932) Beiträge zur Kenntnis der Renntierzucht bei den Juraksamojeden, in *ISK*, Vol. XVI.
(1936) Über die primären ururalischen Ableitungssuffixe, in *MSFOu*, Vol. LXXII.
(1937) Der Tod und die Wiedergeburt des künftigen Schamanen, in *JSFOu*, Vol. XLVIII, No. 2.
(1947) Juraksamojedische Volksdichtung, in *MSFOu*, Vol. XC.
(1956) *Juraksamojedisches Wörterbuch*, Helsinki.
(1959) Eräistä samojedin sanoista (On Some Samoyed Words), in *VD*.
Németh, J. (1913–1914) Über den Ursprung des Wortes šaman, in *KSz*, Vol. XIV.
(Nemirovič-Dančenko, V. I.) Немирович-Данченко, В. И. (1881) Мезенская тундра (The Mezen Tundra), in *Живописная Россія*, Essay No 3.
(1884) *Племена глухих углов* (Tribes of the Wilderness), Saint Petersburg.
Nioradze, G. (1925) *Der Schamanismus bei den sibirischen Völkern*, Stuttgart.
Ohlmarks, Å. (1939) *Studien zum Problem des Schamanismus*, Lund–Kopehagen.
Pais, D. (1949) Reg. A régi magyarság szellemi életének kérdésköréből (The Word *reg*: From the Problems of the Spiritual Life of the Ancient Magyars), in *MNyTK*, Vol. 65.
Pallas, P. S. (1776) *Reise durch verschiedene Provinzen des russischen Reichs*, Vol. III, St. Petersburg.
Park, W. Z. (1938) *Shamanism in Western North America*, Chicago.
(Popov, A. A.) Попов, А. А. (1936) Тавгийцы. Материалы по этнографии авамских и вадеевских тавгийцев (The Tavgi. Materials for the Ethnography of the Avam and Vadeyev Tavgi), in *ТрИЭ*, Vol. I, section 5.
(1948) Нганасаны I. (The Nganasan I.), in *ТрИЭ*, Vol. III.
Prokofjew, G. N. (1931) Materialien zur Erforschung der ostjak-samojedischen Sprache, in *UJ*, Vol. XI.
(Prokofjev, G. N.) Прокофьев, Г. Н. (1935) *Селькупская (остяко-самоедская) грамматика* (Selkup [Ostyak–Samoyed] Grammar), Leningrad.
(1937) Селькупская грамматика (Selkup Grammar), in *Языки и письменность народов севера*, Vol. I, Moscow.
(Prokofjeva, E. D.) Прокофьева, Е. Д. (1949) Костюм селькупского (остяко-самоедского) шамана (The Costume of the Selkup [Ostyak–Samoyed] Shaman), in *СбМАЭ*, Vol. XI.
(1951) Энецкий шаманский костюм (The Enets Shaman Costume), in *СбМАЭ*, Vol. XIII.
(1956) Ненцы (The Nenets), in *Народы Сибири*, Moscow–Leningrad.
(Ryčkov, K.) Рычков, К. (1916) Береговой род юраков (The Coastal Clan of the Yuraks), in *ЗЗСОРГО*, Vol. XXXVIII.
Sauvageot. A. (1927) Az ural-altáji nyelvek egy vallási kifejezéséről (A Word Concerning Religion in Uralic and Altaic Languages), in *MNy*, Vol. XXIII.
Schrenck, A. G. (1854) *Reise nach dem Nordosten des europäischen Russlands, durch die Tundren der Samojeden zum Arktischen Uralgebirge*, Vol. II, Dorpat.
Setälä, E. N. (1902) Zur Etymologie von sampo, in *FUF*, Vol. II.
(1932) *Sammon arvoitus* (The Mystery of the Sampo), Helsinki.
Sprogis, A. (1952) Wörterverzeichnis und grammatikalische Aufzeichnungen aus der Kanin–Mundart des Jurak–Samojedischen. Ed. by Irene N. Sebestyén, in *AL*, Vol. II.
(Starcev, G. A.) Старцев, Г. А. (1930) *Самоеды (ненча). Историко-этнографическое исследование* (The Samoyeds [Nenets]: A Historico–ethnographical Study), Leningrad.
(Stepanov) Степанов (1835) *Енисейская губерния*, II (The Yenisei Government), Saint Petersburg.
(Tereščenko, N. M.) Терещенко, Н. М. (1955) *Ненецко–русский словарь* (Nenets–Russian Dictionary), Leningrad.
Uotila, T. E. (1929) Sammon arvailua (Ideas about Sampo), in *V*.

(Vasilevič, G. M. and Levin, M. G.) Василевич, Г. М.–Левин, М. Г. (1951) Тыпы оленеводства и их происхождение (The Types of Reindeer Breeding and Their Origin), in *СЭ*, No. 1.

Vasmer, M. (1955) *Russisches etymologisches Wörterbuch*, Heidelberg.

(Venjamin) Венямин (1855) Самоеды мезенские (The Mezen Samoyeds), in *ВРГО*, Vol. XIV.

(Verbov, G. D.) Вербов, Г. Д. (1935) Диалект лесных ненцев (The Dialect of the Forest Nenets), manuscript in the library of *МАЭ* (Museum of Anthropology and Ethnography).

(Vereščagin, V.) Верещагин, В. (1848) Самоеды (The Samoyeds), in *Звездочка—журнал для детей старшего возраста*, Vol. VII, No. 25.

(Zuev, V. F.) Зуев, В. Ф. (1947) *Материалы по этнографии Сибирии* XVIII *века* (Contributions to the Ethnography of Siberia in the 18th Century), Moscow-Leningrad.

The Cult of Bear among the Ket (Yenisei Ostyaks)

by

E.A. Alekseenko, Leningrad

One of the ancient cults preserved to some extent by Siberian peoples up to the pre-revolutionary era was the ritual worship of the bear. Probably thousands of years old, the cult of the bear goes back to an early stage of human culture, to the culture of the hunters of the forest areas of Northern Asia and is related to totemistic beliefs.

The cult of the bear among the peoples of Siberia manifested itself in a special attitude towards the bear, in its elevation above the other animals, in the recognition of its kinship with man, in diverse superstitions and symbols, as well as in a number of ritual functions connected with bear hunting and the eating of bear's flesh. This form of bear worship was characteristic of most peoples of Siberia.[1] Another form of this cult, whereby the bear was kept in a cage and the grown-up animal was ritually sacrificed, was observed only among some peoples of the Far East (Ainu, Gilyak, Orochi and Ulchi).

Especially interesting is the cult of the bear among the Ket, who have until lately preserved characteristic traits of the ancient culture of the trappers and fishermen of the taiga (methods and technique of hunting, implements of the trade, means of conveyance, and elements of spiritual culture). Its study not only gives us an insight into the religious beliefs inherent in the archaic economic organization of human society, but helps also to throw light on the ethnogenesis of the peoples of Siberia.

Literature provides very scarce information on the bear cult of the Ket. Mention is made only of the particular role of the bear, of its having a soul like men's, and there are fragmentary notes on some aspects of the worship of the bear and ritual customs observed in the course of bear hunting.[2]

In this treatise, not pretending to an exhaustive description of bear worship among the Ket, I only wish to throw more light on what the literature has to tell about it. This paper is based on the data I collected on a journey I was commissioned to make by The Institute of Ethnography of the

[1] The bibliography on bear worship among the peoples of Siberia is most completely presented by N. P. Dyrenkova (1928); Васильев (1948) and Nahodil (1954).

[2] The Finnish scientist K. Donner, who in the autumn of 1912 took part in a bear festival among the Ket on the Stony Tunguska, devoted two pages of his ethnographical notes on the Ket specially to the cult of the bear. As K. Donner himself pointed out, he did not make out much of what was said during the festival. He described in more detail only the phase of preparing for the feast in the tent of the hunter who had killed the bear. See Donner (1933).

Judging by the data quoted by B. A. Vasiljev (1948), a rich material on bear worship among the Ket was in the possession of N. K. Karger whose manuscripts, unfortunately, are not to be found. After this article was completed, a new publication on the Ket's bear cult appeared — H. J. Paproth (1962).

Academy of Sciences of the U.S.S.R. in 1956 and 1958.[3] Visiting on those occasions all principal places of habitation of the Ket, I had opportunity to verify the collected material and to establish some local differences.[4]

All information has been taken down from the words of Ket hunters, mainly elderly people. At present the Ket observe only certain aspects of the ancient worship of the bear. The Ket today do not have the full complex of ceremonies, and the religious content of the festival has almost disappeared. At the same time the killing of the bear is a festive event among the Ket, evidencing the hunters' courage, skill and endurance. Bear's flesh has a pleasant taste, the skin of the bear has good value, the hunter receives a big price from the purchasing agencies, and it is natural that the people should celebrate the event. Some traditional ritual elements of the festival have been retained for the sake of fun.

*

It is an old superstition, which was still prevailing not long ago, that the bear is the 'king' of the animals living on earth.[5] About the attitude towards the bear as 'god of the earth', M. A. Castrén wrote: 'The Ostyaks (Yenisei Ostyaks is the old name of the Ket.—E. A.) believe that he (the bear.—E. A.) is not an animal like all the others, that his skin is only a cover under which is a being in human shape endowed with divine force and wisdom'.[6] The main feature distinguishing the bear from the beasts, wrote V. I. Anučin, is that he, like man, has a soul (other beasts have no soul).[7]

The attitude towards the bear as a kin to man, a relation of his, appeared also in the name. Speaking of the bear, the Ket called and still call him *qyṕ* or *bat*, which means 'Grandfather', 'Old Man'. The she-bear is referred to in like manner. Characteristically, the Ket *qoi* for 'bear' is used in the Ket tongue also to denote 'stepfather', 'stepmother'. There is another interesting mythological name for bear, *ķaiγuś* 'шерстной человек', 'лесовой человек'[8] (hairy man, forest man), 'one like man, only he has a hide as we have clothes'.[9]

[3] I have received valuable information on the Ket cult of the bear from B. O. Dolgikh, to whom I express my thanks here. The data were noted down by him on an expedition on the Stony Tunguska organized by The Institute of Ethnography of the Academy of Sciences of the U.S.S.R. in 1948.

[4] The Ket live in five groups in the Turukhan district of the Krasnoyarsk Krai along the tributaries of the Yenisei—the Stony Tunguska, Yelogui, Surgutikha, Pakulikha and Kureika rivers. The northern (Kureika) Ket live more than one thousand kilometres away from the southern (Stony Tunguska) Ket.

[5] In Ket folklore there are 'kings' also among birds and fish.

[6] Кастрен (1860).

[7] Анучин (1914).

[8] The Ket gave me information mainly in Russian.

[9] This conventional name for bear, and particularly that placing it on an equal footing with man, is widespread among the Siberian peoples. The name 'Old Man', 'Grandfather' is used by the northern Altai peoples (Tubalars), the Mansi, Nenets, Yakut, north Baikal Evenki, Nanai, Udegei and Orochi. The name denoting the bear as a relation of man's is known also to the Mansi ('forest woman' or 'mountain woman'), the Khanty ('old man with a fur-coat') and others. Васильев (1948) p. 81; Потапов (1928) p. 17; V. N. Černecov's manuscript on the bear festival of the Ob-Ugrians, a lecture at the faculty of ethnography of the Moscow State University (1940), taken from Васильев (1948).

By endowing the bear with a soul, and regarding him as a being akin to man, the Ket invested him with the capability of understanding the speech of all beasts as well as of man. Besides, they fancied that though the bear in summer was dull of hearing because of the rustling of leaves, in autumn or winter, however, it was very dangerous to speak ill of the bear or to boast of successful bear hunting. 'Should you speak badly of him one day or the other, and go hunting and find a good place, then a bear will rise from behind a tree suddenly and grab you with his paw' (Yelogui Ket).[10] When meeting a bear by chance in the forest, the Ket prevailed upon him not to hurt them and to go away. In this case a woman had to say: 'Go away, Old Man, I'm not guilty'. The Stony Tunguska Ket told me that once, while hunting for a bear, the dogs, frightened, ran to hide in the bear's den and found themselves behind him. The hunter did not fire for fear of wounding the dogs. Meanwhile the women were beseeching the bear: 'Old Man, don't hurt our dogs'. And when, mad with rage, the bear began to growl, they kept entreating: 'Quiet, quiet, Old Man, we have tender hearts'.[11] They talked to the bear while hunting after him, and even afterwards, at the feast, when 'feeding' the figure of the bear and 'guiding' him back to the forest (see below).[12]

A reflection of bear worship is the old Ket custom of settling disputes and exposing misdeeds. M. F. Krivošapkin writes: 'When an offence is committed among the Ostyaks, the suspect is given a bear's ear or a bear's tooth which he has to kiss and chew up into small pieces. If the Ostyak does all this without pleading guilty, he is sometimes . . . freed from suspicion . . . '[13]

V. I. Anučin has noted down the term *ķojubat* (*ķoi* 'bear', *bat* 'truth'), equivalent to the expression 'upon my word' or 'honest truth'.[14] G. Miller observed the same custom among the Ket as early as the first half of the 18th century: 'The Ostyaks since olden times used to eat pieces of bearskin to give evidence of their innocence; they hold that whosoever swears falsely shall not live another year, in that time the bear will devour him from within . . . They keep swearing secretly in this way, even though they are christened. If they are obliged to swear or take an oath before a Russian court, they also hold to this superstition and grab the bearskin . . . , fur outside, and the one who swears an oath has to touch it with his lips as if he were to kiss it, for they believe that also in this case the bear in the forest will not fail to punish him who swore falsely.'[15]

Highly interesting is the Ket custom of raising bear cubs. Cases of taming and rearing them were rather frequent. After killing the she-bear, they

[10] Field report of the author, Archives of The Institute of Ethnography of the Academy of Sciences of the U.S.S.R., No. 416, p. 34. (Hereinafter: Field report.)

[11] Field report, No. 435, p. 60.

[12] A similar attitude is shown towards the bear by the Evenki and the Sayan Altaic peoples; this is most manifest among the inhabitants of the forest regions of North Altai—Shors, Tubalars, whose ancient hunting culture was less influenced by the cattle-breeding Turkic peoples. Титов (1923) p.95; Потапов (1928) pp. 16–7.

[13] Кривошапкин (1865) p. 135.

[14] Анучин (1914) p. 20.

[15] L. P. Potapov took these data from G. F. Miller's German manuscript entitled 'Beschreibung des Krasnojarsker Bezirks der Jennissei-Provinz in seinem gegenwärtigen Zustand Anfang 1735', kept in the Central State Archives of Old Documents in Moscow (Vol. 199, No. 9, album 526, sheet 7).

took the cubs home. Cubs were kept usually in childless families. The Ket of our days say that the bear cub has to be cared for and looked after like children. The cub was called *yśkit* 'son' or 'daughter', and was really much cared for. It lived with the foster-parents in the tent, a bed was set up for it as for other people, and it was fed with what the family ate. The cub became a common pet: it was talked to and taken to visit neighbours in their tents, where it was always treated as a guest. The cub wore copper ear-rings and collar. The collar was changed every year as the cub was growing bigger and bigger. In some cases bracelets were put on its paws. For the hunting season the grown-up tame bear was taken to the taiga.[16] He was supposed to give warning if he smelled the presence of any wild bear. The Stony Tunguska Ket say that, sensing the wild bear approaching the hunter or his tent in the taiga, the tame bear grasps a piece of *tiski*[17], thrusts it into the fire and lifts it ablaze, standing on his hind legs. By doing so, as was explained by the Ket woman O. V. Tyganova (Sulomat settlement on the Stony Tunguska river), the bear, as it were, says to his master: 'I shall fight for you, and you make light so as not to miss the mark when you shoot'.[18] The bear cub was kept till three years of age, then it was set free in the taiga, wearing all its ornaments. The neighbours were informed of the release of the son or daughter lest they should chase it. The story is that tame bears sometimes came back to their masters, who received them willingly and, if need be, treated them for their wounds. The Ket never killed a bear they had reared.[19]

The cult of the bear among the Ket manifested itself mainly in a complex of ritual acts connected with bear hunting and the eating of the bear's flesh. Some of these customs have been observed in some places up to recent times.

Before describing the ceremonies related with the worship of the bear, we cite a Ket legend to help us to understand the whole affair.[20]

Once upon a time there lived in the forest a *ķaiγuś*—forest man, son of a woman by a bear.[21] He was six years old, but not yet seven, when he decided to marry and take a human girl for wife.[22] His father objected: 'You are still a child, you must grow up'. The boy insisted. The father then counted out seven steps and commanded the boy to jump to the eighth: 'If you can jump so far, you go to the humans; if you fail, you must not go, the humans will kill you'. Easy to jump for one who is tall, but the boy was small; he

[16] According to the Ket woman T. G. Korotkikh (Sulomai settlement on the Stony Tunguska river), in this case they put 'irons' on the bear's sides, but we know nothing else about this.

[17] Cooked and processed birch bark, used as roofing on the tent, and also on provisional sheds.

[18] Field report, No. 435, p. 61.

[19] K. M. Ryčkov (1923) noticed similar cases of taming the bear cub also among the Sym Evenki. Raising the captured bear cub, as is known, was characteristic of the Ainu and some Amur ethnic groups (Gilyak, Orok, Orochi, Ulchi). But this custom of theirs had specific elements distinguishing the entire ritual concerning the bear as a special aspect of the bear holiday.

[20] Taken down in Russian from the words of the Ket woman V. F. Dorožkina (Surgutikha settlement on the Surgutikha river). Field report, No. 437, pp. 1–4.

[21] It is interesting to note that other Ket legends speak of *ķaiγuś* 'forest hero', 'forest wench', whom no outward signs distinguished from human beings.

[22] Humans, as opposed to forest beings, in this legend are called *ķynereŋ* 'clean folk'

failed, yet he went to the human folk, bringing as gifts a kerchief and clothes. He walked on through the taiga and met some barondukis (*Tamias striatus*). 'Where are you going to?' they asked him. 'To the humans to marry.' 'Don't go; if you don't turn back, the humans will kill you.' The *ķaiγuś* paid no heed, got angry, killed the barondukis with his arrows, and went on. He met squirrels, who asked: 'Where are you going to?' 'To the humans to marry.' 'Don't go ...' and so forth. (The *ķaiγuś* meets also minks, ermine, sable, gluttons and wolves; the same conversation takes place with all of them, and, getting angry, he kills them.) The bear's son (*qoidəγyp*) came to the river where the tent of an old man stood. The daughter of the old man went there to fetch water. The *ķaiγuś* grabbed her and carried her into the taiga. The people became aware that the old man's daughter was gone only when they noticed the footprints of the bear. Next morning at daybreak the old man called the people together, they sharpened their lances and began chasing the bear. The hunt was started: the old man's daughter, wife of the bear, heard the chase going on: 'Father and the men are chasing us'. The *ķaiγuś* decided to offer gifts, pay ransom (*χoŋukś*)[23] to his wife's father, and put a heap of barondukis hides on the way. The men went on chasing the bear and saw the many barondukis hides lying there. They said to the old man; 'This husband of your daughter has left you a ransom, he is not a beast, why chase him?' The old man did not heed and continued the pursuit. The *ķaiγuś* heard the pursuers and left squirrel hides as a present to the old man. (Afterwards the *ķaiγuś* leaves there the hides of minks, ermine, sable, gluttons, and wolves.) The old man continued chasing, not yielding to any persuasion. The bear's son began to grow tired: 'I have left everything to them, and they will kill me all the same' he said. All he had now was the kerchief and the clothes, and he left them behind, too. The men tried to stop the old man: 'You see, it is a man, how many presents he has left you, don't chase him'. The old man did not listen to them and went on chasing. Then the *ķaiγuś* said to his wife: 'They will kill me, whatever I do, because they are scared. If they don't kill me, we shall walk like humans, not like forest beasts. If they kill me, tell them to cut off my right hand and to use it for divination in order to find out how many days shall sit and what kind of food shall I eat. I shall sit for many days, tell them what to give me to eat. Get them to depict me on birch bark and leave only my snout (the skin from nose and lips) of my body. Let them gnaw off my bones, bind my ribs, and carry away head and bones.' He said all this and went to meet the men. The old man killed him and, moving on, saw his daughter standing and weeping: 'Why did you kill my husband?' The old man walked away with his daughter, the people went home. They dragged the flesh along. They threw up the paws, and they drew on birch bark the shadow, the soul, of the bear. They thrust copper spits through the hands, legs and neck—he had said so himself. The girl did not eat of the flesh. The bear was fed for three days running: he was helped to *yukola* (dried fish) and duck's fat. The bones were gathered up, and the ribs bound with the bark of birch-cherry. Then they carried it all to the forest, left it on a tree-stump. When the shade fell on the earth, the bear's son betook himself to his father. On his way

[23] The Ket use the same word to denote the bride-price.

he met all the animals—barondukis, squirrels, minks, ermine, sable, wolverins, wolves—who then rebuked him for not having listened to them. He came to his father's tent (den), dropped down, and his bones were clanking.[24]

Bear hunting took place especially in the autumn months of the 'small round' (first half of the hunting season, from October to early December) when there was little snowfall. The bear was tracked down by dogs, the hunters following the trail and scratches on trees (made by the bear when sharpening his claws). Sometimes it happened that the hunter unexpectedly found a bear's den. Being on the right track, he marked the place, returned to the settlement and let the other hunters know about his good luck by hitting the snowshoes with his stick.[25] Coming to the tent, the hunter made a few jumps, waving his hands. By this he indicated that he had found a den.[26] The men came one by one. When all of them were assembled in the tent, the hunter began to relate how he had gone to the taiga that day, seen the trail, looked at it and found the tent of the Old Man (that is how the bear and his den are called allegorically). He told that the next day would be the time to pay a visit to the Old Man. The word 'hunting' was never uttered, the hunter was forbidden even to invite any one of the men to go with him. The participants had to volunteer. On the morrow some five to ten men went to the den (*kojo*). At a distance of about 200 to 300 metres they built a fire ('that was mandatory!'); in winter the fire was built nearer to the cave.[27] By the light of the fire they fixed their spears, went closer to the cave and addressed the bear in these words: 'Come out, Old Man! Come out and run like the elk!' They goaded the bear with the spears. When the beast still did not come out, they started pushing fir branches or stumpwood into the cave, leaving enough room for the bear to put out his head. Sometimes (when they could in no way lure forth the bear) they filled up the entrance with earth and then dug out the cave again. When they killed the bear they kept telling him not to be angry, inviting him to be their guest.[28] As a rule they hit the bear on the head. After killing him they pulled the carcass nearer to the fire so that the 'bear's eyes may not see his den'.[29] The Stony Tunguska Ket have an

[24] The Ayan Evenki, the neighbouring Evens, and the Orochi also have a story about a woman who happened to give birth to a bear cub in a cave. Her second son, still a young boy, measuring strength with his bear brother, killed him, and the latter, while dying, bequeathed to the men the ritual of the bear hunt and of the burial of the slain bear. Василевич (1957) p. 168.

[25] In the same way, but by striking in a more rapid rhythm, the hunter announced the tracking of other wild game—the elk. The quick blows were the signal for the men to assemble fast before the elk had gone too far.

[26] Assembling for the hunt was accompanied by peculiar pantomimes and dances also among the Mansi, Khanty, Chukchi and Koryaks. Васильев (1948) p. 82. The hunters returning with the slain bear were met by the Evenki jumping and cawing, imitating the crows. Василевич p. (1957) 168.

[27] Fire worship is characteristic of the Ket. Going back, like bear worship, to the ancient tribal cult, it manifested itself in the 'feeding' of the fire, in the placing of the smouldering logs in a definite position on the fire (in the form of a slanting cross, by no means transversely), in a series of taboos concerning the fire, in the use of fire for 'cleaning', etc. Some elements of fire worship can be traced in the cult of the bear as well.

[28] The treatment of the bear (slain or alive) with apologetic or welcoming words has been observed among many peoples of Siberia. Васильев (1948) p. 84.

[29] The custom of dragging the bear's carcass away from the cave existed also among the Altaic peoples.

old story that once a hunter killed a bear, failed to pull away the carcass from the cave, and went back to his mates. Soon thereafter he recalled that he had left the fire-iron at the den, returned, and by that time the bear had come to life. The hunter, beaten up mercilessly by the bear, could just reach his companions and said dying: 'You will live longer if, after killing the bear, you pull him farther away from the cave lest he should see it, or else he comes to life and gets angry'.[30]

As B. O. Dolgikh learned from the Stony Tunguska Ket, they used to slash the slain bear with a small rod.[31]

Before skinning the bear they chopped off (at the wrist) the right foreleg. We have already said above that, in the belief of the Ket, the bear has a soul (or rather a combination of souls). The bear's soul is that of a dead man, who appears to his relatives in the bear's shape. There was a special Ket custom for finding out which of the dead relatives came to them in the shape of a bear. To find this out they threw up three times the cut-off right paw of the bear, and asked the question each time. According to B. O. Dolgikh, the Stony Tunguska Ket hurled the paw at the bear's underbelly, at its right groin. For an affirmative answer the paw had to fall palm up. The first to find out was usually the clan to which the dead person belonged. A representative of each clan threw up the paw three times to know whether the guest had come from his clan. On the Stony Tunguska, where the entire Ket population belonged to two clans, they asked: 'Who are you, *bogdejget* or *ķeńtān?*' This being ascertained, they looked for some special mark on the bear (a certain missing tooth or finger, a crooked claw, a scar, etc.) by which they tried to determine who was before them. If the marks coincided, but the paw still did not fall palm up, they said, 'You still want to deceive us', and continued guessing.[32] They guessed at all relatives, not only at grandfather or grandmother, as was reported by V. I. Anučin.[33] Sometimes, as is claimed by the Ket, it 'proved to be' a very distant relation, who had died very long before and who was not even expected to come back. By tossing up the paw they also determined how many days (usually three) the relation wished to stay as a guest. The host, the killer of the bear, did not go hunting during those days. It is interesting that, when this custom is observed nowadays, the hunter, unwilling to miss the precious time of hunting, directly proposes that the bear stay for one day or even says point-blank: 'You, Old Man, you don't want at all to stay with us as a guest'. Formerly the paw-tossing served also to settle disputes about who had been the first to trace the bear ('To whom did the Old Man show his house?'). Later the right paw was preserved as a talisman.[34]

[30] Field report, No. 435, p. 56.

[31] Cf. I. A. Evsenin's data cited by E. Titov concerning the custom, among the Siberian Turkic peoples, of stroking the slain bear's skin the wrong way with a fir twig. Титов (1923) p. 101.

[32] The Ket custom of guessing is still observed to a certain extent. In the family of V. F. Maksunov, who in the summer of 1958 lived at the estuary of the Pakulikha river, I was told that a bear they had slain a few years before 'proved' to be their departed brother. By throwing up the paw and asking questions they 'found out' in what circumstances he had died.

[33] Анучин (1914).

[34] The custom of preserving the right paw of the bear existed also among the Sym Evenki (Рычков [1923] p. 112) and the Shors (Потапов [1928] p. 19).

The ritual ceremonies over the slain bear among the Ket consisted in a definite sequence of dissection of the carcass. The carcass was ripped up, the head and paws cut off, the skin flayed, and the internal organs taken out. Stomach, lungs, and bowels were buried on the spot.[35] In no case were dogs permitted to lick up the blood or eat the entrails.[36] The body was then cut into pieces—back, ribs and shoulders taken out intact. Cutting of these bones was forbidden.

The Kureika Ket told us of the custom of taking the youngsters on the bear hunt. While skinning the bear, the hunter cut out a bit of the heart, dipped it in blood, and the boy to whom it was given ate it and said: 'You are strong, my father who brought you down is stronger than you; I shall be still stronger, my heart will be stronger than his'. Often the boy was made to get astride the still stirring animal, clutching its ears.[37] When the carcass was carved up, the heart and liver, as well as some fat, were cooked in a copper kettle on the fire laid near the cave prior to the hunt. Every participant of the hunt was given from the spit three bits of fat and a bit of heart and liver. The hunters ate right there, and took the fat with them. They returned home dragging the dissected body of the bear on sledges drawn by dogs or by the hunters.

There was among the Kureika Ket a peculiar custom by which, at the place of killing, they stuck a post in the ground, and every hunter carved in it his sign (*tamγa*)—a little cross, a slanting cross, a cross-like figure in the shape of four triangles with the tops turning towards one and the same point.

In the Ket bear feast we can distinguish three phases: (1) the cooking of the flesh, its distribution and the general celebration in the tent of the killer of the bear; (2) the making of the bear's image (or, instead, the use of the cut-off head, skull or the skin of the bear's nose and lips), the feasting of the bear's 'soul', and magic performances with a view to ensuring success in hunting and the people's well-being; (3) the farewell to the 'soul'—the image and the bones—of the bear.

The rituals took place with the participation of the entire population of the settlement. During those days the killer of the bear and the members of his family did not hunt. Neither did the neighbours on the first day and the last day of the farewell festival, or even all three days.

The number of feast days, determined by predicting from the bear's paw, was carefully observed. The holiday began the day after the carcass had been brought home. In the early morning in the tent of the killer of the bear they began to cook, in a big copper kettle, the main parts—head, breast, ribs, back and paws. They severed from the head the lower jaw, which was cooked in the same kettle.[38] Cooking in more than one kettle

[35] Cf. the custom of the Altaians' burying the heart when dissecting the carcass. Потапов (1928) p. 21.

[36] Among the Evenki it was forbidden to shed the blood of a slain wild beast (elk, wild reindeer, bear). Василевич (1930) p. 64.

[37] Cf. the custom of the boy's fighting with the bear's skeleton among the Evenki of the Lower and the Stony Tunguska basin, as well as the already mentioned mythological theme of the fight between brothers—the boy and the bear. Василевич (1957) pp. 167–8.

[38] The Kureika Ket cooked the head after the three feast days.

at the same time was forbidden, so that when a very big bear was captured it had to be cooked by portions, but always in the same kettle. The bear's flesh was cooked by men only.

When the flesh was ready, the whole community (three to five families), men, women and children, regardless of kinship, gathered in the host's tent. All those present had to consume at least a morsel of the main parts; this was a particularly strict rule. Of the bear's head, however, only the men might eat. Each male participant was obliged to touch, at least with a finger, all edible parts of the head and to besmear his lips. It was taken for a bad omen if anyone of the family of the host fell sick from the bear's meat. Besides the general celebration in the host's tent, every family received a portion of the cooked meat, which the guests took home in flat, trough-shaped wooden dishes called *syq*. Part of the meat, and especially the smoked lard which did not deteriorate, belonged to the host's relatives, no matter how far away they lived.

Fig. 1. Anthropomorphic representation of the bear. Stony Tunguska Ket (Sulomai) settlement, 1948

The collective consumption of the bear meat by the kinsmen has survived among the Ket until the present day. The local Russian population is often astonished at the rapid disappearance of the bear's flesh among the Ket, not knowing that the bear's host is obliged to share out the game among his numerous relations. On the first opportunity the host dispatches the 'presents', or carries them often personally, as far as 50 to 70 kilometres.

The second event of the holiday was the feast offered to the bear himself. The Stony Tunguska Ket started the feast on the first night. For this event they brought the bear's flesh out of the tent and covered the floor of the tent with fresh fir twigs (*koniŋ*). Sometimes they also freshened up the birch-bark beds (*qojatl*). The low tables (*l'am̓*) on which the Ket placed food were carefully scraped down with knives by the women. All this was done in order to remove the 'ghost' of the cooked bear's flesh.

The host took a cleaned panel of birch bark (15 to 20 cm long) carved in oval shape, on which he drew the bear's figure[39] with charcoal or with a powdered and moistened material. According to B. O. Dolgikh, the Stony Tunguska Ket drew a human figure on the panel, which they called *detpinisnim* (Fig. 1). The bear's image was to the Ket the 'shade' or 'soul' of

[39] V. I. Anučin mentions such images executed by carving or painting on small pieces of wood. Анучин (1914) p. 13.

the bear (*ul'vej*). Judging from copies of such figures on birch bark received by K. Donner (1912) and B. O. Dolgikh (1948) from the Stony Tunguska Ket[40], or from the figure made for me by an old Ket living in the village of Surgutikha, all they were of the same type (Figs 2a, b, c).[41] Fastened on the neck and four legs of the figure of bear drawn on the birch bark were thin strips of red copper (cf. the custom of putting copper bracelets and collars on the tame bear).[42] The birch bark was folded lengthwise and covered on top with an oval piece of birch bark folded into equal parts.[43] The two pieces of birch bark were tied with bast or fastened together with four small wooden pegs.

At the place of honour (by the entrance) of the tent (*koteŋ*), near the foremost post (*koteŋdaŋ*), stood a sort of small chest (*ańal*) of fresh-planed cedar boards (the name was noted down by B. O. Dolgikh in the Stony Tunguska region). The chest was left open on top and in front, its boards were put together without nails.[44] The birch bark displaying the figure was put in this chest, the bear's head pointing to the open frontal side, in front of it the snout—a piece of dried skin from the bear's forehead, nose and lips,[45] and behind it the gall and, according to K. Donner, the eyes and the penis of the bear.[46] In the chest was put also a thin stripped cedar twig bent like a ring and tied at its ends (Fig. 3).[47] From the tied-up ends of the twig was suspended a bit of fish-glue (*χit*), and on both sides was fastened a bunch of thin sticks (according to the Ket of the Stony Tunguska, twelve)—the 'ribs' of the bear. Hanging from the ring were a bit of the bear's tongue and a small piece of copper.[48] The ring symbolized, as it were, something holding together the parts of the body.

Copper—the bear's gold—was needed, in the fancy of the Ket, to solder its bones, and the glue to paste together the parts of the body. The Ket of Surgutikha call the cedar ring the 'sledge' of the bear, as if the bear were going to the humans to ask for glue and copper to take home. They explained that it was absolutely necessary to give the bear copper and glue, because

[40] Donner (1933) p. 97.

[41] Sometimes they drew on each side seven slanting strokes—representing the ribs.

[42] On the drawing K. Donner saw they were marked by grass threads. Donner (1933) p. 96.

[43] K. Donner (1933) says that a black stripe bordered the edges of both pieces of birch bark.

[44] The cedar was treated by the Ket with particular reverence. We had occasion to see young cedar trees standing at the place of honour before tents. The Ket call them *qcqścqś* 'sacrificial tree' (*qcqś* is the name for the stuff hanging from trees and other gifts). The worship of the cedar was reflected also in later shamanism—the basic attributes, drum and drumstick, had to be made of cedar wood. Some traditions of the bear festival are also related to the cult of the cedar: the chest was of cedar wood, the ring (see below) was of cedar twig, the bear's figure was hung on a cedar tree when the festival was ended.

[45] The custom of cutting out the skin of the bear's nose was observed also among the Khanty, Tuva, Yakut and other peoples of Siberia, as well as the Sam in the North of Europe. Васльев (1948) p. 88.

[46] Donner (1933) p. 96.

[47] The exact place has not been ascertained; in one case they said it was placed on top of the birch bark, in another case—beside it: K. Donner states that it was placed at the edge of the image. Donner (1933) p. 96.

[48] The custom of cutting out the tip of the tongue existed also among the Tubalars. Потапов (1928) p. 19.

Fig. 2a. The bear's image on birch bark. Stony Tunguska Ket, 1912

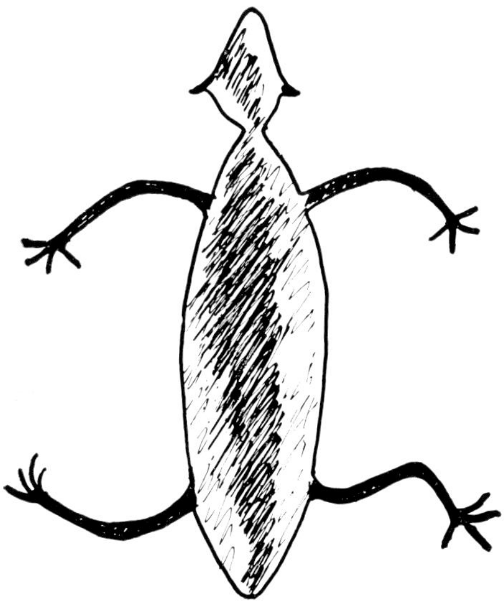

Fig. 2b. The bear's image on birch bark. Stony Tunguska Ket (Sulomai) settlement, 1948

Fig. 2c. The bear's image on birch bark. Surgutikha Ket (Surgutikha) settlement, 1958

the bear would tell the other bears about it to, so that they also should come to the men. Another variant holds that, when after the festival the bear's image is seen off to the forest without copper and glue, then the bear shall come back or send another bear to get the stuff.

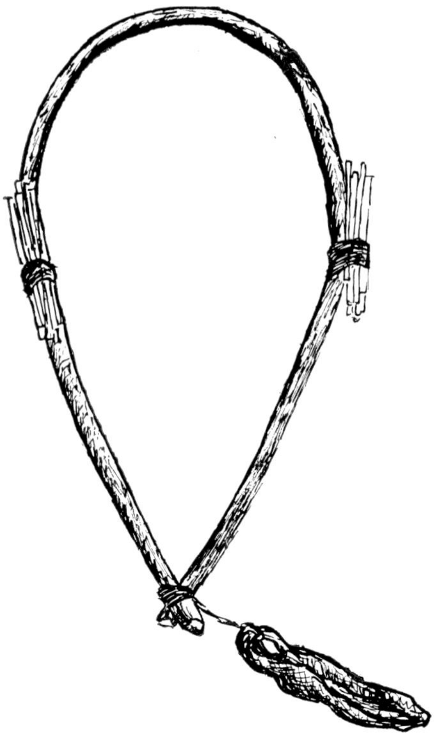

Fig. 3. Ring of cedar twig (made at the author's request by the Ket P. I. Tapkov at Surgutikha settlement, 1958)

During the entire festival the chest stood in a place of honour; for the night it was covered with a white cloth. When all guests were assembled, the host started to tell his story beginning from the day of tracking the bear. He related how he had gotten up that morning, what he had eaten, what he had done before going out to hunt, what preparation he had made and how he had started out, what clue had led him to the 'tent' of the 'Old Man', and that he had decided not to go farther, but to come back for a visit next day, etc. The story-teller sat close beside the chest containing the image, before him stood a *χittyn* (*χit* 'glue', *tyn* 'kettle')—a device for warming up the fish-glue. The *χittyn* is a rod with a funnel-shaped piece of rolled-up birch bark stuck on its cleft upper end. When the story was started the bear's raw eyeballs were put into the funnel. A cedar twig was sharpened to form a small spear 20 to 25 cm long. Getting to the point in the story when the hunters had roused and killed the bear, the narrator suddenly broke the lance, stuck it into the earth floor, jumped to his feet and, shouting '*uχəj!*', swallowed the bear's eyes.[49] The *χittyn* and the broken spear were thrown into the fire. It was held that only a long-lived man could swallow the eyes, and choking over it was regarded as a very bad omen.

Thereafter the tent's owner related what the bear looked like, imitating it with gestures and growling sounds. Meanwhile he held the bear's snout as a mask before his face.[50] While imitating the bear, the hunter whistled like a baronduki (the Ket believe that the bear can whistle like that and does so to catch small game) as though he were calling on the 'animal folk' to come to the humans, too. The hunter time and again took off the 'snout' and warmed it over the fire.[51] The snout was handed to every man in turn, and the men had to repeat the performance. They asked the bear to ensure them, in addition to success in hunting (the bear—the 'animal king',

[49] The taking out of the bear's eyes is known from the rites of several peoples of Siberia. The swallowing of the eyeballs was observed among the Sym Evenki (Василевич [1957] p. 167) and also among the peoples of the Sayan Mountains (Потапов [1928] p. 20).

[50] In another variant the hunter tells the whole story wearing such a mask.

[51] K. Donner writes that it was thrown into the fire and snatched out of it. Donner (1933) p. 97.

in the fancy of the Ket, was able to secure successful hunting), well-being and deliverance from sickness.[52]

Then the 'snout' was put back in its place, and in front of the chest was put a small table, either an old one or a newly made table specially prepared for the purpose. By no means was it permitted to use a table into which iron nails were driven; a piece of copper was fixed on the board (often percussion caps were used for the purpose). According to the Ket of Surgutikha, instead of a spoon they placed on the table a copper plate[53], seven glasess of tea (the bear had as many souls), fish-fat, bread, sugar, etc.[54], all these for the bear. Those attending sat by the other tables. For a certain time no one ate, because the guest was supposed to be eating. Then the host took a sip of each glass. Six glasses were given to the oldest among the attending relations of the bear's host, the seventh was left on the table of the bear. Then the general feasting began. Finishing the feast, they began to guess whom the bear would choose to eat up after him. They guessed at the members of the host's family, throwing up three times the cut-off right paw of the bear. It was held that the one to whose lot it fell to finish eating with the bear would be safe from bears in hunting. The Ket woman O. V. Tyganova (Sulomai settlement on the Stony Tunguska river), who told us about the feast celebrated when a bear had been killed by a member of her family, was very sorry that the honour of eating up with the bear had fallen on her son, a little boy at the time, and not on some grown-up member of the family. On many days following the festival the table stood before the chest. At the time of eating, the food was also served to the bear; after him ate the same man. According to B. O. Dolgikh, with the Stony Tunguska Ket it was always a boy who had to drink up after the bear.

The Kureika Ket have informed us that it was not the bear's image but the cut-off head that was honoured at the feast.[55] On the third day they skinned the head, severed the lower jaw and cooked all in one kettle. The entire male population of the settlement assembled to partake of the head, and no one would miss a morsel of every part.[56] Ket people who now live on the Pakulikha river told me that they did not draw an image of the bear but they put into the chest only the 'snout'. According to N. K. Karger's information quoted by B. A. Vasiljev, the Ket put there the

[52] The same imitation of the bear, the peculiar pantomimes at the bear festival were observed also among the Mansi. It is characteristic that the Mansi hunters while doing so also put on masks, though not of the skin taken from the bear's nose and lips, but made specially for the purpose. V. N. Černecov's manuscript; Васильев (1948) Fig. 1.

[53] Copper played among the Ket a special role, which appeared also in the cult of the bear. Cf. the copper ornaments put on the tame bear and its figure, the use of a copper kettle for cooking bear's flesh, etc. Many loops on the shaman costume were also made of copper. It should be noted that the Ket used the copper dust (scraped off ordinary copper utensils—handles of tea-pots, kettles, etc.) for soldering fractured bones.

[54] K. Donner mentions also bear's fat among the foods. Donner (1933) p. 97. As is confirmed by the Stony Tunguska Ket, the bear's figure was never treated with bear's flesh.

[55] Entertaining the head at the bear festival is observed also among the Khanty, Mansi and peoples of the Far East. Васильев (1948) p. 92.

[56] Obviously, it was the entertaining of the bear's head among the Ket that V. I. Anučin had in mind when he remarked that 'the bear's head is separated as a whole and then serves for a special ceremony'. Анучин (1906) p. 47.

bear's skull, too.[57] This is supported also by the notes of B. O. Dolgikh who says that Ket of the Stony Tunguska sometimes put into the chest, instead of the image, the boiled-out skull with the lower jaw tightly fastened on it.

To see the bear off, again the entire community assembled and a feast was arranged. They asked the bear not to be offended by the bad treat, to send other bears to them, and to fulfil their wishes. As is evidenced by Ket of the Stony Tunguska, the host took copper and glue from the figure and the ring for the bear to come back to get his 'abandoned property'. Before bringing out the figure, they 'cleaned' it—by putting on it the cedar ring, then carried it twice around the fire in the direction in which the sun was supposed to move.

According to B. O. Dolgikh, among the Ket of the Stony Tunguska the bear's host put on gloves, placed the image on the base board of the chest (the sides had been taken away in the meantime), took it in hand and walked around the fire seven times. After that, he carried away the image to the taiga. There, in a secluded place, he selected a young cedar tree and with an axe made a notch on its 'morning' (east) side. The ring and the birch bark with the image were put into the notch or suspended from a branch. The bear's skull was also carried to the forest (if, instead of the image, the head had been honoured at the feast), with the lower jaw tightly fastened on it, and was put on a broken branch or a post staked into the ground.[58] The bones of the bear were carefully gathered; relatives and neighbours, having eaten the flesh, took the bones to the host's tent. All bones were collected, the ribs (the right-side ribs apart from the left-side ones) were bound together with bast of bird-cherry-tree bark. The host, wrapping up the bones in birch bark or sackcloth, carried them to the taiga to 'bury' somewhere in the hollow of a cedar or to dig them into the ground.

The birch bark, straightening with the passage of time, fell to the ground. In the Ket fancy that was the moment of 'resurrection' for the bear to begin a new life.

*

V. I. Anučin, who collected ethnographic material from the Ket in 1905–1908, noted that 'the kind of ritual common to the other aboriginal peoples of Siberia does not exist under the name of "bear festival" among the Ostyaks'.[59]

The material quoted above bears evidence to the contrary of this conclusion. The cult of the bear among the Ket definitely comes to light in the manifold rites observed during the bear hunt and at the celebration of the killing of the bear; this complex of rituals is rightly called the 'bear holiday'. Various elements of the festival—the first and closing phases—are to be found in ceremonies related with the cult of the bear also among other hunting peoples. The first phase (the cooking of the flesh in common, its carving up and the general feast) is a manifestation of the collective form of hunting after big wild game in the past. The closing phase of the festival (farewell to the bear's image and bones) is a brilliant manifestation of the

[57] Васильев (1948) p. 92.
[58] I received these data from Kureika and Stony Tunguska Ket.
[59] Анучин (1906) p. 47.

idea of the rebirth of the slain animal, which characterizes all hunting peoples of Siberia and which can be traced back to the magic rite of reviving the animal. This idea applied to all wild game (elk, wild reindeer, etc.) and was connected with man's striving to secure his food for later times. This also explains why the Ket observe the taboo prohibiting the cutting of the bones of the bear, the custom of cooking of the flesh in one kettle, of placing in a chest bits of the tongue, the penis and the gall of the bear together with the cedar-twig ring and 'ribs', glue and copper to 'solder' and 'glue' the bones of its body. The scenes of imitating the bear, calling the beasts and asking for well-being, all mirror elements of the ancient magic of hunting.

Bear worship among the Ket goes back to the ancient clan cults. Proof of this is, for example, the obligation of sharing out the meat among the relatives. The custom of making the *tamγa* in front of the bear's den must also have been observed by the clan as a whole. The festival became a community event with the participation of all neighbours, not only relatives, and this was evidently due to the disappearance of the clan spirit and to reorganization into territorial communities.

Characteristically, bear worship was coupled with other ancient cults, such as the worship of fire and trees, which goes back to common nature worship.

Some specific traits of the worship of the bear among the Ket, such as the designation of the bear (usually the name *qoi*, used also for 'step-father', 'stepmother') and the custom of raising the bear cubs, can be regarded as a confirmation of the totemistic origin of the cult.

B. A. Vasiljev, who specially studied the bear festival, points out two stages of its development. The first and very ancient stage, called 'Eurasian–American' by the author, is characterized by the following ritual elements: tabooed words and false names, the rituals of assembling for the hunt, the pantomimic bear dances to ensure the successful outcome of the hunt and the multiplication of beasts, the apologetic speeches, the ceremony over the bear's carcass in the taiga (e.g. the ritual act of skinning, the removal of the eyeballs and bones, genital organs, ears, nose, etc.); the phallic ceremonies, the ritual cooking and eating of the bear's flesh, the preservation and the ritual burial of all bones of the bear's skeleton and skull; the fancy of reviving the slain bear and the basic feature of the bear's myth, according to which the bear is a kinsman, a 'metamorphosed' man (or woman), master of the taiga and the mountains, or is in close relationship with anthropomorphic beings of this kind. The second stage, of a more recent origin, the characteristic traits of which are the shutting up of the bear in a cage and its sacrifice accompanied by a festival lasting several days, is called the 'Ainu' stage by Vasiljev.[60]

As can be seen, the afore-mentioned elements of the 'Eurasian–American' stage are characteristic of the Ket bear holiday; what is more, we can say that the most complete manifestation of the event is found just among the Ket.

Bear worship among the Ket revealed also specific features: guessing from the cut-off paw, rites connected with the making of the image, the use of the skin of the bear's nose and lips as a mask, etc. And it is precisely in

[60] Васильев (1948) p. 103.

these peculiarities of the Ket bear holiday that we can recognize the most archaic elements, which other peoples have not preserved.

It is tempting to see a manifestation of different stages of the development of cultural customs in the fact that at the festival among different Ket peoples the bear was personified by its head or skull, or by the skin of nose and lips, or finally by the drawn image. This, however, is only a hypothesis which needs further confirmation.

The similarity of the particular elements of bear worship among the Ket and other peoples of Siberia certainly bears evidences of a widespread hunting culture. The similarities between this cult and some other related ritual customs (notably those among the Evenki of the Sayan Mountains which show some traits similar to the bear holiday's of the Khanty and Mansi) must be taken into account in the investigation of the ancient ethnic and cultural ties between these peoples.

The ancient cult among the Ket has changed into a tradition preserved in everyday life. For a long time past the Ket have not observed all customs when assembling for the hunt and during the chase, nor the custom of sharing out the flesh of the slain bear, and even the strict sequence of the festival is violated. Certain traditional ritual features have longest been preserved by the most isolated Yelogui Ket and the Ket of the Stony Tunguska. It must be said that the bear holiday in the past was practically the only occasion for the Ket to celebrate in common. At the present time the Ket willingly talk about the holiday, just as about the old custom and, particularly, its amusing side. When speaking of the religious purport of the rites, even elderly people often make a reservation: 'This is how our grandfathers believed'. Young people everywhere either know nothing of the custom or talk about it as a sort of fun. A rare and happy occurrence to the hunters—the killing of a bear—is today celebrated by the Ket with a liberal feast given by the host, with the distribution of bear's flesh among the relations, with hunting stories, and general merriment.

The investigation of the evolution of the ancient cult, and its transformation into a customary entertainment which has lost its religious meaning, is of great importance for a study of the cultural development of the Ket—one of the less populous and less studied Siberian nationalities.

REFERENCES

(Anučin, V. I.) Анучин, В. И. (1906) Предварительный отчет о поездке к енисейским остякам в 1905 г. (Preliminary Report on a Journey to the Yenisei Ostyaks in 1905), in *Известия Русского комитета для изучения Средней и Восточной Азии*, No. 6.

(1914) Очерк шаманства у енисейских остяков (Treatise on Shamanism among the Yenisei Ostyaks), in *СбМАЭ*, Vol. II, No. 2.

(Castrén, M. A.) Кастрен, М. А. (1860) Путешествие в Сибирь в 1845—49 гг. (Journey into Siberia in 1845–1849), in *МЗП*, Vol. VI, Part 2.

Donner, K. (1933) *Ethnological Notes about the Yenisei Ostyaks*, Helsinki.

Dyrenkova, N. P. (1928) Bear worship among the Turkish Tribes of Siberia, in *Proceedings of the Twenty-third International Congress of Americanists.*

(Krivošapkin, M. F.) Кривошапкин, М. Ф. (1865) *Енисейский округ и его жизнь* (The Yenisei District and Its Life), Saint Petersburg.

Nahodil, O. (1954) *O původu naboženství* (On the Origin of Religion), Prague.

Paproth, H. J. R. (1962) Das Bärenfest der Ketó in Nordsibirien in Zusamenhang gebracht mit den Bärenzeremonien und Bärenfesten anderer Völker der nördlichen Hemisphäre, in *A*, Vol. 57.
(Potapov, L. P.) Потапов, Л. П. (1928) Пережитки культа медведя у алтайских турок (Vestiges of the Cult of the Bear among the Altaic Turks), in *ЭИ*, Nos 2–3.
(Ryčkov, K. M.) Рычков, К. М. (1923) Енисейские тунгусы (The Yenisei Tungus), in *З*, Vols III–IV.
(Titov, E.) Титов, Е. (1923) Некоторые данные по культу медведя у нижнеангарских тунгусов Киндигирского рода (Some Data on the Cult of the Bear among the Lower Angara Tungus of the Kindighir Tribe), in *СЖС*, Vol. I.
(Vasilevič, G. M.) Василевич, Г. М. (1930) Некоторые данные по охотничьим обрядам и представлениям у тунгусов (Some Data on the Hunting Rites and Beliefs of the Tungus), in *Э*, No. 3.
(1957) Древние охотничьи и оленеводческие обряды эвенков (Ancient Customs of Hunting and Reindeer Breeding among the Evenki), in *СбМАЭ*, Vol. XVII.
(Vasiljev, B. A.) Васильев, Б. А. (1948) Медвежий праздник (The Bear Festival), in *СЭ*, No. 4.

Survivals of Fotemism in the Ancestor Cult of the Kazan Tatars

by

G. V. Jusupov, Kazan

The ancestor cult and, in general, the social and tribal relation of the Kazan Tatars constitute a comparatively neglected field in the ethnography of this region. Thus, as against the vast literature dealing with material culture (chiefly the works of the eminent scholar of the Tatar Autonomous Soviet Socialist Republic, Professor N. I. Vorobjev), scarce are the works devoted to the study of the intellectual culture of this people or to the religious observance of their ancestors, in particular. In the pre-revolutionary literature only a few authors paid attention to this field of research, as e.g. K. Nasyrov,[1] D. Koblov,[2] I. Sofijskij,[3] etc.

What has been done so far for this region fails to satisfy the requirements of descriptive ethnology, and it makes a quite insufficient contribution to the research carried on by modern historiographers especially as regards the origin of the Kazan Tatars, for the elucidation of which it is indispensable to study the intellectual culture of the ancient inhabitants. A rather neglected domain of Soviet ethnography is the study of the origin of ancient religions, particularly '. . . the problems of totemism are still far from being solved and still need considerable attention'.[4] Investigations in this field help us fight against ideas alien to Marxism, and also to awaken part of the people to self-consciousness by showing them that religion is a remnant of the odious past, that 'the helplessness of the exploited classes in the struggle with the exploiters invariably engenders the belief in a better life in the hereafter—just as the impotence of the savage in his struggle with nature engenders the belief in gods, devils, miracles, etc.'[5] and that 'all mythology defeats, tames and shapes the powers of nature in imagination and with the help of imagination; consequently it vanishes as soon as man subjects these powers to his domination in reality'.[6]

The study of the survivals of tribal relationships and particularly of the ancient totemistic beliefs is now becoming more and more difficult among the Kazan Tatars since these survivals have faded away in the minds and disappeared from the daily life even of the old people, let alone of youth.

The fact that so little has survived of the tribal totemism may be explained by the thousand-year-old Islamic rule which has always worked against any manifestation of paganism, and also by the cultural revolution that has almost put an end to these survivals in the formerly backward nationalities of the U.S.S.R. Therefore, relevant ethnographic material can be gathered with great difficulty and only from the old people.

1 Насыров (1880).
2 Коблов (1914) pp. 415–70.
3 Софийский (1878).
4 Толстов (1946) p. 11.
5 Ленин (1935) Vol. VIII, p. 419.
6 Marx.

Not so long ago survivals of animal cult and totemism could be observed among such Turkic nationalities as the Kirgiz, Kazak and Turkmens as well as among the Yakuts and the Altais. For instance, 'the elements of the cult of the lark, whose name is borne by one of the Turkmens clans can be traced among the Kirgiz . . .'[7]

When discussing the survivals of the animal cult and totemism among the Kazan Tatars, it should be stressed that the friendly attitude of the Moslems towards certain animals originates to a certain extent in their dogmas connected with pigeons and spiders, for example, which are considered sacred, because they helped Mohammed to escape when persecuted by the adversaries of Islam. This seems to have been added by later feudalistic ideology.

Traces of totemism among the Kazan Tatars can be detected in (a) the different kinds of prohibitions (taboos), (b) popular medicine and folklore, (c) ceremonies and customs, (d) the archeological material of the region and (e) local toponymy and ethnic nomenclature.

Two kinds of taboos can be distinguished: (a) taboos referring to the animals and plants themselves, and the totemistic subject in general, and (b) taboos connected with the names of animals.

During our 1947 expedition we heard in several villages of the Kzyl-Yul and Atninsk districts that an elk (*Alces alces*) must not be touched by anybody. They mentioned cases when this animal had made its appearance within easy reach of the people of the village and it was not touched by anybody. To kill an elk is considered an injurious act. According to the sayings of the local inhabitants, the village 'Pshanger' in the Kzyl-Yul district owes its name to the fact that an elk (*pši*) was killed there. Therefore its inhabitants had a bad reputation in those parts and were even treated with contempt. In certain places (B. Artash, Kzyl-Yul district) the killing of an elk used to be as great an offence as killing a man. The female elk is called *pši*, while the bull is called *bolan*, and by these two designations some people mean practically two different kinds of animals.

We heard of prohibitions also connected with other animals such as cranes, starlings, swans, frogs, ants, etc.

A most characteristic taboo regarding animal names is the prohibition concerning *qujan* 'hare', which word must not be pronounced when lambs or sheep are present, because they are thought to become sick and perish in consequence. In order to avoid this, it is customary to use fictitious names of which several exist: *qyr täkäse* 'field ram', *ozyn qolaq* 'long-eared', *čal täkä* 'grey ram', *priχ* and *qyrtyj*.

The taboo among the Slavs forbidding the utterance of the name hare is also mentioned by D. K. Zelenin.[8] The Carpathian Ukrainians were afraid to pronounce the name of this animal in the presence of children lest they should become drowsy. The well-known term косой is now a sort of pseudonym for the hare in Russian.

In connection with the name *čal täkä* some authors point out[9] that the Kazan Tatars call the hare by that name for fun. This may be so now. But

[7] Толстов (1931) p. 260. Recently, however, significant works in this line have been published by A. F. Anisimov, D. E. Khajtun, etc.

[8] Зеленин (1929) p. 73.

[9] Рахим (1923) p. 74.

as all pagan ceremonies, in the course of time (parallel with the dissolution of the tribal structure) it lost its meaning. Originally, however, *čal täkä* was the taboo name of the hare. Besides, there are many other such pseudonyms, or euphemisms, as for instance, *qyr aty* 'field horse elk', *qyr syery* 'field cow', *urman syery* 'cow of the woods'; we may include also the words: *qort, qačqyr* 'wolf', *alabuya* 'perch', etc. *qort* meaning 'wolf' still occurs in some Turkic languages[10], but it is no longer used in Modern Kazan Tatar, except in *qortqart* and *qortqa* 'old man, old woman'. Considering the two meanings of *qort* ('wolf' and 'old man') we may safely conclude that the totemic name 'wolf' is tabooed. Similarly the names 'old man', 'old one', 'grandpa', 'granddad' were applied by some Slavic and Turkic peoples to the bear. Perhaps the ethnic name *bašqort*, composed of *baš* 'head' and *qort* 'wolf', may be explained in this totemic sense.

The Turkic name for the perch—*alabuya* literally: 'dappled bull' (*ala* 'dappled, streaked', *buya* 'bull')—should be interpreted in a similar manner.[11] Therefore, obviously also the term *buya* in the Khazar denomination *aq-bugin ulus*, i.e. 'settlement (or clan) of the white bull' should be considered a totemistic-tribal survival.[12]

In the Turkic languages, including Kazan Tatar, the names of some animals end in *an* as, for instance: *arslan, syrtlan, bolan, jomran, tuqran, sazan, čajan, sualčan*, etc. The second member of these names still exists independently in some Turkic languages, but not in Kazan Tatar.[13]

Even today in Altai Turkic *an* means 'wild animal', *anya baru* means literally 'go on wild animals' i.e. hunting. Hence, the meaning of the word *arslan*, consisting of the root *arsyl* 'wild, predatory'[14] and the ending *an* 'animal', is 'wild animal'[15] or 'beast of prey'.

syrtlan—root *syrt* 'mane'—an 'animal with a mane, viz. hyena'; *tuqran*—root *tuq* 'pecking'—'a pecking bird, woodpecker'; *sazan*—root *saz* 'marsh'—an 'animal living in the marsh', i.e. a marsh fish.

This is also true in the case of the word *bolan*, from the root *bol*, conveying in the Turkic languages the notion of 'moisture'. From the same root we have also *bolyn*—meadow, where *yn* (*an*) has already the meaning of 'place'. Hence this word may be interpreted as 'wet place' (the same as луг-луга in Russian). The term *bolan*, with the root *bol*, may be interpreted as 'animal of wet places'[16] or since 'meadow' (*bolyn*) is derived from 'grass', as 'herbivorous animal'. Changing this into 'that which eats grass' we obtain a euphemism, which has a perfect analogy in Russian: 'that which eats honey' (медведь 'bear' < мед ведать 'to know the honey').[17]

[10] *Т-РС* (1931).

[11] Вербицкий (1884) p. 16.

[12] Кудряшев (1948) p. 18.

[13] Рахим (1923) was the first to call attention to the special meaning of these names in Tatar literature.

[14] Радлов (1888–1906) Vol. II, p. 326; Вербицкий (1884) p. 26.

[15] Радлов (1888–1906) Vol. I, p. 182; Вербицкий (1884) p. 21.

[16] The elk willingly goes to graze on marshes where no other animal can survive. *П.Т.* (1947) p. 304.

[17] Реформатский (1947) p. 37.

Finally, the meaning of the Chuvash word *avtan* 'cock' can be explained in the same way: the root *avat*—'sing' (referring to the cock, the cuckoo, etc.)[18] and of *an*, hence, 'singing animal, singing bird'.

We agree with the linguists who consider the particle *an* as an originally polysemantic suffix referring to an object, yet at the same time we feel perfectly justified in assuming as its secondary function that of a substitute for 'animal or bird' as is evident from the examples given above.

A survival of the veneration of animals in the popular medicine of the Kazan Tatars can be discovered in their methods of treatment called *imči*.

When treating smallpox, the Kazan Tatars smeared the eyes of the patient with the blood of a red cock which was, later on, presented to the wise man. Of course, the red cock is a symbol of fire or the sun, as a deity. In various incantations the hare, its paw and skin were used extensively as for instance in the *očan* disease (troubles of the abdomen), in the diseases of sheep, etc.

A method known throughout this country was the healing of abscesses by the application of a dead frog. The skins of hares and mice were also used to heal abscesses. It is known that the legs of geese were used against sore throat. Snake skins were used against eye diseases, crane eggs against jaundice, and wolfskins against skin diseases (called *bure uy* 'wolf's howl'), in which case, the wise man had to imitate the howling of a wolf.

In popular medicine also various parts of the bear were used, particularly its claws, which were sewn everywhere on the clothes and on the headgear of children as talismans—*böti*—against the evil eye.

In some diseases the flesh of certain animals was recommended. For instance, in eye diseases and heart trouble a curative effect was attributed to the meat of the female quail and of the lark.

We know that sacrifices used to be offered and the flesh of the sacrificed animal consumed as medicine in the south-eastern districts of the Tatar Autonomous Soviet Socialist Republic. All kinds of animals such as geese, ducks, chickens, rams, bulls and horses were generally sacrificed until recent times among the nationalities of this region, the baptized Tatars included. It is remarkable that animals of certain colours—black, red and white—were chosen for slaughter. This is attested by names of villages such as *qorman jelγasy* 'ravine of sacrifice', *aq ürdäk čišmäse* 'brook of the white duck', *aq büzat čišmäse* 'brook of the white horse', *aq ügäz čišmäse* 'brook of the white bull'.

We wish in this connection to refer to the considerably earlier custom of offering horses as sacrifices (*žylqy qorbany*), recorded by us during the 1946 expedition in the eastern district of the Tatar Autonomous Soviet Socialist Republic. Thus, for instance, a lengthy illness was explained with the expression *žylqy qorbany tota* ('the horse offering is holding'). In earlier time, to cure such a disease, people boiled a gruel with a right-side rib of a horse, and invited a person of distinction (but not a mullah); or out of the bark of a lime tree they manufactured a bridle without a curb bit called *noqta*, and hanging it up in the barn, undertook to offer a sacrifice. It is not devoid of interest to note that some details of this ritual coincide with the ceremony of the horse offering as practised by the Udmurt (Votyak) and Mari (Che-

[18] Ашмарин (1928).

remis) peoples. For instance, during this ceremony the Mari also pull a bridle made of the bark of the lime tree over the head of the horse.[19]

All these surviving ceremonies reflect, beyond doubt, the ancient pagan custom of testing the body of the totem taken for a deity.

Finally a well-known method of healing is to inhale the air given off in exhalation by animals, like stallions, sheep, etc. This has probably been considered as an inhalation of the spirit of the totem.

The language and folklore of the Kazan Tatars show many traces of animal cult. There is a popular saying, the meaning of which is completely forgotten: When out of egoistic motives somebody tries to occupy an advantageous position among his companions, they say to him: *sin aldanyŋ qašqa täkäseme ällä* 'you are a god's ram with a star'. This recalls another saying with no less a totemistic significance: *allanyŋ bozauyn ašadyŋmy ällä*, 'you seem to have eaten from god's calf'. This exists also in Russian, in the first person: я что у бога теленка съел 'I have eaten from god's calf'.

The meaning of the first proverb reveals itself clearly, when we remember that the Uigurs, until recent times, considered lambs and horses with a white spot on their forehead as animals consecrated to god and, consequently, they bestowed on them all the veneration due to a deity.[20] They showed particular reverence to lambs with brown wool, and bay horses with a white 'star' on the foreheads. These were held in veneration also among the Tatars. The meat and the heart of these animals were used in popular medicine. So, too, was the wool of lambs with a 'star', with which they practised fumigation against fear. In this respect let me refer to the case of a chestnut stallion with a blaze on its forehead sacrificed by the Mari.[21]

The folklore of the Kazan Tatars contains references to fantastic birds, *qara qoš*, wild beasts, many-headed dragons, kind-hearted horses, *aqbüzat*, etc. There are numerous proverbs and common sayings, reflecting, in some way or other, relationship between people and animals.

Their folklore contains a great many examples of various aspects of the idea of kinship and conjugal relationship existing between people and animals and of the transformation of men into beasts. These are typical survivals of totemism. The legend of *saq-soq* (transformations of people into birds) is of special interest here.

The well-known tales of the white wolf, the white queen, the snake with its horns containing wealth, and of red and golden birds, skewbald, striped and white horses (*aqbüzat*), etc., are unmistakable survivals of the universal veneration of animals of unusual and rare colours. Here, too, belongs the proverb *at tanysaŋ ala at al, tanymasaŋ sory at al* 'if you know horses, buy a skewbald one, if you do not, buy a grey one'.[22]

[19] Смирнов, И. Н. (1890) p. 8; Емельянов (1914) p. 7.

[20] Малов (1912) pp. 72–3.

[21] *КЕЛ* (1890).

[22] In this context we may refer to the sacrifice of white ducks, by which—as recorded by Kajum Nasyri—eye-diseases and smallpox were healed. Such sacrifices have been recorded from the village Verchnije Menger, Arsk district TASSR. In the village of Stary Menger there is a well which is called *aq ürdäk čišmäse* 'the well of the white duck'. Obviously the custom used to be practised there.

In this respect the ceremonies and customs of the Kazan Tatars are very interesting. The popular feasts of the Tatars, the social roots of which go back to ancient times, also reveal many traces of animal cult, *narduyan* of the baptized Tatars, with the masqueraders leading a stuffed horse, is closely related to similar customs of the Russians.[23] In the celebration of *Sabantui*, the horse is not only a simple participant but is the central figure. In fact, there horse cult reaches the summit of veneration.

The *ziin*, survivals of the tribal organization of the Kazan Tatars, preserve animal and plant elements in their observances as well as in their terminology. Particularly interesting in this connection are such expressions as *täkä žyeny* 'the *ziin* of the lamb', *qyzyl ajγyr žyeny* 'the *ziin* of the red stallion', *kuk ajγyr žyeny* 'the *ziin* of the blue stallion', *jua žyeny* 'the *ziin* of the wild onion', *tujra žyeny* 'the *ziin* of the young oak', etc. Further research will reveal whether or not these names are the survivals of tribal totemism.

In the ceremonies of these festivals no explicitly totemistic features can be distinguished, as is the case in the *qarγa tuj* of the Bashkirs[24] where the survival of totemism is owing to social and historical reasons, generally better preserved than with the Tatars.

Verbal mansifetations in the spring, at the sight of the arrival of the starlings and wild geese (bliss descending from heaven), are also significant from the point of view of totemistic interpretation.

Nor should we leave unmentioned such customs as the exhibition of horse skulls and of the horns of various animals, practised also by various other peoples. S. P. Tolstov observes the existence of similar customs among the Turkic people of Central Asia. He writes: 'also decorating the *mazars* —these former sanctuaries of tribes—with horns and tails of animals may be classified among these' (i.e. survivals of totemism).[25]

The exhibition of horse skulls, horns and egg shells in gardens next to beehives was already noticed by the 1946 expedition in the south eastern districts of Tataria. In the summer of 1947 in the districts beyond Kazan we recorded cases where the horns of elk were exposed on gate-ways of houses and over windows. Shavings of these used to be sewn on the *böti* or amulets. Cases when the horns of rams were exposed in cottages above doors were observed by us in Russian villages in the Mamadysh district. According to the old people, all these customs were for the purpose of protection from the evil eye, ensuring good luck in household work or farming.

The archaeology of this region also argues in our favour, for we find figures and drawings of animals and animal patterns used in ornamentation. In spite of the thousand-year rule of Islam which severely condemned the representation of animals, animal patterns can still be found in the scenes of daily life among the Tatars. Such are, for example, ornaments consisting of birds above windows, сямяряу at times considerably conventionalized, no doubt. The same holds true for the Trans-Kazanian region, as well as for the villages near Kazan.

[23] Крюкова (1947) pp. 72–3.
[24] Руденко (1955) p. 277.
[25] Толстов (1931) p. 260.

Representations of the elk, bear, snake and horse occupy an important place among the archaeological finds.[26]

As to the widespread veneration of birds by the rural population of the Ural region, Professor A. P. Smirnov makes the following remark: 'Thus, figures of birds occur in the Ural region from the time of the Ananyinski Culture to the 12th and 13th centuries. From the Pyanoborsk period on, they had the likeness of human face on the sternum . . . These figures can primarily be traced back to totemistic notions and associated with the ancestors only at a later stage.'[27]

A great number of figures of various animals and birds may be found also among Bulgarian archaeological remains in spite of the severe prohibitory laws of the Moslem religion.[28] Typical in this respect are Bulgarian door locks in the form of a horse.

Since the horse was sacrificed to the sun deity there is a remarkable abundance of material both archaeological and ethnographical connected to a large extent with agriculture. Evidently, the appearance of the horse in the region must be connected with the beginnings of agriculture, and 'essentially the figure of the horse goes far back to earlier, totemistic, notions that were only vestigial in the Ananyino Period'.[29]

Extremely typical of the Ananyino and Pyanoborsk periods, and even of later periods until the 11th century, are the figures of horses with riders. 'Such figures were found time and again in various parts of the Kama region with objects and coins of the 10th century.'[30]

Finds with similar representations gain much in ethnographical significance in the light of the comparative studies of Professor A. P. Smirnov. Let us refer here to the legend of the imp or goblin (the *šurale* of the Tatars and Bashkirs) widely known among the peoples of the Volga region and also in parts of the northern Russian territories. According to this legend, the *šurale*—a strange anthropomorphic creature—loves to ride a horse which it gallops until it is completely exhausted. To save the horse from the *šurale*, they smear its back with tar and drive it out to the open fields, to which the imp sticks and, thus, is captured in the village. As a punishment it is burnt over a camp-fire.[31] The *šurale* curses the village which, struck by economic calamities and reduced to poverty, is finally depopulated.

Identification of the above-mentioned figure of the rider with the representation of the *šurale*, the spirit of the woods, is reasonable when we consider that, in the representations, the rider is usually surrounded by wild animals and that 'the figure of the rider has no clothes on, is exceedingly disproportionate and makes a comic impression'.[32]

By the figures of the horse with rider, and by the legend of the *šurale*, the activities of the Ananyino and Pyanoborsk peoples and of later inhabitants of the Kama region are reflected as a mixture of agriculture (the

[26] Збруева (1947) p. 38.
[27] Смирнов (1952) pp. 260–1.
[28] Калинин (1945) p. 8.
[29] Смирнов (1952) p. 269.
[30] Смирнов (1952) pp. 267–8.
[31] Коблов (1914) p. 5.
[32] Смирнов (1952) p. 268.

horse) and hunting in the woods (the *šurale*, as the owner of the forest). As for the burning of the *šurale*, it is not difficult for us to recognize in it a later touch of feudalism that spread over to Bashkiria from the end of the 13th through the whole 14th century, when many Bulgarian Mohammedans challenged paganism and preached the Moslem religion far beyond the frontiers of Bulgaria.

The *šurale*'s curse fairly well manifests the attitude of the broad popular masses towards the expansion of Moslem; they adhered to their earlier pagan beliefs and customs for a long time after the Islam invasion. In this respect the inference of the legend is in full accord with the miserable conditions in the subjugated area, as well as the depopulation of the villages that took place as a consequence of enfolding capitalism.

Most of the animals and birds represented in the archaeological material of this region were even in recent times offered as sacrifices (horses, geese, ducks) by the local population or were objects of special veneration (elk, cranes, swans, etc.). This is not without interest from the point of view of ethnogenesis.

Some survivals of totemism in the popular traditions of the Bulgars and their neighbours described by Ibn-Fadlan must be dealt with in this connection.

Speaking of the Bashkirs, he remarks: 'Some groups worship snakes, various kinds of others, fish and still others, cranes'.[33] The cranes protect them and, in wartime, save them with their cries from their enemies. There is a similar legend about the *syŋrau torna* 'singing crane' and the *uγatar* 'the ruler of the birds' known from the poem 'Irkbai'.

Animals considered as possessing the same power were obviously snakes since they occurred in great numbers, but were not killed by the Bulgars.[34] Therefore, the beliefs of the later peoples of this region, particularly of the Kazan Tatars, even after they had undergone social evolution, largely contain survivals of totemism connected with snakes, cranes and fish. This refers in particular to snakes and cranes, as is proved by folk poetry, religious beliefs, ceremonies, archaeology and art.

Survivals of fish totems are the least frequent. We succeeded in recording some that still exist in the Trans-Kazanian region, chiefly among the baptized Tatars, who used some kinds of fish in their commemorative ceremonies. In some places the former existence of the *aq balyq qorbany* (i.e., the 'offering of a white fish') could be demonstrated. In this connection, it is interesting to note that during the excavations at Sholom—a site of sacrificial offerings—large quantities of fish bones were found, representing about one-third of the total amount of bones found there. This sepulchre goes back to the pre-Bulgar or Bulgar periods.[35]

Tree worship, formerly widespread among the neighbouring peoples, is found sometimes in a moderate form in the rites of the Kazan Tatars. For instance, the birch venerated by the Udmurts (Votyak), and the *kläüle qajen* 'sacred birch' of the baptized Tatars are known to the Tatars under the

[33] Ибн-Фадлан (1939) pp. 66–7.
[34] Ибн-Фадлан (1939) p. 72.
[35] Oral communication by B. B. Žiromskij under whose guidance the excavations were carried out.

innocuous name of *jalyyz qajen* 'lonely birch', to which many tales and superstitions are attached.

The tree cult is closely interwoven with the ancestor cult, as is apparent in the Tatar cemeteries, in sacred groves similar to the *kirämät*s of the Chuvash, Mari (Cheremis), Udmurt, whence they probably took their origin. Although no worship of the birch tree in the form customary, in earlier times, among the Finns and Chuvash of the Volga region is known among the Tatars, the survival of the tree cult in the form of *urman žyeny* 'festival in the woods' (offerings of sacrifices in the forest) existed in the not too remote past. We have recorded cases where this kind of a grove (*iske ziarat* 'old cemetery') was worshipped by the Moslem, the baptized Tatars, who called it *kirämät* and offered there pagan sacrifices. Not only the external resemblance of the Tatar cemeteries to the *kirämät* should be emphasized. We should also remember that, in every tree planted on a tomb—all of which form the grove of the cemetery—the Tatar sees his own ancestor. Therefore his devout and pious attitude towards these trees and to cemeteries is not surprising.

The toponymy of Tataria, which has not yet been seriously investigated, is interesting from this point of view. It is worth while throwing a cursory glance at the map of the Tatar Autonomous Soviet Socialist Republic. We will discover dozens of villages bearing the names of animals and plants. Some of these recur in several districts, such as, for instance, Tornaly, Tornajaz, Tolkojaz, Kargaly, Kazyle, Chirshyly, Karamaly, Naratly, Tukmakly-Tokmakovo. Almost all villages named Kargaly are situated in the Kama region and are inhabited mainly by the Mishars.

According to their terminations, these names of villages can be classified into several groups: (a) endings in *-ly* (Kargaly, Yabalakly, Tornaly), (b) endings in *-jaz* (Tornajaz, Tolkojaz), (c) ending in *-sar*, *-sär* (Kainsar, Imänsär), etc.

The question arises whether the ending *-ly*, used also after adjectives, is not an abbreviation of *ile* meaning 'country' as in Kazyle—*kaz ile*. Then Tornaly would mean: *torna ile* 'the country of the crane'; Kargaly—*karγa ile* 'the country of the raven'; Yabalakly—*jabalak ile* 'the country of the owl', etc.

The meaning of the particle *jaz* is intelligible from the name of the Russian village in the district of Tyulyach—Gorokhovoe Pole meaning 'pea-field'. The Tatars of the region call it Borchak-yaz (*Borčaqjaz*), since *borčak* means 'pea', and *jaz* obviously means 'field' or 'clearing'. Evidently, this term is the same as found in the memorial in honour of *Kül-tägin*, i.e. *jazy-* steppe.[36]

We reach a similar conclusion when comparing such names of villages as Tornaly and Tornoyaz. Here the particle *ly*, *ile*, *il* 'country' corresponds to *jaz* 'field, clearing', and is in semantic connection with it. For clarity's sake, let us give another example. In the district of Tyulyach the baptized Tatars village Tolkiyaz is usually called Talkysh (or Tolkish). A comparison of these names shows us that *jaz* 'field' corresponds to *yš*, *iš*, having obviously the same meaning. Some time ago the linguist L. Z. Zaljaj stated that the particle *yš*, *iš* in place-names like Mamadysh or Laish should be interpreted as *il*—'country'. Let us also refer to the Chuvash word *talqqoš*,

[36] Малов (1951) pp. 27, 378, 424.

talkkiš which, as a whole, means 'space, territory, area, volume'.[37] Sometimes the form *qyr* is used for 'field', which—unlike *yaz* and *yš* that have already died out—still exists in the spoken language of the Kazan Tatars.

This is clearly evident from six geographical names used by the Kazan Tatars: Mazar kyr, Begre kyr, Chirmesh kyr, Chypya kyr, Ozyn yar kyr, and Tlau kyr[38], the first three probably containing the tribal names Mazhar, Bulgar, Cheremis.

The same is found in the Tatar place-name Archa kyry 'fields of the Ars' which, in the time of the Kazan khans, the Russian called Арская земя 'land of the Ar-Udmurts'.[39]

Consequently, we can judge the social significance of the particles and words: *jaz, yš, iš, il* and *qyr* which once meant 'land, estate, country' (of the clan, tribe and later of the overlord and of the rural community).[40] Similarly, the suffix *qar* frequently occurring in the place-names of the Finns of the Volga region (Kudymkar, Syktyvkar, Shubashkar, Dondykar, Indokar) means an 'inhabited place, town, settlement', etc.[41] To this group belongs the Tatar village Kyshkar in northern Tataria.

It is not our task to discuss here the origin of the particles and words: *jaz, yš, il, qyr,* but it is clear that these contain traces of Turkic-Finnic clan-tribal interactions, which resulted in the historical birth of the peoples in this region, including the Kazan Tatars.

The above-mentioned place-names Tukmakly, Tokmakovo, are derived from the Chuvash pseudonym of the wolf *tuqmaq* or, as it is still called, *tuqmaq χure,* meaning 'whisking tail'.[42]

Finally the local ethnic names of this region are also of great interest since, in our opinion, they are not devoid of traces of totemism.

It is well known that some family names of animal and plant origin have developed from nicknames, many of which undoubtedly go back to totemistic nomenclature. Naturally, after the dissolution of the clan communities, they remained in use as nicknames for the former members of the clan. Later, at a more advanced stage of social development, they became official family names. (Examples: Sorokiny, Lebedevy, Volkovy, Medvedevy, Berezovy, Jelovy, etc.)

A typical case of such relationship is that one of the clans close to the Selkups (Ostyak Samoyeds) in Siberia was called *karalkup* (*kara* 'crane' *kup* 'man') viz. 'crane people'. At present the most widespread family name in the Ob and Tym regions is Karalkin, originating from that clan name.[43]

Nicknames of animal origin are rather frequent among the Kazan Tatars and the Mishars. We know of families called, for instance, *ajunykylyr*

[37] Ашмарин (1928–[1937]) Vol. XIII, p. 168. The derivate *tolkkišpe* 'in a thick crowd' obviously agrees with the Tatar *qyrylyp* in contexts like, e.g. *kolχozčylar qyrylyp ešlitär ide* 'the kolkhoz peasants are working in a thick crowd'.

[38] *ТрОИТ* (1930) p. 142.

[39] It is worthy of mention that this name, which is a survival, occurs only in the north-easterly districts of Kazan (Валиди [1930] p. 142).

[40] In the context the question comes up whether the Russian word уезд 'district' is not derived from *i̯az* (*i̯az* > *öi̯azd* > *ui̯ezd*).

[41] Егоров (1950) p. 88.

[42] Ашмарин (1928–[1937]) Vol. XIV, p. 109.

[43] Прокофьев (1940) p. 74.

(bears), *čypčyknykylar* (sparrows), *karganykylar* (ravens), or *büre kešeläre* (wolf people), *čäükä kešelär* (jackdaw people), etc.

Although normally these nicknames were not used as family names, we know that some were, as for example: Medvedev (from 'bear'), Volkov (from 'wolf'), Solovëv (from 'nightingale'), Galkin (from 'jackdaw') on the one hand, and Burundokov (from the *burunduk* 'Tamias sibiricus'), Čavkin (from *čäükes* 'jackdaw'), Ajukov (from *aju* 'bear'), Bureev (from *bure* 'wolf'), etc., on the other hand.[44]

Family names of this type appeared among the Kazan Tatars, probably first among the privileged class during the early development of close social and economic relationships with the Russians, and more particularly after the union of Kazan with Russia. This influence is evident in the first group of family names.

We have also found cases where nicknames were used to denote certain villages. For example, the people in the village Alan in the Tyulyach district call the neighbouring village Saush *aju* 'bear', while the inhabitants of this latter mentioned place call their village Alan *pši* 'elk'. The village Chulpych in the Sabinsk district is called in the neighbourhood *kujan* 'hare', etc.

Similar cases have been observed among other Turkic peoples, as the Kirghiz, Kazakhs, Turkomans, etc. 'In the terminology of the clan organization of the Turkomans their totemistic origin appears most distinctly. We can find, for example, such names as: Ak-okuz (white bull), Kara-okuz (black bull), Ajlar (bear), Burkut (hawk), Gaz (goose), Due (camel), Jelan (snake), It-baš (dog's head), Išak (ass), Kulan (wild ass), Käklik (partridge).'[45]

Finally, such ethnic names as Bolgar, Suvar, Mazhar are of utmost importance from our point of view since they were once perhaps totemic-tribal names of the oldest inhabitants of the region. This explanation is strengthened by the meaning of the suffix *ar* 'man'.

There is little doubt that the data presented here in a somewhat schematic form, prove that the cult of animals and totemism were prevalent among the distant ancestors of the Kazan Tatars.

A detailed study of the present question is of vital importance for the understanding of the ethnogenesis of the Kazan Tatars.

REFERENCES

(Ašmarin, N. I.) Ашмарин, Н. И. (1928–) *Словарь чувашского языка* (A Dictionary of the Chuvash Language), Vols I–XVII. Kazan.

(Egorov, V. G.) Егоров, В. Г. (1950) Этногенез чувашей по данным языка (The Ethnogenesis of the Chuvash People as Evidenced by Linguistic Data), in *СЭ*, No. 3.

(Emeljanov, A. I.) Емельянов, А. И. (1914) Языческие моления Черемис (Pagan Prayers of the Maris), in *ИОАИЭКУ*, Vol. XXXI.

(Ibn-Fadlan) Ибн-Фадлан (1939) *Путешествие Ибн-Фадлана на Волгу* (Ibn-Fadlan's Journey to the Volga), Moscow–Leningrad.

(Kalinin, N. F.) Калинин, Н. Ф. (1945) *Булгарское искусство в металле* (Bulgar Metal Art), Kazan.

(Koblov, Ja. D.) Коблов, Я. Д. (1914) Мифология казанских татар (The Mythology of the Kazan Tatars), in *ИОАИЭКУ*, Vol. XXVI.

[44] The Bashkir word is Тойгоновы (*toiγon* 'white falcon').

[45] Толстов (1931) p. 260.

(Krjukova, T. A.) Крюкова, Т. А. (1947) Вождение русалки, в селе Оськине, Воронежской области (The Guiding of the Mermaid in the Village Oskina, Voronezh District), in *СЭ*, No. 1.
(Kudrjašev, K. V.) Кудряшев, К. В. (1948) *Половецкая степь* (The Polovets Steppe), Moscow.
(Lenin, V. I.) Ленин, В. И. (1935) *Социализм и религия* (Socialism and Religion), Moscow.
(Malov, S. E.) Малов, С. Е. (1912) Остатки шаманства у желтых уйгуров (Survivals of Shamanism among the Yellow Uigurs), in *ЖС*, Vol. XXI.
(1951) *Памятники древнетюрской письменности* (Ancient Turkic Documents), Moscow–Leningrad.
Marx, K.: *Zur Kritik der politischen Ökonomie.*
(Nasyrov, K.) Насыров, К. (1880) Поверья и обряды казанских татар (Beliefs and Ceremonies of the Kazan Tatars), in *ЗРГОЭ*, Vol. IV.
(Prokofjev, G. N.) Прокофьев, Г. Н. (1940) Этногония народностей Обь—Енисейского бассейна (Ethnogeny of the Peoples of the Ob–Yenisei Basins), in *СЭ*, No. III.
(Radlov, V. V.) Радлов, В. В. (1888–1906) *Опыт словаря тюркских наречей* (An Attempt at a Dictionary of the Turkic Dialects), Vols I–IV. Saint Petersburg.
(Rakhim, A.) Рахим, А. (1923) Исследование об одном суффиксе в тюркских языках (Examination of a Suffix of the Turkic Languages), in *Магариф*, Nos 9–10.
(Reformatskij, A. A.) Реформатский, А. А. (1947) *Введение в языкознание* (Introduction to Linguistics), Moscow.
(Rudenko, S. I.) Руденко, С. И. (1955) Башкиры (The Bashkirs), in *Историко-этнографические очерки,* Moscow–Leningrad.
(Smirnov, A. P.) Смирнов, А. П. (1952) Очерки древне- и средневековой истории народов Среднего Поволжья и Прикамья (Studies on the Old-age and Mediaeval History of the Peoples of the Volga–Kama Midregion), in *МИА*, No. 28.
(Smirnov, I. N.) Смирнов, И. Н. (1890) Вотяки (The Votyaks), in *ИОАИЭКУ*, Vol. VIII. p. 2.
(Sofijskij, I.) Софийский, И. (1878) Заговоры, заклинания крещеных татар Казанской Губ. (Spells and Magic Words of the Christian Tatars in the Kazan Province), in *ИКЕ*, No. 2.
(Tolstov, S. P.) Толстов, С. П. (1931) *Верования народов СССР* (Popular Beliefs of the Peoples in the USSR), Vol. 1. Moscow.
(1946) Этнография и современность (Current Problems of Ethnography), in *СЭ*, No. 1.
(Validi, D.) Валиди, Д. (1930) Наречие казанских и глазовских татар (Dialect of the Kazan and Glazov Tatars), in *ТОИТ*, Vol. 1.
(Verbickij, V. I.) Вербицкий, В. И. (1884) *Словарь алтайского и аладагского наречий тюркского языка* (Dictionary of Altai and Aladag Dialects of the Turkic Language), Kazan.
(Zbrueva, A. V.) Збруева, А. В. (1947) Идеология населения Прикамья в ананьинскую эпоху (The World View of the Population of the Kama Region in the Ananyino Period), in *ТрИЭ*, No. 1, new series.
(Zelenin, D. K.) Зеленин, Д. К. (1929) Табу слов у народов Восточной Европы и Северной Азии (Taboo Words in the Languages of the East-European and North-Asian Peoples), in *СбМАЭ*, Vol. VIII.

ABBREVIATIONS

КБЛ = *Казанский биржевой листок* (The Kazan Exchange Bulletin), 1890, No. 233.
ПТ = *Природа Татари* (Physical Geography of the Tatarlands), Kazan 1947.
ТрОИТ = *Труды Общества изучения Татаристана* (Papers of the Research Group of Tatar Studies), Kazan 1930.
Т–РС = *Турецко-русский словарь* (Turkish–Russian Dictionary), 1931.

Shamans' Drums of Altaic Ethnic Groups

by

L.P. Potapov, Leningrad

The present paper deals with the shamans' drums of the Altais, and part of the Hakas and is mainly based on my own compilation from 1925 to 1955. Although occasionally I include descriptions already published about these tribes, in principle I confine myself to unpublished material collected during my field work. Therefore I avoid repeating descriptions of shamans' drums previously published by me.[1]

I do not wish to restrict my task to a simple registration of the data I have obtained through personal observation or information on the original pieces described by the shamans and drum manufacturers whom I happened to question. I think the present work will best achieve its purpose if it reveals the general types, and explains the characteristics, of the shamans' drums used by the tribal units or territorial groups of the Altais whose drums have been studied by me for a longer period of time. This objective seems all the more justified since each Altai group had a type of drum of its own, clearly discernible by the form of the handle and by the drawings considered to be characteristic of the respective group. (Some tribes used several kinds of shamans' drums, but recognized only one of them as their own.) Thus, we may state that the general type of the shamans' drum in the Sayan Altai upland included several variants which were distinguished by the peoples themselves not only according to the shaman's functions they were used for but also according to different external features, particularly the form of the handle and the drawings on them. All these variants were considered by the local population as being exclusive to one or the other clan, or regional ethnic group. It follows that there existed among the Sayan Altai tribes, especially among the Altais some kind of classification of shamans' drums, which needs elucidation.

The problem of classification of the shamans' drums has not yet been thoroughly studied in the literature. Researchers who had written about them believed the drums of the Altaian peoples to belong to one single type. Incidentally, they used the term 'Altai peoples' in a rather loose linguistic sense, embracing 'the Turkic-Tartar, Mongolic and Tunguso-Manchurian ethnic groups'.[2]

Only G. N. Potanin made an attempt to classify shaman's drums: He distinguished three types of them among the Siberian peoples according to the form of the handle. The first, called the Sayan Altai type with a view to the region where it was used, includes drums with two cross-pieces—a vertical one made of wood and a horizontal one, of iron. He added that the vertical bar served as a handle and sometimes ended in a human face at the top and in feet, at the bottom. The second type he called Tungusian, on

[1] Potapov (1934a); Потапов (1947, 1949).
[2] Harva (1938)

account of its occurrence among the Tungus. Here, the handle has the form of a cross made of wood or iron, tied to the hoop by a strap or a string. The third, or Samoyed type is characteristic of the Samoyed tribes and, like the Tungusian type, has a suspended handle tied to the hoop by means of straps. Instead of having the form of a cross, it consists of a vertical cross-piece almost equal in length to the diameter of the drum, and of a horizontal bar fitted in the middle of the vertical cross-piece and attached, at its other end, to the hoop by means of a strap.[3] Of the drums of the Altai ethnic groups, G. N. Potanin considered only the round ones typical.[4]

Let us also refer to the compilation of G. Nioradze on Siberian shamanism, in which he divides the drums of the Siberian peoples into two major types: oval and circular.[5] According to this author, the first type exists among the peoples of Eastern Siberia such as the Yakuts, Yukagirs, Tungus, of the Amur tribes: Nanai (Golds), Oroches, Negidals, etc., whereas the second can be found among the peoples of Western Siberia such as the Ostyaks (Khanty), Kets, and the Altais. Such a classification, however, is inadequate and can easily be refuted on the ground of several factual proofs, among others of the Sayan Altai drums.

The fundamental difference in the handles of the drums was noticed by G. N. Potanin. But, like many other researchers interested in Altai shamanism, he failed to discover the obvious dissimilarities which had been known to the Altais themselves. I should like to stress these differences for the benefit of further research.[6]

Beginning with the Shors and the Northern Teleuts, I wish to point out what they themselves consider typical and characteristic of their 'native' drums (as they call them).

SHAMANS' DRUMS OF THE SHORS (*tūr*)

The Shors used only one single type of shaman's drum, which has a wood handle in the form of a vertical rod. The handle is entirely rounded off at its central part where the shaman grasped the object when taking it in his hands. Above and below the middle, the handle thickens; it continues flat at both ends abutting the wall of the framework; a high wooden crest running through the middle dissects each of the planes. The thick portions and the planes on either side are amply decorated with rectilinear carvings in V-shape. Each side of the blades has three or more cuts or holes, triangular, circular, or arched on different drums (Figs 1 and 2). All the Shor shamans agreed in calling the above-described drum-handle *alty köstü ala mars*. They believed it to be the daughter of heaven's supreme divinity, *ülgän*. In addition to what I have published about the Shor drums, their manufacture and the drawings on them, a descriptive evaluation can be found in the work of S. V. Ivanov[7]; therefore I shall confine myself here to a few drawings which best characterize their general and typical features.

[3] Потанин (1893) pp. 13–4.
[4] Потанин (1883) p. 40.
[5] Nioradze (1925) pp. 79–83.
[6] See: Potapov (1957).
[7] Иванов (1955).

A characteristic trait of the Shor drum is that the drawings are concentrated on the drumhead and are sharply divided into upper and lower patterns. In the middle there is a horizontal stripe dividing the drum's surface into two fields of unequal size, the upper being bigger than the lower. The entire upper field is bordered by a curved stripe enclosing triangles and trapezoid figures. These were needed for ceremonies performed in honour of the mountains and the powers above, whereas those of the lower field

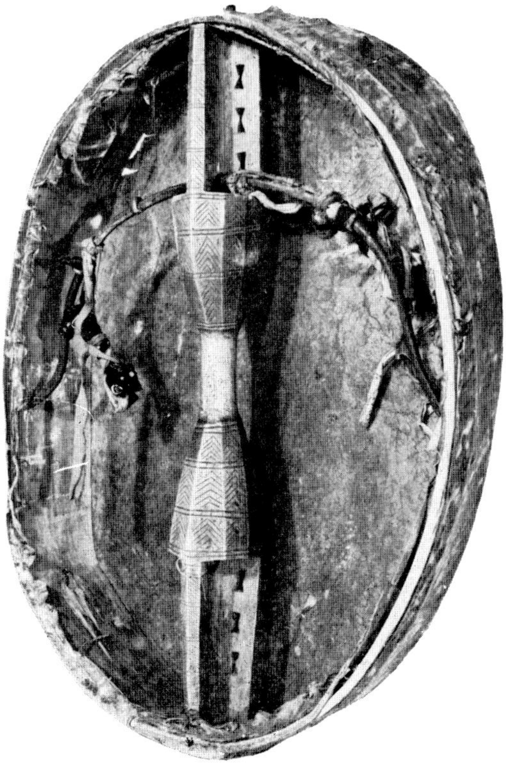

Fig. 1. *mars*, drum of the Shors

played a role during ceremonies for the spirits of the lower regions. Beside the sun, the moon and the stars, birch trees (represented with roots and leaves) and the purple willow tree (with its branches looking on one side) which are arranged in the middle of the drum's left side, there are also drawings of various kinds of horned animals, named *pura*s. The most typical patterns on the upper part of the Shor drum are the anthropomorphic figures with outspread bird-wings, called *qyrlyq*. Similar figures cannot be found on the shamans' drums of the other Altai tribes, but it is very tempting to compare them with the 'winged men' of the Voguls, i.e. the Mansi.[8]

The figure of the frog is frequent not only on the lower field of the drum, as is usual with other Altai tribes, but also on the upper field. This figure

[8] Чернецов (1947).

is named *ülgäniŋ paqazy* '*ülgän*'s frog', i.e. the assistant of the shaman. It is the frog that carries into the sky the sacrificial beer in a cylindrical vessel (туяс). Finally, there are the figures of the horsemen, *pos čorγo,* arranged on the lower part of the drum.

The shamans of the Shors always received their drums from the ancestral shaman mountain, the Mustag, situated at the upper reaches of the Kondoma river, in the Kuznetsk Altai mountain. The drawings on the upper

Fig. 2. *mars*, drum of the Shors

field varied according to the spirits to whom it was consecrated. Polikarp Paldžanov, a Shor shaman, showed me his drum with the figure of a *pura* painted in red and called *baj qoγyštyŋ qyzyl pury* 'the red *pura* of the famous shaman *Qoγyš*'. 'But'—said the shaman—'this *pura* has three names depending on the being the drum is consecrated to'. If the drum was consecrated to *ülgäniŋ uly baj Qyrγys*, i.e. to *ülgän*'s son *Qyrγys*, the drawing was called *adam sardaq*.[9] If it was consecrated to *ülgän* himself, the name of the drawing was *aγaš bažy paqty-χan*, i.e. 'he who lives on the top of the *paqty-χan* tree', (another son of *ülgän*), and this was meant to be the tree under which the

[9] The Abakan Shors use the term *sardaq* for young game, whereas among the Sagai and the Koibals it meant the one-year-old maral-deer.

shaman offers his sacrifices. Finally, if the drum was not consecrated to any of the heavenly beings, the drawing was called *tazy-χan*.[10] Shaman Sandra explained to me that the *pura* received different names because it symbolized different spirits in the shape of the animal *pura*. When I asked whether the *pura* could assume the shape of a reindeer or a maral-deer, Sandra answered in the affirmative.

The drum type referred to can be found also among the Shor shamans who moved, in the middle of the 18th century, from Shoria to the upper course of the Tashtyp river and to the other rivers belonging to the left basin of the upper reaches of the Abakan. D. A. Klemenc describes a drum that was made in 1883 in the Butrakty *ulus*. It belonged to a shaman, who was a member of the *Qobyj* clan living near the Seya river and whose ancestors had come from the region of the Mrassy river. I quote the description: 'The diameter of the drum is 72 cm; the framework on which the parchment is stretched is 15 cm wide; a loop on the upper part permits the drum to be suspended on a nail. Next to it, there are four bosses in a row, where four little rods are inserted into the hoop. The drum and the hoop are covered with the smoked leather of the maral-deer. The iron cross-bar is laid through the handle (*marsy*) placed in the inside of the drum. Three oblong apertures are on each side of the broad scoops of the handle. According to the shaman, upon his incantation the spirits fly into the drum through these apertures on one side, and leave on the other when the ceremony is over . . . The upper part of the drum is called *čalyčyχ*, and is divided by a cross-stripe into two fields, the stripe being called *tiger qurum* "heavenly belt, i.e. rainbow".[11] Painted with red in this stripe are triangles enclosing angles which point towards the opposite side. Along the upper part of the drum, above the stripe, there is a curve with the same ornamentation, also called *tiger qurum*, but meaning the "Milky Way".'

However, judged by the description of Klemenc, the drawings of that drum are not Shor, but rather Hakas. For instance, the following figures are to be found on the lower part:

1. *baj qazyŋ* (in white), literally: 'rich birch tree'. This name is given to birch trees around which the sacrifices are offered every year.
2. *uluγ baj qazyŋ* (in white). According to the *qam* (shaman), these two trees grow in the realm of *ilχan*, under the earth.
3. *aq baγa* 'white frog'.
4. *qara baγa* 'black frog', etc.

Painted in the upper field were the sun, the stars, two birds, serving as envoys of the shaman, as well as the 'bear *tüs*' and others. In other words, it may be said that the upper part of Shor drums is generally decorated with the same patterns as are to be found here on the lower part, and vice versa. From this it may be inferred that the drawings represent the least stable features of the shamans' drums. I have made similar observations about shamans' drums of the other Altai tribes, which will be dealt with later on.

One of the peculiarities of Shor drums was the drumstick, covered with the winter-hide of the white hare.

[10] According to the Teleuts *tazyγan* is the wife of *ülgän*, whereas the Altai Kizhi believe her to be his mother.

[11] Correctly: *täŋäri qury*. L. P.

The northern Teleuts live near the Great and the Small Bachat rivers (the left tributary of the Inya river which flows into the Ob from the east). The drums of their shamans represent a single type and resemble very much the Shor drums but for some details of the drawings and their interpretation. Painted on the exterior of the drums, the drawings represent similar themes and are strictly divided into upper and lower fields. They depict spirits who are helpful to the shaman, as well as beings of the upper and the lower world venerated by him. As among many other Altai tribes, the drawings on the Teleut drums generally represent the shamanistic cosmology with its tripartite form. Separated by a stripe, the lower field of the drum was covered with drawings of triangles and other figures, and represented the lower world. The stripe, which divided the drum into two unequal parts, was interpreted by the Teleuts I talked to as being *jär* 'earth', with the mountains outlined on it. The upper part contained not only the sun, the moon, rainbows and single stars, but entire constellations. These and some other drawings made them easily distinguishable from the drums of other Altai tribes, especially from those of the Shors. Teleut drawings include the 'Book of Fates'—*sabyr pičik*—a compilation of *ülgän*'s laws and verdicts, painted on the upper field of the drumhead, as well as the *quš-tory* a 'net-like sack', or, according to other definitions, a 'cage', used by the shaman to catch the *čamγyl* birds, which are the guardians of some heavenly beings and obstruct the shaman on his way to them. Various representations of the *pura*, although not exclusively found among the Teleuts, are also characteristic.

A specific figure in the lower part is the 'old man of the threshold', i.e. the master and protector of the threshold, the dwelling place. Since the drawings of Teleut drums have been circumstantially described in a number of publications, I shall give here a few references to the description of specifically Teleut drums.[12]

The handle of the Teleut drum is like that of the Shor and is called *ala bars* or *ala mars*. In shamans' rites, it is called *alty köstü ala mars*, just as among the Shors, only the Teleuts translated the word *ala* as 'pied'. They showed me the handle of the drum which was really pied on account of its being covered with carvings and cuts (Fig. 3). Cut into the crests at the end of the handle are seven transverse apertures (*täšik* 'crack, slit') symbolizing the 'earthly apertures' which the shaman happened to pass by during the ceremony. The shaman invites the spirits of the dead 'master tailors' who stitch these apertures together to prevent the shaman from falling through them. Next to the iron cross-bar is a small groove in the handle, where the shaman hides the *jula* 'soul' of the sick man pressing it with the finger of the hand that holds the drum. At an appropriate moment of the ceremony, the shaman lets the captured *jula* enter the right ear of the sick.[13]

Apart from the drawings, there is still another specific difference between Teleut and Shor drums, and this concerns the drumstick. While the Shors

[12] Дыренкова (1949); Потапов (1949); Иванов (1955) pp. 239–43.
[13] About the soul of *jula*: Анохин (1929) pp. 253–8.

covered their drumsticks with the hide of the hare, the Teleuts used the hide pulled off the roebuck's (*quran*) leg for this purpose. Finally, let it be remarked that there were no special drawings seen on the drums of the Teleut female shamans. According to N. P. Dyrenkova, among the Teleuts

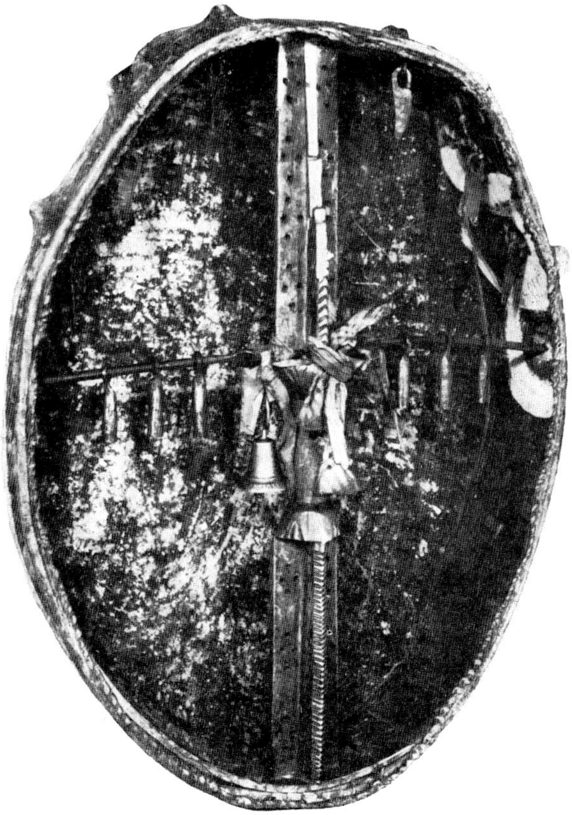

Fig. 3. *mars*, drum of the Teleuts

no woman was allowed to shamanize with *ülgän*.[14] However, the Teleuts I was working with ascertained that women did shamanize with *ülgän*. As an example they mentioned the female shaman Marfa Todyseva and, to confirm it, they reproduced the drawings on the upper part of her drum.[15] It may well be that, in recent times the prohibition against shamanizing with *ülgän* has been broken by Teleut female shamans.

[14] Дыренкова (1949) p. 111.
[15] This drum and its drawings have been published by Потапов (1949).

SHAMANS' DRUMS OF THE CHELKANS (*tűr*)

The Chelkans had two kinds of shamans' drums, one of which was regarded as 'native' Chelkan and the other as Shor.

In general, the Chelkans used the Mongolian term *tűr* for the drums, but each of them had its own 'specific' designation. The name of the specific Chelkan drum was *qalāč* (Fig. 4). This word also means the arrow used for shooting the fish in the water.[16] The Chelkan drum has a vertical handle

Fig. 4. *qalāč*, drum of the Chelkans

made of wood; it is an anthropomorphic figure with two heads placed in opposite directions, like the Kumandin and the Tuba drums, called *qanym* (Figs 5 and 6). According to the personal observations of K. Hildén at the beginning of this century, the handle of the Chelkan drum had no anthropomorphic shape—at least not in every case. It is true that, at that time already, its form reminded one of that of the double-headed *qanym*. The heads did not have the form of human faces but were merely the oval parts of the handle (Fig. 7). At the same time, he notes that there were no drawings on the Chelkan drums.[17]

[16] Cf. Вербицкий (1884) p. 121.
[17] Hildén (1916).

From this it may be inferred that it was only in recent times that the wooden handle of the Chelkan shaman's drum *qalāč* was transformed into a full analogy of the *qanym*, typical of the Kumandin and the Tuba drums. We may also infer that the habit of decorating the drumhead with drawings is not very old either. Although some of the Chelkan shamans possessed their own drums (*qalāč*) as well as the Shor type which they called *alty köstü ala mars* 'six-eyed ala mars', they used the latter for foreigners only,

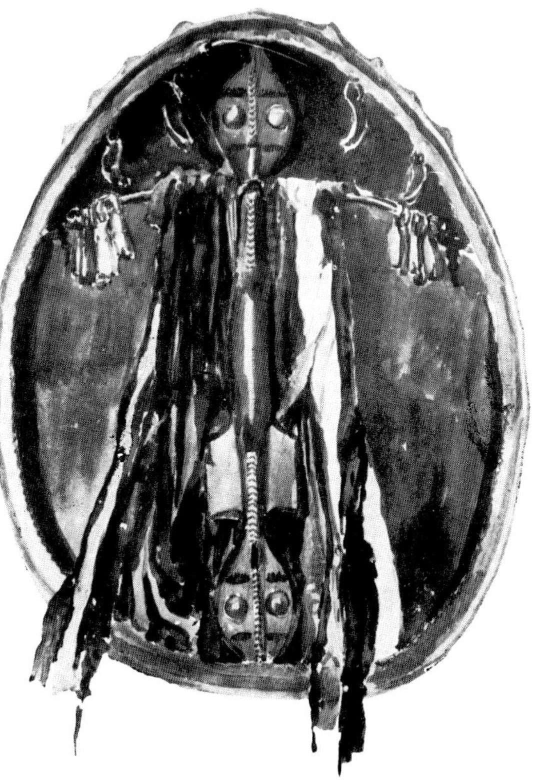

Fig. 5. *qalāč*, drum of the Chelkan shaman Barbachokov

reserving the former for shamanizing for their own tribesmen. The Chelkans maintained that their shamans were far stronger than those of the Shors and made allusions to journeys undertaken by the latter in order to steal their helping spirits.

The Chelkan shamans distinguished two classes of *qalāč* drum depending on the protecting spirit from which they came. The first one, the *täzim*, was received from heaven (*täŋri*) and was consecrated to heaven, while the second, the *očyn*, was received from the ancestral mountains. They definitely told me that these two kinds of drums did not differ from each other. As I was not in a position, however, to see both kinds of drums, I am unable to decide in favour of, or against, this statement. The Chelkans considered the *täzim* to be the superior sort of shaman's drum.

I shall say a few words about this variant since, with the exception of K. Hilden's short report published in a work which is now almost inaccessible, the literature does not contain any description of Chelkan drums. The framework was made of willow, the handle (*tūriniŋ ā* 'master of the drum') of birch. On the framework the leather of a young roebuck was

Fig. 6. *qalāč*, drum of the Chelkan shaman Aspok

stretched, but occasionally the leather of a young maral-deer was used. The iron parts were made by the blacksmith after the instructions of the shaman, if he had not inherited all of them from the drum of the dead shaman ancestor. The iron bar which crossed the handle was made of a twisted rod and called *kiriš* 'bow-string'. It carried thirteen iron pendants (*süŋür*), six on one side, and seven on the other. There were three knife-shaped iron pendants, called *qylyč* 'sabre' at the upper part of the drum, on the inside of the framework, on either side of the handle. Arranged on the outside of the framework was a loop (*čürbäš*) made of a strap permitting the drum to be suspended on a nail or a nipple. Ribbons of various colours were tied on the cross-bar (*kiriš*). Anybody who invited the shaman to shamanize, was expected to fasten a ribbon onto it. The drum had six horns (*müs*) which formed bosses or humps on the upper part of the framework

Fig. 7. Drum of a Chelkan shaman, with a two-headed handle which has not acquired an anthropomorphic form as yet. (From Hildén's book)

and were covered with leather. Small pieces of birch, or willow, were placed under the leather for this purpose. When I was among the Chelkans, in the middle of the 1930's, all the old men, or grown-up men, with whom I spoke, unanimously affirmed that there were drawings on the external side of the parchment. These were made by specialists, according to the shaman's instructions. They used only red paint, obtained from the *qyzyl taš* 'red stone' (ochre) dissolved in water. The drawings on the external part of the Chelkan drum were divided into two fields. The dividing line, running in the middle or somewhat below, was called *täŋri qury* 'belt of heaven'. Among the northern Altai Turks, this is also the name of the rainbow. A typical feature worth noting is that the common name of the entire upper field was *täŋry* 'heaven' and simultaneously *täzim*. The lower field covered with drawings was called *pajana*.[18]

The drawings in the upper field represented the sun (*qojyš*), the moon (*aj*), the stars (*čyltys*), two or nine geese (*qastar*), nine *puras* (*toγus bury*), the birch tree (*qajyŋ*), the bird sitting on the top of the birch tree, called *qaraγuš* or *aj qanatty qaraγuš* '*qaraγuš* with moon-shaped wings'.

The drawings in the lower field showed the frog (*baqa*), the snake (*jylan*), the lizard (*käläskän*), another sort of lizard (*qamnyq*), and the master of the water, the evil *kärpos* (*suγdyŋ äzi aza kärpos*), to whom the shaman applied by saying:

alty astu kär balyq	Six-horned *kär* fish
täbir tištü kär jān	iron-teethed mighty *kär*
jätti äžiktü jär qān	Earth King with seven doors
törtöčöŋä al töspäs	he does not tolerate the road
nojon äžän	*nojon äžän*

The drumstick of the *qalāč* was made of birch and covered with the hide skinned from a roebuck's leg. An iron plate, and six iron rings, three in each row, were attached to the rear part and served as bells (*šaltyraq*).

SHAMANS' DRUMS OF THE KUMANDINS (*tǖr*)

I received the communications quoted below from the shaman Saqan in 1936, at which time he was living in the Surbashin *aul*. He no longer had his drum, since, together with many other shamans, he had officially given up public activity as a shaman but, as he admitted, was still shamanizing for 'himself'.

The Kumandin shamans used three kinds of drums, differing from each other chiefly by the form of the wooden handle carved out of birch. The first was *alty müstü ala mars* 'six-horned pied *mars*'. I have already described this class when dealing with the Shor drums. The Kumandin *qams* made this type of drum only if one of their ancestors possessed a similar instrument and was a descendant of the Shors (usually on the mother's side). The Kumandin shamans received the *mars* drum exclusively from the Mustag mountain, the holy ancestral mountain of the Shors (and of every Shor shaman), situated in the area of Mountain Shoria. Thus, like the

[18] Вербицкий (1884) p. 239. Among the Teleuts and the Lower Kumandins this is one name of the deity: 'good spirit' '*ülgän*'s servant'.

Chelkans, the Kumandins consider the *mars* drum to be of Shor origin. The drawings are the same as those found on Shor drums. The drumstick is made of meadowsweet covered with the hide of a male hare.

The second type (*äki baštyγ qanym* 'two-headed *qanym*') was the most generally used and was considered to be 'genuine' not only by the Kumandins themselves but also by the Chelkans. In fact, as far as external shape and structure are concerned, the *qanym* drum of the Kumandins was the same as the *qanym* drum of the Chelkans. It will be seen later on that the only difference was in the drawings painted on the drumhead. The drumstick was made of birch and covered with the hide of a roebuck's leg.

The third type—the *täzim*—was again like that of the Chelkans. The handle represented an anthropomorphic figure with a head, scarcely indicated shoulders, no hands, and two legs carved out rather indistinctly. It was crossed by a chord-like iron stick at the level of the shoulders. Because of its handle, this Kumandin drum bears a strong resemblance to those used by the Telengits and the Altai Turks living in the basin of the Katun. A distinctive mark, emphasizing, so to speak, that it was a *täzim* drum, consisted of eyebrows and eyes on the anthropomorphic handle, representing the 'drum's master'. They were made of copper instead of being made of lead like the eyes on the *qanym*-drum's head, or of iron, like those on many drums of the Altai Turks of the Katun region.

The Kumandin shamans received this drum not from the spirits living in heaven, but exclusively from their own ancestral mountains: the Solog mountain, near the settlement Turachok, or the Kabak mountain near the Lake Teletsk. But, unlike the *mars* and *qanym* drums, it could not be used for addressing the supreme divinity of heaven—*ülgän*. This is reflected in its drawings which differ considerably from those of the *mars* and *qanym* drums. The drumstick was made of iron and covered with the hide of a roebuck (*quran*). Six small copper rings, serving as bells, were fastened to its rear part (*täs kübä*). The framework of the Kumandin drum was made of purple willow, the handle of birch. The six or seven 'horns' on the upper part of the framework were made from the tinder fungus (*mäškä*) that lives on the birch tree. The iron parts were taken from the old drum placed on the grave of the shaman's ancestor. If no such parts could be found, the blacksmith made new ones modelled on those that had been on the drum of the ancestor who informed the shaman of details in respect of their form and quantity, in the course of a special ceremony. The drum was covered with a membrane made of the hide of a roebuck, of a wild reindeer or, less frequently, of a maral-deer. It was forbidden to use the hide of any other animal for this purpose. Only the exterior of the drumhead was decorated with drawings, the inside was generally painted red. The drawings were furnished by a specialist after the instructions of the shaman. The drawings on all Kumandin drums were exclusively in red (*ochre*) and white paint, made of white clay (*šoyo*) and called *apač*. The paint was dissolved in water and mixed with the resin of the wild bird-cherry (*čärbittig saqazy*) for preservation. The following drawings were characteristic of Kumandin drums: First, the 'belt of heaven' (*täŋäräniŋ qury*), which divided the membrane into an upper and a lower field. The upper part contained the following figures: (1) nine *buras* arranged in a semicircle along the upper border, (2) the sun (*kün*), (3) the moon (*ai*),

(4) the morning star (*šalbon*), (5) a goose-like bird (*qarlyq*), (6) human figures shooting with a bow (*nān täŋri*), (7) birch tree (*baj qajyŋ*), (8) a virgin with three tresses—*ülgän*'s daughter (*üč čürmäčtig ary qys*).

Figures in the lower part were: (1) the great and the small *kärnjajan*, analogies of the Shor *kär-pos*, (2) a frog (*paqa*), (3) two lizards (*käläskän* and *qamnyq*), (4) a snake (*čylan*).

The shaman Saqan gave me a detailed description of the drawings on the 'genuine' Kumandin *qanym* drum. Here, too, the 'belt of heaven' divided the membrane into two fields. The upper part showed the following figures: (1) nine *buras* (*toγus bura*), the helpers of the shaman in attaining *ülgän*, arranged in a semicircle, (2) two birch trees (*baj qajyŋ*), (3) two birds (*qaraquš*) perched on the top of the birch trees, (4) the moon (*aj*), (5) the sun (*kün*), (6) nine stars (*toγus šolban*), (7) two geese (*äki qas*), (8) seven anthropomorphic figures with wings (*qytaj*). They are arranged parallel to the 'belt of heaven' at the bottom of the upper part.

The following drawings were in the lower part: (1) an archer (*atqy*), the helping spirit of the shamans; placed on the left side above, it was represented by a schematic human figure holding a bow (*adar oq*), (2) nine anthropomorphic figures (*čär ajaqtyγ*) 'walking on earth', the shaman's helping spirits, in whose presence he entered the lower world; however, they did not help the shaman in his journeys above the clouds, (3) frog (*paqa*), represented in the middle of the drum's lower part, (4) two images of *kär-čutpa*, placed on either side of the frog, (5) two lizards (*käläskän*) and *qamyq*, arranged behind the *kär-čutpa*, (6) snake.

Some of the Kumandin shamans have a single drum, which, judged by the form of its handle, belongs to the *mars* type, but with the drawings of *mars* and *täzim* drums combined. According to the explanations of shaman Saqan, this happened when the maternal Shor shaman's ancestors ordered the future shaman to make a *mars* drum, whereas the Kumandin ancestral mountains pressed him to manufacture the drum in their honour. Instead of making two separate drums, only a *mars* drum was made with drawings characteristic of the *täzim* type. The *mars* drum was taken as a basis because this type was suitable for addressing *ülgän*, whereas the *täzim* drum was inadequate for this purpose. Thus, this drum became universal, since it could be used for ceremonies connected with any divinity of heaven, earth and the lower world. In case of the above combination of drawings, the membrane of the drum was not only divided into an upper and a lower field, but the upper field itself was cut in two by a vertical line. On the left side were the drawings normally on the upper part of the *mars* drum, while those of the *täzim* drum's upper part were arranged on the right side. The lower part of such drums was common, and not divided into separate fields for *mars* and *täzim* drawings. The upper left side contained the following drawings which belonged to the patterns of the upper part of a *mars* drum: (1) nine *buras*, (2) the sun, (3) the moon, (4) stars, (5) two birch trees, (6) two *qaraquš* birds, (7) two geese, (8) an archer *atqy*.

The right-side drawings, belonging to the *täzim* type,were: (1) five, or seven *buras* painted alternately in white and red, (2) the sun, (3) the moon, (4) stars, (5) four, five or seven *čär ajaqtyγ* figures. The drawings of the combined drum's right side did not include the images of the shaman's assistants who helped him ascend to heaven, to *ülgän*. It was the conviction of shaman Saqan that there must be nine *buras* on a drum if a ceremony was to be performed to address *ülgän*. With a smaller number of *buras* a shaman could not ascend to *ülgän*.

The lower part of the combined drum contained the same drawings as those I have described when dealing with the *qanym* drum.

Such were the main characteristics of the drums of the Kumandins and the drawings on them. There were still other distinctive features—explained to me by shaman Saqan—which were due to the sex of the shaman. As a matter of fact, no female shaman, irrespective of her power or fame, was supposed to address *ülgän* or to ascent to heaven. Therefore—according to Saqan—the drawings on the drums of Kumandin female shamans did not include representations of the *buras*, the chief assistants of the shaman in his heavenly journeys. In the upper field of a female shaman's drum there were seven *čär ajaqtyγ* figures instead of *buras*. This was the main characteristic feature in the drawings of the female shaman's drum.

It should be noted that, among the upper Kumandins (*örö qumandy*) at least, the female shamans received their drums from the same Kumandin ancestral mountains as the men. Unlike the Teleut female shamans, Kumandin female shamans were not allowed to address *ülgän* even though they happened to possess a *mars* drum inherited from Shor (maternal) shaman's ancestors. Saqan told me that female shamans having a *mars* drum did exist among the Kumandins, for instance in the *örö qumandy sök* (*sök* 'clan'), but they could not sacrifice a horse to *ülgän*.

SHAMANS' DRUMS OF THE TUBA[19] (*čalu*)

I found three kinds of drums, differing from each other by the form of their handle among the Tuba shamans. The first of these, named *pars čalu* or *mars čalu* was of the Shor type. It belonged to the highest class: with its help, the supreme divinity living in heaven could be appealed to. By its outward appearance the shamans could tell that this drum was consecrated to *ülgän*'s daughter—*ala bars* 'pied *bars*'. The Tuba called the drum *jān täŋäri bars čalu* 'the *bars* drum of the mighty heaven', *ülgäniŋ qyzy ala bars* 'the *ala bars* of *ülgän*'s daughter'. However, it was very rare among the Tuba, and only shamans belonging to the *čor* clan (*sök*), descendants of the Shors possessed it. The shaman Közerek who lived in the Kayachak settlement of the Choya *aimak*, as well as his son, shaman Tiinci, had one. The handle represented an *ala bars*: it consisted of a wooden bar rounded off in the middle, widening evenly towards its flat ends. It was covered all over with oblique engravings in the form of 'fir trees', imparting a 'pied' appearance to the *bars*. This drum was never covered with anything but the hide of the wild male maral-deer. Regularly distributed into a lower and an upper field, the drawings were similar to those of the *mars* drums as described above.

[19] The Tuba, a group of the Northern Altais, living in the basins of the Sary-Koksha, Kara-Koksha, Pyzha, Uimen rivers (left side tributaries of the Biya river), as well as on the north-western shore of Lake Teletsk. They have also been called Chernevy Tatars (from the word чернь meaning 'dense, dark forest, taiga').

The second type, called *qanym čalu* '*qanym* drum' was the most frequently used and was considered to be of genuine Tuba origin. It could serve for addressing the spirits of heaven, as well as of the lower world. The wooden handle was called *äki baštu qanym* 'two-headed *qanym*' and represented a statuesque carved figure with a head on both its ends, turned in opposite directions. Referring to the double-headed handle arranged on a bar, A. V. Anokhin writes as follows: 'One of the heads can be found in the usual place and is called *čaludyŋ bažy* "the head of the *čalu*" (i.e., the head of the drum—L. P.), the other is called *qamnyŋ bažy* "the shaman's head".'[20] G. I. Gurkin has reported about the *qanym* that 'when the drum is beaten, its lower part talks with *ärlik*, and the upper part with *ülgän* . . .'[21] P. I. Todyžekov, an old hunter belonging to the Alan clan from the Puchak settlement of the Choya *aimak*, said: 'the *qanym* is a certain wild beast'. As has been recorded by A. V. Anokhin from a shaman's blessing (*alqyš sös*), the *qanym* is called the six-eyed daughter of *ülgän–ala bars*, translated by Anokhin as 'pied tiger'.[22] Consequently, the *qanym* handle has become, in the shamanism of the Tuba, the daughter of *ülgän–ala bars*, just as the *mars* handle of the Shor drum has—although, by having two anthropomorphic heads, it differed substantially from the latter.

The Tuba considered the *qanym* also to be the 'master' of all the game that lived in the taiga. We shall return later on to this fact. Let us note here that the handle of the Tuba *qanym* drum (as that of the *mars*) was always made of birch, while the hoop was made of cedar or poplar, but never of asp. This drum was covered with the hide of the wild male maral-deer, or of the roebuck.

The drawings displayed many features that were typical of Tuba drums. First of all, a considerable part, sometimes the major part of the drawings were painted on the inside of the drum. In this sense S. V. Ivanov was right when he said that the Tuba drums were closely related to the drums of the Tuvas.[23] The drawings were made with red paint. Two parallel red stripes, running sometimes also on the inside of the membrane, divided the drum into an upper and a lower field. Between these stripes, there was a painted zigzag line representing triangles. The dividing line ran at the level of the shoulder below the upper head, i.e. of the iron cross bar (*kiriš*). Facing the upper head, the sun, the moon and the stars were on the left side. The Tuba shamans told me that the drawings on the exterior of their *qanym* drums were not always the same. When they wanted to address heavenly beings, they decorated the exterior of the drum with special drawings suitable for that purpose. If, on the other hand, they desired to address the spirits of the lower world, they washed off these drawings and painted on new ones.

The third type, the *täzim* drum, or more precisely, the *täs qabaqtu täzim* '*täzim* with copper eyebrows', was only of restricted use, being prevalent mostly in the areas adjoining those of the Altais (*altaj kiži*). It was covered with the hide of the maral-deer or of the roe (which always had

[20] Анохин (1924) p. 53.
[21] Гуркин (1913) p. 100.
[22] Анохин (1924) p. 144.
[23] Иванов (1955) p. 23.

to be a buck). The handle was a single-headed anthropomorphic figure made of birch, with the eyes, the nose, the mouth and the eyebrows made of brass or copper. The mouth with the moustache and the eyebrows were cut out of a single plate (Fig. 8). The shamans received this drum from the mountains to which it was consecrated rather than to *ülgän* or to his children. Among the Altais living on the right bank of the Katun river, and also among the Tuba, the *täzim* is a sacred mountain—*tajqa*—al-

Fig. 8. *täzim*, drum of the Tuba

though nobody was able to tell me where it was situated. The *täzim* is referred to as a mountain and the protector of shamans in Anokhin's records as well. Thus, for instance, shaman Čujuq was said to have as protecting *tös* the *jäzim tajqa*.[24] The *tös* of shaman Kujbadak was likewise the *jäzim tajqa*. Shaman Abaqaj of the *qaryyl* clan ceremonies addressed not only heaven and *ülgän*, but also *jäzim bij*, called *altaj jäzim bij*. This term may also mean a mountain or a place, since in the folklore of the Altais the word *altaj* frequently meant 'mountain', just as the word *tajqa* means 'snow-capped mountain' or 'high mountain with barren peak'. A shaman's

[24] Анохин (1924) p. 141.

prayer includes a passage saying that *jäzim bij* has seven doors, etc . . .[25] N. M. Jadrincev has given us a description of the drum of shaman Pakjan from the Paspaul region (Isha river, right-side tributary of the Katun). According to the shaman himself, this drum was consecrated 'to the mountain spirit *jezym* (i.e. *täzim*—L. P.) sitting on seven drums'. The researcher was certainly right in stating that 'Pakjan's drum has a closer resemblance to the Altai Kizhi drums than to drums of the Chernevy Tatars. On the inside there is a copper-eyed idol: under the chin, the bar has two thick protuberances looking like hands. On the *kryš* (i.e. the iron cross bar—*kiriš*—L. P.) there are pendants (*sūr* and *sumqyr*). The drawings of the exterior are reminiscent of the Chernevy Tartar drums: a horizontal line divides the surface into two parts; curves and semi-circles divide the upper field into three zones, the first is heaven, the second the rainbow (*soloŋo*); the stars—*čaγan*—are represented in the lower field, the sun on the left side and the crescent-shaped moon on the right side; the drawings include a roe, a small fir tree, other trees and an unrecognizable form called *bura* by the shaman. Some interpreted this word as meaning 'cloud', others said, it stood for 'heavenly horses'; they may be identical, since thunder is called 'horses of heaven'—*qudajdyŋ burγazy*.[26] The female shaman Čiin (in Russian Anna), from the *čaγat-qomdoš* clan, living in the settlement of Artybash near Lake Teletsk, also had a *täzim* drum, which presented the following picture: 'A horizontal stripe divided the drum into two semicircles. In the middle of the upper part there was a tree, driven into a kind of pedestal by its foot. Luminous bodies (the sun and the moon) were on each side. Three hooks emerging from the earth probably represented the *buras*—the mythical shaman's horses. On the left side of the drum was a human figure with one leg shorter than the other: the female shaman called it *toroš* (its full title is *sār pudy qysqa, sār pudy uzun* 'one leg is short, the other long). The broken line which divides the drum into two parts is called *soloŋa* 'rainbow'. Beneath, there are five pendants—*qoŋru*—on the left border and four on the right. All the drawings are painted in red ochre. Under the horizontal stripe the figures of a stone he-goat (*täkä*) and of the female *djumγa* are to be seen.[27]

A Tuba *täzim* drum preserved in the Soviet State Ethnographic Museum (Reg. No. 4332–123. Fig. 8) presents the following figures on the inside of the hide: stars, two rainbows, clouds, taiga, sun, moon, master of the water and the taiga, the sea with the taiga and the rivers flowing into it, the shaman himself (*qam*) shooting with a bow (faded), a maral hind and her fawn, a wild she-goat and her kid. On the outside of the hide there is again the sky with rainbows, stars, clouds, further below, the taiga, still lower the sun, the full moon and the half-moon, the shaman-bird (eagle), the holy white poplar with the *qaraquš* birds at its summit and the shaman shooting with the bow (both drawings are indistinct). It is evident that all these figures formed part of the ceremony addressed to the spirits of the upper world. They were probably washed off in case of ceremonies addressed to the lower world, and new ones were painted on instead. When

[25] Анохин (1924) pp. 136, 141, 142 ff.
[26] Потанин (1883) p. 48.
[27] Гуркин (1913) p. 101.

making this conjecture, I am relying on the authority of what several Tuba have told me: each time the shaman went to a person who had invited him to shamanize, he changed the drawings on the surface of the *täzim* drum, washing off the old ones and painting new ones, as required by the aims of the ceremony.

S. V. Ivanov has given an overall description of the drawings found on the Tuba drums, without considering individual types.[28] I wish to point out some specific traits of these drawings on drum-types which the Tuba themselves consider different. His remark: 'as for the character of their drawings, the Tuba drums resemble those of the Altais and of the Telengits'—is true only for the drawings on the *täzim* drum.

In addition, it should be noted that the Tubas used different kinds of drumsticks for different types of drum. The stick for the *bars* drum was the same as those for the Shor drums. The sticks for the *qanym* and the *täzim* drums were made of birch, covered with hide skinned from the right leg of the *quran*. Six or eight small iron rings were fastened to the rear part.[29]

SHAMANS' DRUMS OF THE ALTAI KIZHI (*tūr* and *čalu*)

The following material refers to the Altais proper, who term themselves *altaj kiži*. Most of my on-site material was gathered from the Altai Kizhi living on the right riverbank in the Katun basin, in the same area where A. V. Anokhin had made his records and observations. I was lucky enough to be present at shamans' ceremonies here and among the Shors, and I studied the drums of different shamans in their own yurtas when out of use. Thus, I did not have altogether to resort to the amount of information I received from the shamans, but I could rely on personal observations. Whenever possible, I have referred to unpublished documentations, relating to Altai Kizhi shamans' drums in museums. I have also made use of various publications.

The Altai Kizhi living on the right bank of the Katun river (now the districts are called Elikmonarsk and Maiminsk *aimaks*) divide their shamans' drums into five categories.

1) Täzim čalu. The shaman called this *täs qabaqtu täzim čalu* '*täzim* drum with copper eyebrows'. I have already pointed out that the handle of this type of drum was an anthropomorphic figure carved of wood. The head of the handle, which was the idol, represented a flat human face with the nose, mouth, eyes, eyebrows made of copper. The hide that covered the drum was not folded back over the lower rim of the framework but was drawn together by a string stitched through the broad margin of the hide. The *täzim* drum was the most current type in the districts mentioned above, and possessed the most characteristic features of all shamans' drums of the Southern Altais. Its handle (called *bars*, *pars* by all Southern Altais) consisted of an anthropomorphic figure made of wood with a well

[28] Иванов (1955) pp. 230–4.

[29] For more about the shamans' drumsticks of the Tubas, Shors, Teleuts, Kumandins Chelkans, see: Потапов (1947).

discernible face. Its drawings contrast with the drawings on the 'original' drums of the Northern Altais where well-defined upper and lower fields can be distinguished. No such partition is found here. The most characteristic feature of the Altai Kizhi *täzim* drum and of the Southern Altai drums in general is that the 'drum's master' (i.e. the anthropomorphic handle, the iron cross bar and the iron pendants) is drawn in a size as large as the drum itself on the outside or sometimes on the inside of the parchment

Fig. 9. Altai Kizhi *täzim* drum

(Fig. 9). Another typical pattern is the arched stripes representing rainbows (*soloŋy*). Descriptions of Southern Altai drums can be found in the literature and need not be repeated here.[30] According to A. V. Anokhin's report, the anthropomorphic figure on the outside of this drum represents the image of the dead shaman (*ozo qam*), in fact, I think, that of the 'former shaman', i.e. the shaman ancestor.[31]

Classified according to their drawings, the *täzim* drums of the Altai Kizhi living on the right bank of the Katun river, belong to the shamans' drum

[30] Анохин (1924) pp. 55–65; Иванов (1955) pp. 244–50.
[31] Анохин (1924) p. 57.

type of the Southern Altais: The patterns of decorations are the same, only the details are depending on the personal ideas of the shaman.

2) Kämčik čalu. Kemchik or Tuva drum (Kemchik is a tributary of the Yenisei river). These drums belonged to shamans whose shaman ancestors were Tuvas, or Soyots, as they were formerly called. The *kämčik čalu* had an anthropomorphic handle made of birch, with a single head. The eyes, mouth and nose were made of iron. They were covered with the hide of the maral-stag. No stretchers were used. The framework was made of cedar wood. The external surface of the drum was divided by a white stripe into an upper and a lower field. All the drawings were arranged in the upper field, the lower one being left empty. The drawings included: (1) three *bura*s with long corkscrew horns like those of the wild goat (*täkä*); (2) the sun, (3) the moon, (4) the stars, (5) the rainbow, (6) *jalbraqtu baj qajyŋ* 'birch tree with dense branches'. The large-size drawing of the anthropomorphic figure, characteristic of the Altai drums, is not found on the Kemchik drums. The drum described above belonged to shaman Čorton (from the *tonžoan sök*), who lived in the Saryskyr of Kuyumsk Selsoviet of the Elikmonarsk *aimak.* I have received my information from Čorton himself, who handed his dum over to the local Soviet at the end of the 1920s, when he renounced public activity as a shaman. A similar drum was in the possession of Sapyr Tujanin, one of the last Altai Kizhi shamans of note, whom I frequently called upon, taking part in his shamanizing. Sapyr (from the *tonžoan sök*) considered his drum to belong to his own clan. He had received it from his shaman ancestors, among whom were also Ruvas. He used it only to shamanize for members of his own clan, and had a *täzim* drum for other persons. He said that he had travelled all over the Altai mountains with this drum.

3) äki baštu čalu 'two-headed drum'. In the Altai Kizhi language its full name was *qajyr qanym čalu äki baštu* 'the two-headed *qanym* consecrated to *qajyr*' (a mountain near Lake Teletsk). The Altai Kizhi themselves believed these drums to belong to 'darkness', i.e. to tribes that lived in the 'darkness' (the taiga)—the Tubas, the Chelkans and the Kumandins. The handle of this drum was the two-headed *qanym* described above. I, personally, have not seen this type. I was told that the drawings on it were like those of the Tuba drums, i.e. partitioned into upper and lower fields. Speaking of this drum, the Altai Kizhi (the former shaman Čorton and Muqulus Tadykin who lived near the Kuyum river) called my attention to the identity of the *qanym* drum (with a two-headed anthropomorphic handle) with the *ala bars* drum (or *mars* drum). They said that the *qanym* drum may also be called *ala barstyŋ čalu*, like the Shor drum (*ala mars*). A. V. Anokhin says the same: the two-headed *qanym* figure was frequently called *ala bars,* meaning (in his opinion) 'pied tiger'. He quotes the text of a shamanistic supplication to *qanym,* to confirm this indication, which begins with the words: *alty köstü ala barys* 'six-eyed pied tiger'.[32]

4) Abuγanyŋ čalu üč köstü 'three-eyed drum from Abugan'. Such drums were made only by shamans who had received them from the holy mountain of Abugan. I was told that it had the same handle—an idol—as the *täzim* drum, only it was not two-eyed like this one but had three eyes, one of

[32] Анохин (1924) p. 144.

which was placed on the forehead. I was unfortunately unable to examine such drum myself. Among the records of N. M. Jadrincev the description of a drum consecrated to Abugan can be found. He saw an Altai Kizhi drum consecrated to *järsu abyγan*, i.e. the mountains and waters of Abugan. Here is his description: 'Inside there is the *bory*, the *kriš* with nine *qoŋrus* and *konces*, i.e. a bow of ribbons tied to the *kriš*. Attached to the framework, on either side of the head of the *bory*, we find a pair of ears—*qulaq*. The transverse stripes on the lateral surface of the drum are called *qočqor* 'the way of god'. The external surface of the drum was decorated with the following drawings: *bory*, *qulaq*, *kriš*, the sun and the moon in the form of concentric circles with a dot in the centre and with rays reaching the circumference. Under the *kriš* was a rainbow represented by a gentle arch, above which there were two anthropomorphic figures—the daughters of god, *istjungunu qystar*.

On one side, beneath the rainbow, there was a drum: on the other, a group offering of a horse sacrifice. Below and above the rainbow were dots, standing for the stars. The shaman said that the girls were the playing daughters of *ülgän* 'coming down to the taiga and rising to heaven'.[33] Nothing is said about the number of eyes on the handle of the drum, the anthropomorphic idol. It is probable that there were only two, for it is hard to believe that a keen observer like Jadrincev would have failed to notice the fact that here were more.

5) pars čalu. Evidently this is the *mars* drum of the Shors. Its handle is of the Shor type decorated with carvings and small holes. The Altai Kizhi, who provided me with information, described this drum as being Shor and said that the shamans called it *alty köstü ala pars, pus tajqadaŋ purun čäčilgän* 'six-eyed *ala pars*, separated (i.e. originating) formerly (or a long time ago) from the ice-covered mountain—the taiga'.[34] The Altai Kizhi whom I questioned considered the *ala pars* to be a wild beast, probably a tiger.

In addition to the above differences emphasized by the Altai Kizhi themselves, another one must be mentioned here. The Altai Kizhi living in the Ust-Kansk, Ongudaisk and Shebalinsk regions (left banks of the Katun river, basin of the Charysh) divided all drums with anthropomorphic (one-headed) handles—the most current of which was the *täzim* drum—into two classes: *aq tüŋür* and *qara tüŋür*, i.e. white (clean) drums and black drums. The white drums were used for addressing the spirits of the earth (mountains and waters—*jär-su*) and those living in heaven, whereas the black ones served for addressing the spirits of the lower world. The *täzim* was a 'white' drum, because it derived from the mountains and definitely had to be covered with the hide of the roebuck. White, blue, red and many-coloured ribbons were tied to its cross bar (*kiriš*). The black drum was covered with the hide of a two-year-old bull. The black drum had the same handle as the *täzim* type, only with the eyes, eyebrows, nose and mouth of the anthropomorphic idol made of black iron. The ribbons tied to the iron cross bar

[33] See: Потанин (1883) p. 47. In this description, the word *bory* means 'handle', *kriš* (bow-string) 'iron cross bar', *qoŋru* 'iron pendant', *qočqor* 'male of the mountain (rock) goat', *qulaq* (ear) 'pendant'.

[34] This is the name of Mustag, the ancestral mountain of Shor shamans.

could be only black or red.[35] As we have said, then, the shamans' drums of the Altai Kizhi living on the left banks of the Katun and in the Ustkansk region were divided into white and black classes, according to whether they were used for conjuring up the spirits of the mountains and of heaven or the spirits of the lower world. The drawings of these drums were not clearly distributed into two fields: the upper with the patterns for addressing the mountains and heaven, and the lower with the drawings for addressing the spirits of the lower world. The drums of the Turks of the Northern Altai (Shors, Teleuts, Kumandins, Chelkans and Tubas) served both of these aims and their drawings were therefore separated into upper and lower parts.

All these shamans' drum types, belonging to the Altai Kizhi living on the right banks of the Katun, had drumsticks made of birch and covered with the hide of the roebuck's leg, or of the maral-deer. From three to nine iron rings were fastened to the rear part. I have not discussed the type of drumstick used with the *pars* drum since it was of the Shor type.

A summary of the various shamans' drums used among the Altai Kizhi shows that there are, in fact, only three basic types, depending on the form and on some features of the handle, characteristic of the entire Altai Region: (1) the Shor *mars* drum, (2) the two-headed drum of the 'darkness', used by the Chelkans, the Kumandins and the Tubas who live in the 'darkness', having a handle with a two-headed anthropomorphic figure, (3) the one-headed Southern Altai type (with a one-headed anthropomorphic figure as a handle). A subdivision of the one-headed type according to the form of the handle into the Kemchik and the Abugan types (having three eyes) does not affect the gist of our classification. My division of the drums into white and black types has nothing to do with differences in the handles: the handle remains the same, only the metal parts of the idol's face (nose, eyebrows, eyes, etc.) are of copper on the white drum and of iron on the black. A division based on traits of an apparently later date can be made by differentiating the Altai Kizhi shamans' drums according to their use for addressing the light (good) or the dark (evil) spirits. Čelekej Apaetov, who furnished me with this information (in the Altai Kizhi language) used the following terms in connection with the two types: *aru töskö tüžär—aq tüŋür* and *qara körmös ärlikä parar—qara tüŋür* meaning 'the white drum issuing from the clean *tös* (protector)' and 'the black drum coming from the evil spirit Erlik'.

SHAMANS' DRUMS OF THE KHAKAS (*tǖr*)

In the year 1946, while staying in the Troyakov *ulus*, I made the acquaintance of the former Sagai shaman, Roman Kaidarakov, a descendant of the *tay-qarγa sök*. Then he was an old man of 67 years. He talked about the affairs of the shamans willingly and told me of many interesting facts. He spoke to me about the drawings painted on the drums of the Kachin shamans, whom he knew and remembered well, although he had not—as

[35] These data are from the Altai Kizhi Čelekej Apaetov (now dead), who lived in the Ust-Kansk district.

he put it—held the drum in his hands for more than fifteen years. Roman told me that the following drawings were on his drum, which he had made according to the Kachin pattern: two parallel transverse red lines dividing the drum into an upper and a lower field, with a white zigzag line painted between the red lines. This was *qam čoly paχparda* 'the remote way of the shaman'. It was crossed by a perpendicular white line, which the shaman called *kiriš* (bowstring) and which was supposed to serve him as orientation when he was going on unknown roads or flying on his drum to the spirits. There were thus four sectors on Roman's drum. Above the upper areas were stripes in dark colour, representing *pulut* 'cloud', which was needed for the calling of rain, since shamans were often invited to shamanize so as to save the harvest from drought. (Roman remarked paranthetically, that the horn of the *tūr müz* drum—a bunch on the framework—was used by him to sail upwards.) He enumerated the drawings of his drum for each sector.

A) Right-hand upper sector: (1) *üs myjγaq*—the Orion constellation (in white). This was necessary to enable the shaman to find his way when shamanizing on earth or in the upper world. (2) *kȫk* 'cuckoo' (in white). This maintained contact between the shaman and the earth, when he was in the heavenly spheres. It was also the *kȫk* that supplied the trees with leaves and pulled the grass out of the earth. (3) *qusqačaq* 'sparrow' (in green). It maintained the contact between the stars, the sun and the moon. (4) *qusqun* 'raven' (in black). Assistant of the shaman on his long journeys. It grasped the ill man's soul (*čula*), when the soul was already playing with the children of the mountain master (*taγ ǟzi*). It took the *čula* with its bill and carried it into the drum of the shaman. (5) *baj qazyŋ* 'birch tree' (in white). It was by inspecting this birch tree that the mountain master found out the *sök* and the name of the sick, where the shaman was shamanizing (apparently this is a survival of the ancestral birch tree cult). Then, again with the help of the birch tree, the mountain master found out the origin of the illness—whether it came from *kürän izyχ* or not. (6) *qam pozy* 'the shaman himself'. (7) *toγus ōl* 'nine young men' (in black). (8) *čätti qys* 'seven girls' (in red), these latter being the children of the mountain master, the assistants of the shaman, through whom he could maintain verbal communication with the audience while he was in trance during the ceremony.

B) Left-hand upper sector: (1) *ai* 'moon' (in white). (2) *kün* 'sun' (in blue or green). (3) *čädigän* 'constellation of the Great Bear' (in red). The moon and the sun illuminated the road of the shaman when he was rising upwards during the ceremony in order to cure the sick. The Great Bear served as orientation.

C) Right-hand lower sector: (1) *pozraχ attyγ kiži-aŋči* 'hunter on a chestnut horse' (in red). This was to protect the shaman against another shaman whom he might meet on the road. (2) *čätti kök püri* 'seven grey wolves' (in yellow). When the shaman was shamanizing for a man whose flock had been devastated by the wolves, he showed these wolves to the mountain master and asked him to 'subdue' the wild beasts. (3) *toγus qara adaj* 'nine black dogs' (in black). These dogs assailed the evil spirits driven out of the ill man by the shaman. (4) *aba* 'bear' (in black). He was the master of the door (*isik äzi*) who did not allow illness to enter the man. He was the friend of the master of fire (*ot äzi*) and they had a 'common language'.

D) Left-hand lower sector: (1) *čylan* 'snake' (in black). (2) *kiläksin* 'lizard' (in green). (3) *somysqan* 'earthworm' (in red). (4) *paγa* 'frog' (in black). The shaman told me that the first three of these drawings were *tüs* (protectors) of the *qoj-izyχ* (i.e. the lamb's *izyχ*) who helped to cure woman's diseases. It was with these assistants that Roman rose up to the mountains and the fields, or even as high as the Caucasus. The frog had a 'common language' with the master of the lake (*köl äzi*) and was ruling the venereal diseases.

He also called my attention to the fact that the female shamans had other drawings on their drums: no perpendicular dividing line, only a horizontal one. On the top there was an arched stripe (*pulut*)—the cloud. No drawings were on the right side of the upper part. On the left side ranged the Great Bear, Orion, the moon, the sun, the sparrow, the cuckoo and the raven. On the right side of the lower part the birch tree, nine boys, seven girls. On the left side the shaman, a snake, a lizard and a frog. The worm was not painted on a woman's drum, it was the assistant of the male shaman only, and helped him cure the sick, whose hand or leg was broken. The shaman sent the worm into the sick, and it pulled the disease out of him.

Then the shaman talked to me about the drawings of the bear drum (*aba-tūr*), covered with the hide of the bear. According to him, the shaman of the Kyzlasov *ulus*. Apsalin who belonged to the *pältir sök*, had such a drum. It had a handle made of wood similar to the handle of the Sagai drums. The ribbons (*čalama*) were exclusively of black colour. On the bowstring—the iron cross bar—were seven iron plates—*qylys* pendants (swords), nine *sünür* pendants and seven copper bells. It had only three horns on the framework (*üs müs*). When the shaman appealed to this drum, he said: *qalyn tajqa tobyr kilgän, qaltar āstyγ qara aba!* 'black bear, arriving through the dense taiga, with a mossy mouth!' The drum was divided into four sectors by a horizontal (the road of the shaman) and a vertical stripe (*kiriš* 'bowstring'). In the right-hand upper sector were three stars.

In the left-hand upper sector: (1) *at pazynča sary-mas* 'the yellow gadfly like the horse's head'. It sucks out the blood of the evil spirit who sent the disease. (2) The shaman, to whom the drum belongs. (3) *coqyr-an* 'glutton'. It must catch the spirits of the disease *ajna* in case the shaman is unable to do so. (4) *aba* 'bear'. The master of the door, i.e. the protector of the house. (5) *syby* 'fir tree'. Roman told me that by the upper reaches of the Abakan, there is the Karatag mountain, with the Karasyn crest, where black fir trees grow (*χara-syby*). All the shamans who go to see this crest, leave their signs (*taŋma*) there, so that the master of the crest may know which of the shamans was there. The crest is very difficult of access and only very strong shamans are able to reach it.

In the left-hand sector of the lower part: (1) seven blue wolves, and (2) nine black dogs were drawn.

Shaman Roman told me that he had made his drum in the Kachin fashion, because he had lived among the Kachins since his childhood and because one of his shaman ancestors had been a Kachin. On the iron bar that crossed the handle of his drum, were seven iron pendants—swords, nine *quŋur* pendants and seven copper bells. The drumstick (*orba*) was made of bone. He inherited it from the drum of his shaman ancestor Tadyk. 'Before beginning to shamanize'—said Roman—'I sat and warmed myself from inside', *idim izirčam* was the term he used, i.e. 'I get into ecstasy'. He added: 'When I get warmed up, I become light and go wherever I want to'.

I have given here something of the material I was able to collect about the drawings of the Khakas shamans' drums from informants. I have also recorded the words of this shaman about how the Kachin shamans made their drums, but the discussion of that material would exceed the scope of the present study.

Now I wish to put forward some personal opinions of general bearing and also in respect of certain details. First let us analyse the Altai Turkic name for the drum. Two terms are used: (a) *tüŋür* or *tūr* and (b) *čalu*, the first being Mongolian, the second Turkic. The drumstick of the shaman's drum is called *orba, orbu, orby* in the entire Altai region. The shaman's name is *qam*—also of Turkic origin. All the tribes of the Altai mountains used the Turkic term for the shaman and for the drumstick, whereas the drum itself was called by both the Mongolian and the Turkic names, the former, however, being more frequently used. This fact cannot be explained without due regard to the development of the shaman's drum from its precursors—the implements used in ancient cults, and naturally, without considering the ethnic development of the Altai tribes. Unfortunately, I myself, for the moment, am not in a position to devote my time to this fascinating problem. But I hope that the material contained in this article will facilitate further research in these lines. I wish to stress that several Altaian tribes use different terms for the drum during the ceremony. This is of great importance for both linguistic and historical analysis. During ceremony, the Kumandins and Chelkans call the drum *aγatan* (*aq-adan*); so do the Bachat Teleuts: *alty örköštü aq-adan* '*aq adan* with six horns'.[36] Among the Chelkans I have also encountered the term *ärbatan* (*ärb-adan*), where the word *adan* can be clearly recognized, as well as the name *paγac* (*paγ-aγač*). During ceremony, the Shors call the drum *ägri adan* (again the word *adan*!) and *aq-čaγal* (*alty örgöštü aq čaγal*) 'six-horned *aq-caγal*'). Hence it is necessary to work out the history of the word *adan*, since it is contained in most of the terms used among the Teleuts, the Kumandins, the Chelkans and the Shors. Judging by the records of A. V. Anokhin, it can also be found in the shaman's incantations of the Altai Kizhi living on the right banks of the Katun river (from a shaman belonging to the *tölös sök*).[37]

It must be noted that all the Altai tribes use *mars, bars, pars, parys* for the handles of the drum although they are of various shapes. In shaman's incantations, even the two-headed anthropomorphic handle of the Chelkan, or Tuba drum is called *mars* or *bars* 'six-eyed *ala mars*'. It is highly probable that also the handle of the Altai drum passed through the well-known process of evolution. Starting from simpler forms—like, up to recent times,

[36] See: Вербицкий (1884) *adan* (В.—К.) 'бубен шаманский' (shaman drum), *egri adan* (В.—К.) 'бубен при шаманском действии' (drum in the shamanistic activity). [В.—К. means an Upper Kondoma dialect, i.e. Shor.]

[37] Анохин (1924) p. 133. He translates this term by 'white father', unaware of its being the name of the drum.

that of the handle of the Shor drum—it gradually received an anthropomorphic shape characteristic of the Southern Altai drums. I am of the opinion that the two-headed anthropomorphic *qanym* figures to be found on the drums of the Chelkans, Kumandins and Tuba have developed directly from the handle of Shor drums, with its thickened or flattened ends. My opinion is supported by the fact that, even at the beginning of the 20th century, the Chelkans had drums with two-headed handles that had not as yet become anthropomorphic. K. Hildén shows photographs of Chelkan drums with such handles. In addition, some two-headed Chelkan drums have handles with vestiges of the carved ornamentation characteristic of the Shor type drum—the *mars*. The fact that the two-headed handle is called six-eyed *ala mars*—as I have pointed out more than once—argues also in favour of this opinion. In this connection I wish to recall the above-quoted information received from the Altai Kizhi Chorton and Muqulus, according to whom the *qanym* drums and the *ala bars* or *mars* drums are one and the same.

From the genetic point of view it is interesting to note that the two-headed anthropomorphic handle of shamans' drums, as well as the two-headed hunter-idol have the same shape and name among the Chelkans, Kumandins, and Tuba tribes.

The Chelkans had a spirit called *qanym*, which was the protector of hunters. They made a two-headed image of it, very similar to the handle of the drum, only smaller in size. One of the heads had a 'beard' made of the tail of a spring squirrel (*sary tïn sayalu* 'having a beard from a yellow squirrel'). I was told by Chelkan hunters that they distinguished two *qanym* protectors: (1) *ȫn qanym*[38] 'wooden *qanym*', (2) *tämir qujaqtyγ qanym* 'iron-armoured *qanym*'. The latter was considered to be the younger one, excelling by his extraordinary impetuosity. The Chelkans believed that both *qanym*s lived in the taiga, on the Southern shores of Lake Teletsk. Before each meal, and especially in the evening, the hunters entertained the *qanym*. For this, they made a *közäm*—a mush made of *talqan* 'crushed barley', mixed up with the crushed seeds of the cedar cone. Then they made a soup from capercailzie and sprinkled the *qanym* with it (small morsels of meat had to be in the soup). The hunters addressed the *ȫn qanym* in the following words:

qan kiži qajraqan!	King of men, saint!
taulu čärgä tajaq bolzyn	Be (my) cane in the mountains
taštu čärgä toloq[39] *bolzyn*	Be a support in the stony area
ägri paraq äš[40] *polo bär*	Bowing tree with dense branches, be a friend
tazyl paraq araš čaqšy bolzyn däp	Be beautiful with your mighty roots, tree with dense branches

[38] *ȫn* 'tree trunk', hence a *qanym* made of it.

[39] *toloq* 'support', against which one can lean, by which one can be supported.

[40] *äš*—a term found also in ancient Orkhon Turkic inscriptions in the sense of friend, one who helps. *paraq*—name given by the hunters to a tree with dense branches (mostly the cedar), which provides a good shelter in bad weather and can be used for making a hunting hut. Generally a large, ramifying tree was chosen for this purpose. This tree is honoured, given food and ribbons are tied to it.

on polyma tajaq bolzyn	Be (my) cane with the right hand
sol polyma toloq bolzyn	Be my support with the left hand
tös taγdyŋ suranyp očor	Solicit the ancestral mountain[41]
ägri paraq äš polo däp	Bowing tree with dense branches, be a friend
tazyl paraq aras bol däp!	With your mighty roots, tree with dense branches, be beautiful!

Many of the Chelkans invited the shaman to perform ceremonies for the success of hunting and for safeguarding their modest harvest of barley in the taiga. In such cases the shaman performed without a drum. In his hand he had a two-pronged wooden fork with a string between the prongs. Pieces of old iron and coal were hanging on this string. The same method of addressing the *qanym* was resorted to when the hunter was ill. The ceremony was performed in autumn, after the reaping of the barley and before the beginning of the hunting season. The *qanym* was offered home-made beer brewed of fresh barley (*orto*) and of barley malt (*uγut*). The beer was poured into big vessels made of birch rind (*tozaq*). Containers intended for the *tämir qanym* were of smaller size.

Fig. 10. *qanym*, idol of the Kumandins

Late in the spring, when the barley was sown, ceremonies were made, when the *qanym* was asked to protect the crops against hail. In short, the *qanym* was the principal protecting spirit for the Chelkans, who led a hunter's life, and a silvan godhead for those inhabiting the mountainous taiga. The Upper Kumandins (*örö qumandy*) also venerated *qanym* as the spirit of the taiga and the protector of hunters. His image was made of a young birch tree and called *äki baštyγ qanym* 'two-headed *qanym*' (Fig. 10). It was flat, and the two heads were placed in opposite directions to one another. One of the heads had a beard made of a squirrel's tail. In autumn, the hunters appealed to the *qanym* and invited the shaman. For this occasion, barley beer (*ortqo*) was made in cylindrical vessels of birch wood (туяс) which were placed in the clean corner of the dwelling house. A stitched net (*aγ*) used in sable hunting was hung on the wall, above the beer vessel. In front of the window a young cedar tree was stuck in the earth. It was called *pažyn čār baj toru* 'honourable, rich with cones', and was decorated with ribbons. The shaman performed with his drum during the whole night. He made the hunter who had invited him repeat the appeals to the *qanym*. The Tuba also knew the protector of the

[41] Literally: 'Sit down, asking'.

hunters under the name of *qanym*. An image of it, collected at the beginning of the 20th century, is preserved in the State Ethnographic Museum of Leningrad (Nos 720–12). The inventory of the Museum contains the following entry in connection with this item: '*qanym*, i.e. the *tüs* of the clan *küzän*, living along the upper reaches of the river Biya. Coarse representation of two anthropomorphic figures without legs, united in one trunk with the heads at the opposite ends. The hands are marked by short cross bars. On the ends, representing the heads, eyes, eyebrow, moustache and chin are indicated with nailheads, resp. drawn with coal. Its length is 33·5 cm'.

According to V. Verbickij, a two-headed hunter idol quite similar to the *qanym* was known in the upper reaches of the Kondoma river among the Shors, living near the Chelkans. It was called *tajγam*.[42] 'It was a two-headed idol made of wood, divided by the gripping part, with buttons instead of eyes.' This is confirmed by a piece from this collection, in the State Ethnographic Museum of Leningrad (Nos 720–9): '*tajγam*, i.e. the *tüs* of Chernevy Tatars, consists of two ovals connected by a bridge. Height: 30·5 cm. Riverhead of Pyzas in the Kondoma system.' This object was received by the Museum at the beginning of the 20th century. At that time, and later too, the riverside of the Pyzas was inhabited exclusively by Shors. Although this image bears a great resemblance to the anthropomorphic *qanym* or *tajγam* (according to the description of Verbickij), the heads are still replaced by the oval forms of the face, i.e. almost in the same way as on the *qanym* drum of the Chelkans photographed by K. Hildén. It is to be noted that the protecting spirit of the hunters called *tajγam*, the image of which is analogous to the two-headed *qanym*, was known also among the Shors living on the river system of the Mrassu (from the *tajaš sök*).

There is, we believe, sufficient material to raise, and to solve, the question of the genetic connection existing between the wooden handles of the Altai Turkic shamans' drums and their wooden idols. The material here published permits us to study the age of the various types of drums, and to ascertain the direction of evolution followed by the ethnic groups of the Altai Sayan mountains. I wish to call the reader's attention to the following significant fact: Among the Kumandins it was not permitted to use the 'original' drum of this people (with two-headed handle) to invoke *ülgän*, the supreme divinity, who was venerated by horse sacrifices exclusively. This seems to imply that the 'original' Kumandin drum had developed when their ancestors were still hunters, when the cult of the ancestral mountains still prevailed—an age, which in my opinion preceded the cult of *ülgän*, the supreme heavenly divinity. Is this not the reason why the hunting deities of the aborigines living in the mountains and the taiga of the Altai Sayan upland gradually have become the daughters or sons of *ülgän* (e.g. six-eyed *ala mars*, i.e. *qanym*)? It is most desirable that these questions should be examined, for they would contribute not only to the study of the history of shamanism, but also to the better understanding of the cultural and ethnographic history of the Altai peoples.

[42] Вербицкий (1884) p. 327.

REFERENCES

(Anokhin, A. V.) Анохин, А. В. (1924) Материалы по шаманству у Алтайцев (Contributions to the Shamanism of the Altais), in *СбМАЭ*, Vol. IV.

(1929) Душа и ее свойства по представлениям телеут (The Soul and its Properties in the Concepts of the Teleuts), in *СбМАЭ*, Vol. VIII.

(Černecov, V. N.) Чернецов, В. Н. (1947) К истории родового строя у обских угров (Contributions to the History of Tribal Organization among the Ob-Ugrian Peoples), in *СЭ*, Vols VI–VII.

(Dyrenkova, N. P.) Дыренкова, Н. П. (1949) Материалы по шаманству у телеутов (Contributions to the Shamanism of the Teleuts), in *СбМАЭ*, Vol. X.

(Gurkin ,G. I.) Гуркин, Г. И. (1913) in *ТТОИС*, Vol. III.

Harva, U. (1938) Die religiösen Vorstellungen der altaischen Völker, in *FFC*, Vol. LII, No. 125.

Hildén, K. (1916) Om Shamanismen i Altai, speciellt bland Lebed Tatarerna (About Shamanism in the Altai, Especially among the Lebed Tatars), in *T*.

(Ivanov, S. V.) Иванов, С. В. (1955) К вопросу о значении изображений на старинных предметах культа у народов Саяно-алтайского нагорья (On the Significance of the Representations on Ancient Cultic Objects of the Peoples in the Sayan Altai Mountains), in *СбМАЭ*, Vol. XVI.

Nioradze, G. (1925) *Der Schamanismus bei den sibirischen Völkern*, Stuttgart.

(Potanin, G. N.) Потанин, Г. Н. (1883) *Очерки северо-западной Монголии* (Papers on the North-western Mongols), collection, Vol. IV.

(1893) О крестообразных фигурах на шаманских бубнах и писаницх (On the Cross-shape Figures on Shamans' Drums and in Scriptures), in *ИВСОРГО*, Vol. XXIV, No. 2.

Potapow, L. P. (1934a) Die Herstellung der Šamanentrommel bei den Šor, in *MSOS*, Vol. XXXVII, Part 1. East Asiatic studies.

(Potapov, L. P.) Потапов, Л. П. (1934b) Лук и стрела в шаманстве у алтайцев (Bow and Arrow in the Shamanism of the Altais), in *СЭ*, No. 3.

(1947) Обряд оживления шаманского бубна у тюркоязычных племен Алтая (The Enlivening Rite of the Shamans' Drum of the Turkish-speaking Tribes in the Altai), in *ТрИЭ*, Vol. I.

(1949) Бубен телеутской шаманки и его рисунки (The Teleut Female Shaman's Drum and Its Drawings), in *СбМАЭ*, Vol. X.

Potapov, L. P. (1957) Samanismus u národnosti Sajansko-Altajské vysocny jako historicky pramen (Shamanism among the Peoples of the Sayan-Altai Highlands as a Historical Source), in *ČE*, No. 2.

(Verbickij, V. I.) Вербицкий, В. И. (1884) *Словарь алтайского и аладагского наречий тюркского языка* (Vocabulary of the Altai and Aladagic Dialects of the Turkish Languages), Kazan.

The Protection of Children among the Shors

by

N. P. Dyrenkova Leningrad

In the southern parts of the Tomsk province within the Kuznetsk Altai, the Shors surrounded by the taiga and isolated by the barricades of the stony Mras have preserved a particularly clear notion of the *ajna*, an evil spirit which constantly shadows the child whether it be a boy or a girl, with the sole purpose of inflicting upon it diseases or death,[1] or 'swallowing the child', as the saying has it. Usually a dead child is not said to have died but to have been swallowed by the *ajna: ajna čyrsaldy*. The *ajna* is addressed as follows: *sän noγa palaγa; qadylčazyn ajna* 'why do you attack the child, *ajna*'; *qazyq pala aγry* 'you make the healthy child ill'.

No sooner is the child born than the *ajna* hides near the yurta and lies in wait for it. That is why, during the delivery, the door of the yurta should, if possible, be, kept closed so as to prevent the *ajna* from slipping in. When coming in or going out, the door is only slightly opened. In the hearth of the yurta a bright fire should be kept constantly burning for three or sometimes more days and nights after the birth of the child. If the appearance of the child is delayed, the women in the yurta throw pieces of burning timbers through the door or through some fissure above the hearth in order to drive away the *ajna*. If the child does not show any signs of being alive, iron cauldrons are beaten, or sometimes a loaded gun is hung in the yurta to scare away the *ajna*.

The rearing of the child is fully centered around the struggle of the parents against the *ajna*, which is always trying to get near the child. If it is frequently ill, its name is changed. In families where children have died, the new-born baby is given a nasty name to make it repulsive to the *ajna* which, therefore, leaves it in peace; or it is named after some animal or bird.[2] The child sometimes receives the name of a very old man so that it should live as long as its namesake. In other cases the name of a man whose own children have been successfully brought up is used.

The colour black attracts the *ajna*. Hence, a new-born baby is never wrapped up in black but in white clothes whenever possible so as to hinder the *ajna* from entering its body. The best is to take the clothes of some very old person to make the child live as long as the latter: *aq qan čažyna* 'to be as old as the White Tsar'. The child must not be awakened lest its soul (*süränäsi*) which is absent while it is asleep should become frightened and fail to return. It might even become insane or die. The cradle (*päžik*), prepared for the child, is smoked with burning coal and juniper (*aračyn*)

[1] The Teleuts and the Altai Kizhi have a similar conception. A. A. Popov (a student of the Leningrad State University), informs us that the Yakuts protect children from the *abasy* ('evil spirit').

[2] In houses where children have died the Altai Kizhi call their children by indecent names so as to protect them. See Radloff (1884) p. 316; Вербицкий (1870) p. 172.

in order to drive away the *ajna* hiding in it.[3] As soon as the child is born it is placed in its first cradle which is made of birch bark. A month or so later the second is made of wood by its paternal grandfather. If the child dies, the cradle is not left in the house nor used for the next child, because the *ajna* is lurking most probably in the cradle which, together with the corpse of the child, is removed from the dwelling far into the taiga. The corpse of the child is wrapt in white linen, never in black, lest is should be kept in darkness for ever.

A rope made of horse-hair or hemp is used to tie the cradle to a tree, preferably to a fir tree. In places where the hanging of dead children is no longer customary, the cradle is hung on the tree near the place in the woods where the body of the child is buried. Most families in which a child has died move to another place. Those in which no child is expected to be born any longer, pass the cradle on to relations or else take it to the woods and hang it on a tree. Nothing that is connected with the cradle should be burned, because if it is, the sons and daughters of that family will have no offspring.

The mother does her utmost to protect her child from the *ajna*. While rocking it to sleep, she first addresses the *qudaj* 'supreme deity' with words to this effect: *äj, qudaj, män palamny alqal* 'O, *qudaj*, save my child'; *män qarayym čažyn (čašyn) tökpä* 'do not let my eyes shed tears'; and, addressing the *ajna*, she adds: *män palamny čeyan ajnanyŋ* 'the *ajna* that swallows my child'—*igi qarayy solzyn* 'should be blind in both eyes'.

The mother, not convinced of the strength of her words, hangs a bear's paw near the door of the yurta or of the house, in order to protect the entrance from the *ajna*.[4] Sometimes the lower jaw of a squirrel, a rabbit's skin, or the shell of *Cypraea moneta* (called *aq tiš* 'white tooth') is hung on the cradle. One or several pieces of this same shell are sewn to the caps of small children when crossing a pass in the mountains. Also the tooth of the maral (*syn tiži*) is hung as an amulet around the neck of small children. Along the lower reaches of Mras-su river the cradle is hung from a hook in the ceiling by means of the collar-bone of a wood-grouse or of a goose to make the child grow strong.

A widespread custom is to hang a small bow and arrow on the cradle of the child or on the wall of the yurta or house, where the cradle is kept, so as to protect the child from the *ajna*. In some places the Shors put in the cradle of girls under the pillow no bow but only an arrow wrapped in the fell of a squirrel, or in a piece of linen made of *kändir* 'hemp'.

The bow and the arrow as means of protecting children is widely diffused also among the other Turkic peoples of Siberia and can be regarded as a common characteristic.

[3] According to Popov, the Yakuts put a tiny wooden poker with one end burnt into the cradle of boys, and scissors into the cradle of girls to prevent the *abasy* from getting in.

[4] Claws near the cradle were found also among the Tuvas. (See Vasilev's collection in the Museum of Anthropology and Ethnography, Leningrad. For the Yakuts, see the paper by Ионов (1915) p. 57, and information given by Popov, student of the Leningrad State University.)

REFERENCES

(Ionov, V.) Ионов, В. (1915) Медведь вопоззрениям якутов (The Yakut Conception of the Bear), in *ЖС*, Vol. XXIV.
Radloff, W. (1884) *Aus Sibirien*, Vols I–II, Leipzig.
(Verbickij, V. I.) Вербицкий В. И. (1870) *Алтайские инородцы* (Altai Nationalities), Tomsk.

The Problem of the Ethnic Homogeneity of Tofa (Karagas) Shamanism

by

V. Diószegi, Budapest

Scientific descriptions of the shamanistic ideas and objects of the Siberian peoples give the impression that a high degree of homogeneity prevails within the various ethnic groups. Earlier publications suggest, for instance, that all the shamans of a given people have the same type of drums, drumsticks, head-dresses and cloaks.

This homogeneity is hinted at not only by such general works on shamanism in Siberia, as are the books by Nioradze and Harva,[1] but also by descriptions that treat the culture of an individual ethnic group.[2] A similar impression can be gained from monographs on shamanism among a specific people.[3]

Theoretical considerations have warned us against adhering to this conception of ethnic homogeneity. Our disbelief found material support in the course of a study of the relevant collections of different museums in Budapest, Leningrad, and in several Siberian towns. We are of the opinion now that the Siberian shamans' paraphernalia show a wide range of differences within one and the same ethnic group.

I decided to study the problem of ethnic homogeneity as part of my field work on shamanism in order to clarify the contradiction that exists between the accessible literature and the objects in the museums. Therefore I sought to make observations among a people that has a comparatively small population and lives in a relatively small and isolated area, where great differences could hardly be expected.

The Tofas, in their own language *topa* ~ *tofa*, satisfied these requirements: they live in a restricted region and their population is only about 430 or 440 all told. So I spent July, 1958 among these people during my second field trip in Siberia.

Today the hunting, reindeer-breeding Tofas live in three villages: Alygdzher, Nerkha and Verkhnaya Gutara, which are quite new, having been brought into existence by the Soviet regime in order to accustom this nomadic people to a more settled life. Before, the Tofas lived in three districts more or less isolated from one another: the western (Gutara river), the central (Uda river) and the eastern (Iya river) parts of the Eastern Sayan mountains (Fig. 1).

The Tofa people consisted of clans (*ńon*), five in all (*beš ńon*) the *aq čoγdu, qara čoγdu, χaš, saryγ χaš* and *čeptej* clans. In accordance with the territorial distribution, they were divided into three groups: (1) To the eastern group belonged the *saryγ χaš*, and *χaš* clans and some of the *čeptej*s,

[1] Nioradze (1925); Harva (1938).

[2] Серошевский (1896); Вербицкий (1893); Лопатин (1922); Богораз (1939); Chodzidło (1961).

[3] Трощанский (1903); Анохин (1924); Агапитов – Хангалов (1883); Широкогоров (1919); Шимкевич (1896); Анучин (1914).

living in the western area of the Eastern Sayan mountains. The *saryγ χaš* carried on their nomadic existence in the region of the Gutara river, the *χaš* lived along the Dzhuglym river, and the western *čeptej*s inhabited the region of the Malaya-Biryusa, the Yerma and the Yaga rivers. (2) In the

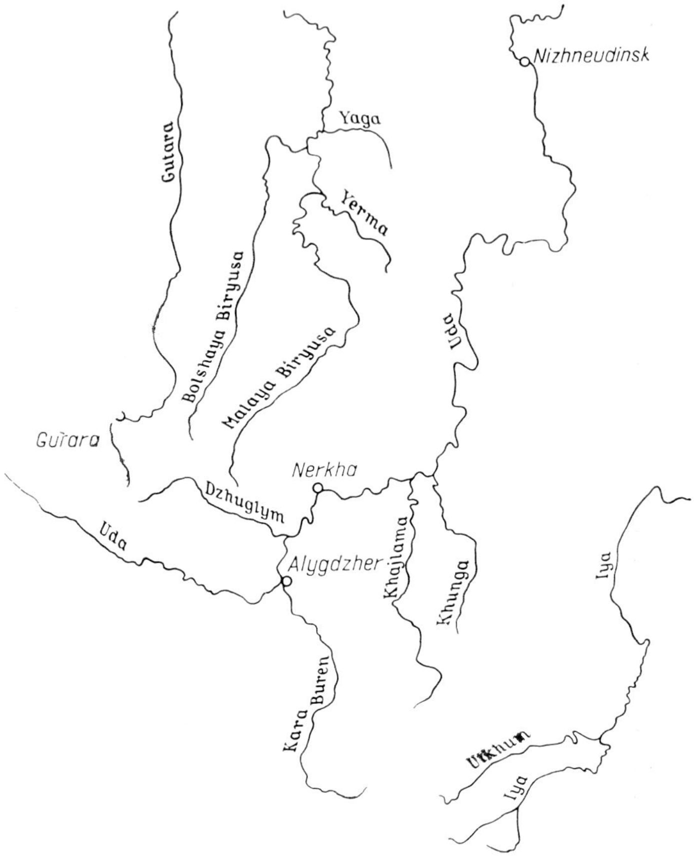

Fig. 1. The habitat of the Tofa people in the first quarter of the 20th century

central area in the region of the Uda, Kara-Buren, Khailama and Khunga rivers lived the *aq čoγdu* clan and the other part of the *čeptej*s. (3) The eastern group of the Tofas, the *qara čoγdu* camped in the vicinity of the Iya and Utkhum rivers.

Officially the Tofas have been Christians for a long time. More than a hundred years ago Štubendorf wrote: 'The Karagas have all been christened, and members of the fourth generation regard themselves as Russian Orthodox.[4] Christianity is, however, merely formal. 'They like to cross them-

[4] Штубендорф (1854) p. 245.

selves and genuflect before the holy icons, but at the first visit to a yurta one notices next to the picture of the venerated miracle-working St. Nicholas, a leather bag decorated with eagle feathers, squirrel-, rabbit- and reindeer-tails, tin disks and leather belts. This is an amulet which is supposed to protect its owner from illness, failure, accidents and all kinds of trouble, and bring him a full bag, a successful journey and good luck with the reindeer herd. Amulets are provided by the shamans, the devil's priests who perform their rites secretly in well-nigh inaccessible places in the mountains. The Karagas refuse to admit that there are still shamans in their midst, and the amulets vanish as soon as a Russian appears. Nevertheless, there is a shaman in every *ulus*, the shaman of the Udinsk *ulus* being the most famous of them all.'[5]

A few decades later Kostrov described a similar state of affairs: 'The Karagas are christened and have considered themselves Orthodox for four generations, but this Russian Orthodoxy is of a most dubious nature. It is true that every Karagas considers it essential to wear a copper cross around his neck, to make the sign of the cross when he enters a wooden house, to cross himself before and after dinner, to keep an icon in his yurta. But all this springs from opportunism rather than from any real inner conviction. Actually the Karagas to this day believe in shamanism. There is a shaman in every *ulus*, and the cloak, which custom demands shamans should don for the magic acts, is found in every yurta.'[6]

According to V. N. Vasiljev, who visited the Tofas at the beginning of the 20th century, the number of shamans had diminished so that he did not find more than three female shamans, and no male shaman at all.[7]

From on-site observations Evsenin concluded that shamanism had not lost ground even by the end of the 1920's. Except for 3–5 people 'the Karagas are christened according to the Orthodox rite. Their idea of God is, however, most confused. Like most of the Siberian natives, they get God mixed up with St. Nicholas . . . On entering a yurta, you will notice in the front corner not one, but often several icons, about which, however, you will not get any information from the Karagas, for they do not know their meaning. There is nothing surprising in the fact that they hang up the images of their own deities next to the icon. Moreover, the Orthodox icons appear in the front corner usually only when Russians arrive at the settlement, otherwise they are kept in a bag. Therefore it is obvious that, although the Karagas are Orthodox, in their case "Russian Orthodoxy" is merely formal. At the same time it would be a mistake to suppose that they have no religion at all. Actually all the Karagas believe in shamanism.'[8]

As B. Čudinov reports, Tofa shamanism was revived toward the end of the 1920's: 'Before the Revolution the Karagas had some three or four old shamans altogether: their number increased rapidly during the Civil War and today it must be in the region of thirty. And this is quite a lot for a community consisting of ninety families! Although during the Civil War there was no actual fighting on the land of the Karagas, they were short of flour, salt and hunting equipment and therefore approached the

[5] Штубендорф (1854) p. 245.
[6] Костров (1871) p. 698.
[7] Васильев (1910) p. 75.
[8] Евсенин (1919) p. 14.

Russian settlements along the Uda river, where discord and typhoid fever were raging. Typhus wiped out many Karagas families. The shamans explained this mass destruction by saying that the people had neglected their former gods—the spirits of the taigas and the mountains—and that the spirits, whose anger had been evoked, were now taking their revenge. The shamans selected good candidates and the number of shamans increased rapidly. The Civil War and the epidemic passed and happier days ensued when life was easier for the entire population of Siberia. The Karagas believed that improvement was due to their original gods. Shamanism spread and grew.'[9]

This sudden recovery of shamanism, however, involved its final extinction at the same time, for to our knowledge no shamans are active any longer in the land of the Tofas.[10]

In the course of my field work,[11] information received from former shamans, as well as the near kin of deceased shamans, gave me a fair picture of the chief particularities of shamanism among the Tofas. The differences were: (a) in clan, (b) in nature (white and black shaman), (c) in degree of power and proficiency (small or minor, middle or intermediate and big or powerful shamans) and (d) individual differences.

THE PROBLEM OF SHAMAN FAMILIES

In the course of my field work I was able to trace 21 Tofa shamans (*χam*). They included men (*χam ašńaq*) and women (*χam qorńaq*) and belonged to different clans (eight to the *aq čoγdu*; five to the *χaš*, four to the *čeptej*; two to the *saryγ χaš* and two to the *qara čoγdu* clan).

In the *aq čoγdu* clan the Amastaev family had provided the largest number of shamans. Three members had been active shamans: Innokentyj, who was so old that he had no patronymic name (at the time of his birth the Tofas did not use the father's Christian name in the Russian fashion), Innokentyj Nikolaevič and Vasilij Mikhajlovič. In the Kholemoev family there had been two shamans: old Ivan (he did not have a patronymic name either) and his grand-daughter: Anna Innokentyjna. There were three more shamans in the *aq čoγdu* clan: Innokentyj Nikolaevič Adamov, Semen Pavlovič Tutaev and Innokentyj Nikolaevič Kokuev.

The *qara čoγdu* clan had given two shamans: Sergej Spiridonovič Bakanaev and Pavel Jegorovič Šipkeev.

The shamans of the *χaš* clan came from two families: the Bokturbaevs and the Saganovs. In the former Innokentyj Semenovič and Innokentyj Alekseevič and his younger brother the 50-year-old Innokentyj Alekseevič.

The *čeptej* clan had four shamans, all of them from the Tulaev family: Innokentyj Abramovič, Aleksander Mikhajlovič, Aleksej Grigorevič and Nikolaj Danilovič.

[9] Чудинов (1931) pp. 102–3.

[10] Persons who used to practise shamanism in the past are today working in kolkhozes like the rest of the Tofas. However, their minds have naturally retained the memory of the shamans' spirits, the different shamans' songs and the details of their dresses, drums and drumsticks. These former shamans were my major informants and sources during my field work. Here again I wish to thank them for their confidence.

[11] Cf. Diószegi (1959) pp. 59–65.

As we list the shamans of the Tofa clans, let us note that it seems as if the gift of shamanism were generally found in certain families and that it was their members who had become shamans. But let us examine this assertion more carefully. Are there definite shaman families? Is the aptitude for becoming a shaman inherited or are there laws governing succession? If we examine the question at a given time, the answer seems to be that there are definite shaman families. All the shamans of the *χaš* clan, for instance, stemmed from two families only: the Saganovs and the Bokturbaevs. Two members of the former and three of the latter became shamans. In the *aq čoγdu* clan the Amastaev family also produced three shamans. All the shamans of the *čeptej* clan derived from a single family—the Tulaevs.

The existence of shaman families is also corroborated, if we examine the question over a period of time:

In the Kholemoev family of the *aq čoγdu* clan not only Ivan, but also his grand-daughter Anna Innokentyjna, and in the *saryγ χaš* clan not only Ilya Spiridonovič, but also his grandmother, great-grandfather and great-great-great-grandfather were shamans.

Also the intermarriages of the shaman families are worthy of mention in this strain. Among the shamans of the Tofas I listed two female shamans: Kholemoeva of the *aq čoγdu* clan and Bokturbaeva of the *χaš* clan. The laws of Tofa exogamy demanded that the men of the *aq čoγdu* clan should take wives from the *χaš* clan and the *χaš*-men from the *aq čoγdu* clan. Shaman Kokuev of the *aq čoγdu* clan married the female shaman of the *χaš* clan Bokturbaeva. Saganov, a shaman of the *χaš* clan, took wife Kholemoeva, the female shaman of the *aq čoγdu* clan. When Shaman Saganov died, his widow married a second time—of course again a *χaš* shaman, Bokturbaev.

It seems evident that shaman families did exist among the Tofas and, what is more, there even appears to have been a deliberate tendency to maintain such families. At the same time this endeavour far from violating the laws of exogamy, is actually subordinated to them.

Is the gift of shamanism hereditary and is there any regularity in succession?

In the Bolkhoev family of the *saryγ χaš*, memory goes back to one of the great-great-great-grandfathers who was a shaman. As far as they can remember none of the great-great-grandparents was a shaman. Among the great-grandparents one great-grandfather, and among the grandparents, the grandfather practised shamanism. The father did not become a shaman, but the son did.

From this example, which is by no means unique in the history of Tofa shaman families, we can conclude that the practice of shamanism is passed down from generation to generation. There was always a shaman among the six generations of the Bolkhoev family with the exception of two generations (great-great-grandparents and parents). At the same time, the thought suggests itself that there was only one shaman in each generation, for among the four grandparents, eight great-grandparents and thirty-two great-great-great-grandparents in the Bolkhoev family, there was always only one shaman.

But let us look at this question once not only diachronically, but also synchronically.

In the Saganov shaman family of the *χaš* clan Jegor and Innokentyj, the two sons of Aleksej became shamans. To the question of inheritance we ought to answer that the gift of shamanism is inherited. But Aleksej did not have only two but five children in all. So, at the same time we should have to add that only two out of his five children inherited his talent.

Or let us look at the *čeptej* clan, where every shaman stems from a single family. In the first quarter of the century four Tulaev cousins, Abram, Mikhail, Grigorij and Danilij—had shaman sons: Innokentyj Abramovič, Aleksander Mikhailovič, Aleksej Grigorevič and Nikolaj Danilovič. Shamanism can be inherited thus in families. However, in the first quarter on the century there were not only four but six Tulaev cousins, and not only four but twenty-six children were living in the related families. Therefore, among the twenty-six second cousins only four inherited the practice of shamanism, whereas the other twenty-two were just every-day people.

Cases of the failure to inherit this gift are even larger in proportion. In fact, we know of a good number of shaman families where no one became a shaman in spite of the many children and grandchildren. This is best illustrated by the Šipkeev and Bakanaev families of the *qara čoγdu* clan where nobody continued the practice of shamanism since the children born in 1891 and 1874, respectively.

In some shaman families none of the descendants carries on the practice. Likewise it happens that the member of a family lacking shaman ancestors becomes a shaman. In the Kokuev and Tulaev families of the *aq čoγdu* clan, Innokentyj Nikolaevič and Semen Pavlovič became shamans without any shaman ancestors.

It follows from the examples given above that the gift of shamanism can be regarded as inheritable in certain families. More than one member of a generation may become a shaman; not all members of a generation can carry on the practice. Inheritance is not necessarily uninterrupted, for it may skip a generation, or the succession of shamans may cease altogether. Not every member of a shaman family becomes a shaman, and descent from a non-shaman family does not preclude the possibility of becoming a shaman. Thus, the shaman's gift does not run in shaman families, but can be possessed by outsiders as well. In conclusion let us stress that while the talent may be inherited, actual advancement to the rank of a shaman is nevertheless independent of inheritance.

THE PROCESS OF BECOMING A SHAMAN

Then what factors determine whether someone does or does not become a shaman? The following examples will give an answer to this question.

'Anjataj was a member of the *saryγ χaš*. Once he fell ill and was sick for three years. His head ached, his arms and legs ached, and he could not move them. In his dream he saw animals, and old men appeared to him. Spirits (*aza*) visited him and told him to become a shaman. These spirits compelled him to practise shamanism. They compelled him with torture. They tormented him for a long time, and finally he slept for three days without waking up. Then he started to practise shamanism and began to feel better. Then he asked his people for a drum and became completely cured.'

How Vasilij Mikhajlovič of the *aq čoγdu* clan's Amastaev family became a shaman was related by his sister in the following words: 'My brother was a shaman, he was 18 years old when he became a shaman. Before he became a shaman, he had fallen seriously ill. Then he had dreams. The fact was that spirits (*χebe, azasy*) appeared to him in his dreams, and they told him to become a shaman. He was ill for a year and lost weight because the spirits tortured him. He did not go out at all, did not even go hunting, just lay in bed at home. Finally he agreed to become a shaman, then he recovered.'

Shaman Kokuev of the *aq čoγdu* clan told me how he had become a shaman: 'I, too, was ill when I became a shaman. First my head began to ache, then my hands. My head still hurts when the moon is full. I was often visited by spirits. When I was asleep, my tongue sang. It sang as the shaman sings. But I knew nothing of this. When I woke up, my mother, my father and my sister told me, "You sang songs, shaman's songs". A few days later I began to feel better. But three or four months later it began all over again. My head kept hurting again all the time and when I slept my tongue sang shaman's songs again. This kept repeating itself every three or four months for three years. One keeps on suffering—suffering and suffering. When one wants to rest or sleep, one's tongue starts to sing. Of course, one does not know these songs, for they are sung by the spirit. But the spirits do not all sing in the same way, some sing beautifully, the songs of others are ugly. The *uluγ* ("great") sings best. I was twenty-seven years old when I first heard the spirit sing. I was visited by the "little one" (*čaryš* = "little spirit"). It flew into my mouth, and then I began to sing shamans' songs. Finally I no longer had the strength to keep on suffering, I did what the spirit told me to do and agreed to become a shaman. I changed completely after I had become a shaman, for a shaman turns into an entirely different kind of a person when he becomes a shaman.'

In other words the shaman, according to the unanimous opinion of the Tofas and the Tofa shamans interviewed, becomes a shaman because of the will of the spirits. Illness and tormenting spirits are mentioned also by N. P. Katanov.[12]

Persons become shamans at an early age. Kokuev, as we have seen, was twenty-seven at the time. Makanaev of the *qara čoγdu* clan 'became a shaman after his marriage. He was still young then'. The Tofas usually became shamans when twenty or even younger. Innokentyj Semenovič of the Bokturbaev family (*χaš* clan) was twenty years old when he started practising shamanism. The two Tulaev boys in the *čeptej* clan must have become shamans at about the same age, since Innokentyj died when he was twenty-three and Aleksander at the age of twenty-two, and both had practised shamanism for several years before their death. Vasilij Mikhajlovič, Amastaev of the *aq čoγdu* clan was eighteen when he followed his vocation. His father was also eighteen years old when the signs of the shamans appeared in him, although he did not actually become a shaman. NikolajDanilovič of the *čeptej* clan's Tulaev family began to practise at the age of twelve.

[12] Катанов (1891) pp. 66, 67.

Ilja Spiridonovič of the *saryγ χaš* Bolkhoev family was even younger. We quote here the relevant part of his shaman song:

'men toran men iχiski čylda	'I was born in '12 [1912]
men χamnan men üšbe ᴅörtbe čylda'	practised shamanism at the age of three or four'

It is exceptional for somebody to become a shaman when older than this. Such, however, were the cases of Anjataj of the *saryγ χaš* clan and Šipkeev of the *qara čoγdu* clan, whom the spirits called at the age of thirty-five and forty-two, respectively.

Many are the spirits that appear before the future shaman.

Anjataj of the *saryγ χaš* saw the 'master of the mountain' (*tajγa ēzi*), and the 'master of the water' (*süge ēzi*). The 'master of the mountain' was translucent, and the 'master of the water' was 'half fish, half man, of medium height, with human hands and feet but the tail of a fish. Its body was covered with scales.'[13] Bolkhoev of the *saryγ χaš* clan was visited by seven spirits: the 'yellow spirit' (*saryγ aza*), the 'blue spirit' (*kük aza*), the 'red spirit' (*qyzyl aza*), the 'blind spirit' (*soγur aza*), the 'wild goose spirit' (*qas aza*), the 'owl spirit' (*χügü aza*) and the 'greedy spirit' (*aqaγ aza*). These spirits lived in the 'yellow light' (*saryγ čyryqta*), where the sky was painted yellow by the sun. For this reason they made their home in the east in the morning, and in the west in the afternoon. Shaman Kokuev of the *aq čoγdu* clan saw also seven spirits appear before him when he was forced to become a shaman. The master of the 'white lake' (*aq χül*), of the 'black lake' (*qara χül*), of the 'holy place' (*ēlig ǰer*), of the 'holy mountain' (*ēlig tajγa*), of the 'rocky volcano' or 'holy volcano' (*χajalyγ sojaq, ēlig sojaq*), of the '*šarm* mountain' (*šarm tajγa*) and of the 'holy sea' (*ēlig dalaj*). 'These spirits', as the shaman Kokuev said, 'had the shape of shamans. Each had a head-dress, a shaman's gown, shaman's boots and a drum and drumstick.'

As we may see, the shaman's spirits are master spirits: master of the water (*süge ēzi*), of individual lakes (*aq χül, qara χül*), groups of lakes (*ēlig dalaj*), of the mountain (*tajγa ēzi*), of individual mountains (*ēlig ǰer, ēlig tajγa, χajalyγ sojaq, ēlig sojaq, šarm tajγa*), of clouds (*saryγ aza, kük aza, qyzyl aza*), of birds (*qas aza, χügü aza*), etc.

The great majority of the spirits were classified by the Tofa shamans themselves, according to quality and degree.

Qualitatively, the spirits were divided into two groups: the 'blacks' and the 'whites'. The former, the 'black spirits' (*qara aza*) or 'black shaman's spirits' (*qara χamnašqyn*) were evil: and the latter, the 'white spirits' (*aq aza*) or 'white shamans' spirits' (*aq χamnašqyn*) were good.[14] Depending on whether he is selected by 'white' or 'black' spirits, a shaman will be a 'white shaman' (*aq χam*), a 'light shaman' (*čyryq χam*), or a 'black shaman' (*qara χam*). Black shamans had the ability to do bad things, and the other shamans were afraid of them. Their garb differed from that of the white

[13] For the mountain and water spirits cf. Петри (1927) p. 32 and (1928) pp. 15–6; Катанов (1907) pp. 615, 628–9, 639.

[14] For the good and evil spirits cf. Евсенин (1919) p. 14.

shamans. Bakanaev of the *qara čoγdu* clan was, for instance, a black shaman. His shaman's gown was made of black cloth and the textile pendants on it were mostly black. Even his drum was covered with the skin of a black bull.

According to the Tofa shamans, the shamans' spirits differed not only in quality but also in degree. There were 'big shamans' spirits' (*uluγ χamnašqyn*), 'middle shamans' spirits' (*ortun χamnašqyn*) and 'small shamans' spirits' (*čaryš χamnašqyn*). The shaman acquired the power of the spirit that picked him. So the shamans, too, belonged to three different classes: there were 'big shamans' (*uluγ χam*), 'middle shamans' (*ortun χam*) and 'small shamans (*čaryš χam*).

The cause of the differences in degree existing among the Tofa shamans was explained to me thus by a former shaman from the *aq čoγdu* clan. On the other hand, the shaman of the *saryγ χaš* clan regarded rather the number of spirits as the reason for the different degrees. 'I was a small shaman because I had only seven spirits', Bolkhoev told me. 'Other shamans were great shamans because they had twenty-seven spirits.'

Every shaman, whether of a shaman family or not, becomes a shaman because the shaman's spirits wish him to become one. How much say the spirits are supposed to have in this matter is amply demonstrated by the belief that the quality of the spirit and the degree of its power, or the number of the spirits, determine the quality of the shaman ('white' or 'black') and his strength ('small', 'middle' or 'big' shaman). This does not depend on whether one of his ancestors was a 'white' or 'black', 'big' or 'small' shaman—or a shaman at all.

Nevertheless, we wish to point out that the spirits which choose the shaman may be 'inherited'. Let us mention in this connection the case of Shaman Vasilij Mikhajlovič of the Amastaev family of the *aq čoγdu* clan as an example: 'Vasilij became a shaman at the age of eighteen. He inherited the spirits of his father, although the father failed to become a shaman. The point is that one must drink blood, a lot of reindeer blood, to become a shaman. His father Mikhail Amastaev asked for reindeer blood, but he was not given any, therefore, did not become a shaman.'

The shaman spirits also have their influence on the shaman's costume, which is made according to their wishes.

Shaman Anjataj of the *saryγ χaš* clan 'was told by the spirit that first the shaman's gown, second the shaman's head-dress, third the shaman's boots and fourth the kerchiefs were to be prepared. It also told him what his drum and drumstick should look like, etc. Later he had his drum made, according to the instructions of the spirits, by a man from the *aq čoγdu* clan.'

How minute these instructions are regarding everything appears from the words of the shaman Kokuev of the *aq čoγdu* clan. 'The *čaryš* (small one, i.e. small spirit) came to me in the shape of a small bird and told me to have a drum, drumstick, roller strips, a shaman's cap, a shaman's gown and shaman's boots made. He said: "Have a drum, drumstick and shaman's costume made for yourself, because you will be a shaman. Perhaps we will be a great shaman." The *čaryš* said: "The drum shall be made of red-deer skin". The *čaryš* said: "The framework of the drum will be of live cedar". The *čaryš* said: "It shall have twenty-one *χabaq* (resonators) on it. Seven shall be of live cedar, seven of live larch and seven of live fir".'

The shaman-to-be, after having agreed to the wish of the spirits, has to procure all the paraphernalia required for the practice of shamanism, and usually asks relatives and neighbours to produce them.

The younger sister of shaman Vasilij Mikhajlovič of the *aq čoγdu*, Amastaev family, gave us the following details concerning the making of her brother's shaman's equipment: 'This took place by the Kara Buren river. Our birch-bark tent was pitched there at the time. Altogether five of us lived there: my brother Vasilij, I, my sister Maria, my sister's husband Aleksej and my younger sister Sura. My brother addressed himself to Maria's husband, Aleksej, for the drum and the drumstick and asked us to make him the rest of the outfit. I, sister Maria and my aunt (my father's sister) made the costume. My father's sister lived in the valley of the Dzhuglym river. We sewed some parts of the costume, the rest was sewn by my aunt in her home. She cut and sewed the shaman gown; we made the head-dress, the boots and the rolled strip pendants, which we made from white and yellow stuff. Altogether twenty-seven strips were necessary for his garment. My brother explained all the details very carefully, so that everything might be in accordance with the will of the spirit.'

The attire of the Tofa shamans[15] consisted of a head-dress (*χamnar ʙört*), a shaman's gown (*χamnar ᴅon*), possibly a separate shaman's breast-cover (*χamnar ᴅüš*) and shaman's boots (*χamnar iᴅik*) (Figs 2 and 3). In addition to these in order to practise shamanism, a drum (*ᴅüŋür*) and a drumstick (*orba*) (Figs 4, 5, 6 and 7) were required.

The shaman had the various items of his attire made in a definite order—according to the Kokuev shaman of the *aq čoγdu* clan as follows: 'At first the rolled strips had to be ordered, then the shaman's gown, the head-dress and the boots and finally the drum, drumstick and drum-case. Then one could begin to practise shamanism.'

The Shaman's Gown

Usually the shaman's gown was made from the skin of tame reindeer. The flaying was done by men, the dressing by women. The skin was stretched on a frame. The frame consisted of two vertical forked rods stuck into the earth and two horizontal cross-bars, one of them laid across the two forks on top and the other tied to the vertical rods at the lower end, at a distance from the upper bar corresponding to the length of the skin to be dried. Over this frame the skin was stretched by means of a single long leather strap. Holes were made in the edge of the skin through which, spaced at equal distances, the long leather strap was threaded, being led to the frame from the skin and then back (Fig. 8). When the weather was warm, the skin dried in a single day, otherwise it remained on the frame for two days.

After the skin had dried, a woman removed it from the frame and, spreading it out on the ground, scraped off the thin layer (*qyrtyš*) on the fleshy side. She pinned the skin down with her foot and peeled off this inner layer with a scraper having a concave elliptical blade (*körgö*) (Fig. 9). Then she spread tanning material made of decayed wood over the fleshy

[15] Соловьев (1926) p. 658; Чудиного (1931) pp. 95, 99; Васильев (1910) p. 73.

Fig. 2. Tofa female shaman in ceremonial costume (front view)

Fig. 3. Tofa female shaman in ceremonial costume (back view)

Fig. 4. Tofa female shaman with drum and drumstick

Fig. 5. Tofa female shaman with drum and drumstick

Fig. 6. Tofa female shaman with drum and drumstick

Fig. 7. Tofa female shaman with drum and drumstick

side of the cleaned skin. Scraps of decayed cedar (*irik*) were kneaded with water and spread over the inner side. Then the skin was folded: First the part toward the neck, then the right and left sides, giving the skin an oblong shape. The upper and lower ends were again folded, yielding a smaller oblong which, in turn, was folded into a square (Fig. 10). The skin was thus left to stand for 24 hours. The next day it was spread out over the ground again. The woman stepped on its tail end and, holding the other edge in the left hand and a scraper (a three quarter crescent blade—*sy*) (Fig. 11) in her right hand, scraped off the decayed pine dust with a motion either to or from the body. When she had finished this work, she pursued her task of thinning the fleshy side of the skin, with another tool. Leaning on a rod split longitudinally in half (*surγuš*) with one end firmly fixed in the ground so as not to slip, and pressing the other end against her loins, she spread the skin with the fleshy side up over the rod, tucking the end between the rod and the loins. With a scraper having a longish blade (*ydrā*) (Fig. 12) she scraped the skin from top to bottom, leaning on it with her full weight. She kept doing this until the skin had dried completely. Now came the smoking of the skin, which was done according to different methods in the Tofa clans.

With the *qara čoγdu* clan three skins were used for the shaman's gown. These were sewn together to form a bag. A pear-shaped hole was dug outdoors, round its mouth some six or eight rods were fixed in a conical form. In this hole a fire was made from the cones of Scotch firs, which the members of the *qara čoγdu* clan thought most suitable for smoking. If Scotch fir was not

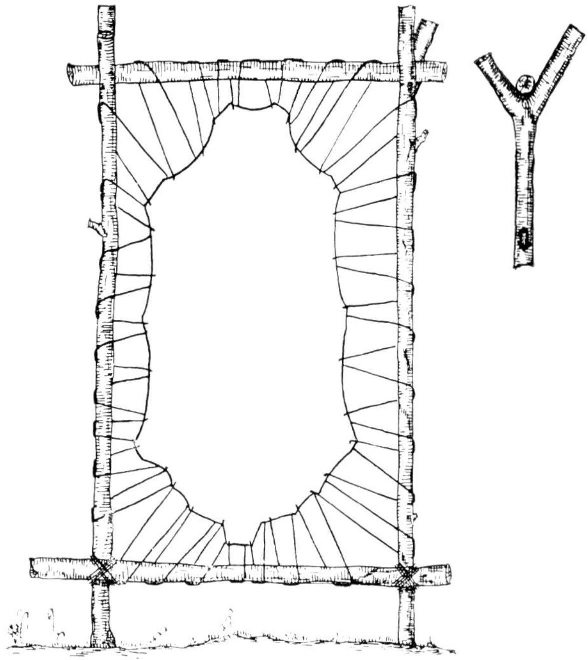

Fig. 8. Drying the flayed skin

Fig. 9. The skin scraper (*körgö*) for removing the upper layer of the epidermis

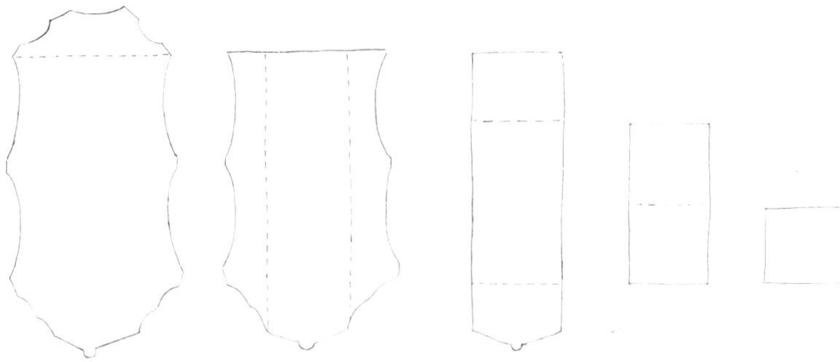

Fig. 10. The folding of the skin smeared with wood dust

available, decayed larch dust was used. The three skins sewn together in the form of a bag were drawn over the rods forming a cone so as to prevent the smoke from escaping through any opening. The 'bag' was creased, since its mouth covered the ground. The skins smoked for approximately an hour became yellow and always remained soft (Fig. 13). With this the process of tanning was ended.

If the hair on the skin was longer than required, it was clipped to the proper length with ordinary shears. It was never removed completely, since it was the inside of the cloak and could not be seen.

Also the members of the *χaš* clan left the hair on the skin to be made into a shaman's gown. They too, smoked the skin, but not according to the *qara čoγdu* practice. They dug three small, ca. one metre deep, holes with round mouths, about one metre wide, next to each other. They stuck three rods forming a cone over each hole and spread the skin to be smoked. Each skin was first separately folded in half and then sewn together along the longer side. Then a fire was lit in the bottom of the ditches, with decayed fir twigs, which the members of the *χaš* clan thought was the best for smoking leather. When the fire was well alight, they drew a skin sewn into a sack over each frame. Generally, they were kept over the smoke for two hours (Fig. 14).

The *saryγ χaš* clan dug three holes side by side each with mouths of 30 cm in diameter widening towards the bottom. Over these they fixed

Fig. 11. The scraper for removing the wood dust (*sy*)

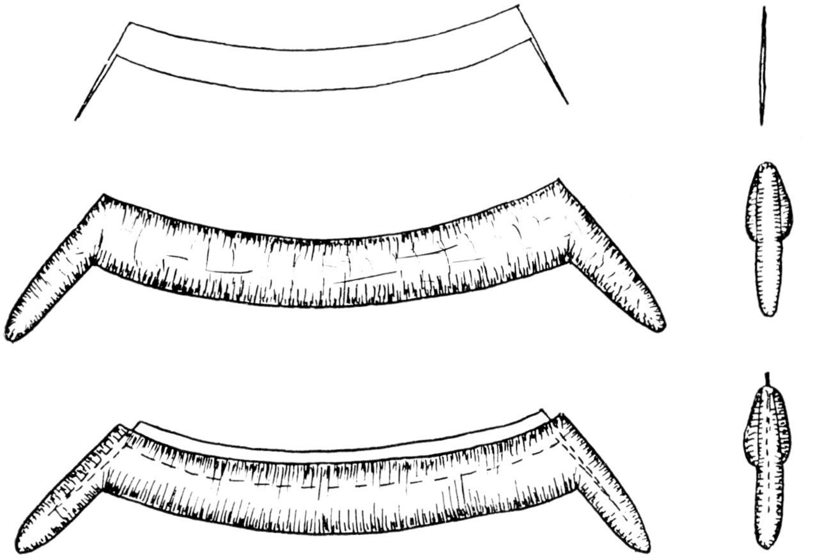

Fig. 12. A long-bladed skin-scraper (*ydrā*)

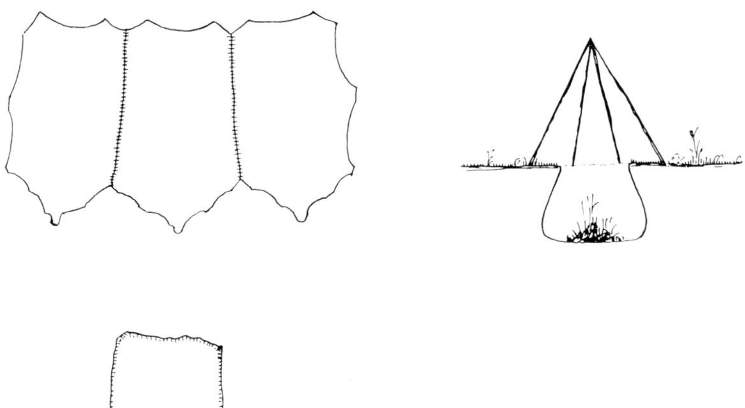

Fig. 13. Three reindeer skins being smoked at once

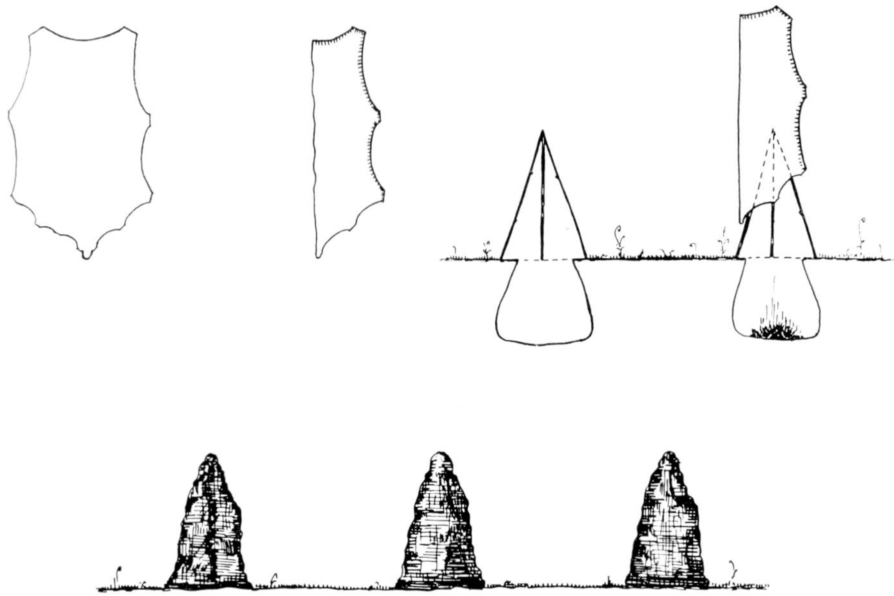

Fig. 14. Three reindeer skins being smoked separately

three rods forming a cone, over which the skin was drawn, to prevent the escape of any smoke. In the holes decayed larch was burnt. The skins were smoked for an hour, by the end of which time they had become yellow.

Then the cutting began. As we said before, for the making of a gown three complete dressed reindeer skins (*ȯčtu̇ geš*) were needed. From the first of these the two front sides of the gown, i.e. the *taľǰyrǰary morna* 'left front side' and the *oŋ ǰary morna* 'right frontside' were made (Fig. 15.) From the second came the back (*arasy*) of the gown (Fig. 16), and from the third, the collar (*ǰaχasy*) and the two sleeves i.e. the *taľǰyr ǰaryn ńine* 'left sleeve' and the *oŋ ǰaryn ńine* 'right sleeve' (Fig. 17).

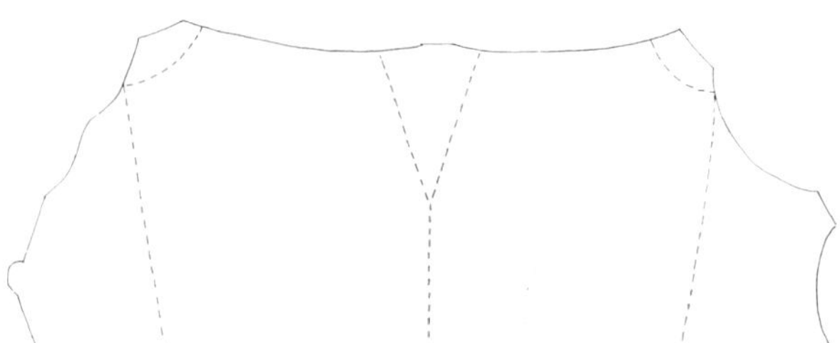

Fig. 15. Cutting the two sides of the shaman's gown

Fig. 16. Cutting the back of the shaman's gown

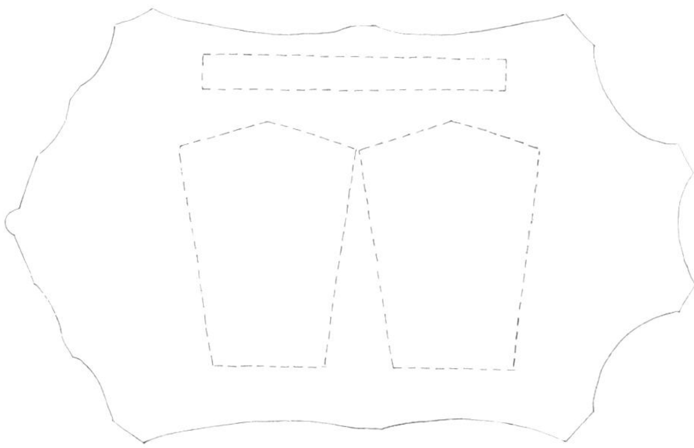

Fig. 17. Cutting the collar and sleeves of the shaman's gown

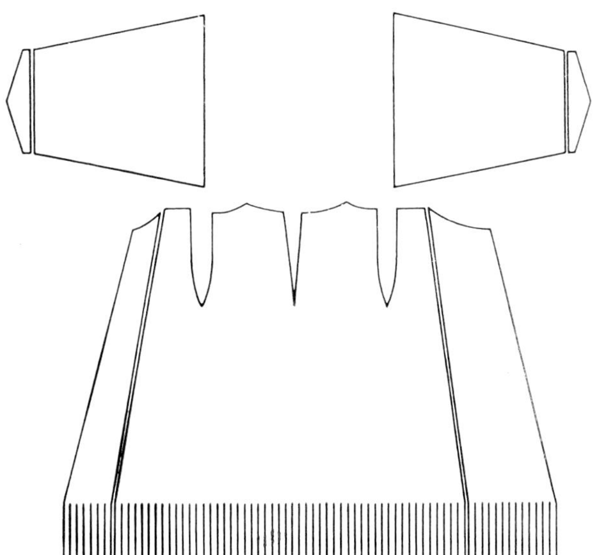

Fig. 18. Assembling the parts of the shaman's gown after they have been cut

We take the Tofa shaman's gown in the Museum of Anthropology and Ethnography (MAE) in Leningrad (inventory number 1480—17) as an example of how the gown was sewn together after being cut out (Fig. 18).

The sleeves, chest and back of the gown were adorned with appliqué or embroidery having different meanings, and with pendants made of textile, leather and metal.

The Textile and Leather Tassels

The textile and leather tassels on the shaman gowns were long rolls (*manǰaq*), made of cotton fabric, or of the skin of tamed reindeer from which the hair had been removed. The material used was always of one colour: blue (*kök*), yellow (*saryγ*), red (*qyzyl*) or white (*aq*).

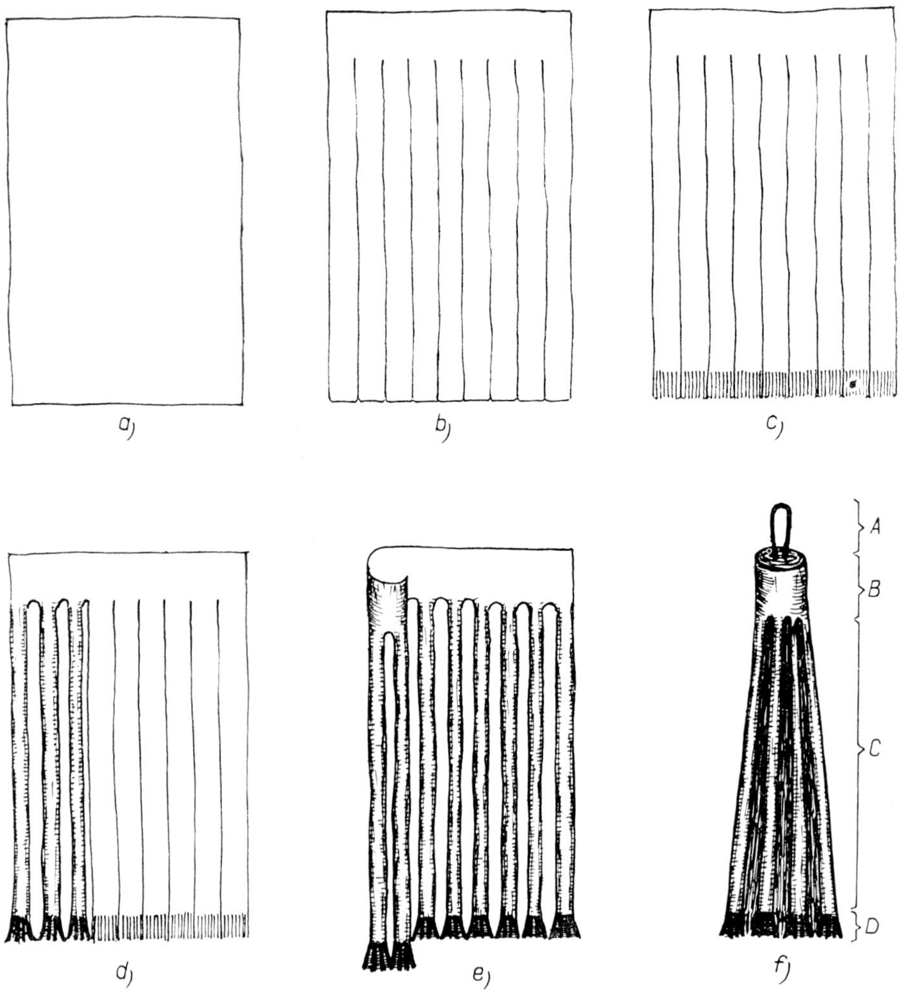

Fig. 19. The process of making a rolled strip (*manǰaq*) (a, b, c, d, e, f)

An oblong piece was cut out of the material (cloth or leather), the length varying between 60 and 100 cm, the width being 20–25 cm (Fig. 19a). The narrower side was cut in so as to obtain nine strips (Fig. 19b). The ends of these strips were again cut to form nine tassels on each (Fig. 19c). Into each of the nine strips a piece of rolled cloth of equal length was placed, and the two edges of each strip were sewn together (Fig. 19d). Now from a narrow horizontal band of textile nine long and thin rolls were hanging, each ending in a tassel of nine strands (Fig. 19e). Then this band of cloth holding the rolls was itself rolled together and a leather-strap was tucked into it by the ends (Fig. 19f), so as to form a loop (A in Fig. 19f), and the whole was thus sewn together. This constituted one bundle. Each bundle always consisted of nine branches. The thick and short roll (*ʙaγ*) which branched into nine long, thin strips, was called *manǰaqʙòžō* 'rolled-strip bundle' (B in Fig. 19f), the branches *manǰaq* 'rolled strip' (C in Fig. 19f) and the tassel (*salvar*) at the end of the branches were termed *manǰaq ʙeχe*, i.e. 'rolled-strip head' (D in Fig. 19f).

Frequently a strip of the white dewlap of a tame reindeer was sewn spirally around each individual branch (Fig. 20a). Sometimes, however, pieces of different-coloured textiles were sewn, in checkerboard pattern, on the branches (Fig. 20b). But there were strips which had not only the end but also the middle decorated with tassels (Fig. 20c).

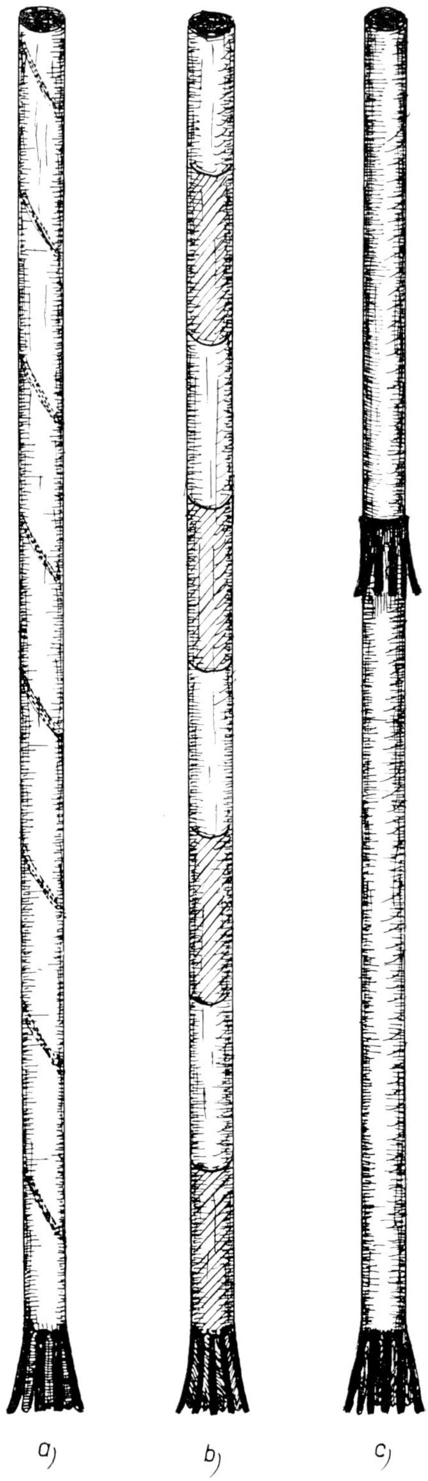

Fig. 20. Variations in rolled strips:

a) *manǰaq* decorated with reindeer hair stripe,
b) *manǰaq* decorated with a checkerboard pattern of textiles,
c) *manǰaq* decorated with tassel in the middle

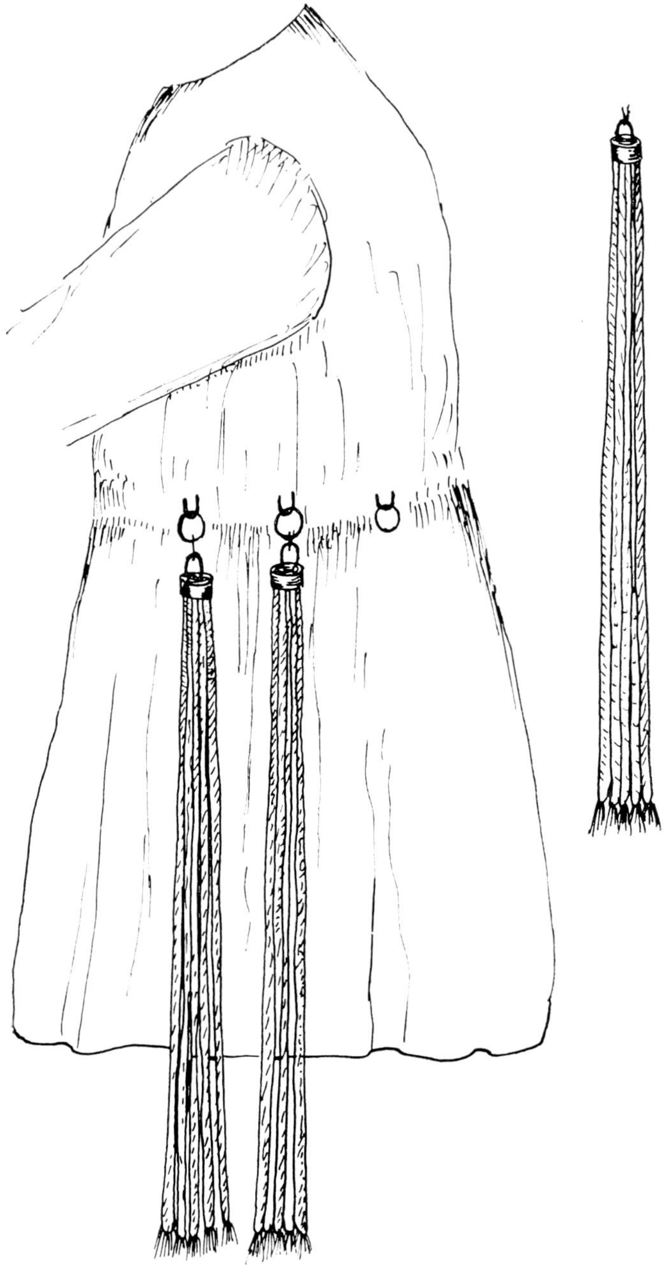

Fig. 21. Attachment of the rolled-strip (*manǰaq*) bundles to the shaman's gown

On a shaman's gown, strips of different colours and lengths might be suspended in varying number. Not even the mode of attachment was always the same: some shaman's gowns had the roll sewn on, that is to say, permanently attached, whereas others had them attached by means of the loops emerging from the bundles—only for the time of actual shamanistic ceremony (Fig. 21).

The Metal Pendants

The shamans' gowns were ornamented with a variety of metal pendants: curved rods (with metal cones and arrows hanging from them), plates and bells. These last were purchased ready-made, but the others were home-made.

The makers of these metal pendants usually bought a new kettle and used its material, since it was the easiest of access. The handle of the kettle

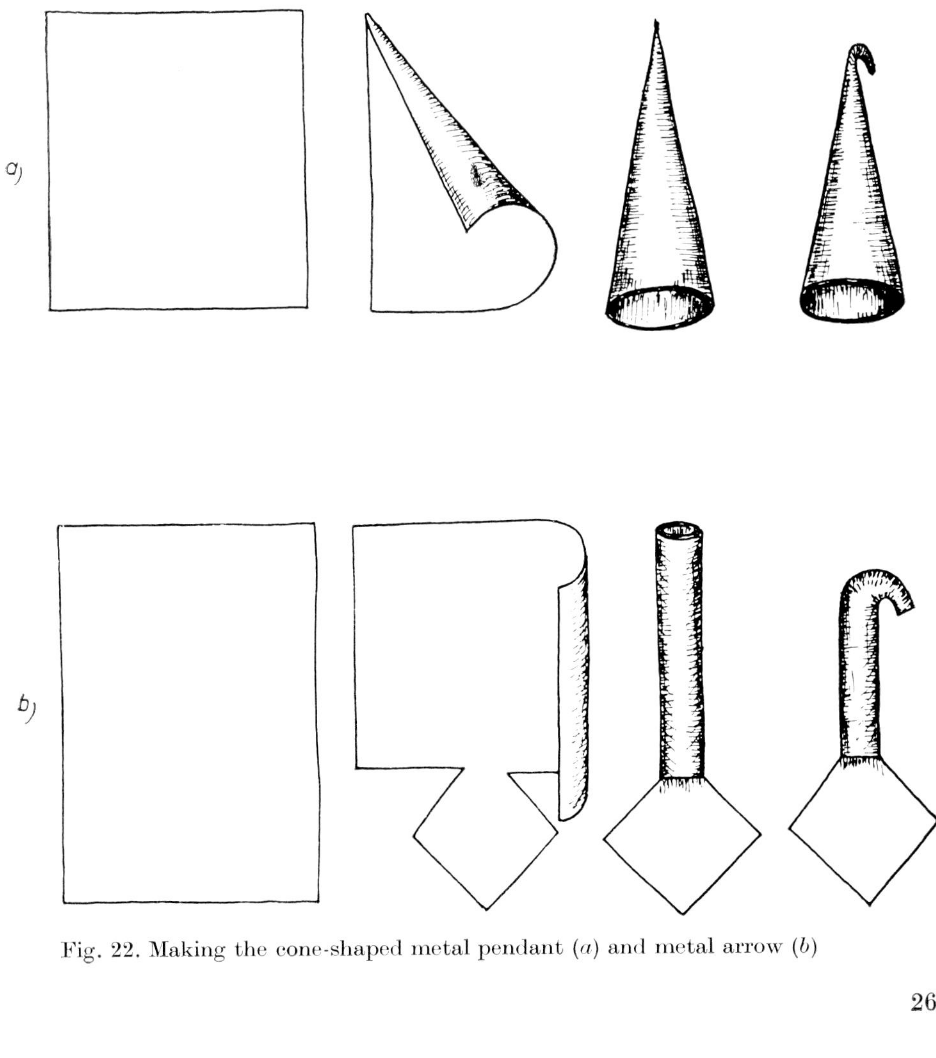

Fig. 22. Making the cone-shaped metal pendant (*a*) and metal arrow (*b*)

was shaped into the curved metal rods, and from its vessel the metal plates were cut out, in elliptical, triangular or trapezoid form, punched with a pointed piece of iron.

The metal cones and arrows were made of tin sheets. Oblong pieces were cut from the tin and rolled into cones (Fig. 22a). Also the arrows were made from similar oblong pieces: a narrower section of the sheet was given rhomboid shape while the rest of the sheet was rolled into a tube, and the arrow was ready (Fig. 22b).

The Head-dress

The head-dress of the Tofa shamans shows striking variety in cut as well as in decoration.

As far as cut is concerned, three types can be distinguished: the head-band, the cap and the crown. The head-band is by far the most customary and the other two types are represented by only a few rare examples.

In the course of my field work among the Tofas, I became acquainted with women who knew how to make head-bands. At my request they sewed two head-dresses, and so we were able to study in detail and photograph the process.

Marija Mikhajlovna of the *aq čoγdu* Adamova family made a feathered head-band as follows: She cut two oblongs of equal size, one out of white

a)

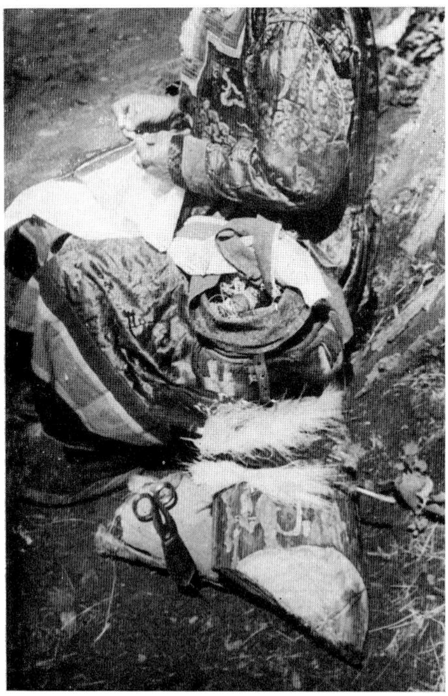

b)

Fig. 23a,b. Sewing a shaman's head-dress (plumed head-band)

Fig. 24a,b.
Sewing a shaman's head-dress (plumed head-band)

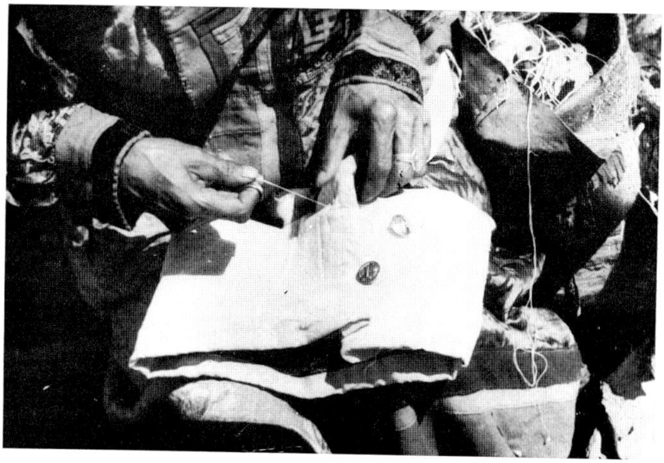

a)

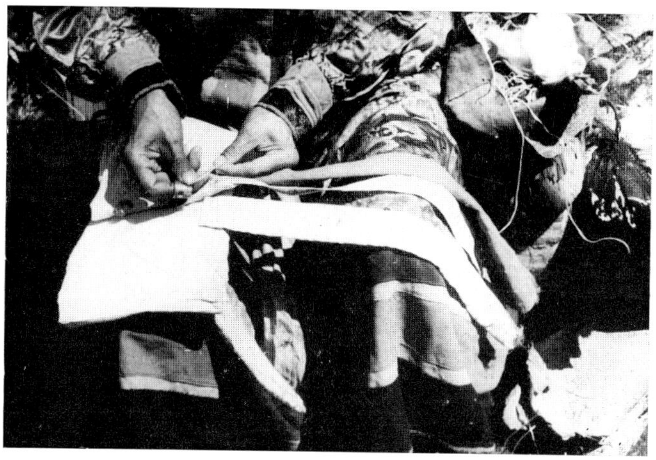

b)

cotton fabric and the other out of a dark-grey material. Their length was 60 cm and their width 18 cm. She placed the longer sides next to one another and sewed them together with cotton yarn (Figs 23a, b). Then she folded the white piece over the dark one, and sewed the two pieces together along the other three sides as well. The white cloth made up the outer side of the band, and the dark and firmer stuff only served to strengthen it. Having done this, she stitched a human face onto the white cotton, applying the long white hair from under the neck of the tame reindeer. She joined some 10 to 15 threads of reindeer hair and wetting them between her lips, laid them on the band which she sewed on by means of white thread with stitches about 0·5 cm apart. When she had finished, she folded the two adjacent ends of the band and sewed two metal buttons on the left end and two loops on the right. Then she attached three ribbons on the back, where the two ends of the band met (Figs 24a, b), and placed feathers into the upper edge of the band.

I also saw how the head-dress of the shaman Vasilij Mikhajlovič of the *aq čoγdu* Amastaev family was made. Upon my request, the younger sister of the shaman, who at the time had sewn the feathered head-band, made it again in the same way as she had originally. Before she began her work, she assembled all the necessary raw materials: white cotton cloth, dark-blue wool, yellow silk, white and red cotton fabrics, black satin cloth, a bundle of reindeer furs, a squirrel-skin and the wing of a capercailzie—and, of course, her sewing kit. First, she cut out an oblong piece of white cotton cloth, and another of dark-blue stuff, both of the same size (52 cm in length and 11 cm in width). She laid them over each other and sewed them together on all four sides. Then she placed the oblong over her knees with the dark-blue side on top, which was later to be the outer side of the band, and pulled out some ten or fifteen hairs from the reindeer fur. Holding these together she moistened them by pulling them between her lips several times. Then she laid the ends of the hairs, adhering together from the wetting, on the textile and fixed them down with stitches about 0·5 cm apart. Where the stitches crossed the reindeer hair, the strips were necessarily narrower, giving the impression of elliptical beads. She sewed the white hair on in the shape of a stylized human face. As she was working, the threads dried up and separated, so she moistened them again with her lips and only then continued the application of the reindeer hair.

After she had finished the face, she sewed three white mother-of-pearl buttons on the left end of the head-band, bringing the thread through the four little holes of the button in such a way that it made the pattern of a square. On the right end of the head-band she made three loops of white thread for buttonholes. She now turned to the yellow silk, red cotton cloth and black satin cloth prepared in advance. Of each of the three, she cut a strip about 1 m long and 8 cm wide. She folded each of the strips lengthwise in half and sewed the two edges together. She then sewed the ribbons thus obtained to the lower edge of the ends of the head-band. To the left end she sewed the black one, in the middle the yellow one, and to the right end the red one. After the long ribbons she began to make the short ones. She cut two oblong pieces from the black satin cloth and two from the yellow silk and one oblong from the white and one from the red cotton (length 20 cm; width 3 cm). She cut in the narrower ends of the yellow and the red fabrics to obtain tassels and then rolled them lengthwise into a long cylinder, stitching down the edges to keep the roll from opening up. This rolled strip ended in a tassel, since the end had been indented. With the white and the black fabrics she did not make a tassel, nor did she roll the ends, but cut three pieces of equal length from the satin cloth, rolled them, and then coiled the white and the two black oblong pieces around them. Here, too, she stitched down the ends, to prevent them from opening. After this she cut out a white strip 3 cm wide and 8 cm long, and a black one, 3 cm wide and 16 cm long, cutting both along their longer sides so as to obtain three fringed strips. Now she rolled these and sewed them onto the ends of the three rolls prepared beforehand, attaching the white to the white and the black to the black. She sewed the finished ribbons onto the lower edge of the left and right sides of the head-band. The order was the following, from left to right as seen by the spectator: On the left side, red, white and black; on the right side, black, yellow and yellow. Then she cut the squirrel skin into straps, 2 cm wide, with which she hemmed all four sides of the head-band. Having completed this squirrel trimming, she pulled out nine feathers from the capercailzie wing and stuck them into the stitches holding the squirrel trimming so that they should be at regular intervals from each other. She attached it long down to the middle. The head-dress was ready at last.

According to the Tofas, the colour and width of the three long ribbons (*čalasy*) on the head-dress have symbolic meanings. The following was said about the ribbons on the head-dress of Vasilij Mikhajlovič of the Amastaev family of the *aq čoγdu* clan:

Mikhail was an orphan, an aunt raised him. When he was eighteen, he began to practise shamanism. He sang songs, shamans' songs. Then his aunt asked him; 'What is it, perhaps you want reindeer blood and that's why you began to practise shamanism?' But she did not give him blood. 'Ask the rich for blood!' Mikhail turned to rich people, but they denied him the reindeer blood. Then he asked for red fabric for the ribbon of the shaman head-dress. This was refused, too. 'What do you need red cloth for, perhaps you want to make the *eɒik* (door of the birch bark tent) out of it?' Then he began to act again like a shaman. He leaned his head on the saddle of a reindeer, spread the skin of a musk deer (less than one year old) under himself, and that was how he slept. And he sang and sang, singing shamans' songs. But he had to sing less and less often, and gradually he stopped acting like a shaman. Then he married. He had twelve children, eleven daughters and one son Vasilij. And when Vasilij became eighteen, he too, began to practise shamanism. His parents were surprised, for his father had also begun shamanism at the age of eighteen. Vasilij did not ask for blood. He did not need it, because his father in his time had been denied reindeer blood. On his head-dress, when it was finished, there was only one little red ribbon—very thin at that, because it had only his own blood. The long and wide ribbon on Vasilij's cap was yellow. The head-dresses of the shamans of other clans had usually a long and wide red ribbon, because they were rich clans and could give much reindeer blood to their shamans.

The Boots

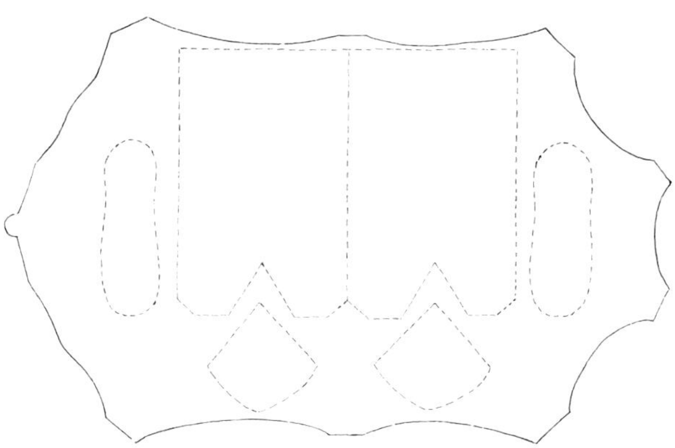

Fig. 25. Cutting the shaman's boots

The shaman's boots were sewn out of a complete dressed skin, the right boot from one half and the left one from the other half. Each boot was composed of three parts: the top which was cut separately (*iᴅik χanǰasa*), the vamp (*iᴅik ʙeχe*) and the sole (*olduŋ*) (Fig. 25). The top of the boot was trimmed in front and on the two sides, either with a strip of cotton fabric or with a strip from the long hair under the neck of tame reindeer. Strips made of similar material were also applied to the vamp. Often a long and thin iron plate was fixed to the outer side of the boots.

The Drum

The shaman found out from the spirits what kind of wood the framework (cylinder) of his drum was to be made of, and what kind of skin was to be used to cover one side. The shaman's drum (*ᴅüŋür*) is more or less round and is made in different sizes: male shamans have larger drums (the diameter being 80–95 cm), the female shamans have smaller ones (the diameter being about 60–70 cm). The framework (*ᴅüŋür ńeǰe*) was carved from fir (*šibi 'Pinus abies'*), cedar (*ʙüš 'Pinus cedrus'*) or resinous forest pine (*qaᴅy 'Pinus sylvestris'*). The handle (*tuta*) was cut out of cedar, forest pine or birch (*qaᴅyŋ 'Betula fruticosa'*). The two cross-bars (*χümündrük* and *quᴅurγa*) were also of cedar, forest pine or birch branches. Three-knobbed resonators (*χabaq*) lined the outer side of the Tofa drum frames. There were 9, 12 or 21 of these. The resonators were usually made from various kinds of wood. Such as: cedar, larch (*ᴅyt 'Pinus larys'*), fir, spruce (*čojγan 'Abies pectinata'*), forest pine, birch (*qaᴅyŋ 'Betula fruticosa'*), white poplar (*ujγut*), willow (*qaty sôske 'salix'*), privet (*numurut 'Ligustrum vulgaris'*) or rowan (*ergus 'Sorbus aucuparia'*). A piece of deer skin (*saryγ aŋ 'Cervus elaphus'*), or that of wild goat (*teχe*), roe, red horse (*qyzyl āt*), cattle or gray or black bull, was stretched over the frame.

The shamans resorted to the skill of experts by whom they had their drums made, irrespective of whether the maker was a relative of theirs or not. The Tofa people said that Vasilij Nikolaevič of the Adamov, Semen Innokentyjevič of the Mukhaev and Aleksej Kojtovič of the Ungustaev family were good drum-makers.

The normal practice was to make the drum at home, although sometimes it was carved and constructed at the place where the tree was felled. The drum belonging to Innokentyj Nikolaevič Adamov of the *aq čoγdu* clan was made where the wood for the frame had been obtained.

'The first step in making a drum', declared shaman Kokuev of the *aq čoγdu* clan 'is to carve the resonators. When my drum was made, I was living by the Aganit river. Some four or five other tents were pitched next to mine. The men from every tent helped to make my drum. It is always the shaman who gives instructions as to the kind of wood the resonators of his drum are to be made of.

'Mine were made of three different kinds of wood: cedar, fir and larch. For when I had sung songs when becoming a shaman, I had sung about the necessity of having seven cedar, seven fir and seven larch resonators for my drum. For this reason my first neighbour fetched cedar, the second, fir and the third, larch.

'These trees were felled. From their trunks, about 10 to 12 cm thick, every man brought home only as much wood as was needed for the making of the seven resonators. The same man, who had fetched the wood, carved the seven resonators. It took two days to complete them, because it took one day to find and fell out the tree and bring the wood home, and another day to carve three times seven resonators.'

The man who made the resonators hacked the cylinders out of the trunk in pieces of the right size and then began to work on each piece, making two V-shaped incisions in the wood (Fig. 26a) with his axe. Later he used his knife to blunt the edges of the teeth (the heads of the resonator) and carve an arc out of the wedge-shaped incision (Fig. 26b). Turning the resonator upside down, he carved the bottom in the shape of an arc (Fig. 26c). Then he cut the sides to the proper thickness. Finally he made a perpendicular incision in each of the three heads (Fig. 26d). On the average the resonators are 13 cm long, 2·5 cm thick at the bottom and 4 cm high.

When the resonators were ready, three men went to the forest to get wood for the framework of the drum. Shaman Kokuev of the *aq čoγdu* clan describes the scene as follows: 'In my case it happened like this. I sang of this, too. I said that on the *ʙetik taγ* mountain, on its peak, there would be a cedar from which the framework of my drum should be made. When I wanted to have it manufactured, I set out with two men to get the wood. We climbed the *ʙetik taγ* mountain, and at the top we found the cedar we were looking for. I pointed out the cedar, and said: There it is. Then my companions cut a piece from the trunk, because the tree from which the wood for the framework of a shaman's drum is taken must not be cut entirely. If the tree is felled, that is, if it is destroyed, the shaman dies, too. Therefore they cut out only a piece of the proper length from the trunk so that the tree could live on (Fig. 27). If for some reason or other the tree perished—for instance, if it dried out, or if the wind uprooted it, the shaman would die too.'

As the *aq čoγdu* people, so, too, the members of the *χaš, saryγ χaš, čeptej* and *qara čoγdu* clans cut out the wood required. For example, the wood for framework of the drum made for the female shaman Aleksandra Nikolaevna Andalaeva was removed from the trunk in a way as to permit the tree to continue living. The trunk was cut with an axe (*süge*) at the top and bottom first, and then with wedges vertically so that a segment could be split off.

Shaman Kokuev of the *aq čoγdu* clan continued: 'The piece of wood that had been cut out, was carried home by two men who hung it up in my tent to dry. It has to be dried a little. They attached straps to each end and hung it horizontally on the two tent poles, opposite the door. The wood was kept there drying for three days. This coincided with the three days during which the skin for the drumhead was drying in the hunter's tent. On the fourth day they began making the frame.'

After two men had cut out the piece needed for the framework from the trunk of the tree, all three of them started for home. On the way they hacked out also from a fir a piece just like the piece cut out from the cedar. This was for the two 'tendons' (*sīr*), that is, the two narrow strips of bark encircling the top and bottom of the framework. Also this piece of fir was tied to the tent poles to dry.

a)

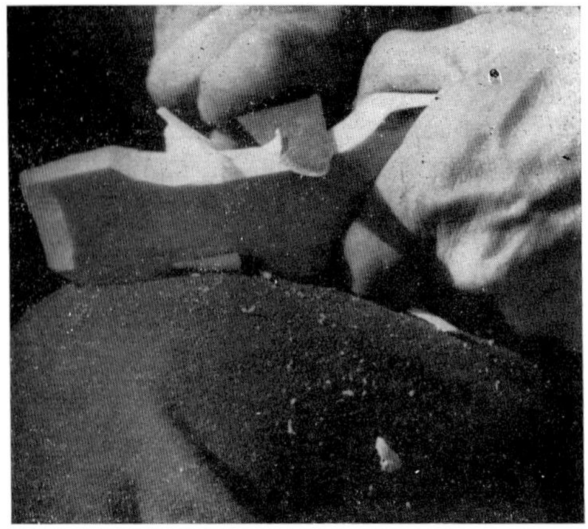

c)

d)

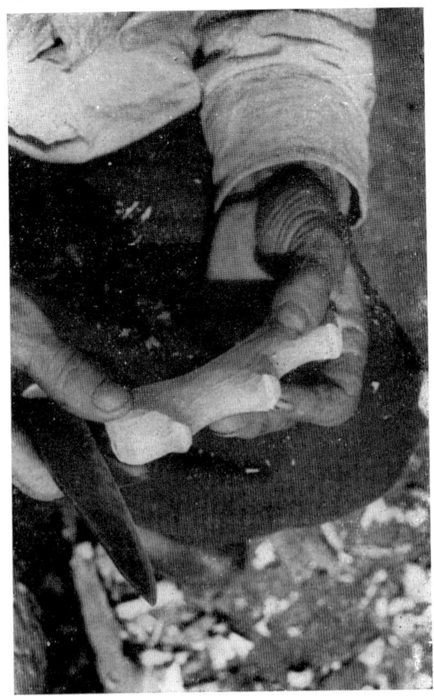

b)

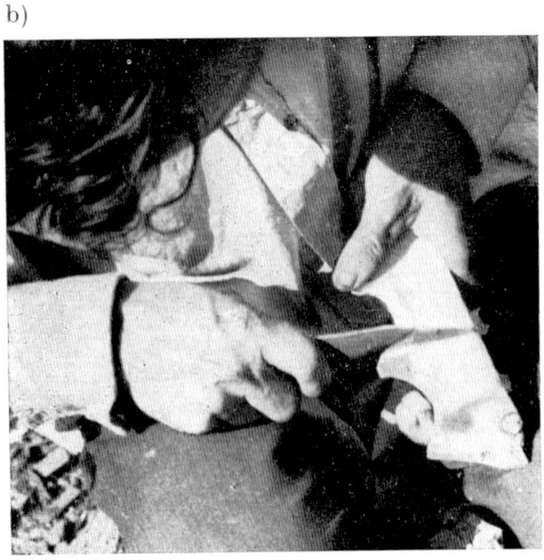

Fig. 26a, b, c, d. Carving the resonator (*χabaq*) of the shaman's drum

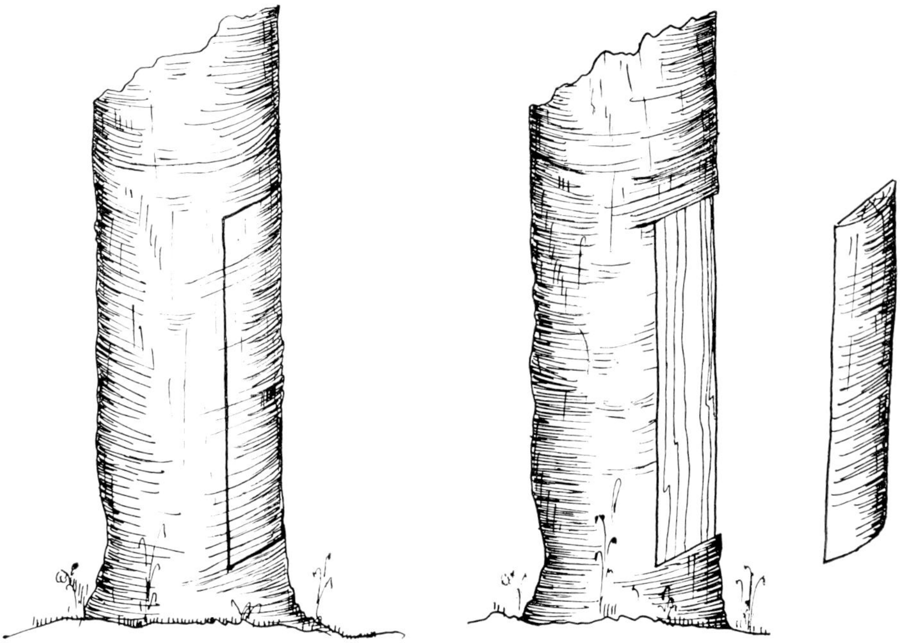

Fig. 27. Cutting the frame of the drum from the tree trunk

While the shaman and his two companions were looking for material for the frame, two other men went in search of wood for the handle and the cross-bars. Quoting shaman Kokuev: 'For the handle of my drum cedar was needed and for the two cross-bars, white poplar. Again it was I who told them what kind of wood was required for these parts of the drum.'

Trees from which wood was obtained for the handle and the cross-bars were felled just like those which provided wood for the resonators. All pieces of wood were hung horizontally on the tent pole to dry. On the left side hang the fir cut for the wooden ribbon, and below it, the cedar cut for the framework of the drum. On the right side the cedar necessary for the handle was suspended horizontally and beneath it, the white poplar for the cross-bars. These two pieces of wood were cylindrical, since the whole trunk had been cut, not only segments, as it was not essential that the tree should remain alive. They were kept drying in the tent for four days.

On the same day the hunter went out to kill the animal whose skin was to furnish the drum-head. Shaman Kokuev of the *aq čoγdu* clan says in this connection: 'Before I became a shaman, I was ill. Then I began to sing. I did not know anything about this. The relatives who lived with me understood it. I sang about the Nerkha river in the mountains, and that somewhere it was joined by another river, the *tottyγoj*, by whose upper reaches lived a deer, which would be my drum. Then a hunter went to the upper course of the *tottyγoj* to shoot it. This is very easy, because when the hunter sets out, the animal itself goes to meet him. The hunter killed the deer and flayed the animal on the spot, cut up its flesh, but left it there and went

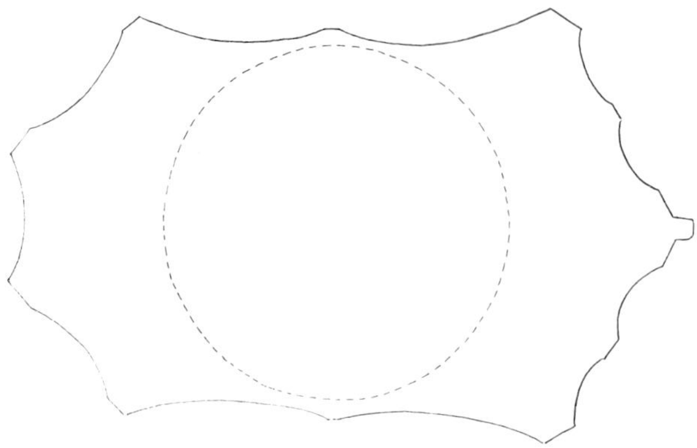

Fig. 28. Cutting from the flayed skin the leather disc to be stretched over the frame of the drum

home for reindeer to carry the burden. He started back for the animal with three reindeer on which he loaded the skin and the meat. So he started homeward. The meat of the animal was cooked in the hunter's tent, all the neighbours were invited and ate it together. I ate of it, too. While the meat was being cooked, the hunter began to dry the skin. He took it into his tent and stretched it out near the side opposite the door, binding it to the tent poles with straps. The skin was left like this for three days. On the third day he unfastened the skin and gave it to those who were to make the drum for me. A shaman usually receives the skin for his drum free of charge from the hunter. The drum-makers put the skin into hot water where they left it soaking for one day. Then they took it out and scraped off the hair. They cleaned it with an ordinary knife (*вišek*). The man sat on the ground, and laid the skin on his left shoulder with the hair up and part of it hanging over his knees. It was always this part that he cleaned and scraped with the sharp knife. He did not touch the fleshy side of the skin, which had to remain as it was, because otherwise the drum later would not sound good and the skin might become injured from energetic beating.' (After the hair had been removed, a circle large enough for the frame of the drum was cut out of the skin and was sewn on to the framework which in the meantime had been completed. Fig. 28.)

The members of the *qara čoγdu* clan dressed the skin used for the drumhead in the same fashion. 'Nothing has to be done in advance with the skin that is to cover the frame of the drum. If the inner side has a large lump, for instance of flesh still clinging to it, this is cut off with the knife. Again a disc of proper size is cut out of the skin with the knife. Then it is ready to be stretched over the frame of the drum.'

The skin to cover the drums of the *χaš, saryγ χaš* and *čeptej* shamans was dressed in the same way.

The drum was made on the fourth day. The frame was carved out of one of the segments drying in the tent, from another the two 'tendons' were made. From one cylindrical piece of wood the handle was prepared and from the other, the two cross-bars.

The framework of the drum for the shamans of the *aq čoγdu* clan was made as follows: the maker first split a plank from a half-log with an axe, smoothed it out with a knife, and then began the bending.

Fig. 29. Bending the frame of the drum in the slit made in a tree trunk

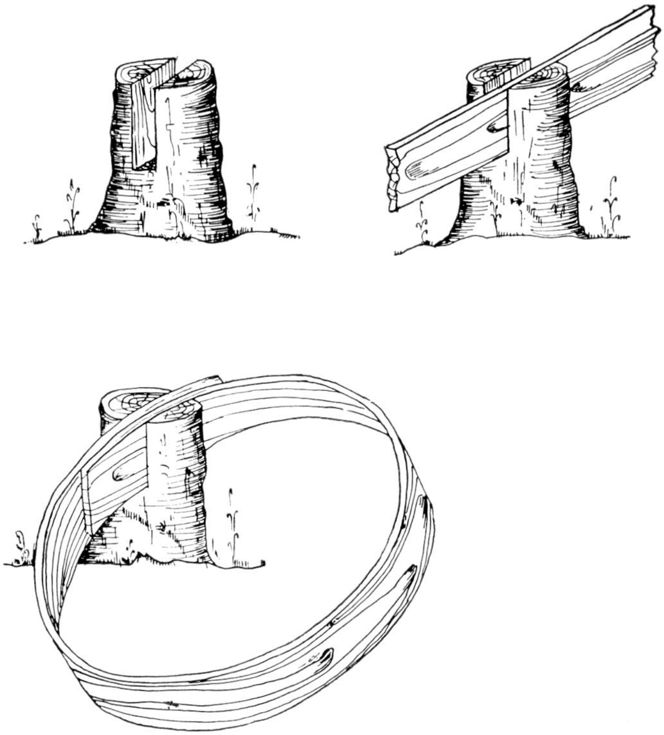

When the drum was made for the female shaman Aleksandra Nikolaevna Andalaeva, the wife of shaman Sipkeev of the *qara čoγdu* clan, the wood for the frame was first thinned down with an axe, then the segment, cut out of the trunk, was carved into the shape of a plank and thinned with a plane (*noŋu*). The plane was a grooved piece of wood in the shape of an ellipse with a longish slit in the middle. An ordinary knife was inserted into the slit, and the plank was thinned with this. When the board was of the right thickness, it was bent in the shape of a circle around a structure made especially for this purpose on the ground. Near the tent a tree was cut, its stump left waist-high. A U-shaped groove was carved vertically in the trunk. The width of the groove was somewhat larger than twice the thickness of the board, and the depth was equal to the height of the frame. One end of the board was inserted into this slit, the board was bent around the outside of the trunk in a circle, and its end placed into the groove next to the other end. Later, three holes were drilled through the overlapping ends of the board and on either side of the trunk and the two ends were attached by leather strips threaded through these holes (Fig. 29).

Among the *saryγ χaš* people the frame of the drum was made in a different way. On the ground a circle was drawn and a stake (*örgen*) driven into a point on the circle. The frame was laid down edgeways outside the stake, and another stake was driven on the other side. At a hand's span from it, another stake was driven on the inside. The frame was folded over this and on the outside one more stake was driven into the ground. In this

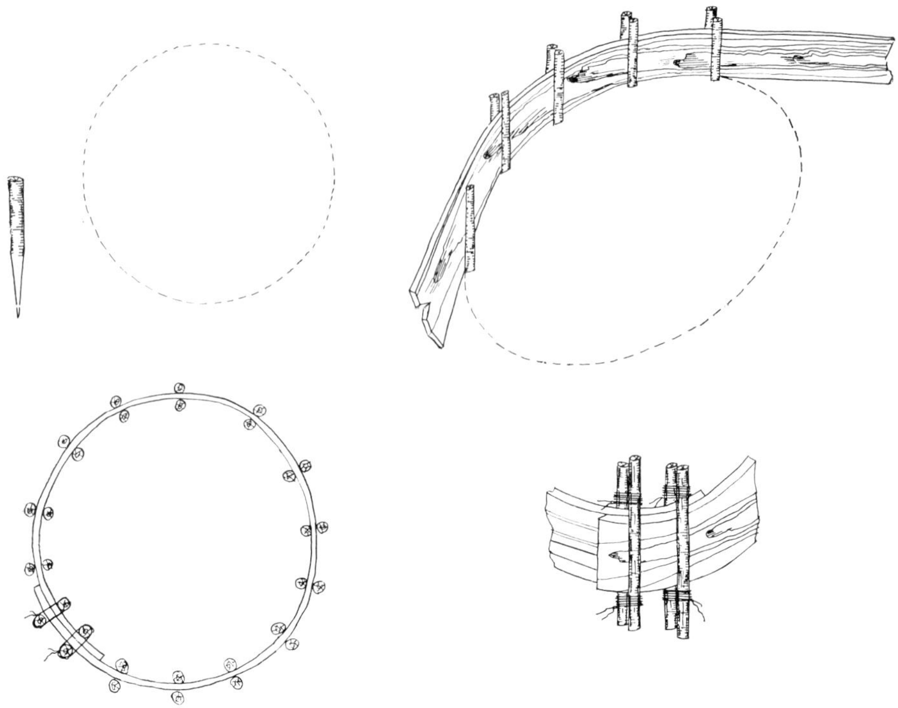

Fig. 30. Bending the frame of the drum by driving in two posts one after the other

fashion the work was carried on: first always the inside post was fixed and then the outer stake was driven in, to press the frame against the inside one. When the frame had become quite round, the two ends overlapped. Here two stakes were fixed on each side—two inside and two outside—and their two lower and two top ends were strongly tied to keep the frame together. Now in each of the overlapping ends of the frame three holes were driven in a vertical line. The two ends of the frame were then sewn together with straps pulled across the holes (Fig. 30).

When making the drum of the *χaš* shaman, nine or twelve stakes were driven into the ground along a circle. Since a male shaman's drum is bigger than that of a female shaman, twelve stakes have to be used for making a drum for the former and only nine for the latter. The drum-maker propped the middle of the board, still straight, against the stake on the opposite side of the circle and, seizing both ends at once, bent it around the stakes. When the two ends of the frame overlapped, he first pressed them with his legs against the stake on the nearest point of the circle and then fastened them together with a wooden clip (*qysyp qaqdar*). This was made from a cylindrical piece of wood, in the middle of which a U-shaped opening was cut. Its breadth was somewhat larger than twice the thickness of the frame and its depth was the same as the height of the frame. When the clip pressed the two ends of the frame well together, they were bound with leather straps and sewn in one piece through three holes on each end (Fig. 31).

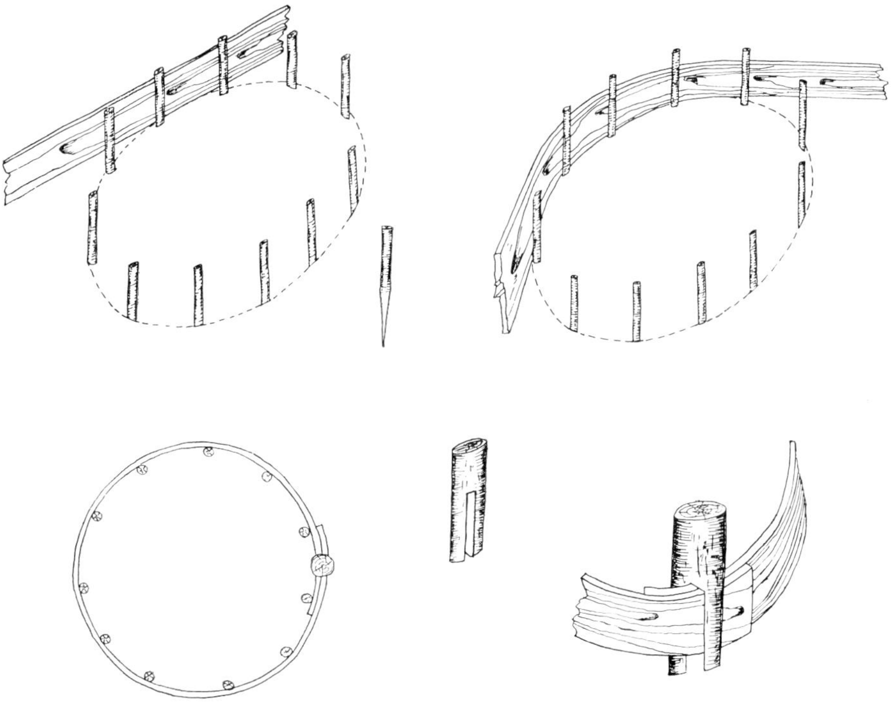

Fig. 31. Bending the frame of the drum in the form of a hoop by means of posts driven into the ground

The members of the *qara čoγdu* clan drew, on the ground, a circle which was as large as the drum they wanted to fabricate. Then they laid stakes along the line of the circle, about 15–20 cm apart. The closer the stakes were to each other, the easier it was to shape the frame. Then, next to one of them, they put the end of the frame against the outside of the stake, and opposite the inner stake they drove another on the outside. Then they folded the frame gradually around these stakes always driving a raw stake on the outside, opposite the inside one. When the two ends of the frame met, they were folded over each other. However, the first outer stake was removed to permit the two ends to overlap directly and another outer one was driven into the ground so that the overlapping ends were caught between the two. The overlapping ends were drilled in three places and tied together. This completed the operation of bending the frame (Fig. 32).

For Andalaeva, the wife of the *qara čoγdu* shaman Šipkeev, the frame was bent around a peculiar structure. Two shorter boards were placed edgeways on the ground, their breadths corresponding to the height of the frame. Between the two boards two short rods were placed to keep the boards at a proper distance one from the other. It was around this structure that the frame was bent. When the two ends met, two posts were driven in both on the outside and inside of the overlapping ends, the four posts

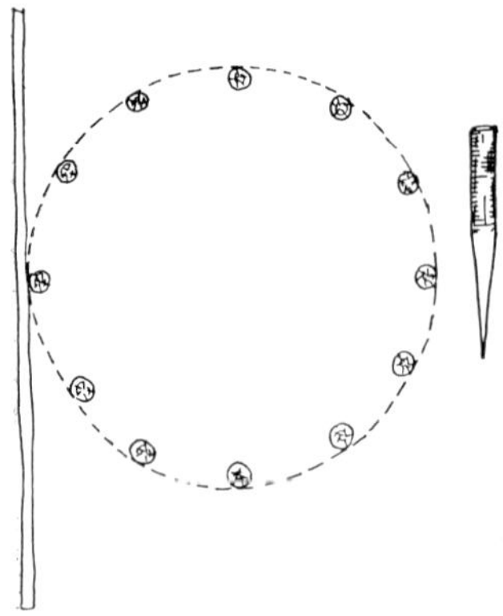

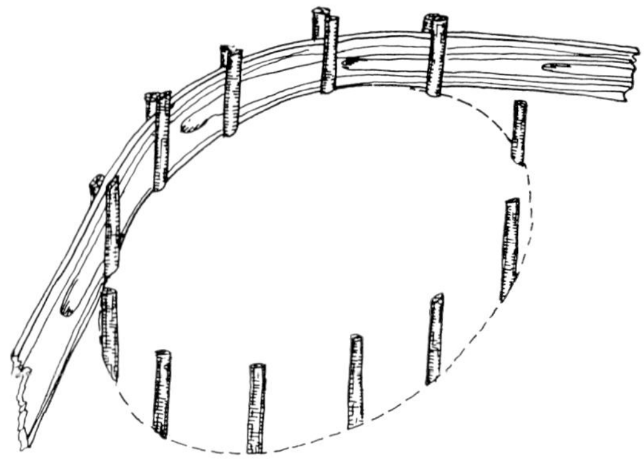

Fig. 32. Bending the frame of the drum by means of posts driven outside a circle of posts

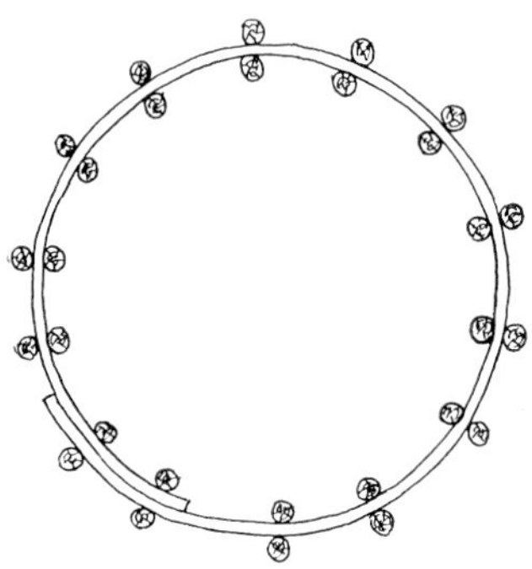

pressing the two ends tightly together. Three holes were now drilled through both ends, which were tied together with thin straps pulled through the holes. After this it was possible to remove the stakes and the framework remained circular (Fig. 33).

While the framework of the drum was being bent, other workers were busy making the other parts of the drum. As we have seen, the wood for the two 'tendons' on the *aq čoγdu* shaman Kokuev's drum were cut from thin-trunked firs. The log was first carved with an axe, and, when it had reached a reasonable size it was made even thinner with a knife until a 2 or 3 mm wooden sheet was obtained. This strip was split lengthwise into two. First, however, a thin line was traced down the middle of the thin board with the point of the knife so as to keep the splitting straight. Then it was split with the point of the knife. The entire operation was performed on the hard ground of the tent, on which two finished thin wooden sheets had been laid.

The two tendons for the drum of the female shaman Andalaeva were made of the same wood as the frame itself. Two strips were split each about an inch wide.

The handles for the drums of the Tofa shamans were manufactured by a different man. He first shaped a piece of board out of a log with an axe, then he smoothed it down with a knife, carved a protruding peg at each end, made the board thinner and cylindrical in the middle and cut two grooves in the top oblong part, and one or sometimes two oblong grooves in the lower oblong part. He drilled a round hole with a red-hot pointed iron rod (*šüвüǵej*) above the two openings made on top (Fig. 34a).

A third man made the upper and lower cross-bars (the *χümündrük* and the *quturγa*), each consisting of two parts. A cylindrical log was split lengthwise and the two halves were split again. The four pieces were first fashioned with an axe into quadrilateral rods and smoothed down with

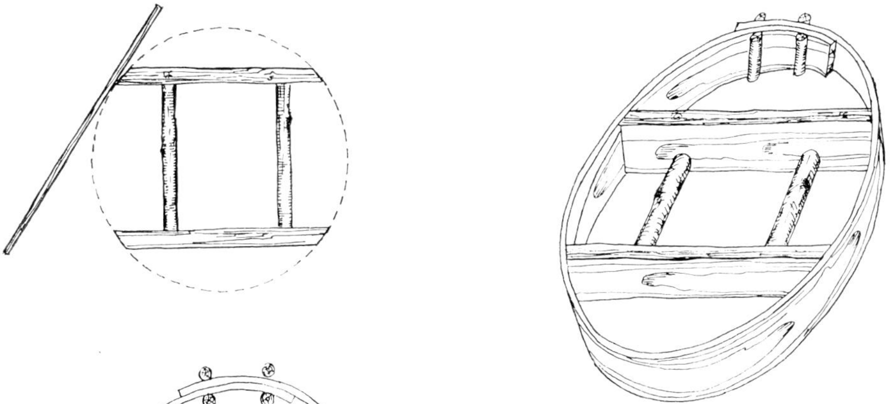

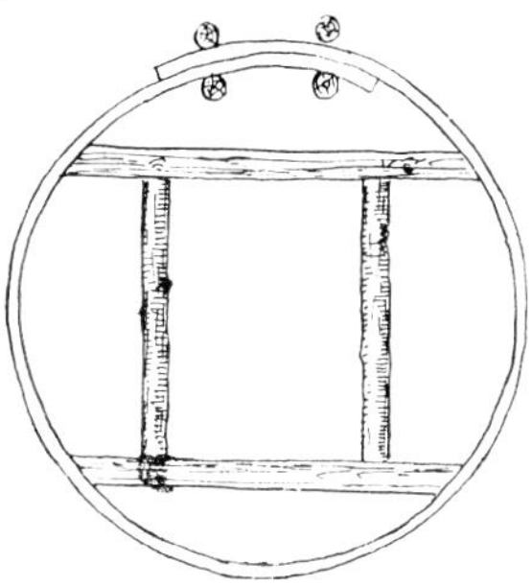

Fig. 33. Bending the frame of the drum by means of a structure consisting of four pieces of wood

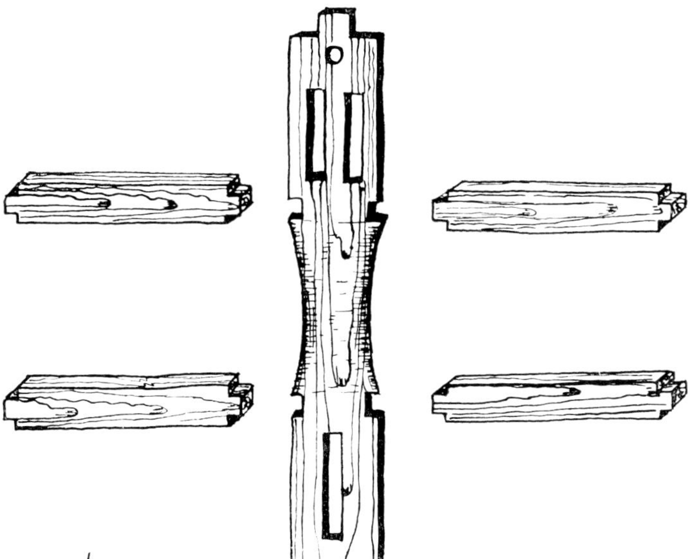

Fig 34. The handle of the drum (*a*) and the top and bottom cross-bars of the drum (*b*), each consisting of two rods

a knife. On both ends of each of the four sticks protruding pegs (*uǯu*) of the proper size were cut out (Fig. 34b).

The next step was to assemble the drum. This was done by an old man skilled in this type of work. First he attached the two thin wooden ribbons to the frame. Then he laid out one strip on the ground, put the lower edge of the frame on the end pressing the strip against the edge of the frame with his thumb, and rolling it along the strip. When the 'tendon' had been wound around, he made the two ends overlap and, drilling holes into them, tied them to the frame with a strap. The upper strip was attached to the frame in the same manner.

After the two 'tendons' had been attached, the man cut openings (*tutuq*) with a knife in the frame for the upper and lower pegs of the handle. After having put in the handle, he attached the right and left parts of the upper cross-bar in between the handle and the frame. The ends of these also had pegs for which openings had to be cut into the frame and into the right and left sides of the handle. The maker put in first the right and then the left side of the upper cross-bar, and inserted the two parts of the bottom cross-bar in the same way.

In the meanwhile another man was making the 'ears' (*qulaq*) out of wire. 'For each one he bent a thin metal rod into a broad U, hammering the ends pointed. He bored two holes in the frame at a proper distance from the handle to the right and to the left. Later he pulled through the ends of the 'ears', folded them back, and bent them with a hammer against the inside of the frame.

The attaching of the 'ears' was followed by the making of the 'ear-rings' (*syrγa*). They were in all respects like the metal cones on the shaman's gown and were rolled in the same fashion from oblong metal plates. Two or three 'ear-rings' were placed on each 'ear'.

The resonators were attached when all this had been successfully accomplished. For this the drum was laid on the ground, a strap was bent around it with its two ends held together. All the resonators were then inserted under the strap. The strap was always loosened a little after a resonator had been added, as the drum got a little thicker. When all the resonators had been inserted under the strap, the two ends were bound together. Then reindeer tendons were placed around the upper heads (*χabaqtyŋ вэχe*) of the three-headed resonators, the two ends of the tendon being pulled tight and tied together. The procedure was the same for the lower bosses of the resonators when these were once firmly attached to the frame, the strap which had kept them pressed against it, was loosened. Finally the tendon was tied also round the wedge-shaped grooves of the middle heads of the resonators.

The next step consisted in stretching the skin (the drumhead) over the drum. With a red-hot pointed iron rod holes about 5 cm apart were bored along the upper rim of the frame. The skin was placed over the frame with its hairy side out. After it had been measured so that it should cover the whole outer side of the frame with the resonator and could be folded back over its lower rim, to leave a border of 1·5–2 cm width, the skin was cut round with a knife. Then, through holes drilled into the frame, it was sewn to it with reindeer tendon threaded through a primitive large-eyed needle. The skin was not pulled tight, but was left rather loose at this point.

'This operation closed the day. They started to dry the drum the next day. My drum was dry in five hours' said shaman Kokuev. The drying of the drum of shaman Šipkeev of the *qara čoγdu* clan was done in the following manner: One side of a log about 30 cm long was carved so as to assume a slightly convex shape, in the centre of this a gentle depression was carved (Fig. 35a). A stick about 30–40 cm long and 4–5 cm thick was thinned down over a periphery of 3–4 cm close to its end (Fig. 35b), and was then placed perpendicular to the handle, with its thinned part against it. The log with the convex side was placed on the other side of the stick to fit the depression and was then tied with a cord to the handle (Fig. 35c).

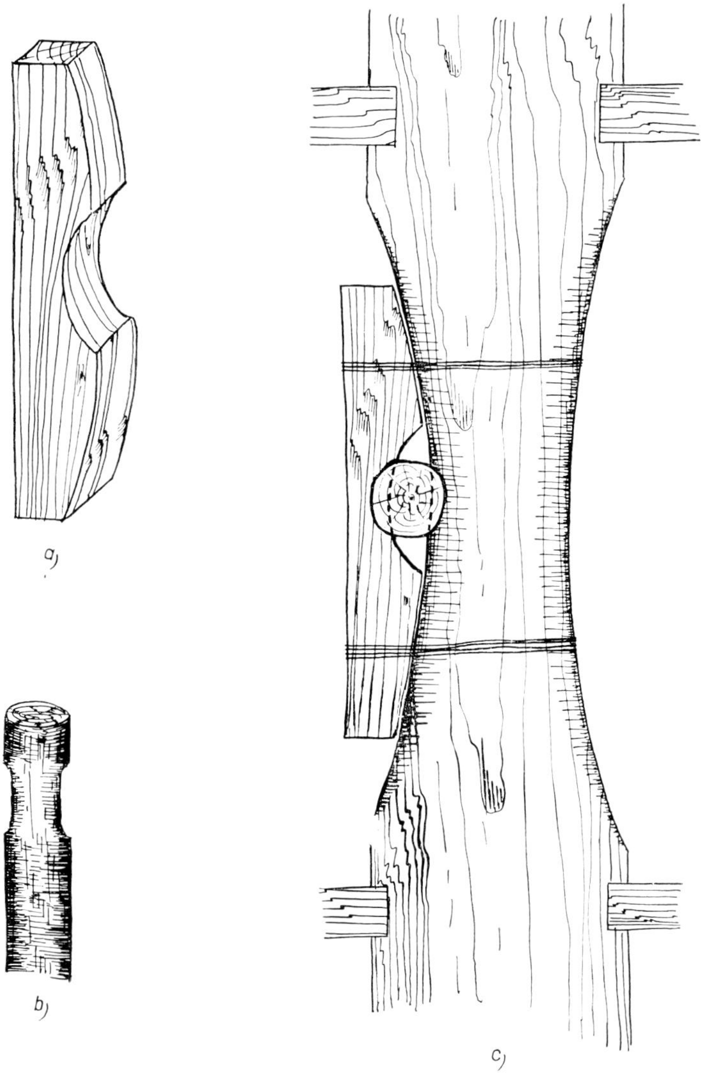

Fig. 35. Preparation of the handle for drying the drum

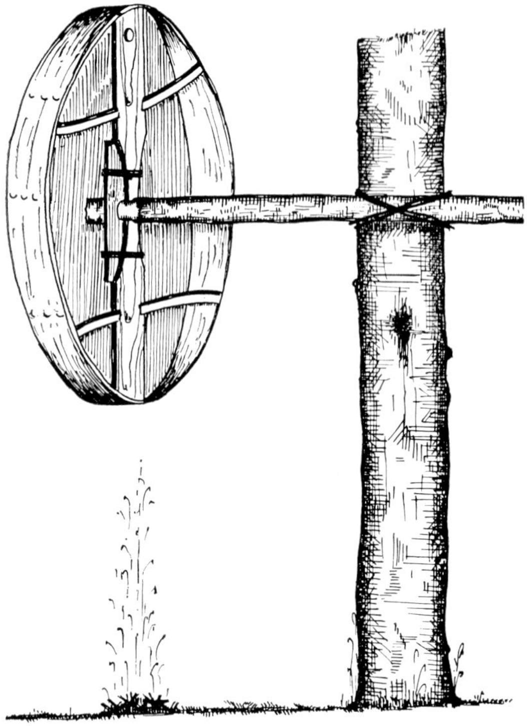

Fig. 36. Drying the drum

This having been done, a vertical stake was driven into the floor of the tent at a distance of 30–35 cm from the fire-place in the centre. The stick holding the drum was tied to this stake (Fig. 36). A fire was lit and the drum was spun over it for about one and a half to two hours.

The members of the *saryγ χaš* clan dried the shaman's drum in a similar fashion. To the handle of the drum belonging to shaman Kokuev of the *aq čoγdu* clan not a carved-out stick but a flexible branch was attached: the two ends of the bent twig were bound to the handle, but the middle part was left loose, forming an arc, i.e. an opening between the handle and the branch. It was through this opening that the drum was strung on to the horizontal rod above the fire, where it was given a fast twirl with the fingers. The fire was fed with juniper branches (*artyš*), the smoke of which reached the drum. Time and again a burning branch was picked out of the fire and the skin of the drum was beaten with it. Since the skin was moist it did not get scorched, but when the hair caught fire, it was scraped off with a knife. So gradually all the hair was removed. The originally loose drumhead tightened, too. For this reason it had to be readjusted sometimes, as the resonators occasionally slipped a little. Care had to be taken to keep them perpendicular to the frame end at the right distance from each other.

After the drum had been dried over the fire, it was hung up in the shaman's tent on the side opposite the entrance. 'My drum dried there for three more days' shaman Kokuev informed us.

Once dry, the drum had to be painted. 'I heard that some people painted the drum. Mine was not painted' shaman Kokuev told us. 'In fact, I have not even seen such drums.' It is a fact that our informant could not have seen painted drums, for the *aq čoγdu* shamans did not use them. The *qara čoγdu* shamans practised likewise with unpainted drums. Only the *χaš*, *saryγ χaš* and some *čeptej* shamans used drums painted with natural dyes such as red clay (*šibit*), which they obtained from the mountains. They dissolved it in milk, and painted the skin of the drum all over from the outside with this solution.

They attached two or three rolled strips to the handle of the drum. These were made of leather or textile. The leather strip was made from the hairless skin of wild reindeer (or sometimes of deer, *'Cervus elaphus'*) and the textile strip from linen. Their colour was white, red, blue or black and were made like the rolled strips of the shaman's gown.

The Tofa shamans kept their drums in a special case (*ᴅüŋür qaby*), made from the skin of tame reindeer, and having the shape of a broad and low cylinder. One half of the upper disc-shaped lid was not sewn to the vertical side so that it could be opened. This part had to be buttoned.

The Drumsticks

The sticks with which the shaman drums were beaten (*orba, χam orbasy*) were made either of horn (*mȳs*) or of wood: namely of the antlers of deer (*saryγ aŋ*), the maral deer (*uluγ aŋ*), the wild goat (*teχe*), or of the wood of birch or cedar. The stick consisted of a cylindrical stem (*tutar* 'handle') and an elliptical head (*orba ʙeχe* 'head of drumstick'). The bottom side of the latter part was slightly convex and was covered with leg skin (*ʙyšqaq*) from the reindeer, maral deer, deer, wild goat or even from the bear (*üles aŋ*). The other half of the head of the drumstick was usually ornamented with metal rings (*ᴅümür syrγa* 'iron ear-ring'). Some drumsticks had one, others two or three metal rings. Frequently three metal cones hung from the ring. Often a metal rod was attached along its longitudinal axis, and three rings were strung on it. The stem was composed of two parts: the neck of the drumstick (*tuqpoǰo*) next to the head, and the actual grip (*tutar ǰere*) at the end of the handle.

To the end of the handle of the drumstick the loop of a narrow strap (*ʙaγ*) was attached in many different ways. An L-shaped (i.e. rectangular) hole was drilled in the end of the drumstick, and the two ends of the strap were pushed right through the longitudinal and transverse parts. The two ends of the strap which thus came out of the transverse part, were tied in a knot large enough to prevent them from slipping back (Fig. 37a). Ends of other drumsticks were bored only in a transverse direction. First the strap was looped round the handle, and then the two ends were led through the boring from the top downwards. The two ends were then tied together (Fig. 37b). Sometimes the strap was simply pulled through, and the two ends, tied together without previous looping (Fig. 37c). The ends of some drumsticks had T-shaped holes. The two ends of the strap were moved through the longitudinal hole, then one end was pulled out to the right and the other to the left through the transverse parts. Both ends were knotted to prevent their slipping back (Fig. 37d). Finally, an entirely different method of attachment was used: A metal ring was affixed to the end of the handle and the strap was strung through this.

I had the good fortune of meeting an old man among the Tofas who knew how to make drumsticks. Vasilij Nikolaevič Adamov of the *aq čoγdu* clan (born in 1893 in Alygzher) had made two in his life. The first was for the shaman Nikolaj Danilovič Tulaev of the *aq čoγdu* clan in 1920, when he lived with him in the Nerma valley. He made it out of cedar. The shaman himself had told him to do so. Adamov covered the wood with skin from the leg of a wild reindeer according to the shaman's wish. The handle of

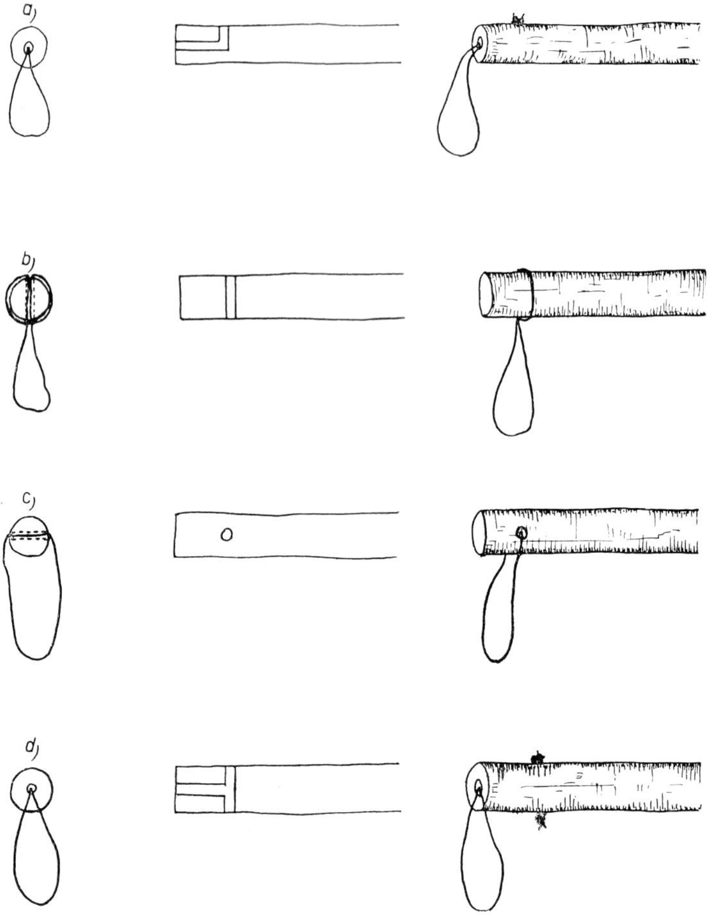

Fig. 37. Ways of attaching the leather-strap to the handle of the drumstick

the drumstick bore three ribbons, all white. This, too, had been the shaman's wish, who paid him 20 rubles. Had he not paid in cash, he would have given a bottle of brandy. It took him two days to make the stick.

Adamov made the second drumstick for the shaman Amastaev of the *aq čoγdu* clan, who was carrying on a nomadic life in the valley of the Karaburen river. He was a relative of Adamov's. This drumstick, carved of wild-goat horn and covered with the leg skin of the same animal, was made in 1930. The shaman himself had told him what to make it of. The handle had also three ribbons, all white. Also Amastaev paid 20 rubles for the making of this stick. Vasilij Nikolaevič Adamov made a drumstick at my request and we were able to follow and photograph the whole process.

First he laid out an animal skin on the ground and sat on it while he was working. Then he produced a three-pronged reindeer (*ibi*) antler, a saw and an axe (Fig. 38). He started by removing the tines. Sitting on the animal skin on the ground, he rested the antler on his left instep and pressed its end down with his right foot. Also his left hand provided a fixed support to the lower tine which he was going to remove first, with the saw held in his right hand. After he had sawed off the tine, he stopped to see whether he had removed it properly; then he spat on the blade of the saw to make it slippery (Fig. 39). Again he fixed the antler and sawed off the second tine as well (Fig. 40). The third tine followed. This was not done in the same way as the others, but was made by sawing off the entire end of the antler. After having sawed the end of the antler almost through, he broke off the part to be removed. Now he had got the length required for the head of the drumstick. The antler ended in a straight line, and therefore he had to saw off the two corners, i.e. 'round them off'. Later he 'rounded off' the other end of the antler, where the handle was to be, so as to obtain the proper length for the handle. Adamov carefully examined the piece of the antler from the point of view of length (Fig. 41). Now he concentrated on the width of the antler. He sawed off the stubs of the tines. He spat on the blade again while he was sawing to make it slide and continued working with his saw until he had thinned the antler to the required width through its whole length. Having finished this, he made a few movements imitating the beating of the drum: He was judging the weight of the drumstick, whether it was too heavy or too light, and whether it would beat well. He found that a little more should be taken off the width. This operation he now performed with an axe. Then he checked what he had done. The next step was to thin off the two sides of the head of the drumstick and the end (Fig. 42). After this had been done, he checked his work again, later he bored holes into the tapering edges of the head of the drumstick 2–2·5 cm apart. This he did with a pair of scissors and a drill: He made little holes with the point of the scissors (Fig. 43), and broadened and deepened them with the drill (Fig. 44). After having drilled the antler through in about ten or twelve places, he bored three evenly-spaced holes in the axis of the head of the drumstick, and inserted a metal ring into each of these holes. The next step was to make a hole through the end of the handle. First he made a longitudinal bore (Fig. 45) and then a transverse one perpendicular to the first. The two holes formed an L-shaped channel. He checked whether they really joined by blowing into the vertical opening (Fig. 46), thereby also clearing the hole of the shavings inside. Then from the bag in which he kept his sewing kit, he took out a needle of proper size and a thin tendon to sew the skin on the antler. Skin from the leg of a bear was used to line this particular drumstick made before us. Adamov covered the lower part with the skin and cut off the superfluous pieces with a knife. After this he sewed on the skin through the holes bored into the edge of the antler (Fig. 47), and attached a loop of narrow strap into the boring, so concluding his task.

The drumstick was the last step in making the complete outfit of the Tofa shamans, all of whom had their head-dresses, gowns, boots, drums and drumsticks.

The pieces of the shaman's costume and the objects of this equipment, however, showed a great variety according to clan, category (white or black shaman), eminence (big, middle, or small shaman). In addition to this, there were also individual differences.

Fig. 38. Preparations for making the drumstick

Fig. 39. Spitting on the saw to make it slip

Fig. 40. Sawing off the second tine of the antler

Fig. 41. Checking the length of the drumstick in preparation

Fig. 42. Thinning the edges of the antler

Fig. 43. Boring through the edges of the antler with scissors

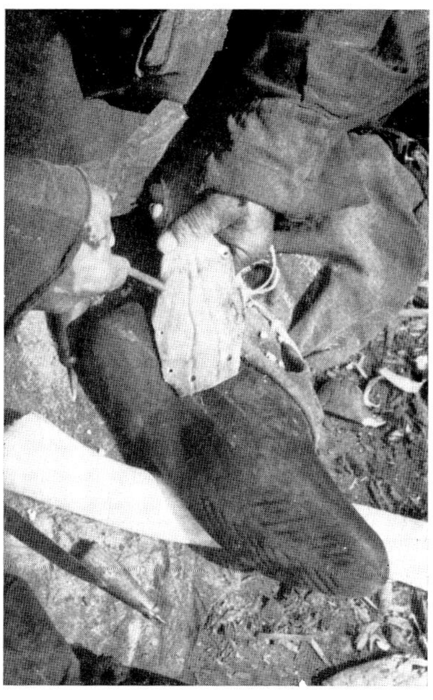

Fig. 44. Boring through the edges of the antler with a drill

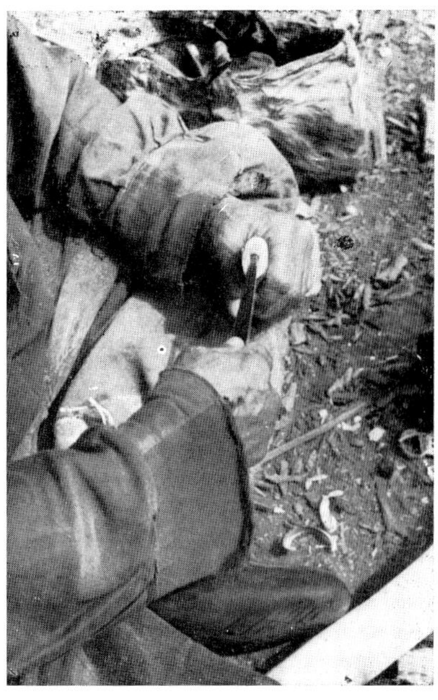

Fig. 45. Making the longitudinal bore in the end of the antler

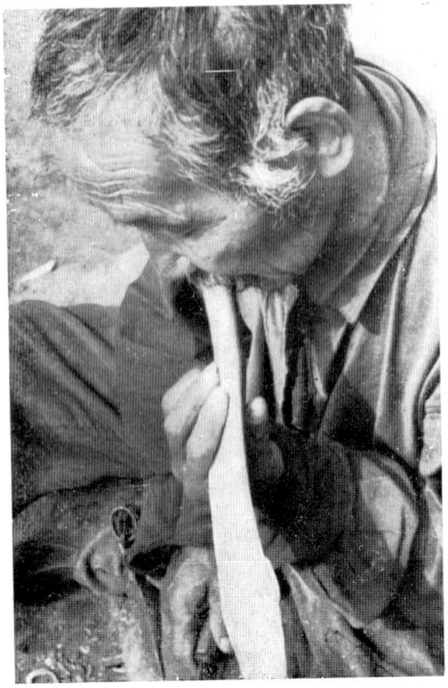

Fig. 46. Testing if the two bores meet

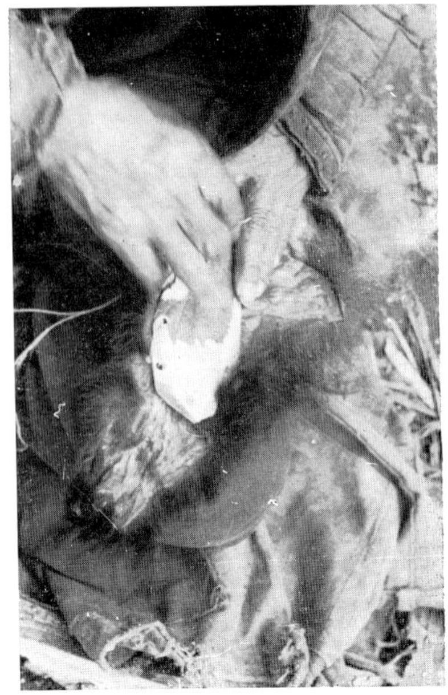

Fig. 47. Sewing the skin to the antler

DIFFERENCES IN THE SHAMANS' COSTUMES

The Head-dress

As we have mentioned before, the head-dress of the shamans shows a great variety in shape, decoration and colour. The survival of so many variations is fairly well accounted for by the fact that a large number of head-dresses were needed in the shamanistic practice. The Tofa shaman, for instance, used in the course of his lifetime not one, but two, three and sometimes even more head-dresses, each more or less different from the one used previously.[16]

We may divide the head-dresses into three groups depending on their shape: (a) the head-band, (b) the cap, and (c) the crown.

The head-band occurs most frequently, being worn equally by the *aq čoγdu*, the *qara čoγdu*, *saryγ χaš*, *χaš* and *čeptej* shamans. It was made of textile (usually cotton) 28–56 cm long and 10–15 cm wide. It consisted of either a single strip of the same colour (white, red, blue or yellow) or strips of different colours, such as: blue, white and red; white and red; white, yellow and blue; or red and yellow.

[16] The Tofa shamans had several head-dresses made in the course of their lives, which they designated as 'first', 'second', etc. They changed head-dresses during the ceremony (as described below).

Usually the head-band and a hem (*ɢeš*) of the fur of white rabbit (*χodan*), squirrel (*ʙīŋ 'Sciurus vulgaris'*), lynx (*ùs* 'lynx'; *ùstùŋ ʙùgù* 'lynx fluff') or wild-duck feathers (*qas* 'wild duck'; *qastyŋ ɢeši* 'wild-duck fur').

Feathers were stuck in the upper hem of the head-band. All the Tofa clans used feathers from the long-eared owl (*χùgù 'Strix bubo'*). Yet other feathers were also used, differing according to clans: the shamans of the *aq čoγdu* clan used capercailzie (*qara quš 'Tetrao urogallus'*) or hazel-grouse (*paryqan 'Tetrao bonasia'*). Those of the *qara čoγdu* clan wore eagle (*ēzir*) feathers, the *saryγ χaš* shamans had wild-goose (*qas*) or crane (*turuńa*) feathers and the *čeptej* shamans decorated their head-band with hawk (*tiligēn*) feathers. There were three, seven or nine feathers either all of equal length, or with the middle one long or all different, shortening gradually towards the two sides.

On each side rolls (*čaryš manǰaq*) were sewn to the lower edge of the head-band. In addition to the general term given in brackets they were also called *čaš* and *syrγa*, or *börtüŋ syrγa*. They had an average length of 15–20 cm. There were either three, or seven, on each side, black or yellow. They were made not out of cotton, but occasionally also of reindeer skin. These rolled strips ended in tassels called *čaryš manǰaq ʙeχe* or *salbar*.

To the two ends of the head-band one or three long ribbons (*čala* or *čalazy*) were sewn at the bottom. They were white, black, blue, yellow and/or red in colour, 30–80 cm in length, and 6–8 cm in width.

After having put the head-dress on his head, the shaman buttoned the two ends of the band at the back. There were two or three buttons (*òk*) on the left end, and the same number of loops (*òk ebsise*) on the right end.

The head-bands may be divided into four groups according to their decorations: those adorned with (1) faces, (2) geometric figures, (3) shapes of birds and (4) the unadorned. The ornaments were either embroidered or appliquéd. The embroidery was generally stitched with the long dewlap hair of tame reindeer (from under its neck) or with thread; the appliqués were of cotton fabrics.

Head-bands decorated with faces were the most frequent and were used by all the five clans.

The head-bands we collected had the following motifs stitched on them on the part above the face: in the middle, two horizontal or arched lines, the *χabaq;* under them, two ellipses, the *qaraq* (sometimes a bead, for instance, a blue bead, called *kùk ńinǰe* was sewn in the centre); between them, a vertical lines of stitches, the *χaj*; and under it, a horizontal line or ellipse, the *ās* or *aγsy*; and on each side, at the height of the eyes, a half-moon looking toward the motif (the *qulaq*).

The head-bands decorated with geometrical designs were covered with discs and with zigzag lines stitched on with reindeer hair or with thread. It should be noted here that the head-bands covered with discs (a geometrical design) belong, in fact, to the group of head-bands adorned with faces, for the discs represent stylized faces. The Tofas called the individual discs: eyes, nose, mouth and ears.

Head-bands decorated with figures of birds are rare. Such one was the second head-dress of the big shaman Kholomoeva, which was ornamented with three appliqués cut out of blue cotton, each representing a crane.

The second type is the high cap. It is made of reindeer hair, covered with cotton fabric, and sewn together from triangular pieces. We find this type only in the *χaš* clan. The cotton fabrics used for each are of three colours: white, blue, and yellow (see Fig. 61a, b). Long ribbons of the same colours hang from the hem of the head-dress. The peak of the cap bears wild-duck (*ȯᴅu̇rek*) and the edge is trimmed with a gray rabbit fur.

The third type, the crown or *qumsar bȯrt* was worn by shamans of the *aq čoγdu* and *čeptej* clans. Although it is rare, it is not an individual invention but occurs sporadically true to type. As a matter of fact it is very much like a head-band, only the two ends are sewn together, and to the upper edge of the circular band two narrow iron bands are attached crossing each other at right angles in two arcs over the head. A tuft of plumes emphasizes the point of crossing. From the lower edge of the head-band are suspended nine black rolled strips on each side, and three long and wide ribbons hang in the back.

The Breast-cover

The shamans of some Tofa clans, namely of the *aq čoγdu*s, *qara čoγdu*s and *čeptej*s, had breast-plates, or rather breast-covers (*ᴅuš χu̇nǰu̇k*), worn over their gown.

Although these breast-covers were all more or less oblong, there existed certain differences in shape. Some were arched at the neck and at the sides, others were arched only under the neck with straight sides, the front either widening or narrowing evenly (see Figs 62, 77).

The mode of attachment varied, too. For instance, the right edge was sewn to the gown and the left, buttoned on with two buttons; or else three leather straps equally spaced were sewn to both the right and left edges. Sometimes only two straps were sewn to the upper edge of the breast-cover, one on each side, tied behind the neck so that the breast-cover hung from the neck. This garment was ornamented with appliqués or stitchings and also with textile tassels.

A characteristic ornament of the breast-covers was the vertical stripe running down along the middle axis, whence narrower strips branched off to each side. Also the pattern and the fabric of the breast-covers differed, being sometimes made of white or blue cotton and decorated with slanting white reindeer strips parallel to each other; or else the vertical strip consisted of nine cotton squares in a vertical row. This design was called *ᴅu̇š sȫgi* 'breastbone'.

From the vertical strip, parallel strips branched off to the right and the left: nine white cotton strips on each side slanting downward, straight or in a gentle curve, or sometimes seven straight white cotton strips slanting downward on each side, or nine straight strips running upward obliquely stitched with reindeer hair. The name of these branch-strips was *ēgi* 'rib', or *ēgi sȫgi* 'rib bone'.

In addition to the above ornaments *šu̇ᴅ* ('chest'), the breast-covers were decorated with other appliqué and embroidery designs as well.

Marking the upper end of the breastbone, a disc of reindeer hair was sewn on to the breast-cover of the *aq čoγdu* shaman. It was called *oru*, 'hole above the breastbone'.

The same garment had, at the height of the collar-bone, a V-shaped appliqué of white cotton strip. This was called *munuj sōgi* 'collar-bone'.

There was an acute-angled triangle of appliquéd, white cotton fabric, sewn to the same breast-cover, marking the lower end of the breastbone. This was the *ᴅüš qudrū* 'the lower end of the breastbone', or literally 'the tail of the chest'.

Over the nipples a disc of white reindeer hair was sewn on each side on the *qara čoγdu* shaman's breast-cover. Its name was *emi* 'nipple'. This garment had another disc at the height of the navel,—the *χin*, or 'navel'. Under the navel the *aq čoγdu* breast-cover was adorned with a bundle of nine white ribbons (*čaryš čaqyn manǰaq*), and the *qara čoγdu* garment had three parallel strips of reindeer hair (*qasy*).

In the case of shaman's costumes with no separate breast-cover the above appliqué and embroidery designs were on the gown itself.

The Gown

The gown (*ᴅon*, *χamnar ᴅon*, *χam ᴅonu*) of the Tofa shamans was usually made of the skin of reindeer stag (*čary*) or doe (*ibi*) killed in summer. Sometimes the skin was covered with a fabric, as for instance, blue cotton. Summer skin was preferred because the hair on it is shorter. The hair was never scraped off, but simply cropped short with scissors when necessary. The gown had two folds, meeting either in the middle, or on the right side. Its length varied: beside the usual knee-length gown, some reached the middle of the leg, others, below the knees, and still others, ended at the middle of the thighs.

A tall, standing collar about 2·5–3·5 cm high (*čaχa*) was a characteristic feature of the gowns, which were trimmed with strips of the fur of lynx, lamb, squirrel or rabbit. Along the fur trimming, or even when the fur was absent, the edge of the cloak was adorned with tassels (*salbar*, *čaryš manǰaq*), usually made of reindeer skin, or, rarely, of yellow cotton. Their length ranged anywhere between 6 and 20 cm.

The shaman's gown of the Tofas was adorned with embroidery and appliqué, textile, leather and metal pendants, and bundles of feathers.

We shall begin classifying these designs here according to the nature of the appliqué and embroidery.

Characteristic is the ornamentation on the chest, which is not restricted to any one clan, but is used by all the shamans. In some cases it decorates the gown itself and in others it appears on a separate garment, which is the breast-cover.

Along the midline of the gown a vertical line runs in front from the neck down to the end of the breastbone. This strip called *ᴅüštüg sōgi* consists of an appliqué of white cotton or of stitches of white reindeer hair.

To the right and left of the vertical strip nine downward-slanting parallel strips branch off, also of white cotton appliqué, or white reindeer hair stitchs. The branches are called *ēgi* 'rib', or *ᴅüštüg ēgisi* 'rib of chest'. The name of the whole design (the vertical strip and the branches together) is *ᴅüš* or 'chest'.

Some of the Tofa shamans' gowns had other designs than those representing bones and parts of the body on the part covering the chest.

The gown of a *qara čoγdu* shaman had three parallel horizontal cotton-strip appliqués on the chest. The top strip was black, the middle one white, and the bottom one black again.

From the collar down to the lower edge, two vertical cotton strips were appliquéd on the gown of the *χaš* shaman Bokturbaev. The left strip was blue, the right one yellow.

The sleeves of the Tofa shamans' gowns were also covered with appliqué and embroidery designs. At each shoulder a white cotton strip was sewn, around the sleeve of the middle shaman Amastaev of the *aq čoγdu* clan. This was the *ǰüsü* 'shoulder-joint'. Decoration of the sleeves with horizontal strips was more customary. The number, colour, material and place of the strips showed great variety. Some gowns had sleeves adorned in front, others, on the back with one single white cotton strip. Some had three cotton strips sewn on them: a red one on the top part of the sleeves, a blue one in front and white on the lower part; or white, yellow, and blue. Others had only a single white strip of reindeer hair in front on the sleeves called *qol* 'arm' or *qol sȫgi* 'arm bone'. On some gowns the strip running on the sleeves was broken by a white disc at the elbows. The name of the strip closer to the shoulder was *üstü*, of the disc *ǰüs* or 'joint', and that of the part from the elbow to the hands was *qary* 'its arm'.

At the ends of the sleeves five thin strips (called *ergek* 'finger') branched off from the thicker strip on the front or back. These were embroidered with white reindeer hair.

Some shamans' gowns had sleeves not decorated by bones of the arms. For instance, a thin strip of reindeer hair spiralled round the sleeves from the shoulders to the extremities. According to some members of the *qara čoγdu* clan, it was this lack of adornment that characterized the gown of their shamans.

The back of the gown of the Tofa shamans was also covered with appliqué and embroidery. A strip ran along the middle following the backbone from the neck to the rump-bone. The material and shape of the appliqué varied: Some gowns had a white cotton appliqué, a simple strip about three fingers wide; others had a row of nine trapezoids. Still others had white reindeer hair following the line of the backbone, sewn on with slanting black thread stitches. Sometimes the white textile strip was broken into nine sections by eight discs embroidered with white reindeer hair. On some gowns the vertical appliqué was made of white or blue cotton three fingers wide, and on each side of the strip there were seven or nine semi-discs stitched on with white reindeer hair. The strip is called *ōrγa* 'backbone', *ōrγa sȫgi* 'spine-bone', and the semi-discs, *ōrγa ǰüsü* 'dorsal vertebra'.

From the strip running along the spine parallel bands branch off slanting downward, varying in material, colour, and number. They may be of white or yellow cotton or reindeer hair, and are usually nine, seven or five in number on each side. They are either short, or so long that they span the entire back and even reach over to the front part of the gown. The name of these bands is *ēgi* 'rib'.

On one shaman's gown there were four very short parallel strips made of white reindeer hair branching out to each side slanting downwards. These are called *orγa χanaty*, or 'back wing'.

The gowns of the Tofa shamans were ornamented with a large number of accessories, most of which were textile pendants.

Up in front, at the height of the shoulderbone, one gown had one rolled-strip bundle with three branches (red, white, blue) attached on each side. This was called *uzun manǰaq* 'long rolled strip'.

Textile pendants attached in front, at the height of the waist, were frequent. Some attached at the height of the nipples had nine branches. Other gowns were ornamented in the same place with seven- or nine-branch bundles on each side, or with three nine-branch bundles on each side. Some had a white belt (*qur*) sewn to the waist-line, from which hung nine bundles with nine branches each. The colour of the rolled strips varied on the different gowns: one had only blue strips, the second had red, yellow, green, or else red, yellow, blue and white strips; the third had red, yellow and blue strips. Others had red, brown, pink, yellow, light-blue, dark-blue, green strips.

Textile pendants were attached also to the sides of the gown: either three with three branches on each side, or a seven-branch strip on each side, or else three nine-strip bundles on each side. The colour of these rolled strips showed great variety. The textile pendants hanging from the sides of the gowns were called *uluγ manǰaq* 'great rolled strip'.

Also the backs of the Tofa shamans' gowns were decorated with textile pendants. In some cases the rolled strips were attached to the shoulders on the back, either in the form of one three-branch bundle (blue, white, red) or two single rolled strips (white, with some parts coated with red) on each side. The term used for them was *usar čaqpa*.

Waist-high textile pendants were frequent on the back. The big female shaman of the *aq čoγdu* clan had a gown decorated with four nine-branch bundles on each side. The big shaman of the *qara čoγdu* clan had one ornamented with five seven-branch bundles and their middle shaman had a gown embellished with one nine-branch bundle on each side. The gown of the *χaš* big shaman had fourteen nine-branch bundles and that of the *čeptej* middle shaman, nine nine-branch bundles. The colours used were: red, brown, lilac, yellow, blue, light- and dark-green, and white.

Another variety of textile pendants was the short tassel (*salbar*), which is rarely found on the dress of the Tofa shamans. The sleeves of an *aq čoγdu* gown had tassels of this type under the elbows.

Kerchiefs—another form of textile pendants—hardly ever occurred. We know of an *aq čoγdu* shaman's gown which had two metal rings on each side, each bearing twenty-seven kerchiefs, called *čalγyn* 'wing'.

Leather pendants were also found as decoration on the Tofa shaman's gowns. On *qara čoγdu* gowns two nine-branch rolled-strip bundles were attached at the height of the nipples, one on each side, below the armpits and on the back on each side at shoulder-height.

Not only rolled strips, but also tassels were made of leather. Under the chest part of the gown, a tasselled reindeer skin, 20 cm long on each side, was sewn on the gown of the *saryγ χaš* shaman Bolkhoev. It was called *ᴅüštüg manǰaq* 'chest rolled strip'. The hems of the sleeves were ornamented with reindeer skin tassels on the gown of the *qara čoγdu* middle shaman.

Usually leather provided the material for a typical pendant (the *quᴅruγ* 'tail') on the Tofa shaman's gown. A piece of reindeer skin rolled up into the shape of a long cylinder was attached at the height of the rump-bone, with a trifurcated end. The three branches were long on some gowns and short, on others.

The gowns of the Tofa shamans were also decorated with metal ornaments. Between or at the height of the shoulder-blades always an odd number of metal cones or arrows hung on one, or several, curved metal rods. Some gowns had one metal rod bearing seven or nine metal cones, others had two rods attached to them either next to each other or one above the other. In the first case each rod bore seven metal cones, and in the second case, the upper one had seven and the lower one nine metal arrow-heads. Some were known to have three metal rods placed under each other, each with seven metal cones. The metal bar was called *čele,* the cone *χoŋora,* the arrow *oq.*

A metal plate was attached to the gown over each shoulder-blade. The pendant was fixed at one point only, so as to permit it to swing as the wearer of the gown moved. The shape of the pendants was elliptical, triangular or trapezoidal, the sizes varied. The pendants were called *ǰaryn* 'shoulder blade', or *ǰarna* 'blade of the shoulder'. All of them had punched ornamentation.

Bells called *qaχinōr* ~ *qoχinōr* were also often attached to Tofa shaman's gowns. At the same place some had bells instead of the 'shoulder-blades' mentioned earlier. Several had bells on their gowns under the arms. Some had bells attached to the elbow joints. 'The shaman listens to the tinkle of this bell to find out certain things', shaman Bolkhoev of the *saryγ χaš* clan told us.

There was still one more metal accessory on the gowns of the Tofa shamans: The elliptical metal plate sewn on the hem of the sleeves between the 'arm-bone' and the 'fingers', it was found on an *aq čoγdu* shaman's gown and was called *χalbur χaj.* Apart from the textile, leather and metal accessories, the Tofa shamans' gowns had also feather ornaments. Two bundles of feathers were attached to the gown of a *χaš* shaman at the height of the shoulder-blades. Each bundle consisted of nine owl feathers. The fine, downy feathers from the sides of the bird were used.

The Foot-wear

The boots of the Tofa shamans (*iᴅik* 'boot', *χam iᴅi* 'shaman's boot', *χamnar iᴅik* 'boot for shamanistic practice') were made of the skin of tame reindeer. The uppers ended in a horizontal line under the knees. The legging and the fronts of the boots were decorated with appliqué or embroidery. The upper edge of the legging was often trimmed with squirrel or lynx fur. Sometimes red and white cotton strips were sewn below the fur trimming. In this place some boots had a strip of white reindeer embroidery, or white cotton. Others did not have trimmed edges at all. The name of the appliqué or embroidery is *χon ǰüsü* 'knee-joint'. A similar white appliqué is found round the legging at ankle-height. This is the *ʙut ǰüsü* 'ankle'.

Vertical cotton strips also ornamented the legging. It was usually white in front, or sometimes on the outside and inside; or white in front, blue on the inside and yellow appliqué on the outside. The name for these is *čoᴅa* 'leg', *čoᴅa sō̆gi* 'bone of the leg'. The strip running vertically along the front of the legging fans out into five branches in the forepart, called *ergek* 'finger', or *ʙut ergē* 'finger of foot'. This ornament is typical of the foot-wear of all the clans.

The long metal strip attached to the outer sides of the legging is always present. This is the *qoš čoᴅa* 'small leg-bone' (literally: 'paired leg-bone'), sometimes simply called *čoᴅasy* 'leg-bone'.

THE MEANING OF THE DECORATIVE MOTIFS ON THE SHAMAN'S COSTUME

The ornamental elements of the Tofa shamans' gowns constitute an integral unit. Already S. V. Ivanov pointed out the fact that, as a whole, they symbolized the human skeleton, consisting of the backbone, the chest, the shoulder-blades, the pelvis and bones of the arms and legs. 'The motifs on the fur-trimmed head-bands decorated with straight feathers symbolized parts of the human face and are probably the natural complements of the skeleton depicted on the clothing.'[17]

This is even more clearly proved by the parts of the shaman's costume which we have illustrated, as well as by the names of the decorative elements and pendants on the gowns.

The head-dress bears the eyebrows (*χabaq*), the eyes (*qaraq*), the nose (*χaj*), the mouth (*ās, aγsy*) and the ears (*qulaq*). The rolled strips hanging from it represent the hair (*čaš*), or ear-rings (*syrγa, bȯrtüŋ syrγa*). Various parts of the human anatomy were also marked on the costume. In front the chestbone (*ᴅüš sō̆gi, ᴅüštüg sō̆gi*), the ribs (*ēgi, ēgi sō̆gi, ᴅüštüg ēgisi*), the chest (*ᴅüš*), the hole above the chestbone (*oru*), the collar-bone (*munuj sō̆gi*), the lower part of the sternum (*ᴅüš quᴅrū*), the nipples (*emi*), the navel (*χin*), the shoulder-joints on the sleeves of the gown (*ǰüsü*), the arms (*qol*), the arm-bones (*qol sō̆gi*), the upper arms (*üstü*), the elbow-joints (*ǰüs*), the palms (*χalbur χaj*), the fingers (*ergek*), and in the back the backbone (*ōrγa, ōrγa sō̆gi*), the dorsal vertebrae (*ōrγa ǰüsü*) and the ribs (*ēgi*). The accessories also added reality to the portrayal of the different parts of the body, for on the back exit shoulder-blades (*ǰaryn, ǰarna*), and on the sleeves of the gown were palms (*χalbur χaj*).

The ornaments on the foot-wear also fit nicely in the picture: The uppers of the boots carried the knee-joint (*χon ǰüsü*), the ankles (*ʙut ǰüsü*), the leg bone (*čoᴅa, čoᴅa sō̆gi*), the fibulae (*qoš čoᴅa*) and the forepart carried the toes (*ergek, ʙut ergē*).

Some pendants pointed to connections with the animal world. Using material furnished by V. I. Vasiljev[18], already S. V. Ivanov stressed this point. 'The leather plaits and leather ribbons which were sewn to the shaman's costume in large numbers (three hundred), were called the "tails" and "wings" of the bird.'[19] Some of the costumes we collected, had bunches of feathers on

[17] Иванов (1955) p. 250; cf. Иванов (1954) p. 679.
[18] Васильев (1910), p. 76.
[19] Иванов (1955) p. 250.

the back of the gown, rolled strip bundles called 'wing of the back' (*ōrγa χanaty*), and the customary 'tail' (*quᴅruγ*).

Since except for the museum descriptions of S. V. Ivanov[20] and W. Hartwig[21] the costumes of Tofa shamans are unknown in ethnographic literature, I wish to publish part of the drawings I had the opportunity to have made in the course of my field work. These objects are no longer in existence: their owners destroyed them when they stopped practising shamanism. They were drawn and described from memory by ex-shamans or their relatives who still clearly remembered these costumes.

In order to have all the clans and different degrees of shamans represented, we present the gowns of the following subjects: The big female shaman Anna Innokentyjna Kholemoeva of the *aq čoγdu* clan (Figs 48–55); the small shaman Innokentyj Nikolaevič Kokuev of the *aq čoγdu* clan (Figs 56–58); the big shaman Sergej Spiridonovič Bakanaev of the *qara čoγdu* clan (Figs 59–62); the middle shaman Pavel Jegorovič Šipkeev of the *qara čoγdu* clan (Figs 63–65); the big shaman Innokentyj Semenovič Bokturbaev of the *χaš* clan (Figs 66–69); the small shaman Ilya Spiridonovič Bolkhoev of the *saryγ χaš* clan (Figs 70–73); and the middle shaman Aleksej Grigorevič Tulaev of the *čeptej* clan (Figs 74–78).

THE DEDICATION OF THE SHAMAN'S EQUIPMENT

When all the pieces of the equipment of a new shaman were finished they could not be used immediately. First a dedication ceremony called *ᴅüŋür ōredir*, that is to say 'enlivening of the drum', had to be held, attended by all the adults living in the community.

For this ceremony in honour of the female shaman Kholemoeva of the *aq čoγdu* clan, a reindeer was killed, and it was a white one, too! It was killed in the same way as usual: with the dull end of a hatchet. Then it was cleaned, its bowels thrown out. The female shaman drew three glasses of the blood of the reindeer, for it was said that she had to drink three glasses of the blood of the white reindeer that had been killed. Then the flesh of the reindeer was cooked.

The preparations were completed by about 6 o'clock p.m., when the real ceremony began. The drum hung in the middle of the wall, opposite the entrance of the tent. Animal skins were laid under the drum to raise the level of the ground slightly. White cloth was placed on the skins and three plates were placed on the cloth. Pieces of meat taken from the various parts of the reindeer were put on the plates. In front of and behind the three plates, double rows of three cups (a total of twelve) were aligned. The three cups closest to the wall had unsalted tea (*šaj*) whereas the three placed before these contained milk (*süt*). The three cups in front of the plates contained reindeer blood (*ibi qan*), and the three cups in front of these were filled with vodka (*araqa*). The food dishes placed in front of the drum were called *ᴅüŋürnüŋ ᴅoj*.

[20] Иванов (1954) p. 679 and (1955) pp. 250–5.
[21] Hartwig (1957) pp. 37–50.

After the food offered for the drum and for the shaman's costume had been properly placed, all present started to eat of the meat of the white reindeer. After the feast they began to practise shamanism. First seven men picked up the drum one after the other, and finally the female shaman herself. After she had shamanized, the food and drink made for the drum were also consumed.

Shaman Kokuev of the *aq čoγdu* clan described his drum dedication ceremony as follows: 'When everything was ready, i.e. both the shaman's costume and the drum, then, at five in the afternoon, everybody collected in my tent. There were a lot of people, perhaps about thirty or so. They had killed a white reindeer and were cooking its meat in three large kettles in three different tents. Tea was brewed and wine was brought. The shaman's drum hung on the side opposite the opening of the door, under it, the shaman's gown was on the *ʙarba* (reindeer bag)[22], and in it the rolled strips. Next to it lay the drum-cover. Before it, my wife had placed a small table, on which she had put three plates along the midline with pieces of the flesh of the white reindeer in them. They included: a little piece of the lung (*ȯkpe*), of the kidneys (*bēr*), of the heart (*čürek*), of the last rib (*qyʙyγ ēgi*), of the breastbone (*ʙüš*), of the flesh below the skull, the flesh from the end of the tail, the right shoulder-blade, and of the right foot. A small piece of each. Before the three plates seven small cups were placed (on the side by the fire), and each of them was filled with brandy. Behind the three plates (toward the side closer to the drum) three cups were set, tea (*qara šaj* "black tea" without salt!) being poured into the first (from left to right if you looked at it from the threshold), tea with reindeer-milk (*aq šaj* "white tea") in the second, and reindeer milk (*süt*) in the third. The food prepared for the drum was called *ʙüŋürnüŋ ʙoj* "the feast of the drum". When my wife had put all these things on the table, we, too, began to eat. Having finished, my brother took the drum from its place and began to play on it. He beat the drum and sang to it. My brother was no shaman and knew no real shamanistic song, he just made up something which he sang. He sang for about five minutes. He put on the shaman's head-dress and donned the shaman's gown, but did not pull on the shaman boots. After about five minutes he leaned the drum against the table, on which were placed the dishes prepared for the drum, and took off the shaman's head-dress and then the shaman's gown. Then he asked: "Who wants to practise shamanism?" Then someone else stood up, who also donned the shaman's head-dress and the shaman's gown, picked up the drum and the drumstick, and began to practise shamanism. Three people acted as shamans altogether. After the third, my wife hung the drum again in its place, folded up the shaman's gown and laid it on the reindeer-bag. In the middle, between the two poles of the tent, a strap was stretched (this is always stretched there before the shaman begins to practise), and she put the shaman's head-dress on this. It was also my wife who stretched the strap. Then I put on the boots, the gown and the cap, I picked up the drum and the drumstick and began to practise shamanism. When I first laid my hand on the drum I sprinkled

[22] Oblong bags sewn of the skin taken from the leg of the reindeer. They were generally 45 cm long, 17 cm wide and 25 cm high.

Fig. 48. The first head-dress of the big female shaman Kholemoeva of the *aq čoγdu* clan

it with milk and vodka. I practised shamanism until four o'clock in the morning.'

In the *aq čoγdu* clan the drum-livening celebration for the female shaman Andalaeva was as follows: 'When the drum was completed, they killed a reindeer—a white reindeer. After its meat had been cooked, a piece was cut out of every part (heart, lungs, etc.) and placed under the drum. The drum was hanging from the wall opposite the opening of the entrance. In front of it, on the ground, a rug made of skin from the legs and head of a reindeer was spread, and the pieces of meat were placed on this. In addition to the plates containing the meat, there was tea in three cups,

Fig. 49. The second head-dress of the big female shaman Kholemoeva of the *aq čoydu* clan

and vodka in three others. Then those present began to practise shamanism under the instructions of the husband of the female shaman. He handed the drum to someone in the congregation, at random. That person put on the shaman's gown, placed the head-dress on his head and began to beat the drum. He chanted some words, too, but they were not real shaman's songs. He practised for about 15–20 minutes, after which time he gave the drum back and removed the shaman's gown and the head-dress and passed them on to the person next to him, who also donned the shaman's gown, put the head-dress on, took over the drum and began to practise. In this particular case seven persons, altogether, practised shamanism. I have heard, however, that sometimes nine, or else only three people are engaged. In the case of the female shaman Kišteeva, for instance, nine people practised shamanism. When the seventh person had tried the drum, they all began to feast. The party lasted until late in the night, when the female shaman herself began to practise on the new drum. This she did until morning, because she had to make many journeys. First of all she had to find the place where the deer, whose skin was used for the drum, was born. Then one after the other she had to visit all the places, where the deer had been in the course of its life. She had to walk along all the rivers where it had been, seek out all the places where it used to drink and graze. She had to follow all the trails of the deer, and trace them to the place where the hunter had killed it. The female shaman told also who had shot it and how the deer had been killed.

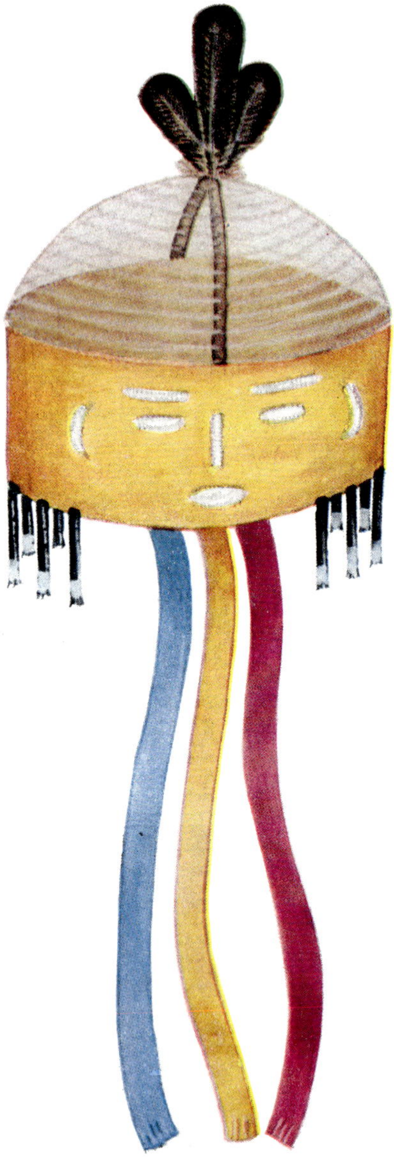

Fig. 50. Third head-dress of the big female shaman Kholemoeva of the *aq čoγdu* clan

But, since two hunters were together when the animal was killed, she also related what the hunters had been talking about at the time. She told the age of the animal, named the person who had flayed it, who had cut the skin and related how it had been sewn on the drum. Not only did she relate these things, but she also imitated all the movements made by the drum-makers while making the drum. After finishing with the reindeer, she began to speak about the tree which had furnished the wood for the framework of the drum. She said that the tree had been very old, three hundred years old. She told us where it had grown and even gave an account of its life explaining when nourishment had been plentiful and when not, when the rain had been abundant and when scarce, when the tree had been shaken by savage storms, etc. She also related and showed how the men had cut the wood for the drum out of the trunk and what they had been talking about when doing so. After she had told all about the life of the reindeer, whose skin covered the drum and the story of the tree which had furnished the frame for the drum, she was busy until morning with the deer, because that animal refused to carry her, so the female shaman had to break it in, just like the reindeer when it is first saddled. She kept on spurring the restive animal promising it that they would ride to the places where it had been in its life and which it liked. At last, she finished her activities by morning, when they all began to feast again, consuming the whole reindeer except the parts set before the drum. The female shaman practised shamanism every night for seven days. Not everybody, however, does it as she did. Nikolaj Danilovič, for instance, practised shamanism for only three or four evenings, while other shamans continued for nine days.'

THE STORING OF THE SHAMAN'S EQUIPMENT

After the drum-livening ceremony was over, the candidate was a qualified shaman and now he or she could contact the spirits at will, by simply dressing in the proper attire and beating the drum. The necessary objects were

Fig. 51. Front view of the gown of the big female shaman Kholemoeva of the *aq čoγdu* clan

kept in the shaman's tent, at a definite place. This was considered a sacred point in the conical tent. It was the same for all the Tofa clans: the side opposite the door. But the manner in which the objects were placed there differed with the individual shamans.

In the tent of the shaman Kokuev of the *aq čoγdu* clan, thick branches were laid down parallel to each other, before the wall. About twenty reindeer skin bags were lined up on this 'support'. In or on the two bags standing in front of the side opposite the entrance stood the shaman's equipment. The bags contained the rolled strips and on them the folded shaman's gown and the drum in the drum-cover with its uncovered side

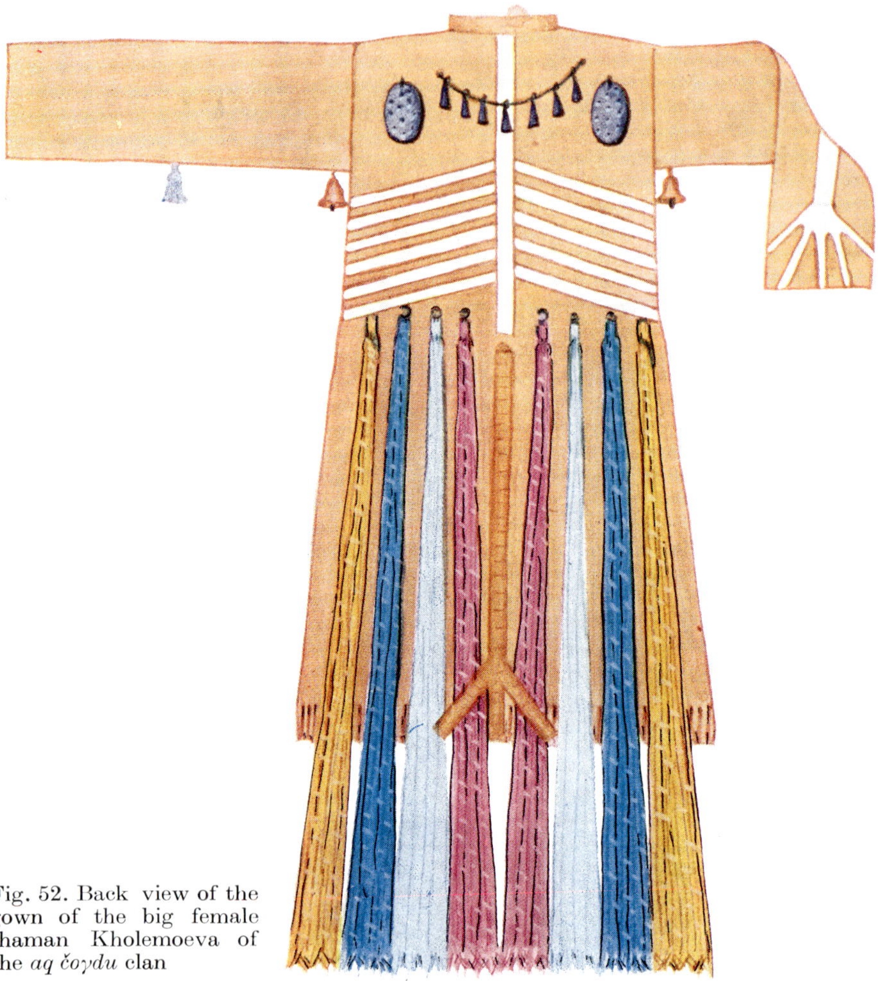

Fig. 52. Back view of the gown of the big female shaman Kholemoeva of the *aq čoγdu* clan

upward and contained the head-dress. The drumstick was kept in the same reindeer bag as the rolled strips.

In the tent of the shaman Innokentyj Semenovič Bokturbaev of the *χaš* clan, the bag placed opposite the entrance contained the rolled strips. On it lay the shaman's gown folded; surmounting this was the drum in its cover. The head-dress and the drumstick were in the drum.

The objects of the shaman Ilja Spiridonovič Bolkhoev of the *saryγ χaš* clan were situated on the side of the tent opposite the entrance. The rolled strips and the head-dress were kept in the bag standing there. Spread on the bag were first the boots and, on top of them, the shaman's gown. The drum lay on these in its cover. The drumstick was placed in the drum.

Aleksej Grigorievič Tulaev, a *čeptej* shaman, kept his equipment in two bags opposite the entrance. The bag closer to the wall contained the rolled strips and the boots, the head-dress and the drumstick; the other contained the shaman's gown and the breast-cover. On top of the two bags lay the drum in its cover. In the tent of the shaman Pavel Jegorovič Šipkeev of the *qara čoγdu* clan, the shaman's costume was in the two bags opposite the entrance. The first contained the rolled strips; the second, the shaman's gown, the head-dress and the boots. On top of them, in the drum-cover lay the drum, the drumhead turned downward, and the drumstick inside.

With the Tofa shamans not merely the storage but also the transport of the costume was regulated by strict rules. When he moved or had to go somewhere to practise shamanism, the shaman Bolkhoev of the *saryγ χaš* clan laid the two bags containing the rolled strips and the head-dress on the back of a reindeer, placed the boots on top and above them the shaman's gown and finally the drum in the drum-cover, containing also the drumstick, tying them with a long strap to the back of the reindeer.

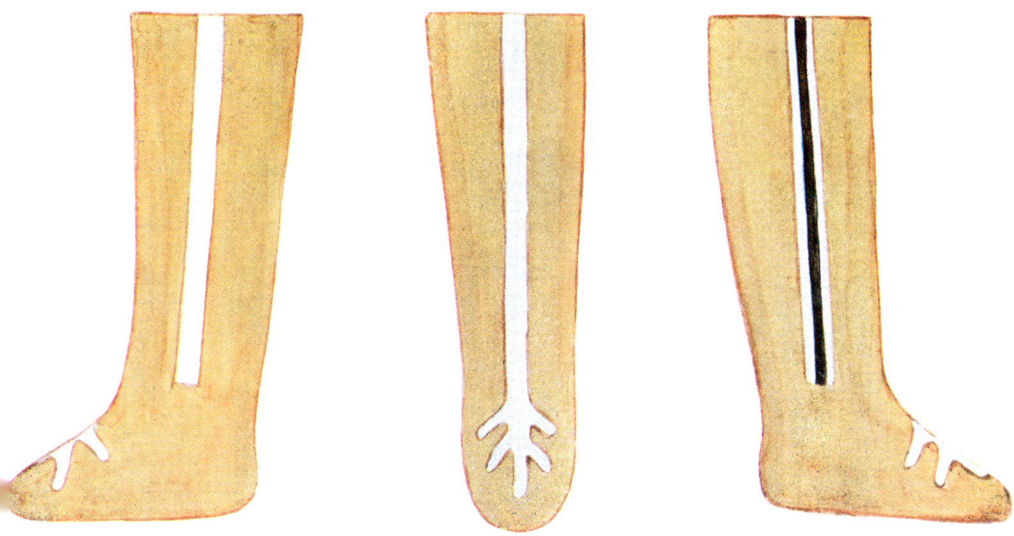

Fig. 53. The boots of the big female shaman Kholemoeva of the *aq čoγdu* clan

THE SYMBOLIC MEANING OF THE SHAMAN'S DRUM

The Tofas called or visited the shaman for several reasons: (a) when someone fell ill; (b) if something got lost (for instance, a reindeer); (c) to find out how successful the hunt would be, or what the future held in store. But often the shaman practised his art for its own sake, and travelled to places 'on his drum', where in a trance he met the spirits.

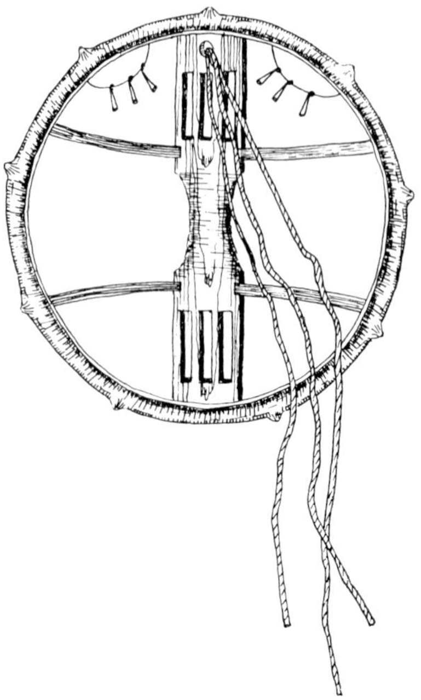

Fig. 54. Drum of the big female shaman Kholemoeva of the *aq čoγdu* clan

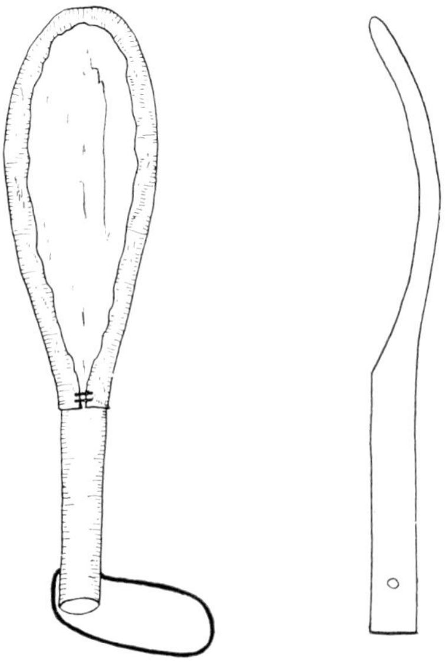

Fig. 55. Drumstick of the big female shaman Kholemoeva of the *aq čoγdu* clan

For it was the belief of the Tofa people that the shamans 'rode' their drums. The fact that it symbolizes their saddle animal was already brought out in the description of the ceremony attached to the newly made drum—the *ᴅüŋür ōredir*, 'training of the drum', which is, in fact, the symbolic breaking-in of an animal which had not yet been saddled. Not only this ceremony but also the names of the drum and its parts, as well as the beliefs attached to it prove this point.

I was told by the shaman Bolkhoev of the *saryγ χaš* clan that the shamans referred to the drum in their songs as *aqāt* 'white horse' and *čary* 'reindeer bull'.

The names of the different parts of the drum also indicate that the drum symbolized an animal, in fact an animal suitable for riding.

To the inside of the frame, on the top part of the drum, a ⌣-shaped metal rod is attached to either side of the handle, the *qulaq*, i.e. 'ear'. There is a round opening over the ornament above the handle, through which always three leather or cotton rolled strips are laced. The members of the *saryγ χaš* clan have special words to denote these strips. According to the shaman Bolkhoev, the names of the two lateral ones are *ȯkpe* ('lungs') and the name of the middle one is *čürek* ('heart'). According to his explanation they really symbolize the lungs and the heart of the drum.

As we have seen, varying numbers of resonators are attached to the outer side of the frame. The knots of these resonators have wedge-shaped incisions through which a reindeer ligament is threaded around the framework; this is the 'artery' of the drum. According to the younger sister of the *aq čoγdu* shaman Vasilij Nikolaevič Amastaev 'the blood of the animal, (i.e. of the drum) runs in these'.

Thus, the drum has, according to Tofa belief, a variety of organs such as ears, lungs, heart and arteries.

As we have seen, there are two cross-bars in the drum. The top pair is termed *χümündrük* or *χülündrük*, and the bottom, *quᴅurγa*. These two words are technical terms used not only for shamans' drums, but also in

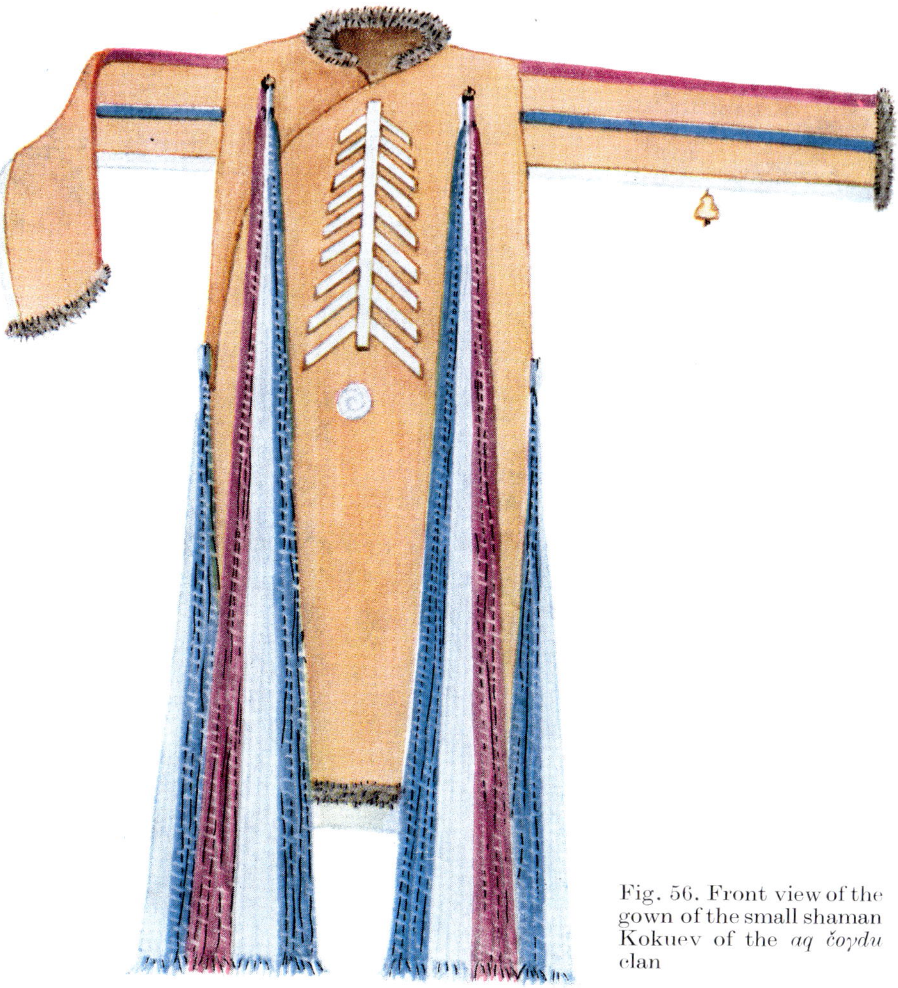

Fig. 56. Front view of the gown of the small shaman Kokuev of the *aq čoγdu* clan

every-day life, in connection with reindeer keeping. *χümündrük* means the breast-harness and *quᴅurγa*, the tail-case of the saddled beast.

We have mentioned above that there are three rolled strips on the drum-handle. While the *saryγ χaš* clan call these 'lungs' and 'heart', the other clans use the name *ᴅüŋürnüŋ ᴅyna* 'reins of the drum'.

Therefore it may be said that the drum is equipped with the breast-harness and tail-case holding the saddle, and with the reins essential for guiding the saddle animal.

According to the belief of the Tofa people, the shamans actually rode the drums as if they had been saddled animals. 'When they practised shamanism without a drum' —as I was told by a Tofa—'they were on foot, but with a drum, they were mounted. When beating their drums with the drumstick, they were whipping their animal, on which they could travel very far.'

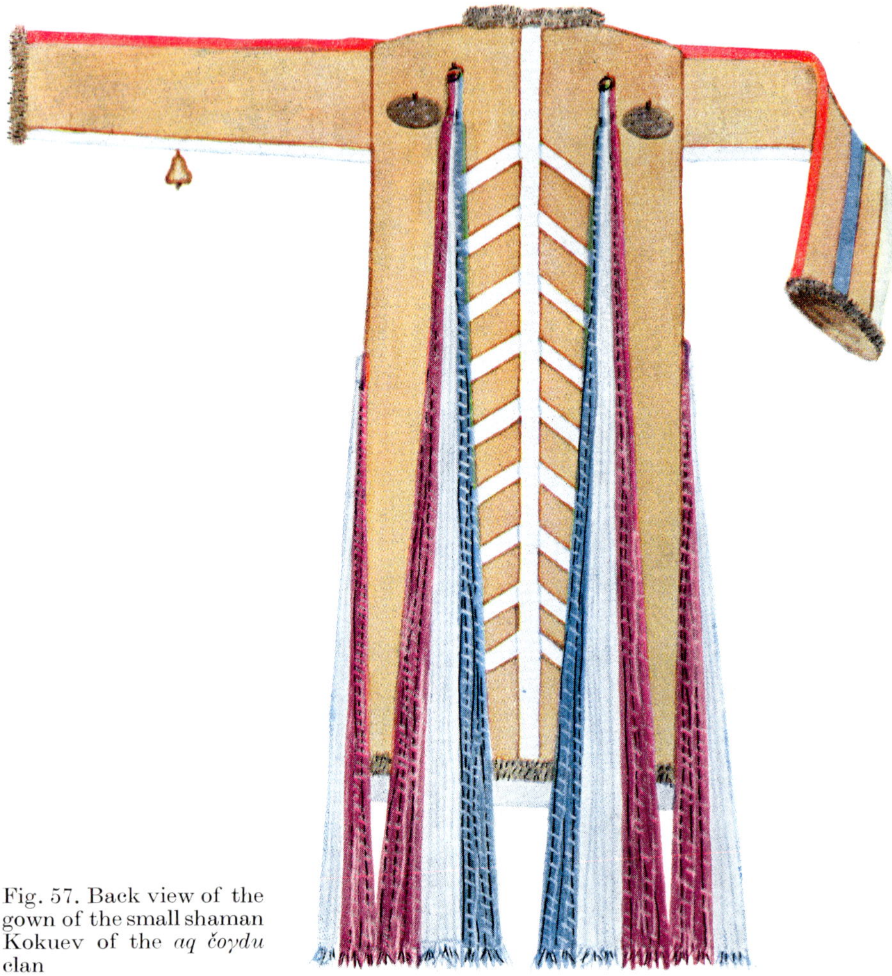

Fig. 57. Back view of the gown of the small shaman Kokuev of the *aq čoγdu* clan

The requirements of riding were decisive in connection with the skin that was to be stretched over the frame. 'For a drum, the skin of a middle-aged animal is the most suitable. The drum is the shaman's saddle animal, and for this reason the skin of a very old or very young animal is not suitable. If the animal whose skin was stretched over the frame was very young, it would tire out very soon, and if too old, it could not ride sufficiently fast.'

With the Tofa people, it was the age of the animal whose skin was used for the drumhead that determined the durability of the shaman's drum. 'Not all drums are suitable for use for the same amount of time' said the shaman Kokuev of the *aq čoγdu* clan. The duration depended on the age of the animal whose skin was put on the frame. An animal runs well only till it is about 7–9 years old. Therefore, the drum could be used only for as many years as separated the animal at the date of its killing from the age of 7–9. After this the skin on the frame had to be replaced. 'I, too, had two drums' continued the shaman Kokuev 'because I played the first one, which was of deer skin, only for three years'.

A similar opinion was expressed by members of the *čeptej* clan. 'The shaman knows how old the animal that furnished the skin for his drum was. If, for instance, it was covered with the skin of a deer, it was thought that one 14–15 years old was already too old for riding because it got soon tired. Therefore, the shaman had a new drum made at the time his deer would have reached the age of 14–15, had it been left alive. This is the reason why one shaman used his drum for five years, another for seven, and a third one for nine years. It sometimes happened that the drum had to be replaced after one or two years. All this, of course, depended on how young or old the animal was from whose skin the drum was made.'

THE ACTIVITY OF THE SHAMAN

This central role of the drum as saddle animal will seem quite natural if we are acquainted with the essence of the shamanistic activities which consisted of travelling over many regions and reaching different places.

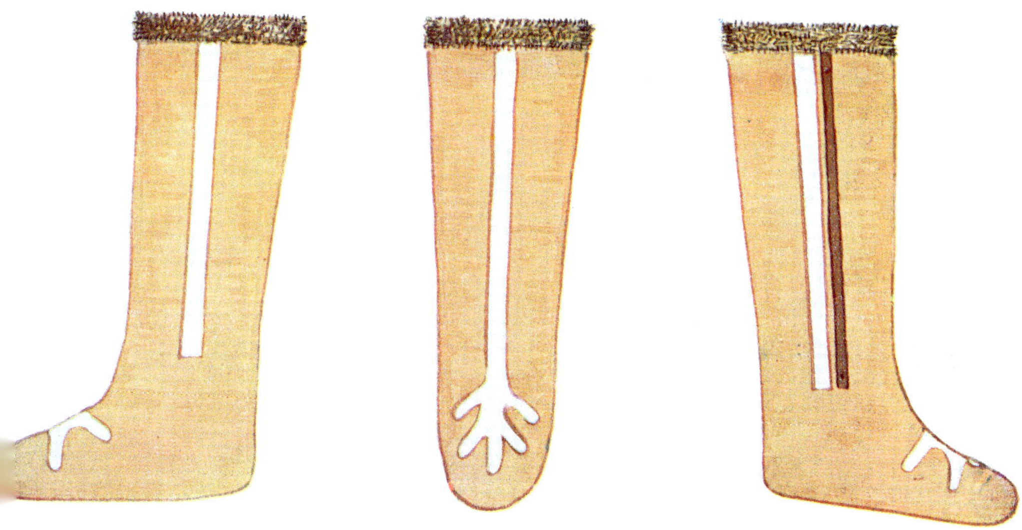

Fig. 58. Boots of the small shaman Kokuev of the *aq čoγdu* clan

Fig. 59. Second head-dress of the big shaman Bakanaev of the *qara čoγdu* clan

Let us, for instance, listen to the shaman Bolkhoev of the *saryγ χaš* clan, who declares: 'I began practising shamanism in the evening. It was always in the evening, because the spirits come more readily at this time of the day. I would do it for two or three hours. First I picked up the drum and held it over the fire. I kept drying it over the flames for about ten minutes. In the meantime I tried it out repeatedly to see whether its tone was sharp enough. When it sounded all right, either I or my younger sister attached the rolled strips to the shaman's gown. If my sister attached them, she did so while I was drying the drum. The rolled strips were on the shaman's gown in a few minutes. Then I pulled the boots on my feet, donned the shaman's gown and put the shaman's head-dress on my head. Now I took the drum in my left hand and picked up the drumstick with my right one. Then I asked someone of those present to hold juniper branches in the fire. This is how we summoned the spirits. It was all done in a few minutes. Then I

Fig. 60. Front view of the gown of the big shaman Bakanaev of the *qara čoydu* clan

slowly began to beat the drum and to sing. I summoned the spirits. I always called all the seven spirits. In their company I journeyed to sacred places. The shaman is in the tent, beats his drum, but his soul is riding the drum and his spirit is travelling to holy places. My holy places were the mountains Ug Taiga, the Ulug Taiga, the Kaibiti and the Shibikhaya. It was only gradually that the spirits showed me the way to these sacred places. After

I became a shaman, in the first year I travelled to the Ug Taiga. Here on the top-most spot lived the "master of the earth". This is a very holy place. The master of the earth (*ǰarniŋ ēzi*) is as tall as a tent. In the second year I frequented the Ulug Taiga. Here lived the elder brother of the "master of

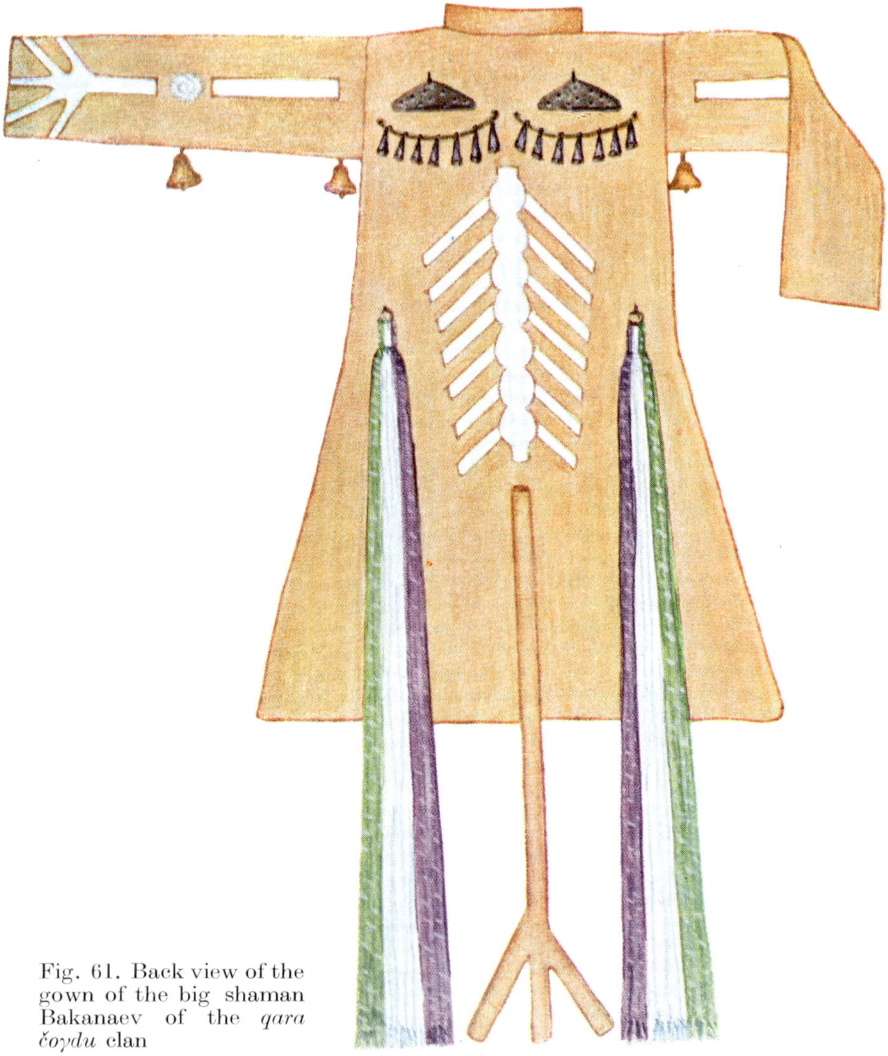

Fig. 61. Back view of the gown of the big shaman Bakanaev of the *qara čoγdu* clan

the earth", the "noble-capped man" (*kidis ʙörtüg ašaq*). In the third year I was guided to the Kaibiti mountain. This is by the upper reaches of the Kazyr river, and is the home of the "bald-headed otter man" (*čülünüg baštyγ tilgimej ašaq*). The fourth year I journeyed to the "fir rock" (*šibi χaja*). That is the place where the Alandzu river joins the Lower Gutara, the place of the "great shaman master of the *saryγ χaš* clan" (*saryγ χaštyŋ uluγ χam ēzi*). From the spirits living in these holy places I found answers to all the questions the others had asked me. Such as: what is troubling a sick man, and how he should be treated, where is the lost reindeer and where one had to look for it? In the fifth year I no longer went anywhere, because I had to surrender the drum to the authorities. I was a small shaman so I had travelled low. I flew at a height of only eight or nine metres. The great shamans rode very high, they flew in the clouds.'

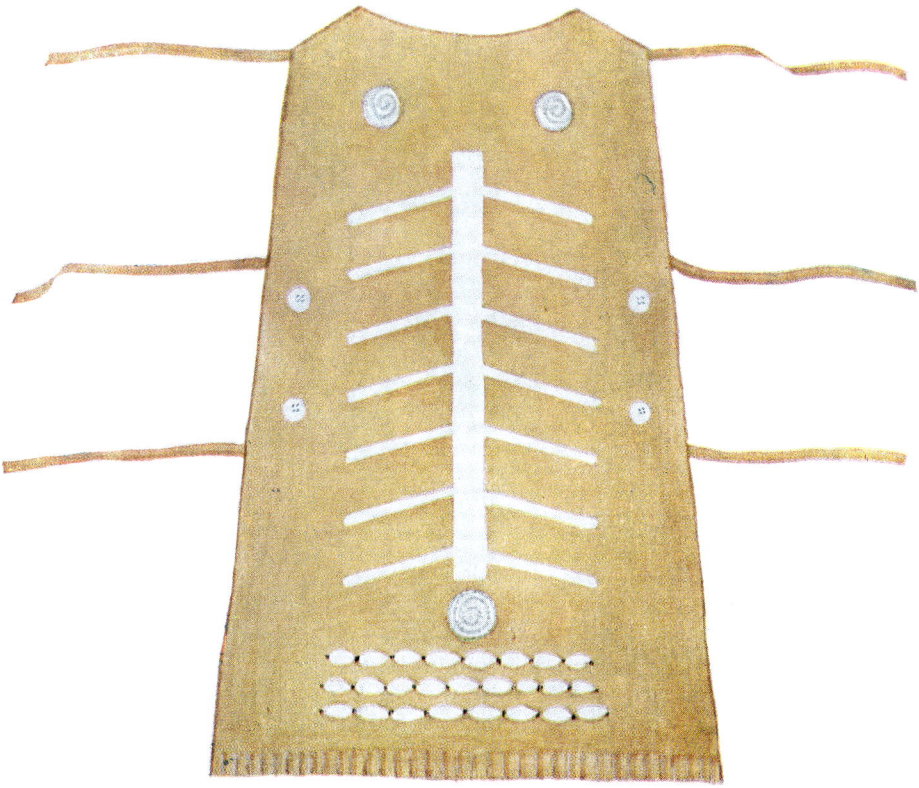

Fig. 62. Breast-cover of the big shaman Bakanaev of the *qara čoγdu* clan

Shaman Kokuev of the *aq čoγdu* clan practised his art as follows: 'When I began, first I took the drum from its cover and held it above the fire to dry a little. Then I first pulled on the shaman's boots. Only then did I put on the shaman's gown. The rolled strips were separate from the shaman's gown. I put these on only when I practised shamanism. When I began, one of the relatives (*ulusu*) attached the rolled strips to the gown. When I had finished, again one of the relatives took them off and put them back in the bag. To the side of the gown three metal rings were sewn, each with a small leather strap, at a distance of 3–4 cm one from the other. There was a ring on the end of each rolled-strip bundle as well. When I began to practise shamanism, the rings of the rolled-strip bundles were tied with pieces of leather strapping to the rings on the gown. Finally I put on the shaman's head-dress. Then I took the drum in my left hand and held the drumstick in my right and began to beat the drum. In the first year the spirits always took me to the Ak Kul (or *ēlig aq χùl*)—a lake situated in the land of the Tuvas. I learned the way to the Ak Kul. This was where the "master of

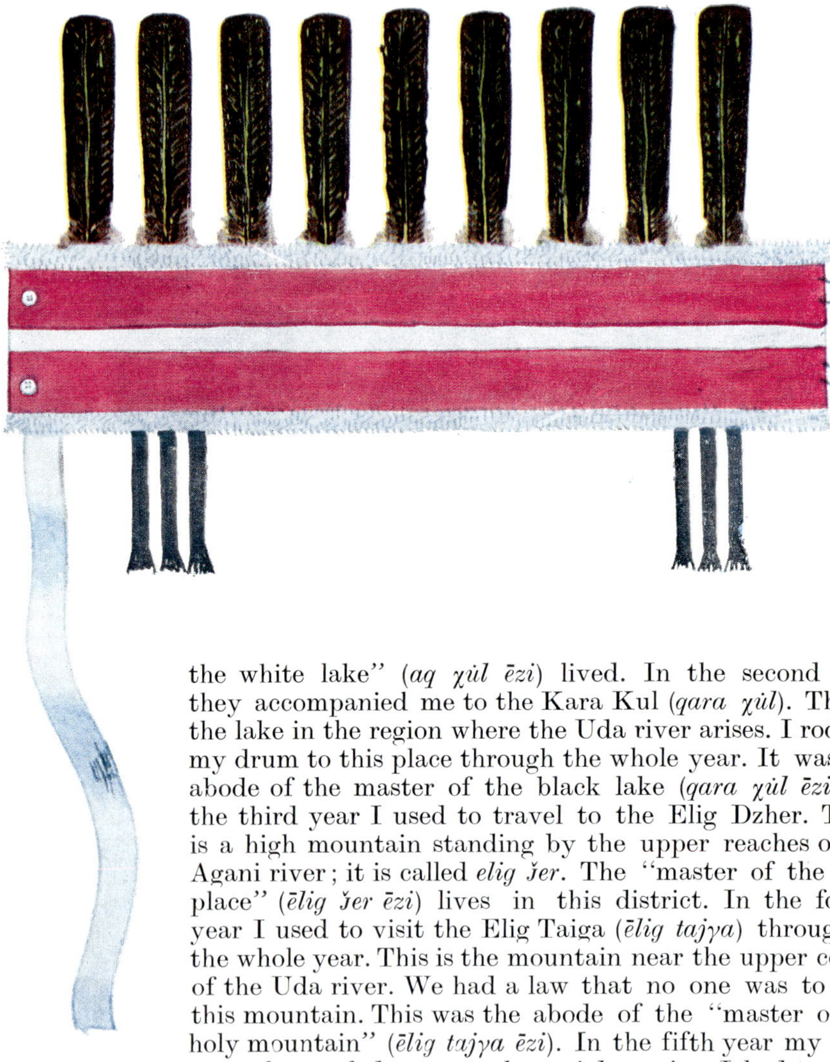

Fig. 63. Head-dress of the middle shaman Šipkeev of the *qara čoγdu* clan

the white lake" (*aq χùl ēzi*) lived. In the second year they accompanied me to the Kara Kul (*qara χùl*). This is the lake in the region where the Uda river arises. I rode on my drum to this place through the whole year. It was the abode of the master of the black lake (*qara χùl ēzi*). In the third year I used to travel to the Elig Dzher. There is a high mountain standing by the upper reaches of the Agani river; it is called *elig ǰer*. The "master of the holy place" (*ēlig ǰer ēzi*) lives in this district. In the fourth year I used to visit the Elig Taiga (*ēlig tajγa*) throughout the whole year. This is the mountain near the upper course of the Uda river. We had a law that no one was to visit this mountain. This was the abode of the "master of the holy mountain" (*ēlig tajγa ēzi*). In the fifth year my journey always led me to the *χajalyγ sojaq*. I had to go to this place, in order to learn the road to the residence of the "master of the holy snow-capped mountain" (*ēlig sojaq ēzi*) or of "the master of the rocky snow-capped mountain" (*χajalyγ sojaq ēzi*). In the sixth year my trips took me to the *šarm tajγa* in the land of the Tuvas.

Fig. 64. Front view of the gown of the middle shaman Šipkeev of the *qara čoγdu* clan

I came to know the Tuva shamans and learnt the road leading to the "master of the Sharm Mountain" (*šarm tajγa ēzi*). In the seventh year I rode my drum to the *ēlig dalaj*. In the land of the Tuvas there are several lakes close to each other and we Tofa shamans call these *ēlig dalaj* ("holy seas"). Thus I came to know the road leading to the "master of the holy sea" (*ēlig dalaj ēzi*), too. By the eighth year I knew all the roads, and I could ride to any place at will, to any "master", to find out

Fig. 65. Back view of the gown of the middle shaman Šipkeev of the *qara čoydu* clan

from him whatever I wanted to know or my companions wanted me to find out. In the ninth year I no longer rode to any place because I gave up practising shamanism.'

'The shamans ride at various altitudes. I used to ride in the lowest altitude, because I was a small shaman. The big shaman flies about a thousand metres, the middle shaman about seven metres, and the small shaman quite low, only about four metres, from the ground. I used to go only about four metres from the ground.'

Thus, whatever the purpose of his act, the shaman first summoned his spirits and then, accompanied by them, visited other spirits. Of course, he stays at home all this time, only his soul rides the drum. The shaman beats his drum and sings during all this time.[23]

Fig. 66. Head-dress of the big shaman Bokturbaev of the *χaš* clan (front view)

The activities of the Tofa shamans are not limited to singing and beating the drum. Nikolaj Danilovič Tulaev of the *čeptej* clan, for instance, often released his drumstick, which then swung freely, suspended by the strap at the end of the handle, into the loop of which he had inserted his four fingers. Shaman Kokuev of the *aq čoγdu* clan attached the drumstick to the third finger of his right hand. He put his third finger through the

[23] A separate paper will be published on the written record of the songs of various Tofa shamans.

loop (Fig. 79, on p. 325), and then began to turn the drumstick to have the loop twisted (Fig. 80, on p. 325). He turned the drumstick until it became completely twisted (Fig. 81, on p. 326) and then held firm the handle in his palm (Fig. 82, on p. 326).

Fig. 67. Head-dress of the big shaman Bokturbaev of the *χaš* clan (back view)

The different shamans had different ways of attaching the strap of the drumstick to their hands. Some suspended it from their wrists, others wrapped it around their thumb, and still others on their middle finger.

A shaman smoked and drank tea during the ceremony. He laid the rolled strip bundles on the front of his gown in his turned up palm and so stretched out his arm toward the person who happened to be smoking and who then put his pipe into the hand of the shaman. When the shaman had smoked the pipe, he threw it into the drum. When he wanted to drink some tea, also holding the rolled strip bundles in his palm he stretched his hand toward someone who was drinking tea. The latter immediately handed his cup to the shaman. The shaman finished the tea and threw the empty cup also into the drum.

Usually only one shaman practised at times. Shaman Kokuev of the *aq čoγdu* clan and his wife Kokueva (née Bokturbaeva), however, often per-

Fig. 68. Front view of the gown of the big shaman Bokturbaev of the *χaš* clan

formed together. An old Tofa recalled that once he had seen four shamans practising in a birch-bark tent.

The shamans, as we said before, often engaged in healing. In fact, this was their most characteristic activity. For this purpose, the shaman often made, or had somebody else make, amulets (*ēren*) which had to be hung over the patient lying in the birch-bark tent. The shaman ordered that when the members of the family were drinking tea, they should sprinkle some on the amulet and should set a cup filled with tea before it.

Many kinds of amulets were in use among the Tofas. To mention only a few at random: *us ēren* 'lynx amulet', *āt ēren* 'horse amulet', *ibi ēren* 'reindeer amulet', *χodan ēren* 'hare amulet', *ᴅīŋ ɢeš ēren* 'squirrel skin amulet'. All these were made of the skin of the animal whose name they bore. In addition, amulets made of textiles were also used. Such were, for instance, the *aq ēren* 'white amulet', *qyzyl ēren* 'red amulet', *torγu ēren*

Fig. 69. Back view of the gown of the big shaman Bokturbaev of the *χaš* clan

'silk amulet', *qyzyl torγu ēren* 'red silk amulet', etc. They were made of white or red cotton or silk. There were amulets attached to sticks, for instance, the *tajaqtyγ ēren* 'amulet with stick'. It was always the shaman who determined which kind of amulet was needed in a given case and what kind of *ēren* was to be made.

The shamans often performed magic tricks. Shaman Bolkhoev of the *saryγ χaš* clan related the following story:

'Even the father of my grandfather was a shaman. He was the greatest shaman. He did with people what he wanted. For instance, a person was

Fig. 70. First head-dress of the small shaman Bolkhoev of the *saryγ χaš* clan

just sitting there, and if my great-grandfather so desired that person would just stand up, facing the wall. I heard it was said that when he practised shamanism he was able to fly over rocks. But he would often jump from one rock to the other, for instance, from the *surar χaja* to the *soγny χaja.*' These two cliffs were by the Lower Gutara river, about half a kilometre apart.

The shaman also played a part in the burial ceremonies. The dead were buried after having lain at home for two or three days. While the deceased

Fig. 71. Front view of the gown of the small shaman Bolkhoev of the *saryγ χaš* clan

was at home, his relatives mourned for him. On the day of the burial there was a feast and seven days later a memorial celebration was held in the evening. Half a month or one month after death had occurred, the shaman practised shamanism. But not everybody addressed himself to the shaman in such cases. It is true that only those people did not call a shaman who did not believe in him. There were some who commemorated the deceased one year after death, in recent times there was a commemoration after forty days, too—under Russian Orthodox influence.

Fig. 72. Back view of the shaman gown of the small shaman Bolkhoev of the *saryy χaš* clan

THE RENEWAL OF THE SHAMAN'S EQUIPMENT

In the life of every shaman who practised regularly there came a time when some items in his equipment had to be renewed, or when he received a second piece of a part of his equipment. Shaman Kokuev knew only by hearsay that in the past the shamans had possessed two sets of equipment at the same time. Once they practised with one set, then with the other. But he knew from experience that the Tofa shamans had to renew their drums, and received often second, third, etc. head-dresses in addition to their first one.

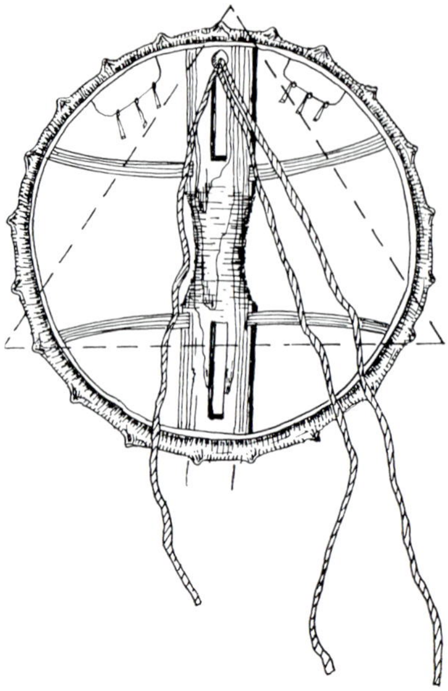

Fig. 73. The drum of the small shaman Bolkhoev of the *saryγ χaš* clan

Fig. 74. Head-dress of the middle shaman Tulaev of the *čeptej* clan

With the passage of time, the shaman's drum became 'superannuated'—that is to say, the animal whose skin was stretched over the frame of the drum had reached the age of 7–9 years and, hence, could not 'run' fast

Fig. 75. Front view of the gown of the middle shaman Tulaev of the *čeptej* clan

enough, or grew tired very soon. In such cases the drum was renewed. 'When the second drum was made, only the skin had to be replaced.' Shaman Kokuev said, 'the wooden frame of the drum remained unchanged. The skin taken off the old drum had to be taken to the same place in the woods where it hung on a cedar. It was always taken there by someone else, never by the shaman. Therefore it was not I who took the skin off my first drum to the forest. In fact I don't even know where it was hung.'

For the second time, usually not the same animal's skin was put on the frame of the drum. The drumhead of the *qara čoγdu* Bakanaev's drum was, for instance, made at first with the skin of a bull and later, with the skin

Fig. 76. Back view of the gown of the middle shaman Tulaev of the *čeptej* clan

of a deer. The *aq čoγdu* shaman Šipkeev also had bull-skin on his first drum and roe-deer skin on his second. Over the *aq čoγdu* shaman Kokuev's drum the skin of a red deer was stretched first, and of a maral deer later.

In spite of the fact that only the skin was new in the second drum, it had to be 'trained' once again.

We have heard what follows about the dedication ceremony of the second drum of Shaman Nikolaj Danilovič of the *čeptej* clan, which took place in 1924 when the Tofa people were still leading a nomadic life. At that time Nikolaj Danilovič Tulaev dwelt by the upper reaches of the Todott river, which is Tuvan territory. It was in July, and for this reason the tent was

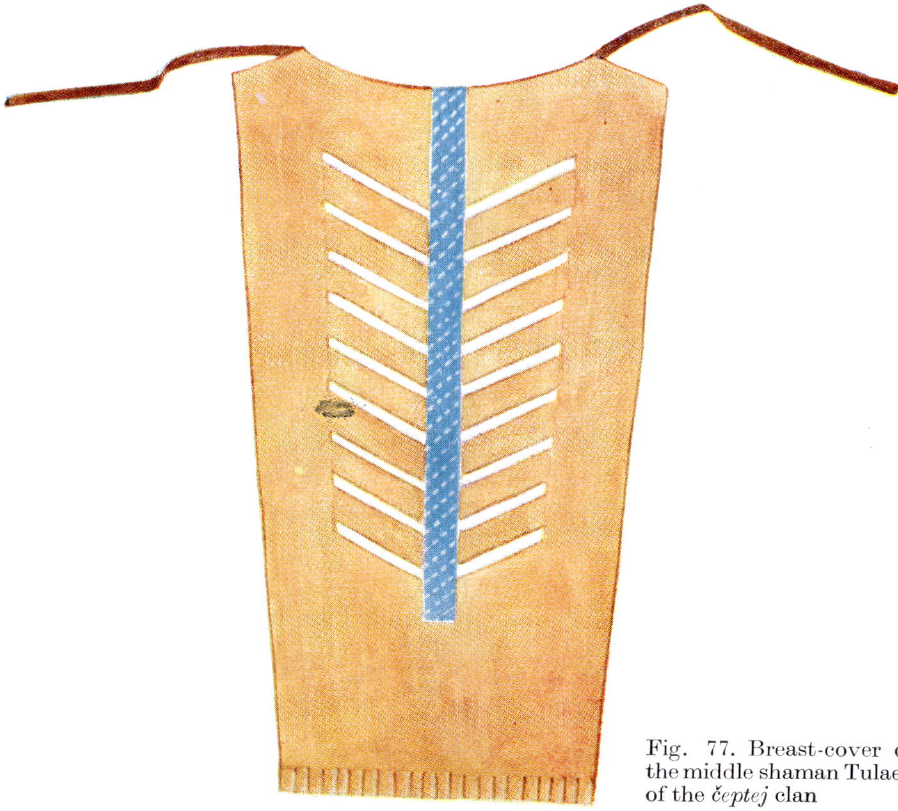

Fig. 77. Breast-cover of the middle shaman Tulaev of the *čeptej* clan

of birch-bark and not of leather. Both the dress and the drumstick were old, only the drum was new. The preparations for the drum dedication ceremony (*ʙüŋür ȫredir*) began around noon. Seven people beat the drum, taking turns. They were not relatives; for instance, Vasilij Nikolaevič Adamov was there. Both in its main features and in its details the ceremony was the same as in the case of the first drum.

Not only a second drum, but also a second head-dress was often made for the Tofa shamans. But, whereas one shaman has always only one drum, for he has the framework of the old drum covered with new skin, he may have several head-dresses at once, for he does not destroy the first one when the second is made. Sometimes he practises in the first, at others in the other.

'I'—the *aq čoγdu* shaman Kokuev continued—'had two shaman's head-dresses. In the first part of the night I played wearing the first one, and in the second part, the other. I kept them both in the drum case.'

The Tofa shaman performed a special ceremony in connection with this second head-dress (*iχiški вôrt*).

'In the centre of the tent stood a post and the new shaman's head-dress was tied to this post. An old shaman practised shamanism for the sake of this cap, because an old shaman had been called, and he himself tied the new head-dress to the post and it was he who placed it on the head of the new shaman. The old shaman belonged to the *saryγ χaš* clan and the young one to the *aq čoγdu* clan. This took place in 1893, here at Alygdzher. The ceremony was held in the birch-bark tent of the new shaman, and that was where the old shaman had been called to. It began in the evening, and we practised shamanism until morning, only in connection with the cap, because it was the shaman's second headgear.'

Some of the Tofa shamans had even a third head-dress made in the course of time. When the shaman becomes a shaman, he receives his first head-dress. Three years later, he gets a second one, and another three years later even a third one (*ùčùški вôrt*). The shaman who had three head-dresses 'performed' for two hours in the first one, for two hours in the second one, and for two hours in the third one' said the shaman Kokuev of the *aq čoγdu* clan. 'To be more exact, he wore the third one until morning, that is, until they had finished the shaman's act.'

The shaman Bolkhoev of the *saryγ χaš* clan said in connection with his three head-dresses: 'I had three shaman's head-dresses: all the same, only the feathers were different. The first had wild-goose feathers, the second crane feathers and the third one owl feathers. I wore the head-dress with the wild-goose feathers when the wild geese lived with us, in other words, from May till October—for they migrate to this region in May and leave early in October. I practised shamanism in the head-dress with the owl feathers when the wild geese were no longer here, that is to say, from the

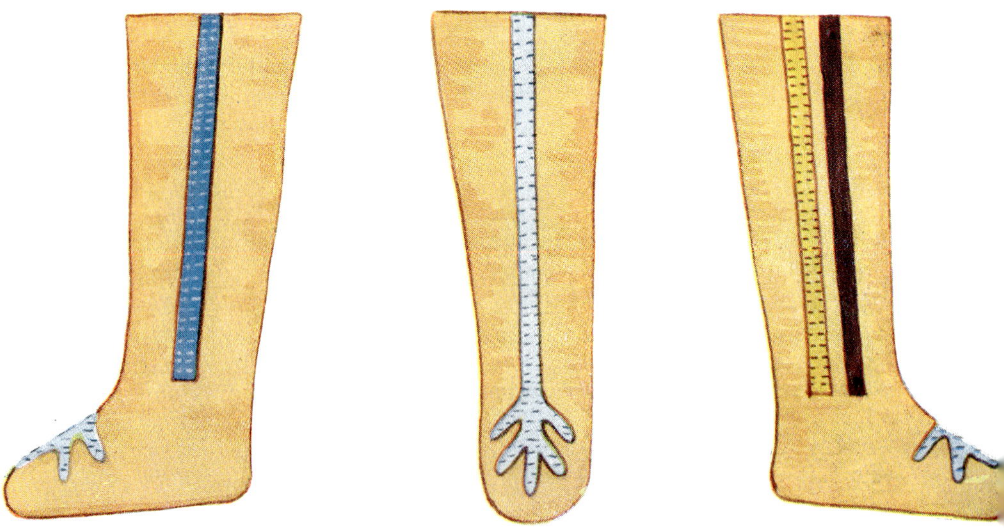

Fig. 78. Boots of the middle shaman Tulaev of the *čeptej* clan

Fig. 79. Attachment of the loop on the end of the drumstick to the fingers

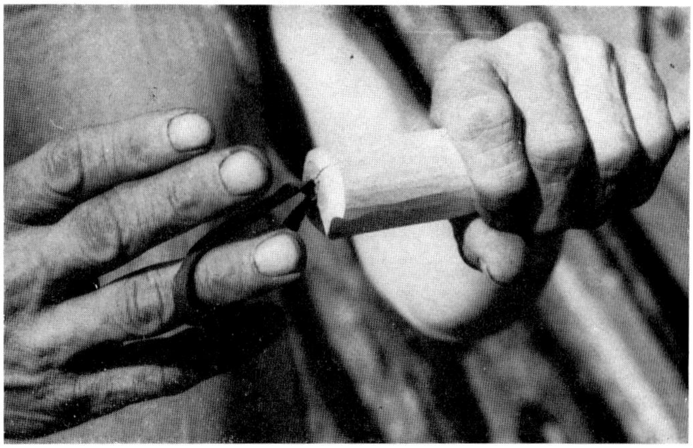

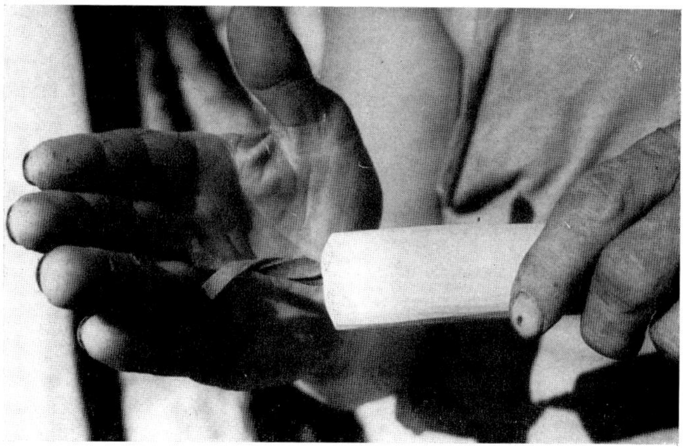

Fig. 80. The shaman turned the drumstick in order to twist the loop

early part of October until the first part of May. I wore the head-dress with the crane feathers all the year round, but only during the second part of the shaman's act. I used this because the cranes can fly very high. The other two head-dresses were in the tent when they were not needed. Between the two poles of the tent a strap was stretched from which hung the two caps not in use. It was my sister who had made all three head-dresses. An old shaman, Amastaev (Khabastar) of the *aq čoγdu* clan told me this. He was a big shaman. He died in 1922.'

But the three shamans' head-dresses were by no means always so similar one to the other as they were in the case of Bolkhoev.

For instance, the third head-dress of the big female shaman Kholemoeva of the *aq čoγdu* clan differed from her first and second head-bands, and were even quite different from any of the types of Tofa shaman's head-dresses described so far. Her third head-dress was—as we have shown—not a head-band with feathers, but a 'grilled' head-dress. This type was much rarer than the head-band with plumes, but nevertheless it cannot be regarded

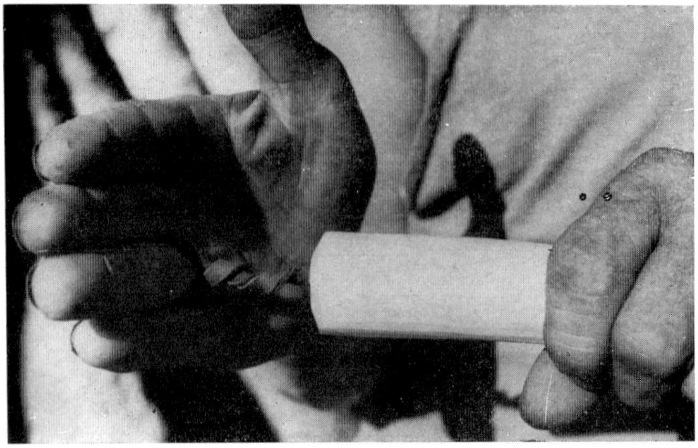

Fig. 81. The shaman kept on turning the drumstick till the loop was fastened to his finger

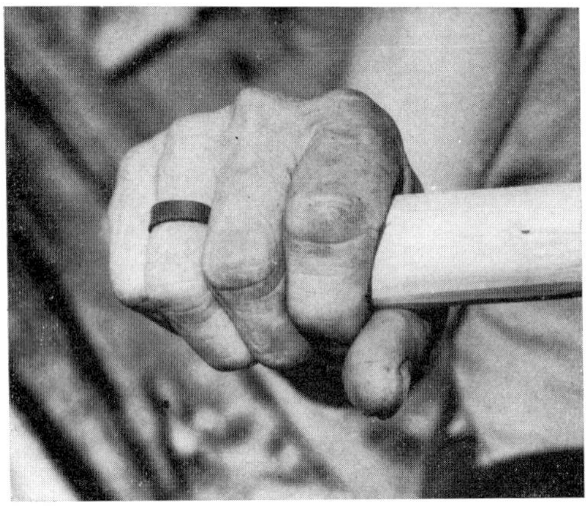

Fig. 82. Then the shaman could grapple the handle

as unique, since other shamans used it as well. Nikolaj Danilovič Tulaev of the *čeptej* clan, for instance, had one like this. A 'grilled' cap, shaped like the tall Russian fur caps, was, for instance, possessed by one of the north-eastern Tuva shamans, too.

Three head-dresses were no rarity among the Tofa shamans, nor was the possession of more than three an unusual occurrence. For there were shamans who received shaman's head-dresses even after the third one. The fourth was called: *dörtiški ʙört,* the fifth, *bečiški ʙört,* the sixth, *altyšqy ʙört,* the seventh, *četiški ʙört,* the eighth *seχeziški ʙört* and the ninth head-dress, *doχozyšqy ʙört.* The shaman received new ones every three years, but none was allowed to have more than nine. The shaman who had nine head-dresses wore each of them for approximately one hour. Only in winter when the nights were long could he use all the nine. If a shaman had nine head-dresses, he rolled each of them up separately and placed them above the rolled strips in the reindeer-skin bag.

Among the Tofas few were the shamans who had nine head-dresses. Most of them got their ninth only toward the end of their lives, and many did not live to have so many, for the shamans died relatively young. Bakanaev, for instance, died at the age of 55; Šipkeev at the age of 47; Aleksej Grigorevič Tulaev, Nikolaj Danilovič Tulaev and Jegor Alekseevič Saganov did not live to be forty, whereas Innokentyj Abramovič Tulaev and Aleksander Mihajlovič Tulaev died before their 25th birthday.

THE DEATH OF THE SHAMAN

According to a Tofa belief, the life of a shaman was threatened by more and severer dangers than those faced by ordinary mortals. If, for example, the tree which provided the wood for the frame of his drum perished, the shaman died too. If the skin covering the drum was torn during practice, the shaman or one of his relatives had to pay for it with his life. Often the shaman died in a struggle with other shamans in his trances.

After a shaman's death the skin on his drum had to be destroyed, and the damaged drum was hung on a tree in the woods—as Shaman Kokuev of the *aq čoγdu* clan explained. In the *saryγ χaš* clan, too, the drum and equipment of a shaman who died were hung up somewhere in a forest. These places were situated at a distance of about three kilometres from the grave. The members of the *qara čoγdu* clan took the shaman's paraphernalia in some unfrequented place of the woods after his death. They hung every item on a tree—the gown, the boots, the head-dress, the drum and the drumstick, but first they broke the skin of the drum.

SUMMARY

In the foregoing paper we have examined shamanism among the Tofas and its manifestations in the objects and spiritual life of that people. In respect of ethnic homogeneity, we found that there was a strong similarity in certain general patterns, but sharp differences in the details.

Identity was evident when we studied the basic and general phenomena of shamanism. Some of the common characteristics are: The selection of the shaman by a spirit, the illness of the would-be shaman or the facts that every Tofa shaman practised shamanism in a special attire and with a drum and that his tools were kept in a place of honour in the tent, etc.

On the other hand, within these general features, practically every detail is different, from the duration of the shaman's illness to the mode in which the strap on the end of the drumstick is attached to the hand. These differences stem from various sources and some of them are individual peculiarities such as: The beginning of becoming a shaman (within the generally valid limitation of 'youth'); the shorter or longer duration of the shaman's illness; the individual elements of his attire; the variations in the shaman's songs; the manner of their performance, etc. Some phenomena can be explained by the quality of the shaman, as: the black dress of the black shaman; his black drum, etc. Differences in the categories or degrees of the shamans are also a source of variation. The small shamans 'fly' just above ground, the big shamans in the clouds, etc. Other differences may be accounted for by the distinctive customs of the individual clans. Every clan had, for example, a different way of bending the frame of the drum. Similarly, the skin stretched over the frame differed in each clan: the *aq čoγdu* drums were covered with deer skin, those of the *qara čoγdu*, with bull skin, those of the *χaš* with the hide of the bay horse, those of the *saryγ χaš* with mountain goat skin, and the drumhead of the *čeptej* shamans was made of the skin of deer dyed red. Manifestations characteristic of the individual clans may be seen in the material, colour and ornamentation of the attire.

Naturally, most of the differences apparent in Tofa shamanism cannot be regarded as the outcome of fortuitous development. In these special traits the historical evolution of the clans is reflected, making them more or less different from others. For example, their settling in a region with a substratum of alien people; contact with a strange ethnic group, life free from more decisive foreign influences, etc.

A more exhaustive explanation of the many deep-rooted differences in Tofa shamanism requires further analytic work.

REFERENCES

(Agapitov, N. N.–Khangalov, M. N.) Агапитов, Н. Н.–Хангалов, М. Н. (1883) Материалы для изучения шаманства в Сибири. Шаманство у бурят Иркутской губернии (Materials for the Study of Shamanism in Siberia. Shamanism among the Buryats of the Irkutsk Province), in *ИВСОРГО*, Vol. XIV, Nos 1–2.

(Anokhin, A. V.) Анохин, А. В. (1924) Материалы по шаманству у алтайцев (Materials on the Shamanism of the Altai Turks), in *СбМАЭ*, Vol. IV, No. 2.

(Anučin, V. I.) Анучин, В. И. (1914) Очерк шаманства у енисейских остяков (Essay on the Shamanism of the Yenisei Ostyaks), in *СбМАЭ*, Vol. II, No. 2.

(Bogoraz, V. G.) Богораз, В. Г. (1939) *Чукчи* (The Chukchi), Vol. II, Leningrad.

Chodzidło, T. (1961) Karagasi—Tofalary Monografia etnograficzna (The Karagasy and the Tofas. An Ethnographic Monograph), in *Lud*, Vol. XLVI.

(Čudinov, B.) Чудинов, V. (1931) *Путешествие по Карагасии* (Journey to the Land of the Karagas), Leningrad.

Diószegi, V. (1959) Bericht über eine Forschungsreise nach Südsibirien, in *Sociologus*, Vol. IX.

(Evsenin, I. A.) Евсенин, И. А. (1919) *Карагасы* (The Karagas), Krasnoyarsk.

Hartwig, W. (1957) Gedanken über ein Schamanenkostüm (nach Notizen von J. A. Jewsenin), in *JMVL*, Vol. XV.

Harva, U. (1938) *Die religiösen Vorstellungen der altaischen Völker*, Porvoo–Helsinki.

(Ivanov, S. V.) Иванов, С. В. (1954) Материалы по изобразительному искусству народов Сибири XIX– начала XX в (Materials on the Graphic Arts of the Peoples of Siberia in the 19th and at the Beginning of the 20th Centuries), in *ТрИЭ*, Vol. XXII.

(1955) К вопросу о значении изображений на старинных предматах культа у народов Саяно-алтайского нагорья (On the Problem of the Significance of Representations on Ancient Objects of Cult among the Peoples of the Sayan-Altai Upland), in *СбМАЭ*, Vol. XVI.

(Katanov, N. F.) Катанов, Н. Ф. (1891) Поездка к карагасам (Journey to the Karagas), in *ЗРГООЭ*, Vol. XVII, No. 2.

(1907) Наречия урянхайцев (сойотов), абаканских татар и карагасов (The Dialects of the Uryankhai, the Soyots, the Abakan Tatars and the Karagas) in *Образцы народной литературы тюркских племен*, Vol. IX. Saint Petersburg.

(Kostrov, N.) Костров, Н. (1871) Карагасы (The Karagas), in *ИГ*, No. 4.

(Lopatin, I. A.) Лопатин. И. А. (1922) *Гольды амурские, уссурийские и сунгарийские* (The Golds of the Amur, the Ussuri and the Sungari Regions), Vladivostok.

Nioradze, G. (1925) *Der Schamanismus bei den sibirischen Völkern*, Stuttgart.

(Petri, B. Ê.) Петри, Б. Э. (1927) *Оленеводство у карагасов* (Reindeer-breeding among the Karagas), Irkutsk.
(1928) *Промыслы карагасов* (Occupations of the Karagasy), Irkutsk.
(Seroševskij, V. L.) Серошевский, В. Л. (1896) *Якуты (опыт этнографического исследования* (The Yakuts [A Preliminary Ethnographic Study]), Saint Petersburg.
(Solovjev, D. K.) Соловьев, Д. К. (1926) *Основы охотоведения* (Game management), Moscow.
(Šimkevič, P. P.) Шимкевич, П. П. (1896) Материалы для изучения шаманства у гольдов (Materials for the Study of Gold Shamanism), in *ЗПОРГО*, Vol. II, No. I.
(Širokogorov, S. M.) Широкогоров, С. М. (1919) Опыт исследования основ шаманства у тунгусов (Essay on the Foundations of Shamanism among the Tungus), in *УЗИФФВ*, Vol. I.
(Štubendorf, E. F.) Штубендорф, Е. Ф. (1854) О карагасах (About the Karagas), in *ЭО*, Vol. IV; *ВРГО*, Vol. XII, No. 2.
(Troščanskij, V. F.) Трощанский, В. Ф. (1903) Эволюция черной веры (шаманства) у якутов (The Evolution of the Black Art [Shamanism] among the Yakuts), in *УЗКУ*, Vol. IV.
(Vasiljev, V. N.) Васильев, В. Н. (1910) Краткий очерк быта карагасов (A Short Sketch of the Daily Life of Karagas Women), in *ЭО*, Vols XXXIV–XXXV.
(Verbickij, V. I.) Вербицкий, В. И. (1893) *Алтайские инородцы* (The Natives of the Altai), Moscow.

The Tuvan (Soyot) Shaman's Drum and the Ceremony of its 'Enlivening'

by

S.I. Vajnštejn, Moscow

So far very little attention has been devoted to the religious conceptions of the Tuvas and in particular to their shamanistic beliefs. The scholars who have written on the subject (E. K. Jakovlev, Felix Kon, etc.) touch only on a few aspects of shamanism.[1] Ceremonial acts pertaining to the transformation of a Tuva into a shaman, as, for instance, the enlivening of the drums, have been rather neglected. Nothing has been written about shamanism among the reindeer-breeding Tuvas (Todzhans) living in Todzha, in the north-eastern Sayan mountains.

At the beginning of the 20th century F. Kon describing the Tuva region wrote: 'In the west among the Kemchik Soyots (Tuvas) and in the east among the Todzhans, shamanism was still fully flourishing'.[2] Nowadays, however, there are no more active shamans in Todzha.

In the present paper we shall describe the rite of enlivening a shaman's drum among the Tuvas of Todzha, relying on material collected during our field work in the Todzha region in the years 1951–1955. Our informants were former shamans. Their names were Kol Čenzer (born 1892, clan *saryg sojan*), Sojan Kadanaj (born 1876, clan *aq sojan*), Baraan Čilopsan (born 1895, clan *aq todut*) and Ak-Daržaa (born 1894, clan *dargan*).

Shamanism among the Tuvas was inherited in either the paternal or maternal line. In most cases not all children became shamans but one of them only. Sometimes a person whose parents were ordinary people became a shaman. Men as well as women could become shamans.

A shaman is recognized by typical attacks of a special 'disease' called *albystār*. The person in question 'goes off his head', utters inarticulate sounds, breaks dishes, leaves the house and rooms about in the taiga, twitches in hysterical convulsions, is seized with nausea and rends his garments, etc. The 'invalid' repeats that the soul of a shaman *ištig irgäk* has moved into his body, that the spirit urges him to take a horse (by which the drum *düŋür* is meant) and clothes (*ala χujaq*) and to become a shaman. The *albystār* generally appears at a young age: between 20 and 25 years in men and from the 10th to the 15th years in women. There are, however, cases when the shaman 'disease' comes on at a mature age.

Baraan Čilopsan, whose parents were not shamans, for instance, became 'ill' when he was 40 years old. According to his story, his 'illness' lasted three months, during which he was unconscious. He did not remember what he had shouted, sung or done. The parents of the 'invalid' sent for a shaman of the clan *dämči* whose name was Pušku. This man said that the

[1] For the various aspects of Tuvan shamanism cf. Яковлев (1900) pp. 54–8, 109–20; Потанин (1883) pp. 40–78; Кон (1934) pp. 67–87; Катанов (1903) pp. 30–6 and (1907) pp. 190–202.

[2] Кон (1934) p. 53.

soul of a shaman who had died earlier without leaving children had entered the diseased person: Čilopsan had breathed in the spirit when he was hunting in the taiga. Pušku told the relatives that the 'illness' would last until Čilopsan himself had become a shaman. Then the members of the clans manufactured a shaman's outfit for him. 'As soon as I had become a shaman the illness ceased'—said Čilopsan.

The other former shaman Kol Čenzer of the *saryg sojan* clan told us that he had 'contracted the disease' when 24 years old. He was constantly shivering and all his bones ached. When somebody entered his tent, he fainted and lost his senses. When he was alone, he felt quite well and even went hunting. The shaman who was sent for said that the soul of a shaman—the brother of the diseased man's father—had moved into the body of Kol Čenzer, and that this was a black spirit (*qara buq*) now sitting in him as a bear.

The 'sick' man made efforts to expel the spirit by special trances, and sent for several shamans, but nothing seemed to help. His 'illness' lasted for 4 years. All shamans invited to heal him declared that if Kol Čenzer were not willing to become a shaman, he must die. By order of the shaman Aldanaj of the *balyqčy* clan the relatives manufactured for Kol Čenzer a shaman's staff and clothes.[3]

The first requisite of shamanhood, which the Tuvan shaman received was the staff (*dajaq*), in contrast to the Altais where it was a drumstick.

As his 'powers are increasing', the shaman, on the suggestion of his predecessor, asks his relatives to give him a drumstick (*orba*) and a drum (*düŋür*). During his trance the shaman turns to his relatives with a request, as for instance:

χoj qaranyŋ bädik sivirindän	From the crest of Khoi-Kara[4]
ādym- χölüm ālyry män,	I intend to take my horse,
aq ajnyŋ on bäštä	On the fifteenth of the white month
ādym- χölüm bälätkäŋär,	Prepare my horse,
ala χujaq därigläŋär.	Prepare also my '*ala χujaq*'[5].

As a rule, if the new shaman obtained the *dajaq* in the autumn, they made the drum for him either in the winter, in the middle of February, or in spring, in mid-May.

A few days before the middle of February two or three people, usually relatives of the shaman, set off to the place indicated by the shaman to manufacture the drum.

Having chosen a suitable cedar tree for the cylinder of the drum, they let the tree know of the desire of the shaman, in words like these:

yŋāžyq χam silärdän mynar	The shaman *Yŋāžyq* requests
āt dilädip čorān čüvä ijyn	From you a saddle horse.

[3] Анисимов (1952) pp. 229–31 gives a most interesting explanation of the social character of the process of becoming a shaman described here and common among most peoples of Siberia.

[4] The name of a site. The shaman indicates the place where the cedar tree should be felled for manufacturing the horse, i.e. the drum.

[5] The name of the shaman's cloak.

Then they felled the tree and to the remaining trunk fixed a gift consisting of a long white ribbon, which they wound several times around the log.

With the help of a special tool (*čoŋu*) they prepared the wooden frame of the drum on the spot, in the course of one day.

Returning to the *āl* (camp) the relatives fixed to the frame a crosspiece to serve as a handle, later the iron parts, and then as a rule they took the skin of an elk they had killed—(it had to be a male one)—cut out a piece of the

Fig. 1. The shaman's drum of the Tuvan reindeer breeders. (Front view)

skin from the chest and belly region, carefully removed the fur and (without any other kind of processing) covered the drum with it.[6]

The drum of the Todzhan shaman was of a circular, somewhat oval form, with a diameter of 60 to 80 cm. The drum preserved in the Museum of Tuva is 64×67 cm and the breadth of the rim is 21 cm (Figs 1, 2).[7]

The drums of the Todzhan shamans have no drawings on them; this is also characteristic of the Tofa drums, unlike the shamans' drums of the

[6] Among the Tuvas living in the steppe region the skin of the maral and also of the red deer and musk deer were used to cover the drum.

[7] The description is that of a drum belonging to the shaman Šančimas of the *aq čōd* clan. It is now in the Tuvan Regional Museum (No. 1139).

Khakas, the Altais and the Tuvas living in the Western districts of Tuva.

Resonators (*χavā* 'front') of cedar wood were applied to the frame of the Todzhan drum. Each resonator had the form of a 2 cm wide lath with three protrusions. To the inner side of the drum along the greatest diameter a vertical, flat wooden handle (*tuda*) was attached. On the upper and lower parts of the handle there were two incisions. Often two ribbons, called *dyn* ('driving-reins') were added to the handle. The lath was generally adorned with simple geometric patterns (Fig. 3). The smooth handle of the drum of the Todzhan shamans is similar to that of the Tofas (Karagas) and the Selkups (Ostyak Samoyeds).[8]

Along the small diameter of the drum here described, across the handle was fixed a cylindrical wooden lath (*čažā*) to which was attached a bent yoke of iron crupper (*qudurγa*).

To the inside of the wall of the drum two arch-shaped bent iron rods (*syrγa*) were fixed, on each of which three conical brass tubes (*χoŋura*) were suspended. The drum was kept in the tent on the wall opposite the entrance.

In Todzha also a shaman's drum with two horizontal laths was used, which was very similar to the type widespread among the Tofas.[9]

The characteristic features of the Todzhan drum (the shape of the handle, the structure and shape of the resonator, the width of the rim, etc.) distinguish it from the drum common among the Tuvas of the steppe regions and rather compare it with the drums of the Tofas[10], Kets[11] and Selkups[12]. In connection with a detailed description further research is required.

Fig. 2. The shaman drum of the Tuva reindeer breeders. (Lateral view)

The skeleton of the drumstick (*orbu*)[13] was manufactured from the horns of the elk whose skin was stretched on the drum. This was then covered with skin taken from the leg of a bear (in exceptional cases with the skin of other animals, e.g. of elks). At the end of the drumstick (*orbu*) they fixed a noose (*ildirgä*) made from the skin of the elk. To the *ildirgä* they attached black-

[8] Collection of the MAE (Museum of Anthropology and Ethnography, Leningrad) No. 3871–5a.

[9] According to data obtained from V. Diószegi.

[10] Collection of the MAE No. 1480—12.

[11] Collection of the MAE No. 4034—153.

[12] Collection of the MAE No. 3871—5a.

[13] Among the Altais the drumstick, called *orbu*, is made from the twigs of the meadow-sweet and covered with skin from the legs of a deer. The Altais also used to bind ribbons of the spirit *orbu āzi* to the *orbu*. In the prayers of the Altai shamans the drumstick sometimes is called *taš orbu*. See Дыренкова (1949) p. 33. According to the information kindly communicated to us by V. Diószegi, the Tofas manufacture the drumstick in the same way as the Todzhans.

and-white ribbons (*čalama*)[14] for the spirit of the drumstick (*orbu āzi*). In their supplications, the shamans often called the drumstick *daš orbu* i.e. stone drumstick.

Fig. 3. The shaman drum of the Tuvan reindeer breeders. (Back view)

As soon as the drum and the drumstick were ready, the relatives set themselves the task of enlivening the drum. At the beginning of the 20th century the Tuvas used to invite to the ceremony the relatives of the shaman, as well as his neighbours in the camp; but—according to the statement of the old people—in the past century only the members of the clan could participate. The relatives slaughtered a saddle reindeer (*čary*), belonging to one of the persons present, but not to the shaman. Its flesh was cooked and eaten. The rite of enlivening the drum was a clan festival (*düŋür toj*). The ceremony was regarded as the process of 'taming and training the horse'. The Tuvas take the drum for a horse and the drumstick for a whip. It is very interesting to note that also the Mongols[15], Buryats[16] Yakuts[17] looked upon the drum as a horse. The assumption of L. P. Potapov

[14] Among the Shors these ribbons are called *čalaba*.
[15] Потанин (1882).
[16] Агапитов—Хангалов (1883) p. 19.
[17] Приклонский (1886) p. 134.

that the Tuvas called the drum a horse only on account of its 'function as a riding animal, but really meant a maral *(Cervus maral)* with the skin of which the Tuvas covered the drum'[18] is not confirmed by material evidence.

When the festival opened, they brought the shaman his ritual clothes sewn by his relatives. He then addressed a request to those present saying for instance: *ādymny ōredip bäriŋär* 'Will you tame my horse'.

Each of the attendants took the drumstick in his hands, one after the other, and struck the drum a few times. When performing these acts the person in question jumped about, shouted incoherent words, and, in general, imitated the gestures of a shaman when in a trance.

Women do not attend the consecration of the drum. It is not excluded, as stated by L. P. Potapov, that the prohibition for women to assist at the festival is a survival of their former status in the clan of the husband where they were considered strangers. The shaman whose 'horse was tamed' did not attend the ceremony either, but was represented only by his observer. He was not even allowed to take the drum into his hands.[19]

At the *düŋür toj* also other shaman relatives were present. When all persons who were not shamans had struck the drum, it was handed to the 'weakest' of the shamans present, who gave it to a 'stronger' one. The last man who took the drum into his hands was the greatest shaman (*uluγ χam*), who, striking the drum said:

adyn-χölüŋ	Your horse,
ala χujat	Your pied clothes
bičä-bora syryγlyγ	*bičä-bora syryγlyγ*,
amy kirip botaŋan-dir,	Is now coming to life,
	Has found a master
āttäni-dā bärgän-nä-dir.	And is overgrown with flesh.

The second and concluding phase of the enlivening of the drum began at nightfall when nobody remained in the tent, except the shaman, who then went into a special trance, which lasted until daybreak. He called the spirit of the drum and addressed it as follows:

čüs aldy saj sögünnü	All hundred and six bones
čümüläp bälüp alγaš,	Are gathered and united by us.
borodaj säni, syryγlyqty,	You, *borodaj—syryγlyqty,*
bottandyryp alȳlynān.	Let us make you overgrown with flesh
āttändirip alȳlynān,	Let us make you a living creature,
āläp-dā alȳlȳnan	We will ride you
uzun kyska yjaštarnyn	Over the tops of low
užu-bilä, bažy-bilä	And high trees
užup byjyladyp orālynān.	We shall fly, smoothly fly.

The shaman called the drum (*düŋür*) by the name of the horse, e.g. *bičä-bora syryγlyγ*.

[18] Потапов (1947) p. 179.

[19] Among the Teleuts, when a similar ceremony was performed, the drum is struck only by men; the shaman himself is not allowed to do so: see Дыренкова (1949) p. 117.

The shaman prepared a special treat for the 'horse' when he made the first ceremony with the new drum: He brewed a very thick tea and poured it into a wooden cup while pouring *araγa* (milk-brandy)[20] into a second cup and reindeer milk into a third. Before falling into trance, the shaman smeared the drum with these liquids, fumigated himself and the full cups with smouldering juniper tree and among other things pronounced the following words:

bo mynarny män am	This harnessed [horse] I,
bo χündän ägäläp	From this day on,
ättändirip, bottandyryp,	Make overgrown with flesh
äki öredip alyjn . . .	And tame it . . .

In his subsequent ceremonies the shaman often addressed the drum—his horse. He said, for instance:

bičä-bora χörtküjü,	Poor *bičä-bora*
aštap-suksap moγavyty.	Feels hungry, has grown tired.
bičä qada syrγγlyqty	We shall feed the poor thing
ašqaryp čämgärip alāly,	Here, on the road.
sularān-dyr, χōrtqujnu	Poor fatigued creature
suγγāryp, otqaryp alālym.	We give you to drink . . .

Then the shaman, declaring that the horse had grown tired, handed the drum to one of those present, who took it and 'dried' it at the fireplace. Often when in a trance the shaman wanted to refresh the drum, his 'horse', he offered him his pipe, or moistened the outside of the drum (*arty*) with milk and *araγa*, which he spread over with the drumstick.

When after some years the drum got damaged, the shaman would say to an old man of his clan: 'A new horse is needed, call him here!'. For the cylinder of the new drum they felled a cedar in the same part of the forest where the cedar for the first drum had been felled. When taming the new 'horse' the ceremony described above was repeated. The old drum was taken out by the shaman in the taiga and addressed in words such as:

qyryp qalγan bičä-bora	Worn-out *bičä-bora*
čärlār čärin čärläzin,	May you graze to your liking,
čorūr čärīnčä čorāj-dā ān.	May you go where you want to.

As a rule the shamans had one or two drums in the course of their lives: only very powerful shamans (*uluγ χam*) had three or four.

The enlivening of the drum, then, was done in two phases. In the first phase all adult members of the clan took part (except the women), but the owner of the drum did not attend the first part of the ceremony. In the sec-

[20] The reindeer-breeding shamans obtained the milk-brandy from the stock-breeders of the Upper-Yenisei valley.

ond phase the drum, which had by now become the exclusive property of the shaman, was subject to a concluding ceremony, performed by the shaman alone. Subsequently nobody could use the drum except the shaman himself, for the spirit of the drum might cause the death of the person who took it.

The ceremony of 'enlivening' of the drum reflects clearly the evolution of shamanism from a clan activity to a profession monopolized by a few persons. The first phase of the ceremony is a survival of an early stage, when this function could be assumed by any member of the clan.

The second part stems from a later period when, parallel to the disintegration of the clan, the shamanistic functions devolved on certain specific persons—the professional shamans. This seems to be verified by the Altai data.[21]

REFERENCES

(Agapitov, N. N.–Khangalov, M. N.) Агапитов, Н. Н.–Хангалов, М. Н. (1883) Материалы для изучения шаманства в Сибири. Шаманство у бурят Иркутской губернии (Contributions to the Study of Shamanism in Siberia. Shamanism of the Buryats in the Irkutsk District), in *ИВСОРГО*, Vol. XIV, Nos 1–2.

(Anisimov, A. F.) Анисимов, А. Ф. (1952) Шаманский чум у эвенков и проблема происхождения шаманского обряда (The Shaman Tent of the Evenki, and the Problem of the Origin of Shaman Rites), in *ТрИЭ*, Vol. XVIII. — English transl. in *Studies in Siberian Shamanism*, ed. H. N. Michael, University of Toronto Press, 1954.

(Dyrenkova, N. P.) Дыренкова, Н. П. (1949) Материалы по шаманству у телеутов (Data on Shamanism among the Teleuts), in *СбМАЭ*, Vol. X.

(Jakovlev, E. K.) Яковлев, Е. К. (1900) *Этнографический обзор инородческого населения долины Южного Енисея и объяснительный каталог этнографического отдела* (Ethnographic Survey of the Aboriginal Population of the Lower-Yenisei Basin and an Explanatory Catalogue of the Ethnographic Section of the Museum), Minusinsk.

(Katanov, N. F.) Катанов, Н. Ф. (1903) *Опыт исследования урянхайского языка с указанием главнейших родственных отношений его к другим языкам тюркского корня* (About the Uryankhai Language with Reference to its Most Important Relations with Other Languages of Turkic Origin), Kazan.

(1907) Наречия урянхайцев (сойотов), абаканских татар и карагасов (Uryankhai [Soyot], Abakan-Tatar and Karagas Dialects), in *Образцы народной литературы тюркских племен*, Vol. IX.

(Kon, F.) Кон, Ф. (1934) *За пятьдесят лет* (After Fifty Years), Vol. III, Moscow.

(Potanin, G. N.) Потанин, Г. Н. (1882) Громовник по поверьям племен Южной Сибири и Северной Монголии (Thunder Calendar in the Popular Beliefs of the South-Siberian and North-Mongol Tribes), in *ЖМНП*, Jan.–Febr.

(1883) *Очерки северо-западной Монголии* (Sketches of Northwestern Mongolia), Vol. IV, Sammlung.

(Potapov, L. P.) Потапов, Л. П. (1947) Обряд оживления шаманского бубна у тюркоязычных племен Алтая (The Enlivening Rite of the Shaman's Drum among the Turkic-speaking Tribes of the Altai), in *ТрИЭ*, Vol. I.

(Priklonskij, V. L.) Приклонский, В. Л. (1886) О шаманстве у якутов (On Shamanism among the Yakuts), in *ИВСОРГО*, Vol. XVII, Nos 1–2.

[21] Потапов (1947) р. 182.

The Acquisition of Shamanistic Ability among the Evenki (Tungus)

by

G. M. Vasilevič, Leningrad

In contrast to the abundant material dealing with the acquisition of shamanistic ability among the Turkic tribes,[1] there are only a few sporadic traces referring to this subject in connection with the Tungus peoples and particularly to the Evenki. In his monograph on the Tungus of the Yenisei,[2] Ryčkov does not furnish any concrete material, although he correctly associated hysteria with the acquisition of shamanistic ability among the Yenisei Tungus. S. M. Širokogorov[3] mentions some facts about the Evenki of the Middle Amur. Like Ryčkov, he too connects inherited shamanistic ability with hereditary neurosis—the *mänärik*. Among the Evenki of Manchuria the same means of acquiring shamanistic ability was complemented by a method, borrowed from the Manchus, of teaching the young shaman the knowledge necessary in his profession. This method can be found also among the Tungus tribes of the Lower Amur such as the Nanai (Golds), the Ulchi, etc.

Before going into details concerning the different ways of becoming a shaman, let us glance at the terminology used in this respect in the Evenki and other Tunguso-Manchurian languages.

The Evenki language has four expressions for 'shaman' i.e. for 'to act as a shaman':

shaman	to act as a shaman
1. *jajan*	*jaja-*
2. —	*nimŋan-*
3. —	*səwənčə-*
4. *saman*	*samaldī-*

Among these, *jajan*, *jaja-* is the most ancient and the most generally used. Originally it meant singing—appealing to the fire for good luck. The shamans made big fires in the open air and performed their magic around them. Evenki legends still contain reminiscences of this. A ceremony of this kind was called *jandrapka*, which is derived from *jandrɪ* (Nepa subdialects *d'aŋə*) 'flame'; literally: 'ceremony' (*-pka*), of appealing to the clan (*jandra*), hence *jandɪ* 'to shamanize by the camp-fire'. The words *jaja* 'to act as

[1] See Дыренкова (1930).

[2] Рычков (1922) '. . . among the Tungus, the idea of spirits is connected with hysterical and epileptic fits which used to seize the men . . . On account of this, such persons, being in intimate relation with spirits, may dispose of powerful and special qualities . . . First of all, the shaman has to be *tui* (epileptic)'. This refers to the shamans of the Ilimpeja group settled in the region of the Lakes Yessei, Nekengde and Khuringda.

[3] Shirokogoroff (1935); Широкогоров (1919).

a shaman' and *jajan* 'shaman' are preserved not only in the Tunguso-Manchurian ethnic group,[4] but also in the Paleo-Asiatic languages.[5]

It is indisputable that all of these words are derived from a single root which existed in the ancient tongues before the original languages of the Altaic family were separated. The existence of this word, with a meaning restricted to the object ('drum') and to the act of playing on it, in the Paleo-Asiatic languages shows that the root had been borrowed by these languages from those of the neighbouring Tungus prior to their dispersion. The limited meaning of the word *jajar* (referring to the object and the action), shows that it is a loan word in the Paleo-Asiatic languages. This is confirmed by another Tungus word standing for 'to act as a shaman'—*nimŋan-*, which is used in a double sense in all Tunguso-Manchurian languages except Manchu proper[6] and has one single meaning in the Paleo-Asiatic languages.[7]

The meaning of *nimŋa(n)* lends itself to different conclusions. (1) On the early stages of shamanism the transmission of the entire body of knowledge and traditions formed part of the functions of the shaman. That is the reason why there are many common features in the transmission of the tales and in the ceremony. (2) When there still existed a connection between the ancient Tungus and the Siberian ancestors of the Koryaks and the Chukchi, this function belonged probably only to those who were in possession of the traditional knowledge and who knew the folk-tales—but not to the young shamans, or else one or the other meaning of the word would have had something to do with shamanism in the Paleo-Asiatic languages. (3) The restricted meaning of the root in these languages, as well as its productiveness in word-formation in the Tungus languages, show that the former had borrowed it from the latter. (4) The absence of this root in Manchu (so far, at least, it has not been traced) implies that it must have been adopted in the ancient Tungus language after the separation of the first groups which formed the Tungusic basis of the Manchus.

The third expression for 'to act as a shaman'—*səwənčə-*—is typical only of the Tunguso-Manchurian languages.[8]

[4] Evenki (every group of the Yenisei and the Upper Aldan regions), Oroch, Ulchi, Nanai *jaja*, Manchu *jan-* (Ulchi also *jaja-* ~ *jajala-*) 'to act as a shaman' Evenki, Ulchi *jajan*, Even *n'aja*; Oroch *jaja* 'shaman!' Manchu *jajki* 'shaman's song'; Mongolian *jaŋ* 'tune, song'; Manchu *jaγa* 'camp-fire'; *jaŋa* 'torch made of resinous branches'; Ulchi *jampa* 'shaman's belt'. Phonetic variants: Evenki *ǯaŋnə-* ~ *ǯanŋu-* 'to burn with flames', Dahur, Khalka *ǯanda-* 'to sing'; Kirgiz *ǯan-* ~ *jan-* 'to burn'. Nenets *ja* 'coal' (*jadas-* 'to forge'); Koybal, Tuvan *jal* 'flame'; Old Turkic *jalvy* 'shamanistic ceremony' Uigur *jalbar-* 'to request', Teleut *jalbar-* 'to sacrifice'.

[5] Chukchi: *jarar*, Koryak *jajaj* 'drum', *jajajtko-* 'to beat the drum'; Chukchi *jyn-jyn* 'fire'.

[6] Evenki: *nimŋa(n)-* 'to act as a shaman', *nimŋaŋki* 'shaman's drum' *nimŋakān* 'myth, story, legend, tale', *nimŋakat-* 'to perform a tale in song', *nimŋamāt-* 'to swallow shaman's assistant spirits during ceremony, invitation of spirits'; Solon *nimŋamat-*, Nanai, Ulchi *niŋman-* 'to act as a shaman, to tell a tale'; Evenki *nemkan*, Nanai *neŋma*, Ulchi *niŋma* 'tale, story', Even *nemkan-*, Nanai *neŋman-*, Udegei *nimas-* 'to tell tales'.

[7] Chukchi and Koryak: *lymŋyl* 'tale, puzzle', Chukchi *lymnylkyn* 'he tells tales'.

[8] Evenki *səwən* (*šəwən* ~ *həwən*) '(a) traditional meal made of bear meat and fat for the bear-eating ceremony, (b) the name of the bear-eating ceremony itself, (c) the assistant spirit of the shaman', *səwə-* 'to melt bear lard', *səwənčə-* 'to act as a shaman'

The meaning of the word *səwən* is evident in the following. (1) This word was known to the ancient Tungus prior to the separation of the groups which constituted the Tungus basis of Manchu. (2) The word referred to a part of the rites in the bear cult and was related to the melting of grease. In the meaning 'censing with grease' it has been preserved among the Evenki, as well as in shamanistic ceremonies. (3) This tradition was handed down and developed by the shaman (*səwən*, *səwəkī*, *səwək*, *səwənčə*) and finally has become one of their most important functions (*səwənčə*, literally: 'to be in a state of *səwən*', i.e. 'to inhale the assistant spirits, to cure the sick, to look for the spirit of the sick').

The fourth word for 'acting as a shaman'—*samanil-*, *samalda-*, as well as the word *saman* (*šaman* ∼ *haman*) can be found in every Tunguso-Manchurian language[9] and also in some Mongolic and Turkic languages.[10]

The following reasons support the assumption that the word penetrated the ancient Tungus languages from the south. (1) The word 'to act as a shaman' (*saman-i-l-*, *sama-ldɪ-*) is a derivative in the Evenki and the other languages, and the root *saman* has only one meaning (whereas true Tungusic words are characterized by polysemantism). (2) In the Evenki language it does not indicate the ceremony as such, but only the movements of the shaman (jumping, dancing, etc.)

The Ilimpeja and other groups had two words side by side: *haman* and *hirurī*.

Thus, the terminology for 'shaman', 'to act as a shaman' shows:

1. Inheritance of ancient, pre-shamanistic traditions, by the shamans (appeal to the fire; eating of, and censing with, grease; transmission of the whole tradition and folklore) and a certain southern influence in the development of Tungusic shamanism.

Attributes of the shaman, such as a scoop similar to the shaman clapper (find from Anosovsk) and anthropomorphic representations on the chest of the skeleton (find from Ust-Udinsk) permit us to speak of a nascent shamanism in tge Glazkov period of the Neolithic around the Lake Baikal.

2. This terminology indicates that the appearance and evolution of shamanism among the Tungus are connected with general conceptions of spirits, i.e. the *səwən* which, later on, began to form different categories: *səwəkī* (or, among the Amur Evenki, *buγa*)—the master spirit of the upper world, *səwən*—the assistant spirit of the shaman, *musun* ∼ *muhun*—the master spirit of certain natural phenomena, *hargi*—the assistant spirit of the hostile shaman, etc. As far as the origin of the conceptions referring to spirits is concerned, we agree with L. Ja. Šternberg's statement that it was given

---

(literally: 'to be in the state of *səwən* '(i.e. to contact helping spirits, *səwəkī* 'master spirit of the upper world, god', *səwək* 'reindeer under taboo, carrier of a part of *səwəki*'s force'. The word *səwən* meaning 'assistant spirit of the shaman, master spirit of heaven and earth, protecting spirit of the house', to whom the shaman appeals, is typical of the Solons (*səwu*), Udegeis, Orochi, Oroks (*səwə*) and the Manchus (*soku*), in the latter meaning. The word *səwəkī* means 'god' in the languages of the Evens, the Negidals, the Orochi, and 'anthropomorphic figures hung on the shaman's costume' in the languages of the Evens, Orochi, Udegeis and Nanai.

[9] Even *haman*, Solon *sama*, Udegei *samo*, Negidal, Nanai, Ulchi, Manchu *saman*, Evenki *samanil-*, *samalda-*, Even *hamalči-*, Manchu *samaša-*, *samda-*.

[10] Dahur *saman*, Mongolian *samadi*, Uzbek *šomon*, Uigur *samaty*.

'in the idea of primitive clan coalition, on one hand, and in ancient natural philosophy, on the other'.[11]

As L. Ja. Šternberg states, the term 'master' (husband, host) stood initially for a person of outstanding skill, strength and courage in hunting, of excellent memory and sound judgement—qualities which brought material welfare and help to the family-clan society. In peacetime, such a man provided food for the clan and administered justice, in war he was the chief. When the belief in the master spirits developed, this interpretation was transferred to the sphere of the surrounding natural phenomena.

The idea of the assistant spirits was created and developed by the shamans. These spirits included, first of all, the birds of the early mythology (birds which participated in the creation of the earth: loons, golden-eyes, ravens), quadrupeds which were the main objects of pre-shamanistic hunting rites (bears, reindeer, elks, dogs) and finally, various animals which were chosen in later periods and were needed for the ceremonies of the shaman. A major group of assistants consisted of the spirits of the shaman ancestors. The most ancient ones of these, belonging to their own or to a foreign tribe, together with the female shamans, formed the circle of the ancestral family spirits and were endowed with various supernatural qualities. Of certain ancestral spirits, not too far removed in the past, even biographical particulars were remembered by the shamans.

After having developed beliefs in different kinds of spirits, the shamans distributed among these such functions as the care and protection of the souls belonging to a certain clan and, later, to a given regional group. In this manner, they kept in peace the souls of the clan members. Gradually, the shaman became an indispensable member of outstanding importance in the clan and later, in the regional ethnic group. In the clans which consisted of several generations, living in different places and no longer remembering the names of the first ancestors, there were several shamans.

As long as the shaman lived, the spirits of the ancestors were in contact with him, while the assistant spirits were always at his disposal, constantly protecting the clan against the spirits of hostile shamans. But when the shaman died, the Evenki believed that the spirits 'scattered all over the world' and ceased to protect men. Defenseless, they might then be exterminated by the spirits of the hostile shamans. S. M. Širokogorov observed among the Evenki of Transbaikal and the Middle Amur that the death of the shaman and the belief in the dispersal of the spirits led to hysterical cases among the neurasthenic people and the younger generation, intimate with the shaman. And if among those who fell ill, nobody felt that the spirits had passed on to him, the disease spread to others. As a result of this abnormal state, accidental deaths increased in number. But as soon as one of the sick people 'began to become a shaman', so strong was the belief that the new shaman gathered all the spirits together and was able to fight with the hostile spirits, that the disease gradually came to an end.[12] On very rare occasions when nobody declared that he would be a shaman, the oldest of the clan, or the old shaman of another generation, found an appropriate person who had to replace the dead one. This person was taught the knowledge of the shamans.

[11] Штернберг (1936) p. 217.
[12] Широкогоров (1919) pp. 89, 99.

According to the Evenki, the spirits of the shaman ancestors, being members of the clan, could not go over to a foreign clan, nor could the assistant spirits which were connected with them. It follows from this that the shamanistic power, that is to say, the capacity of mastering the spirits, also had to remain within the clan. If we now consider the extraordinary nervous and mental constitution of all those who had been shamans throughout the subsequent generations, as well as the point at issue that this disposition was hereditary, we may naturally conclude that the shamanistic activity among the Evenki was depending on certain specific hereditary dispositions and conditions.

Our own observations and the data recorded from old men who were former shamans among the various groups of Evenki indicate that in most cases the persons belonging to the clan of the shaman—whether they were living with him or elsewhere—suffered from extreme nervousness and susceptibility, some of them being even *mänäriks*. They were very sensitive, had a sharp memory and a vivid imagination and were able to relate the events of their life in an artistic way. From this class the most talented tale-tellers and song improvisers emerged, famous not only among their own tribesmen but also in the neighbouring ethnic groups. We were able to check on site L. Ja. Šternberg's theory concerning sexual selection in shamanism.[13]

All inquiries as to how a person became a shaman are always answered to the effect that 'the spirit whispered or sang into his ears the words he had to sing', or 'a spirit moved into his brain and urged him to utter the words'. We could not find a single tale telling of some sexual motive, although such motives did occur in the relations between the hunter and the master spirit of the hunting-ground who sometimes called on the hunter at night in the shape of a woman.[14] Nor was there any reference to sexuality in the stories about the ancient shamans.[15] In the folklore of the various Evenki groups, there are no themes on sexual intercourse between men and animals (except for the myth of the Okhotsk Evenki and Evens about the girl who spent the winter in a bear's den and gave birth to a bear's cub) though these seem to be widespread enough among the Tungus-speaking tribes of the Lower Amur.

As the shamanistic ability was regarded as heritable through the male line, the assistant spirits and the spirits of the shaman ancestors were considered to remain within the clan. In some cases, however, they could pass down through the mother's side to members of another clan that had family ties with the shaman's clan. Time and distance were of no importance: succession might skip three or four generations and cover over 400 kilometres.

[13] Šternberg has based this idea on the instances of sexual selection observed among the Nanai (Golds), Yakuts, Shors, Selkups and Chukchi. In his article (Штернберг 1927), reprinted in his book in 1936, he examines the sexual factor in selection from the lowest up to the highest form during the whole course of history of religions.

[14] See Васильевич (1930) p. 63.

[15] See e.g. the unpublished stories about the first shamans existing among the descendants of the Angara Evenki (Василевич 1936) about Curivul (pp. 45, 51), Murivul (pp. 59, 64, 151–2, 241). The story about Unyany and Kerendo (pp. 41–4) was also recorded by Tretjakov in the first half of the past century, as a tale about the ancient shaman.

The tradition concerning the inheritance of shamanistic ability and spirits was often carried through connections between kinsmen living far away, but visiting each other from time to time, and thus maintaining the memory of their related genealogy.

Among the vast number of genealogies (some running into 5–14 generations) which have been registered among the various Evenki groups, we have found 48 cases where it was remembered from whom the shaman had inherited the dead shaman's spirits. The general tendency was from grandfather to grandson and great-grandson or from father to son, with a few isolated exceptions.

From father, to son	11 cases
From father, to daughter	1 case
From grandfather, to grandson	16 cases
From grandfather, to granddaughter	1 case
From great-grandfather, to great-granddaughter	2 cases
From great-grandfather, to great-great-grandson	1 case
From great-grandfather, to grandson of great-grandson	1 case
From brother, to brother	3 cases
From brother, to brother's son	1 case
From brother, to brother's grandson	1 case
From mother's father, to her son	1 case
From mother's father, to her grandson	1 case
From mother's father, to her daughter's son	1 case
From mother, to son	1 case

In six of the cases, the spirits had not been inherited by anybody, and in two cases of these six, spirits of female shamans had been involved.

It was believed here and there that the spirits were passed from generation to generation (from father to son and then to grandson, or from father to daughter and then to her son), or that the spirits of two dead shamans were inherited by one shaman as, for instance, in the case of Evgraf Lopoko (Stony Tunguska basin). To him came the spirits of his mother who belonged to the Momol clan and those of the son of his paternal grandfather's brother belonging to the Lopoko clan. To Kepete Ngangagir (Vitim basin) came the spirits of his great-grandfather and of his first aunt, both belonging to the Ngangagir clan.

In reply to the question as to what happens to the spirits that had not passed over to a successor during the period of time between the shaman's death and the days when these records were made (1925–1935), the Evenki said: 'They wander about in the taiga and will eventually meet the person whom they are to enter'. Such was, however, rarely the case and even then no more than three generations had passed since the shaman's death. According to records of I. M. Suslov (up to 1926), succession on the mother's side was predominant among the Evenki living north of the watershed of the Lower Tunguska and the Stony Tunguska. Širokogorov has recorded cases among the Evenki of Nercha and Chita where the grandson was bound to inherit the spirits from the grandfather; when they were to pass from a woman to another, it was only in regard of the spirits of her own clan.[16]

According to G. Spasskij, a new shaman belonging to the mounted Evenkis of Transbaikal declared: 'The dead shaman had appeared to him in his dream and ordered him to be his successor'.[17]

The person 'selected by the spirits of the shaman ancestors' could be of any age—in fact, even a child. The essential thing was to have a deep sense of vocation. Thanks to P. Tretjakov we possess a fairly accurate description of a shaman's personality: 'His natural sensibility and susceptibility were coupled with a vivid imagination, as well as with the belief in the spirits and the secret communion with them . . . abandoning himself to the figments of his own imagination, he became tormented and timorous, especially at night when his head was filled with dream visions. On the day when he had to act as shaman, the visions stopped, he fell into a trance and stared for hours at an object. The pale and worried man, with his piercing look, made a peculiar impression.'[18]

After his dream, the chosen man became uneasy and timid, began to meditate, did not answer when he was addressed and frequently heard 'words whispered into his ears' which he had to sing, *ǯariǯačān* (literally: 'to repeat'). That was the moment when, in shamanistic terminology, 'the spirits entered his ears and brain', telling him the words of the song. Sitting in his tent and meditating, he would begin to rock from one side to the other and to sing. Sometimes he went into the taiga, sat there all day long in the same place and repeated from time to time the words whispered to him by the spirits. The spirits might suggest to him, in his dream, that he should kill one of the shaman's birds (the loon among the western tribes, the cuckoo in the eastern regions), to cut it to pieces and eat a piece every day. This meant that the bird of the assistant spirit 'had flown already into the young shaman'. The Sym Evenki called the first dreamt instruction, as well as the instructor *burunī* or *ildıdı*; the Evenki near the Lower Tunguska also used the word *ildıdı*. They believed that the choice itself and the transfer of the assistant spirits to the novice were directed by the master spirit of the upper world. The first words heard by the chosen person were instructions, such as: *əŋnəkəl ərūt n'əkəmī butıləwə ələ ajnāŋnəkəl ilədılə əŋnəkəl iləčərə bi uγiriktət bišim šinə šiŋnilim əŋəšijəš šindu buǯəm* (Sym), 'do not commit evil things, only cure the sick. As for the *ıləs*[19] (Evenki), leave them in peace. I shall be above you, and take care of you, and give you strength.' In his dream, the young shaman could see the shaman ancestors who also gave him instructions. Then other spirits came 'whispering into his ears'. The *burunī ildıdı* constantly checked whether the future shaman really complied

[16] Широкогоров (1919) p. 90.
[17] Спасский (1822) pp. 39–40.
[18] Третьяков (1869) pp. 209–10.
[19] In this address, it is interesting to note how he compares himself (he uses the term *əvən*—from the Angara region) with the other groups of the Lower Tunguska and the Stony Tunguska region). Quarrels were frequent between these groups and it is only natural that this should find its reflection in the shamanistic beliefs too: not only their shamans were enemies, but also the assistant spirits. *ılədı* means literally: 'who belongs to the *iləs*'. Therefore, the instruction not to touch anything that is *ilə* is fully justified since offending them might involve similar acts from the *ilə* shamans' spirits.

with the instructions he had received. He could punish him (*buwkāniŋnəčən*, literally: 'force him to die'), if he did not carry out the orders. Sometimes the novice went into the taiga, where he fell into a trance that made him think he was an animal (a glutton, a wolf, or a deer) and he returned home perfectly imitating the animal.

According to information received from certain shamans (Lower Tunguska—river-head of Vilyui), the ancestral spirits of dead shamans sometimes invaded the novice during the preparation period: they lacerated his flesh and sometimes even roasted him on the gridiron, drank his blood, they shot at him or cut his hands off and threw them on a heap. Imagination or auto-suggestion of this kind often caused morbid symptoms such as flatulence, or heavy breathing resembling asthma, or epileptic fits.

After this period of preparation the candidate made a drumstick with the figure of a human head at its end for his future drum. This he put into the earth, sat down and rocking to the right and to the left, began to sing shamanistic songs. So far he had no shaman's clothes, but his head was always covered with a cloth reaching down over his face. After one or two years, he saw in his dream a reindeer whose hide was destined for his drum. He told his kinsmen where the deer was to be found and what it looked like. The kinsmen (formerly the mother's relatives) tracked down the animal and killed it. Sometimes the hunt lasted for a whole year. The would-be shaman selected girls who prepared the hide for the drum, while he himself made the wooden frame and the cross-bars. The initiation of the drum was performed by means of ceremony with the assistance of another shaman who belonged either to the same clan or to the clan of the candidate's mother. This rite was generally connected with another one the purpose of which was mostly curing. For the initiation ceremony a tame reindeer was killed, the drum smeared with its blood, while the hide, and formerly also the head and the antlers, were hung on a tree as a sacrifice to the spirit by which the novice had been chosen. Later, the hide was replaced by a stretched band which had the shape of a hide. Now, the new shaman was at last given the right to act as a shaman whose normal functions he took over (forecasting, prediction, curing). He gradually acquired his costume: first came the bodice—*həlmī* or the shoulder piece—*arkalān* (Amur-Zeya), followed by the gown—*samasik*, *lombolon* and the shoes, finally the headgear called *šərgəmī* by the Sym tribes and *awun* by the others. Originally this was made of hide, later of metal. In addition, the shamans of the riding Evenki had sticks symbolizing horses. By gradual stages, metal flaps were procured for the drum and for the clothes. Each of the new pieces had to be 'initiated', i.e. smeared with the blood of the sacrificial reindeer. Among the Evenki, this was generally connected with one of the main ceremonies. Before the preparation of the costume or of the drum, the shaman had a dream in which he received instructions concerning the hunters of the reindeer to be killed and its whereabouts.

The Evenki of Mankovsk and Chita called the rite for becoming a shaman *turani*. Širokogorov[20] has observed cases among the reindeer-breeding groups of the aforesaid Evenki, where the candidate was nominated by the clan. The old shaman or the oldest members of the clan selected and educated in

[20] Shirokogoroff (1935) pp. 351, 352.

a special way a boy whose nervous state, behaviour and keen attention indicated the faculties necessary for a shaman. The clan provided opportunities of 'contact with the spirits', consisting of interpretation of dreams and forecasting.

The most important criteria for determining whether the candidate could really become a shaman, in the opinion of the Evenki, were his instructions concerning the animal to be killed for the drum. During the ceremony, when the novice communicated his dream, the members of the clan offered a sacrifice and implored the spirits of the ancestral shamans to help them find the reindeer. Then a clan member went out hunting for a long time—sometimes for several months—and tracked down the animal described by the candidate. If the hunter was unable to find the required animal (reindeer, elk), the novice was allowed to repeat his request to the spirits of the ancestral shamans and to ask them for a more precise dream. Finally the animal was killed. Always girls prepared the hide and sewed the costume. After the drum and the clothes were finished, all the members of the clan assembled, sacrificed a full-grown reindeer and dressed the novice in shaman's clothes. He shamanized with the drum and became an accomplished shaman. The other pieces were gradually provided later on.

According to the records of Širokogorov, the rites for becoming a shaman among the Biras of the Middle Amur resembled those of the Manchus and the Dahurs. The most important event was the ceremony in the course of which the novice received the first spirit who had to bring him in contact with the many shamans' spirits of the clan. This was followed by a probation period during which the novice was educated by an experienced shaman, through talks. The novice was called *asaran* 'the pursued' (literally: 'he will be followed'), the old shaman *səbu* 'teacher' (from *səwən*); but they called each other *dowɛj*. A small ritual place was furnished for this education. On both sides of the entrance to the tent a *turu* (∼ *turu* ∼ *turi* ∼ *turə*)-tree was set up with its lower branches lopped but the crown left untouched. These trees were linked up crosswise by horns. The novice, dressed in the clothes of the old shaman, took his place between the trunks. A few yards from the left *turu*, a third *turu* was erected—this was the place of the old shaman. These two *turus* were tied together by a strap or rope—*sidım*—furnished with removable rings (*ǯuldu*), cloth bands and bird figures[21] symbolizing the road where the spirits assembled. Around the *turu* the pictures of the ancestors' spirits were placed, called *an'an* (literally: 'shadow' or 'soul').

While the new shaman was sitting and beating the new drum, the old shaman invoced the ancestral and assistant spirits, and, pushing the rings on the strap, sent them one after the other to the novice, who pushed the rings back to prevent the spirits from moving into him. During this time, the elders 'examined' the novice, asking him what each of the shaman's spirits was like, what he was when he was still alive, where he lived, which shaman he had served, when that shaman had died, etc. This 'test' could last from 5 to 9 days. After it had been successfully passed, the first animal sacrifice was offered, the *turu* and the *an'an* smeared with blood and the

[21] Similar equipment was characteristic of the sacrificial places of the shamans of almost all the Eastern Evenki who call themselves *oročɛn*.

new drum decorated with drawings or strips. The novice was then recognized as a shaman and allowed to shamanize on his own.

Other ways of becoming a shaman were also known. Širokogorov describes the following data he had collected among the mounted Evenki of Manchuria: 'An Evenki of about forty went into the taiga and lost his way. He did not know how long he wandered about, but one night, among the rocks, he was frightened, and fainted. The *kadar burkan* had moved into him and declared him the chief *səwən*. He acted as a shaman for a few days on the spot, regained his conscience and returned home where he told what had happened to him, and the clan assembly . . . recognized his rights.'[22]

It follows from all what has been said so far that, according to the general beliefs of these peoples, the shamanistic ability and the shaman's spirits almost always remained within the clan[23] and rarely passed over to another, related one.

The continuous tension resulting from the struggles with the spirits of hostile shamans and from the strain with which they listened to and followed the guidance of their own assistant spirits, as well as watchful and timid behaviour towards the surrounding shamans, developed a mistrustful reticent and secretive nature in the shamans, many of whom lived in seclusion.

Since it was the main task of the shaman to protect his own tribesmen against any machination of the spirits of hostile shamans, he had no right to accept any reward from his kinsmen for his performances. The sacrifices offered to the master spirit of the upper world: the new scarfs and hides hung upon the *turu* were left on the spot, or distributed among the disabled members of the clan. If he were in great need, however, the shaman could take his share of these. On the other hand, he was entitled to remuneration from members of foreign clans, consisting at the utmost of one or two bottles of brandy. It was only at the end of the 19th century and the beginning of the 20th that he was allowed to accept reindeer meat, occasionally a reindeer or furs.

As he was obliged to spend most of his time in communicating with spirits and shamanizing upon request, the shamans had little time left for hunting or fishing. Therefore many of them had to endure hardships, especially those who lived in solitude.

Side by side with these shamans, there cropped up at about the turn of this century the 'false' shamans (as the Evenki called them). Certain persons, looking for easy profit (fees received from foreign tribesmen for ceremonies), learnt the songs and activities of the shamans, made drums and declared themselves shamans. They only foretold, and were sometimes

[22] Широкогоров (1919) pp. 92–3.

[23] Širokogorov found and interesting case among the Evenki of Manchuria concerning the relationship that existed between the shaman and the spirits in the imagination of the Evenki. 'A shaman had been persecuted by a spirit for a long time. His contact with other spirits proved unsuccessful. At last he decided to take an energetic step: he made a small figure into which he placed the harmful spirit and slashed it with his knife. After having returned home and reached his own house, he suddenly became very angry and again with his own knife, inflicted a mortal wound on himself from which he died.' (Широкогоров 1919, pp. 102–3.) The Evenki gave the following explanation: The spirit of the shaman, wishing to get rid of his disagreeable master, challenged the shaman, took his place first in the figure and then in the shaman himself.

called to the sick if no 'real' shaman was available in the region. They did not participate in the rites connected with hunting and reindeer-breeding. As they always performed their ceremonies in one and the same way, they could not gain any real recognition from their communities, and they were only very exceptionally asked for help. They were in contact with the kulaks, and hindered, in every way, socialist construction and cultural progress in the 1930s.

REFERENCES

(Dyrenkova, N.P.) Дыренкова, Н. П. (1930) Получение шаманского дара по воззрениям турецких племен (The Acquisition of Shamanistic Ability in the Beliefs of the Turkic Tribes), in *СбМАЭ*, Vol. IX.

(Ryčkov, K.M.) Рычков, К. М. (1922) Енисейские тунгусы (The Yenisei Tungus), in *З*, Vols I–II.

(Spasskij, G.) Спасский, Г. (1822) Забайкальские тунгусы (The Transbaikalian Tungus), in *СВ*.

(Širokogorov, S.M.) Широкогоров, С. М. (1919) Опыт исследования основ шаманства у тунгусов (An Essay on the Foundations of Shamanism among the Tungus), in *УЗИФФВ*.

Shirokogoroff, S.M. (1935) *Psychomental Complex of the Tungus*, London.

(Šternberg, L. Ja.) Штернберг, Л. Я. (1927) Избранничество в религии (Election in Religion), in *Э*, No. 1.

(1936) *Первобытная религия* (Primitive Religion), Leningrad.

(Tretjakov, P.) Третьяков, П. (1869) Туруханский край (The Turukhan Region), in *ЗРГООГ*, Vol. II.

(Vasilevič, G. M.) Василевич, Г.М. (1930) Некоторые данные по охотничьим обрядам и представлениям у тунгусов (Some Data on the Hunting Rites and Beliefs among the Tungus), in *Э*, Vol. XI.

(1936) *Сборник материалов по эвенкийскому (тунгусскому) фольклору* (Collection of Material in Evenki [Tungus] Folklore), Leningrad.

Shamanistic Songs of the Evenki (Tungus)

by

G.M. Vasilevič, Leningrad

Although they maintain the structural form of improvised songs, the songs of Evenki shamans are not without characteristic peculiarities.

From the point of view of structure, they have the following common features: (1) An obligatory introduction which sets the time and introduces the rhythm of the song. The introductory words, as well as the rest of the improvised lays, have little to do with the material of the living language. In certain cases, some connection can be stated between these words and the name of the clan (e.g. *waku-waku* introductory words and *wakuwa-γir*, name of a clan) and in legends (e.g. the introductory word *kimənin* is the name of the clan *kimə, kimənɪ*; the introductory word *kildɪnərəj* is the name of the hero *kildɪnɪkān*). On this same pattern it may be assumed that the opening words of the shamanistic songs were the names of former shamans who had become spirits. (2) Generally, each word of the sentence formed a line with the refrain, which was the last word of the introduction. (3) If the words did not have a sufficient number of syllables for the rhythm, extra ones were inserted, as: *-γu, -γa, -γə, -γi, -wu, -wə, -wa, -wi, -ja, -jo, jə,* (e.g. *akī-γi-ŋasa-ja-γa-mī buru-γa-ǯəkildu-γu-k-tən*... 'my late uncle from the place of the fall, i.e. from the realm of the dead'), where three syllables *-γi, -ja, -γa*—can be found in the first word and two, *-γa, -γu,* in the second.

The peculiarities of the shamanistic songs are due to the antiquity of their syntax. Many of the forms are no longer used in modern speech, have not been recorded in any dialect and are missing in the vocabulary of the Tunguso-Manchurian languages. This is the reason why a modern Evenki cannot very understand the texts of the shamanistic songs which obviously, for that reason, make a deep impression on him.

During the ceremony, the audience together with the shaman (*ǯari-ǯawkīl bičətin*) recited the words of the refrain. And only among certain groups of the Amur Evenki the shamans had assistants (*ǯariči-mŋū* literally: 'singing companions').

We have already published several shamanistic songs with their literal translations made with the aid of modern Evenki dialects.[1] These include songs performed to lead the souls of adults into the realm of death and those of children into the upper world (No. 103, recorded in 1930 from J. Bajaki, in the region of the Sym-Chirembo river, a left-side tributary of the Yenisei; Nos 105, 106, 107, recorded in 1931 from A. Kombagir in the settlement Ngekengne, in the middle course of the Lower Tunguska; No. 104, an ancient song of the first shaman, recorded in 1930 from Š. Kima, region of the Sym river; No. 108, samanizing song for the success of hunting, recorded in 1925 from N. Kongnokogir, upper course of the lower Tunguska

[1] Василевич (1936).

and Tokma rivers). E. N. Širokogorova has recorded the tune of one of the shaman songs by ear.[2]

The Archives of the Leningrad Ethnographic Institute contain on-the-spot records made by N. P. Nikuljšin[3] in 1940, from different Evenki belonging to the Sym group, who still remembered the ceremonies. (The last ceremony—the spring hunting ceremony, *ikənipkə*—was performed in 1930, during our visit there.)

The texts were taken down from dictation, without translation. Those who dictated them, knew only their general content, which they related as explanations of the songs. N. P. Nikuljšin has recorded from three informants (one of them, I. Bajaki, is the same from whom we took records in 1930) the variants of the songs of the *ikənipkə* ceremony, the variants of the shaman's song, recited during the *ildrəməčipkə* (the ceremony of 'searching for the souls of the sick'), and during the *šiŋkələpkə* ('the ceremony for obtaining hunting luck').[4]

Here, we shall present only a small selection of the songs belonging to the above-mentioned ceremonies and rituals. The persons who dictated the quoted examples were, at that time, illiterate. For the sake of comparison we present the introductory part of the songs performed at the *ikənipkə* ceremony we recorded in 1930 from Šomirčā Kima, who was my informant and could read. The variants of his songs are nearer to the spoken language and can be translated, although the old shaman's song, in its literal translation, requires considerable revision to be understood.

*

1. Informant: I. Bajaki, 23rd July, 1940

Sym river. Recorded by: N. P. Nikuljšin. Song of the shaman during *ildrəməčipkə* ceremony called 'searching for the sick man's soul'.

The shaman starts downwards, along the shaman river together with his assistant spirits.

Gēkarōkal! Gēkarōkal!	Иди иным путем! Иди иным путем!
Gēkarōkal! Gēkarōkal!	Иди иным путем! Иди иным путем!
Činəkəl umčətin,	Птички-чинэкэ отправили
ənəl ulənmušnə,	Не крутясь,
ajat dōγujkānčā,	хорошо внутрь спустился,
əjəďəšinǯənrən,	вниз [по реке] направился
ələ hutəmilwəw	сюда детей моих [эвенков] привел.

Go another way ! Go another way !
Go another way ! Go another way !
činəkə Birds have sent [him off]
Without circling
He went well inside
Down [along the river] he started
Here he led my children [the Evenki]

[2] Shirokogorova (1934).
[3] Died during World War II.
[4] Василевич (1957).

kuŋ̍akār čāškīwər	ребята дальше
ńōšiklāǯ̆əpunaš?	зачем вперед идете?
əl hutəmīlwɛ!	Сюда дети мои!
əši tālā ōmī	Сейчас там душа
dundʳa šurkakīdūn.	На острой вершине [горы] земли.
Tādū turkutlilǯ̆ap	Там встретимся
dundʳa kiləǯ̆aktūn	на плохом месте земли
Šukarkī ōǯ̆akin	там, где острая вершина
dundʳə dulkuktaďūn.	в самом центре земли.
ǯ̆ūrdū kultišəldū	У двух порогов [на реке]
ilītin kultišəl	на третьем пороге
minə təwəlǯ̆ənə,	меня задерживают [ловят],
diγītin kultišəl	на четвертом пороге
minə təwəlǯ̆ənə,	меня задерживают,
tuŋītin kultišəl	на пятом пороге
minə təwəlǯ̆ənə,	меня задерживают,
ńuŋītin kultišəl	на шестом пороге
minə təwəlǯ̆ənə,	меня задерживают,
nadītin kultišəl	на седьмом пороге
minə təwəlǯ̆ənə,	меня задерживают,
ǯ̆apkītin kultišəl	на восьмом пороге
minə təwəlǯ̆ənə.	меня задерживают.

Children farther
Why do you go forward?
Come here, my children!
Now there is a soul
On the sharp peak [of the mountain] of the earth
There we meet
On the wretched place of earth
There where the sharp peak [stands]
In the very middle of the earth.
At the two waterfalls [of the river]
On the third waterfall
I am held up [they catch me]
On the fourth waterfall
I am held up,
On the fifth waterfall
I am held up
On the sixth waterfall,
I am held up
On the seventh waterfall,
I am held up
On the eighth waterfall
I am held up.[5]

[5] The shaman was unable to go down the shaman river further than eight shallows: he, who went down, died—this was the explanation for the shaman's death during ceremony.

Kutur! Kutur! Kutur!	Поворот! Поворот! Поворот! [передышка]
Durəl! Durəl! Durəl!	Вверх! Вверх! Вверх идите!
Čiwinǯaldəl, duramnīlwi	Собаки мои, бегуны мои
ənəkəkər šuptilčaldra!	Не отставайте!
inəŋīkən təγinŋīlən	По направлению к свету
išičaǯakšōga . . .	бегите, не разбегаясь...
Timānīkān inəŋīlən	К началу дня
ənīwi-timānī	Матери-утра
čuninčəǯəwurtin	[всякие] звуки производят.
akīja šagditkār	Дяди старшие
čuninčəkīčiltin	[всякие] звуки производящие
ənīlə-timānī	к матери-утру
šudinčaǯawurtin	перенесут [нас]
ǯapkundū tumnəjəl	На восьми местах спуска
dōnnal, lukīčilwi!	Спуститесь [сядьте] мои стрелки!
šaduŋəkər daričimnilwi	Мелкая тварь—моя охрана
Šēŋašəl darilwi	Налимы—моя охрана,
kəštərəl darilwi	Рыба *kəštər*—моя охрана
ǯabǯarija-ənməjə	Змей—моя древняя мать—
tulunašalǯaŋā	светом горящей бересты
minə ńuŋnīlǯana	мне указывая,
išigošiktəwəw	то место, до которого дойду,
tulunak talǯana.	светом бересты освещая.

Turn round! Turn round! Turn round!
Upwards! Upwards! Upwards you go!
My dogs, my fast ones
Don't fall behind!
In the direction of light
You run, without dispersing . . .
By the beginning of day
Of the morning-mother.
They utter [all sorts of] sounds.
The older uncles
Uttering [all sorts of] sounds
To the mother morning
They carry [us]
At the eight places of the descent
Descend, my arrows!
Small creature—my protection
Burbots—my protection,
kəštər fish—my protection
Snake—my ancestress—
At the light of burning birch bark
Showed me.
The place which I reach
Lighted with the gleam of the birch bark.

ǯabdarjə ənmārjə	Змей—моя древняя мать,—
amigul Iwəlčə̄	Отец Иволчо
dōwu biŋəšiwi	Когда я был внутри
ərkə ikə̄šəlim	Так вот пел.
Ajanun ganirkil	С духом-болезни
ikə̄mə̄tcə̄wūrtin	Пели соревнуясь
dund^r*əl biŋəšiltin*	в то время, когда земли были,
jaŋramrajagdiltin	Камланье у костров устраивали
gušumrijagdiltin	. . . .
tuktipkəjagdilǯi	. . . .
jaŋrapkāwi ləkšilkānmān	Камланье—узнавание будущего
ikə̄ljəpunəŋnəm	пою я.
ējaŋan gundə̄wi	Что-нибудь сказать
əši-tālā gunǯəm	сейчас там скажу
šulakītkān čipkān	лисенок—зверенок
əši tālā ōmī!	сейчас там души!
galir biǯəǯənd^r*i*	. . . .
Goldijāgə! Goldijāgə!	Остановка! Остановка!
(goldiran əšīlə̄, ətərən jajaǯəmī, ələ̄ndə jajald^r*an).*	(Теперь остановился и перестал шаманить. По движению солнца начал шаманить).
ərə̄! ərə̄ umilgəwi,	Ой, ой, моя сова
bakučəmālākəl!	Посмотри на будущее!

The snake—my ancestress—
Father *Iwəlčə̄*
When I was inside
Like that I shall sing
With the spirit of disease
We sang competing
At the time when there were fields
They shamanized near the camp-fire
. . . .
. . . .
The ceremony—the knowledge of the future
I am singing.
Something to say
I say it now
Fox-cub—young game
Now the souls are there!
. . . .
Stop! Stop!

(He paused and ceased to shamanize. Then he started to shamanize again turning round according to the course of the sun.)

Oh, oh, my owl
Look into the future!

Hargija, məlkəjə
tuktipkə̄jə əgdindū!
tārwəwəkānŋilā
əši tālā ōmi (biši)
Tāduk turkutlilǯap
walaja! walaja!
ə̄tə̄m ŋidimərə
šinduk ŋidimərə.
Ašuja-wal bikəl
ǯəlšukjagdəwəš
kamaya tətilmə,
kamaya ətulkə,
əši šinəjəwə
ǯapkunǯi šaplalil
iniŋilāšinǯam
Dundre šunkərkədūn
tādū turkutlilǯap.
əši tālā ōmi (biši)?
əjə̄nlīl, əjə̄nlīl
əjə̄nlīl, əjə̄nlil
(hərgiški šururəu)
ənīlə̄ luŋuri
šipǯənkittulə̄

Дух-*харги*, дух-*мэлкэн*
Подымись на крутой откос!
вот туда
нет ли там души?
Потом сценимся в борьбе
Дух-*вала!* Дух-*вала!*
Я не меньше
не меньше тебя
Как бы силен ты ни был
Хитрость [вертлявость] твою,
дух болезни *тэтылмэ,*
дух-болезни *этулкэ,*
Сейчас я с тобою
восьмю саблями
биться буду.
На самой острой вершине горы
там мы схватимся в борьбе.
Нет ли там души?
Вниз по течению, вниз!
Вниз по течению, вниз!
(вниз отправился)
К матушке-ночи
к месту ныряния [к месту выхода в нижний мир]

hargi spirit, *məlkən* spirit
Rise on the steep slope!
Behold there!
Are there no souls?
Now we match our strength in battle
wala spirit! *wala* spirit!
I am not less,
Not less than you.
However strong you are
Your cunning [slyness],
Spirit of the *tətilmə* disease,
Spirit of the *ətulkə* disease
Now with you,
With eight swords
I will fight.
On the sharpest peak of the mountain
We clash in struggle.
Are there no souls there?
Down along the river, down!
Down along the river, down!
(He started downwards)
To little mother-night
To the place of dive [to the outlet of the lower world]

ičəšipǯənikɛ
jūləǯəktūn dundrə
tawar ičəwuldrən.
umūkən kultikən
tawar ičəwuldren,
gēkārkān kultikən
tawar ičəwuldrən
ilītin kultikər,
diyītɪn kultikər
tɪŋŋītin kultikər
tawar ičəwuldrə.
(šaman hərgiškī ŋənəǯərən buŋīkīttikī, tar šəwəšəl nəŋətčərə)
N'uŋītin kultišəl
tawar ičəwuldrən.
ǯapkītin kultišəl
tawar ičəwuldrən
(ororwo ilanma uŋnən čāškī bunīkīttikī, walāwān kukipkənən orordū. Bəjətpi əšila kuturərən inəŋitkī)

. . . .
На земле, где есть выход,
там показался,
один порог [на реке];
там показался,
второй порог,
там показался
третий порог,
четвертый порог,
пятый порог
там показался.
(шаман двигается вниз к миру мертвых, духи-помощники отправлены на поиски).
Шестой порог
становится видным.
Восьмой порог
там показался
(шаман отправил трех оленей-духов-помощников дальше к миру мертвых; заставил оленей поймать [ртом] духа-*вала*. Сам всем корпусом лег в направлении на юг)

. . . .
On the earth, where there is an outlet.
There appeared
A waterfall [in the river]
There appeared
The second waterfall,
There appeared
The third waterfall,
The fourth waterfall,
The fifth waterfall.
Appeared there

(The shaman went down into the realm of the dead, and sent assistant spirits out on the search.)

The sixth waterfall
will be visible.
The eighth waterfall
Appeared there

(The shaman sent three reindeer assistant spirits in the direction of the realm of the dead, he made the deer catch the *wala* spirit [with their mouths]. He himself was lying with his whole body turned to the south.)

Inəŋikə̄n šujukəwə̄	Светлый день
išičəššilǯalim	Попытаюсь я увидеть
ǯapkunma kultišəl	Через восемь порогов
əjə̄nmə̄tčə̄kilwi	через быстрое течение
gašalil-amtilwi	птицы-*гаша*—мои отцы
minə dəgdəktəlnəl!	меня перевезите!
Timānīkān šujurənmən	До начала утра
išičaššilǯalim.	Попробую дойти.
ənīwə-timānī	Матушку-утро
tawar əlbənəldʳən.	тот начал закрывать
tɛrwə mōkānŋilā.	столбом-деревом.
Timātnə ōǯakin,	Когда утро настанет
tādū ičəčilǯəp	там увидим
ənīwə-timātnə.	Матушку-утро.
Čuninčəǯəksōga	Производить звуки станем
dapkašilkīttuwar	на том месте, где плотно соединяются
timānī ōdakin,	когда утро настанет,
timākta ōdakin	когда утро настанет
ǯapkuna inəŋil	За восемь дней
išičaǯaksōgdɛ! . . .	дойдем мы...
(šamanmə uγirrə uγiskī	(Шамана подняли вверх, бубен его
uŋtuwūnmān duwuldʳa . . .	перетащили... Шаман пришел на
Šaman əmərən inəŋitki,	юг, сел. Начал курить. Покурил.
təγərən. Tāldʳən. Tānən.	Принесли три посудины с мясом.

On a bright day
I try to see
Through the eighth waterfall
Through the swift course
gaša birds—my fathers
Carry me
Until the break of the day
I try to arrive.
Little mother morning
He began to lock up
With a wooden log.
When morning arrives
There we shall see
Little mother morning.
We shall utter sounds
On the place where they fuse closely
When the day breaks
When the day breaks
In eight days
We shall arrive . . .

Ilanma tiγəlwə uldrəwə əməwrə. Mušuγəldʳən, horolišnən umū-kə̄nmə tiγə. Šaman, šəwəšəlin musulgəldʳe, šāplətpi minə̄kə̄t-čənə, čokčigakātčina. hišakil-watin oror šimičildʳən šaman, šəwəšəlin. Jajaldrən ńāni. ete-rən. Sunmi luktan, həlmīwi luktan, šərgəmiwi luktan, ijəg-wi luktan. Šuwildʳa tāləγiwə. Inməkildū šuwšinə. ətərən jaj-aǯamī.)

Он обошел вокруг одной чаши. Шаман и духи-помощники начали воевать, саблей рубить, подрубать под корень. Шаман и его духи-помощники, лопатки оленей начали грызть. Опять начал шаманить. Кончил. Снял кафтан, снял нагрудник, снял ровдужный венец, снял наспинную пластинку с рогами *иег*. Развязал опояску. Все уложили в сумку. Кончил шаманить.)

(The shaman was lifted up, his drum carried over . . . The shaman arrived at the south, he sat down. He began to smoke. He lit his pipe. Three vessels with meat were brought. He walked round a drinking vessel. The shaman and the helping spirits began to fight, cut with swords, stab under the root. The shaman and his helping spirits began to gnaw at reindeer's shoulder-blades. He started again to shamanize. He finished, took off the caftan, took off the breast plate, took off the head-cover made of hide, took off the back-cover with antlers [*ijəg*], untied the belt, put all this into the bag. He finished the performance.)

2. Informant: Legurča Bajakī, July 1940
Sym river, recorded by: N. P. Nikuljšin. Song recited during the ceremony for success in hunting *šiŋkə̄lə̄pkə̄* or *šinkə̄lə̄wūn*, in the period called *uwun*—when the squirrels were hunted onfoot.

Šiŋkə̄līlǯə̄kilwi
ulukīl əkīntin
ənə hunaktaldʳa,
nimŋakān biŋəši-n
itiwūkākunin.
Tar-da-gla totonin
ši-də jogo ōǯakal,
minə nuŋnilǯəkəl.
nudur kamījawar!

Обряды поисков охотничьей удачи
тетка белок
без задыхания [от бега]
в старину
установила.
Тогда же добычу
ты же предскажи,
мне укажи
зверя, которого добуду!

The ceremonies for hunting luck
Auntie Squirrel
Without being out of breath [from running],
In days of old
Were fixed.
Then the catch
Must be predicted by you,
Show me
The game I shall kill!

Nimŋakān gučəǯən
mēlilǯaŋāšiwər,
dundrə himirdukin
hata bagar bičə̄n
mēlilǯaŋāšiwər.
šiŋkə̄n čuŋurə̄wə̄n
əjə̄činiliŋnəm.
Ējaŋan gundə̄wi!
jandraljapuŋaŋnəm

čuŋurə̄lə̄n dundrə
ənīlə̄-timānī
čuninčəkīčilwi
ejaŋa gundə̄wi!
Ikə̄ljəpuńəǯəm
jandrapkājəgdilǯe
tintipkə̄jəgdilǯe
tɛrwəwakānŋila.

Amīγu Dulinča
tawar šarilānən
tɛrwəkānŋilā

В старину завещено
в то время, когда [все] оживало,
после того, когда появилось земля.
Всегда так было
в то время, когда [все] оживало.
Самую суть удачи
начинаю я просить.
Что-нибудь скажи!
Шаманить [обращаться] у костров хочу
к центру земли,
к матери-утру.
Писком животных
каким-нибудь отзовись!
Я шаманить хочу
камланьем у костров
отпусканием в разные стороны,
в дали, где растет любимая оленья трава.

Отец-предок Дулинча
тот отправился узнать
туда, где растет любимая оленями трава

It was deemed in olden times
When [everything] came to life
After the earth had appeared.
It was always like this
When [everything] began to live.
For the essence of success
I begin to ask.
Say something!
At the camp-fire I want to shamanize
[to appeal to you]
To the centre of the earth
To mother morning.
With the search of animals
Manifest yourself somehow!
I want to shamanize
Doing magic at the fire in the open
With releasing to different sides
Far away where the reindeer's favourite grass grows.
Forefather *Dulinča*
He went to discover
Where the reindeer's favourite grass grows.

Amar Šinājətən	предок Шинаетэн
mānəkājə gulkəl!	сам скажи!
irkəkudušalwaw	Добычу
minə nuŋnilǯənə	мне указывая,
išigəšiktəwəw	места, куда я дойду
abaderokōkun!	покажи!
ətəγāt! ətəγāt!	Прекратим! Прекратим! [Остановился для передышки].
Amar goldišilnəl	Предки, двигаясь по кругу,
ējagda gulǯəlmī	что-нибудь если и скажете,
əšim ilbən biši	я не из пугливых [меня не вспугнете]
dulbaralǯakaldrə!	тихо идите!
Hargijal, məlkojəl	Духи-*харги*, духи-*мэлкэны*
goldišiǯakaldrə!	двигайтесь кругом!
ējagda guldilmi	что-нибудь если и сказать
goldija, goldija!	кругом, кругом!

Our ancestor *Šinājətən*
You yourself must tell me
The catch
By showing me
The place I shall reach,
Show it!
Let us stop! Let us stop!
[He stopped to draw breath.]
Ancestors, circling around,
Do say something,
I am not easily frightened
[You do not frighten me]
Come quietly!
hargi spirits, *məlkən* spirits,
Circle around!
Whatever you say
Around, around!
[The shaman finished the ceremony.]

3. Informant: N. I. Bajaki, July 1940
Recorded by: N. P. Nikuljšin. Song recited during the hunting ceremony, *ikānipkā*

durulə, durulə	
durulə, durulə	
melija, melija	
melija, melija	
melija, melija	(18 lines are not translated)
melija, melija	(after a pause to take breath)

čuŋkeš, čuŋkeš	
čuŋkeškald^{r}*ə*	
čuŋkeš, čuŋkeš	
čuŋkeškald^{r}*ə*	
harandijan, harandijan	
harandijan, harandijan	
gujadakān, gujadakān,gujadō!	
gujadakān, gujadakān, gujadō	
gujadakān, gujadakān, gujadō	
jadu, jadu, jadugō!	
jaduje, jaduje,	
jaduje, jaduje	
əlātinǯəm, əlātinǯəm,	По солнцу начинаю движение в круге,
əlātinǯəm, əlātinǯəm,	По солнцу начинаю движение в круге,
elātinǯəm, əlātinǯəm	По солнцу начинаю движение в круге,
mawudikān, mawudikān,	все, как аркан, выравнявшись,
mawudikān, mawudikān,	все, как аркан, выравнявшись,
mawudikān, mawudikān,	все, как аркан, выравнявшись,
ulkunǯajal, ulkunǯajal,	все вместе сцепившись руками,
ulkunǯajal, ulkunǯajal,	все вместе сцепившись руками,
ulkunǯajal, ulkunǯajal,	все вместе сцепившись руками,
kutušilgə, kutušilgə,	По кочкам болот-топей,
kutušilgə, kutušilgə,	по кочкам болот-топей,
kutušilgə, kutušilgə,	по кочкам болот-топей
hənrijəjəl, hənrijəjəl,	быстро бежит [небесный олень]
hənrijəjəl, hənrijəjəl,	быстро бежит [небесный олень]
hənrijəjəl, henrijəjəl,	быстро бежит [небесный олень]
ǯurandājəl, ǯurandājəl,	быстро несется,

I begin to turn round according to the sun,
I begin to turn round according to the sun,
I begin to turn round according to the sun,
All are straightened in a line like a lasso,
All are straightened in a line like a lasso,
All are straightened in a line like a lasso,
All are linked by the hands,
All are linked by the hands,
All are linked by the hands,
On the moor of the marsh,
On the moor of the marsh,
On the moor of the marsh,
Runs quickly [the heavenly reindeer],
Runs quickly [the heavenly reindeer],
Runs quickly [the heavenly reindeer],
It runs quickly,

ǯurandājəl, ǯurandājəl,	быстро несется,
ǯurandājəl, ǯurandājəl,	быстро несется,
čumurkə̄γə̄, čumurkə̄γə̄,	по кочкам пойдем!
čumurkə̄γə̄, čumurkə̄γə̄,	по кочкам пойдем!
čumurkə̄γə̄, čumurkə̄γə̄!	по кочкам пойдем!
Kutušilgə̄, kutušilgə̄	По болотым, топям
gilgəjə-čipkājə	олень-*гилгэ*—зверь лесной
hənrišindəkjəgdən.	длинными прыжками-галопом бежит.
Nikākār, nikākār	Малыши, малыши [к присутствующим]
turgandīnānənəl	будьте все вместе,
ənəl girānərə!	Не разбегаясь [не расходясь]!
Kiragin oŋkojikol!	на косогоре [олень] пусть есть мох!
Nikājə, jambukūkān	Малыши, потише! [не вспулните]
gilgəjə čipkaja,	бычка зверя,
dulbakšilčā.	тише пошел.
	[четыре дня гонятся за воображаемым небесным оленем]
ələ̄tinǯəm, ələ̄tinǯəm.	Двигаюсь по направлению движения солнца
Nikāje ələ̄tilnəl!	Малыши, по солнцу, двигайтесь!
ələ̄nmuǯəm	По солнцу идти хочу
ələ̄nmuǯəm	По солнцу идти хочу
Nikāǯəl, šintalnal	Малыши, охраняя

It runs quickly,
It runs quickly
Let us go on the moor!
Let us go on the moor!
Let us go on the moor!
Across the swamp, in the thicket,
The *gilgə* reindeer, the forest game
Runs with long strides, galloping
Children, children [to those present]
Keep always together,
Do not scatter,
On the hillside let him [the reindeer] eat moss!
Children, more quietly! [Don't frighten him]
The wild bull-calf
Went quietly.
[For four days they run
after an imaginary heavenly
reindeer]
I go in the direction of the sun's course
Children, go with the sun!
I want to go according to the sun
I want to go according to the sun
Children, who guard,

ələtilǯəkəldʳə!	двигайтесь по солнцу!
ənīlə-timanī	К матушке-утру
ələtikšonīwar!	направьте свой путь!
Tawar ajakādɛnǯi	Так хорошенько
duligiǯəŋātpər!	тихо, смирно [двигайтесь]!
Tawar gilgəlǯənəl	Так гонясь за оленем-гилгэ
hargijal-məlkəjəl	духи-мэлкэны,
hargijal-dolbojol	духи-ночные
duligiǯəŋātpər	тихо, смирно [двигаясь]
haričāŋnakaldʳə!	следите [за оленем]!
ədikər, ədikər!	Ветры, ветры! [название духов-помощников]
əši dundʳa ōčāl	Теперь земля стала
šuwušin kɛlbumə.	вихрь страшный.
(šuγurə inəŋitki)	(пошли на юг)
ələtinǯəm, ələtir	По солнцу двигаюсь, по солнцу
Nikākār, nīkākār	Малыши, малыши!
ələγətinǯəkəldʳə!	двигайтесь по солнцу!
ələtinǯəm, ələtir	По солнцу двигаюсь [по солнцу] *элэтыр*
Inəŋīkən ələtir	светлый день *элэтыр*
šiγuranmān ələtir	солнца свет *элэтыр*
šonmukšalǯakaldʳə	Хорошенько увидьте

Go according to the sun!
To little mother morning
Direct your steps!
Like this, gently
Quietly, taking care [you must go!]
A-chasing the *gilgə* reindeer
məlkən spirits
Nocturnal spirits
[Going] quietly, taking care
Follow [the reindeer]!
Winds, winds! [Name of the ancillary spirits]
Now the earth appeared,
A terrible storm.
(They went south)
I am going according to the sun, [with the sun]
Children, children!
Go with the sun!
I am going according to the sun
[with the sun *ələtir*]
Bright day *ələtir*
The light of the sun *ələtir*
Look at it closely,

šintalnal hurilwi	оставшиеся дети мои
timānīkān ələ̄tir	утра *элэтыр*
mujələšpən ələ̄tir	границу *элэтыр.*
ənəkəkər ələ̄tir	
mikəlmirə ələ̄tir	Не потеряйте *элэтыр*
tarwāgəwə kōnnilā	в том самом месте догнал
tawargilā ələ̄tir	в том самом месте *элэтыр,*
timānīkan əlˉtir	когда утро *элэтыр*
šujunrəkin ələ̄tir	зародится, *элэтыр*
tarkalagla ələ̄tir	там вот *элэтыр*
kēlbanaldran ələ̄tir	светить начнет *элэтыр.*
hargi bəjur ələ̄tir	Лесной олень—зверь *элэтыр*
dunaraŋinran ələ̄tir.	след оставил *элэтыр.*
Ašulaŋan ələ̄tir	Небольшой след *элэтыр,*
tawaldāwər ələ̄tir	чтобы дышать *элэтыр,*
əŋəŋədū ələ̄tir	но в погоне *элэтыр*
tirgawuŋnam ələ̄tir	задержались *элэтыр*
hargi bəjun ələ̄tir	олень-зверь *элэтыр*
durgaŋimran ələ̄tir	от испуга бросился бежать *элэтыр.*
Tārwāɣəwa konila	Того вот
tarwāgə konila ələ̄tir	того вот *элэтыр*
hitalanin ələ̄tir	вдали еще заметил *элэтыр*
Durundəšəl, durundo	. . . . [запев]

My children who stayed here,
Of morning *ələ̄tir*
The boundary *ələ̄tir*
Don't lose *ələ̄tir*
On the same place
I caught it up
On the same place *ələ̄tir*
When morning *ələ̄tir*
Is born *ələ̄tir*
Behold, there *ələ̄tir*
Begins to shine *ələ̄tir.*
The forest reindeer — the wild *ələ̄tir*
It left a foot-print *ələ̄tir*
A small spoor *ələ̄tir*
That it should pant *ələ̄tir*
But in the chase *ələ̄tir*
We remained behind *ələ̄tir*
The reindeer—the wild *ələ̄tir*
Frightened began to run *ələ̄tir*
Behold him
Behold him *ələ̄tir*
I still noticed him far away *ələ̄tir*
. . . . [refrain]

durundəšəl, durundo	....	
durundəšəl, durundo	....	
Hargi bəjun durəŋinrə	Оленя зверя погнал,	
ašulaŋan təwuldāwər durando	где же догнать	
durundəšəl durundo	....	[запев]
durundəšəl durundo	....	
tāʳwəwəkān durundo	Вот там *дурундо*	
hitalanin durundo	мелькнул *дурундо*	
īdūgilə durundo	где же *дурундо.*	
īdūgilə təwuldə̄wər?	Где же догнать?	
inəŋədū tirgawunam	К свету двигаюсь я,	
irjə(kə̄g)lə̄, irjə(kə̄g)lə̄	где же, где же	
durundušəl, durundo	....	[запев]
Nika, tutuščəlnəl,	Малыши, двигаясь	
durundušəl, durundo	....	[запев]
timānīkān durundo	Утра *дурундо*	
igandulān durundo	к свету *дурундо*	
ədāwər durundo	чтобы не *дурундо*	
mìkātčərə durundo	ошибиться *дурундо*	
hargi bəjun durundo	олень зверь *дурундо*	
duraŋildʳən durundo	быстро пошел, *дурундо*	
Tirgakšan durundo	Полдень настал *дурундо*	
Nikal tāŋəǯəγāt!	Малыши, давайте передохнем!	

. . . .
. . . .
I chased the wild reindeer
Where shall I catch him?
. . . . [refrain]
. . . .
Behold there *durundo*
It flashed *durundo*
Then where *durundo*
Then where to catch it?
I go towards the light
Then where, then where
. . . . [refrain]
Children, moving
. . . . [refrain]
In the morning *durundo*
Towards light *durundo*
In order not to *durundo*
Make a mistake *durundo*
The wild reindeer *durundo*
Moved quickly *durundo*
Noon has come *durundo*
Let us rest, children!

4. Informant: I. P. Bajaki, July 1940
Recorded by: N. P. Nikuljšin. The shaman's song during the *ikǝ̄nipkǝ̄* ceremony.

oldur, oldur, oldurəl,	.... [запев]
oldur, oldur, oldurəl,	....
ilər, ilər, ilərəl,	По ходу солнца, иди,
ilər, ilər, ilərəl,	По ходу солнца, иди,
gidu, gidu, gidurəkəl!	двигайся, двигайся!
šiŋə, šiŋə, šiŋərə̄ǯəm	....
gārəša-al, gārəša-al,	Филины, филины [духи]
gārəša-al, gārəša-al,	Филины, филины [духи]
gārəša-al, gārəša-al,	Филины, филины [духи]
hiru, hiru, hirukəld^{r}*ə,*	заговорнте!
hiru, hiru, hirukəld^{r}*ə*	заговорнте!
jand^{r}*a, jand*^{r}*a, jandrakald*^{r}*ə,*	Начните камлять у костра!
jand^{r}*a, jand*^{r}*a, jandrakald*^{r}*ə,*	Начните камлять у костра!
hirukəl, hirukəl,	Говори, говори!
hirukəl, hirukəl,	Говори, говори!
hirukəl, hirukəl	Говори, говори!
nororol, nororol, nororoga-al,	Поднимайся кверху!
nororol, nororol, nororoga-al,	Поднимайся кверху!
nororol, nororol, nororoga-al,	Поднимайся кверху!
čulga, čulgataγā,	даль синеет
čulga, čulgataγā,	даль синеет

. . . . [refrain]
. . . .
Along the course of the sun, go,
Along the course of the sun, go,
Move on, move on!
. . . .
Brown owls, brown owls [spirits]
Brown owls, brown owls [spirits]
Brown owls, brown owls [spirits]
Secret allies!
Secret allies!
Begin to shamanize at the fire in the open!
Begin to shamanize at the fire in the open!
Speak, speak!
Speak, speak!
Speak, speak!
Rise high!
Rise high!
Rise high!
The distance grows blue
The distance grows blue

čulga, čulgataγā,	даль синеет
ńēkši, ńēkšišagā,	[спустился] вниз [дух]
ńēkši, ńēkšišagā,	[спустился] вниз [дух]
ńēkši, ńēkšišaga,	[спустился] вниз [дух]
čuŋurǯəl, čuŋurǯəl, čuŋurǯel,	кубарем, кубарем,
čuŋurǯəl, čuŋurǯəl, čuŋurǯəl,	кубарем, кубарем,
čuŋurǯəl, čuŋurǯəl, čuŋurǯəl	кубарем, кубарем
nəkərəl, nəkərəl, nəkərəγə̄!	. . . .
ədir, ədir, ədiril, ədiril,	вниз, вниз
čuŋurjəkə̄n, čuŋurjəkə̄n, čuŋurjəl.	кубарем, катится.
tunturjəkə̄n, tunturjəkə̄n, tunturjə	Подымаются [духи]
golta, golta goltoko-ol	пересекая пути друг у друга
uwəldiškān, uwəldiškān	по волнам вверх,
təmilgə, təmilgə, təmiliγə̄-ə̄	по занесенной снегом [земле]
kurəlǯākən, kurəlǯākən	ругая дух болезни
kurəlǯa, kurəlǯa,	. . . . [припев]
kurəlǯa, kurəlǯa	. . . .
girijalil uldigō,	Шагая вдоль сухого леса,
girijalil uldigō,	шагая вдоль сухого леса,
girijalil uldigō	шагая вдоль сухого леса,
dušələnǯəm, dušələnǯəm,	По узкой долине иду я,
dušələnǯəm, dušələnǯəm	по узкой долине иду я
durəl, durəl, durəlkəldrə	по берегу реки идите

The distance grows blue
Down [came the spirit]
Down [came the spirit]
Down [came the spirit]
Like a peg top, like a peg top,
Like a peg top, like a peg top,
Like a peg top, like a peg top,
. . . .
Down, down
Like a peg top, it is whirling.
They [the spirits] rise
Crossing each other's way
Upwards on the waves,
On the snow-drifted [earth]
Cursing the spirit of disease
. . . . [refrain]
. . . .
Walking down the dry forest,
Walking down the dry forest,
Walking down the dry forest,
Along a narrow valley I am going,
Along a narrow valley I am going,
You go along the riverside

šabgalilǯal, šabgalilǯal	по следам ....
šabgalil, šabgalil,	....
šabgalil, šabgalil	....
gēkārəkəl, gēkārəkəl,	по другим следам иди,
gēkārəkəl, gēkārəkəl,	По другим следам.
gēkarəkəl, gēkarəkəl,	По другим следам.
kiləǯakən, kiləǯakən, kìlə,	По одному следу [пошли],
kiləǯakən, kiləǯakən, kìlə,	по одному следу [пошли],
kiləǯakən, kiləǯakən, kìlə	по одному следу [пошли]
kurnalāšəl, hirugal,	К свету пошел, говоря...
kurnalāšəl, hirugal,	К свету пошел, говоря...
kurnalāšəl, hirugal ...	К свету пошел, говоря.....

Following the footprints ...
....
....
Follow other footprints,
Other footprints
Other footprints
One of the footprints [was followed]
One of the footprints [was followed]
One of the footprints [was followed]
Came to light, while he was saying ...
Came to light, while he was saying ...
Came to light, while he was saying ...
etc.

5. Informant: Šomirča Kima, June 1930
Sym river. Recorded by: G. M. Vasilevič. The song of a shaman of Sym during the *ikōnipkō* ceremony.
(On the first day of the ceremony)

Tākiləl! Tākiləl!	Кузнецы! Кузнецы!
Adī tawičimnīlwi	Сколько кузнецов у меня
tākilǯəčāl bəjəlwi?	ковавших мужиков?
ēlaw? — Dagujiləw.	Что ковали для меня?—Наспинные рога
Adī kurgəmnīlwi	Сколько раздувателей мехов,
ǯašanal, ǯajačāl	Изготовителей металлических частей

Blacksmiths! Blacksmiths!
How many blacksmiths have I
Who forge men!
What have you forged for me? Horns on my back.
How many bellows,
How many manufacturers of metal parts?

ǯaša, ǯaša, ǯašāləl
urāšəmrə ǯašāləl
tāksəmigǯem ǯašāləl
əmirdiptəǯənə̄ ǯašāləl
ǯašā, ǯašā, ǯašāləl!
Adī ǯawaŋkīlwi,
ənəl mušunmurə!

Adī halkarwi ǯašāləl?
Adī əjurgəlwi,
ənəl mušunmurə?

ōraǯačal ašāl
Iŋašalwa ōǯačāl
kiktakārə ōǯačal
dəwəkšə, jəndrəγi
ənəl hučulkurə
talunašinǯačāl . . .

əlōr, elōr, əkəl-əl-əlul!
ŋinakir šiŋtalnal

Железо, железо, железо [припев]
Железные опилки *дяшалэл*
соберу я *дяшалэл*
делая их более острыми *дяшалэл*
. . . . [припев].
Сколько клещей у меня,
которые не поддаются духам *мусун?*

Сколько молотов у меня? дяшалэл
Сколько клещей у меня,
которые не поддаются духам *мусун?*

Работавшие женщины,
швы с подшейным волосом делавшие
швы с закусом зубами делавшие
сурик и черную краску
не портя,
размягчая...
[На второй день обряда]
[Призыв двигаться в круге по
солнцу и гнать оленя].
Собаки, не останавливаясь,

Metal, metal, metal [refrain]
Iron filings *ǯašāləl*
I am gathering *ǯašāləl*
Making them much sharper *ǯašāləl*
. . . . . [refrain]
How many tongs have I
Which do not yield to the *mušun* spirits?
How many hammers have I? *ǯašāləl*
How many tongs have I
Which do not yield to the *mušun* spirits?
Women who have worked,
Making stitches with dewlap hairs
Making stitches with bites of their teeth
The red lead and the black paint,
Not spoiling,
Softening . . . .
[On the second day of the ceremony]
. . . . . .
[Summons to move around according to the course of the sun and to chase the reindeer.]
Dogs, not stopping,

ələ̄lilǯəkəldʳu.	двигайтесь [гоните] в направлении движения солнца
Hargija-bəjujə!	оленя лесного, зверя!
Diγinrə̄čə̄n bičə̄n	Четырехдневным был
əŋnəkə̄ndin očā̄n,	с теленка стал,
ēpkāčāndin očān	с двухголовалого оленя ростом стал.
ələ̄ntiǯə̄ktin dundʳə	Земля, по которой гонят
ulgənməγiwə̄n	среди ряда деревьев
ələ̄lilǯəkəldʳu! . . .	Гоните!... [и т. д.]
	[перед перерывом, предлагается устроить окружение]
əγəlgə, əγəlgə!	Окружение! Окружение!
əγəlgəǯəkəldʳu!	Окружайте, окружайте!
hargiwa-bəjunmə	лесного оленя—зверя
əγəlgəǯəkəldʳu!	окружайте!
Ičəwlə̄mīl bəjəl	Бодроствующие мужики
əγəlgəǯəkəldʳu . . .	в окружении держите!...
	[на пятый день обряда]
Dundʳə, dundʳəkəkūn!	Земля, землище!
Šuwušin, tamnakša	Вихрь ты мой, туман
gēmakāǯe, gēmakāǯe!	изменись, изменись!

Go ahead [chase] along the course of the sun
After the forest reindeer, the wild!
It was four days old,
From the calf it became,
From the two-headed reindeer it became full-grown.
The earth, where it is chased,
Among the multitude of trees
Chase it! . . . [etc.]
[Before the pause, circling
is to be begun]
Encirclement! Encirclement!
Encircle him, encircle him!
The forest reindeer, the game
Encircle him!
Men awake
Keep him in the circle! . . .
[On the fifth day of the ceremony]
Earth, big earth!
My storm, my fog
Change, change!

Ičətmī-nun, irgədəšukə̄nmə
ǯəlgəšunŋijəwər
Məlkušujəwər-bəjəkə̄kūr
ǯəlgəšunŋijəwər
irgədəšukə̄nmə
hirurəŋnəktin!

Так, чтобы увидеть где-либо
близко друг к другу стоящих
духов *мэлкэнов*-мужиков
близко друг к другу стоящих
где-бы то ни было,
пусть заговорят! [пусть подадут голос!]. . . .

So that we may see something,
Standing close to each other,
The spirits of *məlkən* men
Standing close to each other
Wherever they are,
Let them begin to speak! [let them hear their voice] . . .

REFERENCES

Shirokogorova, E. N. (1934) Folk Music in China, in *The China Journal of Science and Arts*, Vol. II, Shanghai

(Vasilevič, G. M.) Василевич, Г. М. (1936) *Сборник материалов по эвенкийскому (тунгусскому) фольклору* (Collection of Material in Evenki [Tungus] Folklore), Leningrad.

(1957) Древние охотничьи и оленеводческие обряды эвенков (Ancient Hunting and Reindeer-Breeding Rites among the Evenki), in *СбМАЭ*, Vol. XVII.

A Nanai (Gold) Tale about the Fortieth Brother and His Wife, a Washbear (*Ursus lotor*)

by

V.A. Avrorin, Novosibirsk

The Nanai (Gold) tale (*ńiŋmã*) published below was recorded by me on July 8, 1948, in the settlement of Dairga, Nanai District, Habarovsk Region. My informant was an eighty-year-old Nanai, Pintuka, of the Oninkan family, who had lived in this settlement ever since his birth. Judging by the peculiarities of the speech of my informant, he spoke the Naikhin dialect which underlies the literary Nanai language yet displays some characteristic traits of the dialects (Bolon and Garin) spoken further north. It almost completely ignores the diphthongization of suffixes in the oblique cases of the personal possessive declination.

The tale clearly shows the remnants of shamanism connected with animistic conceptions, and the belief in various miraculous metamorphoses. These elements of obsolescent ideology are interwoven with later elements evincing a critical attitude toward shamanism expressed with undisguised irony at the expense of the professional shamans. Such an interlacing of beliefs and scepticism is characteristic of many creations in Nanai folklore.

The *mergen* is a strong, brave, clever, quickwitted young man. He is always the most skilful and successful hunter and fisher. Corporeal beauty is united with high moral qualities in him. By hyperbolizing these human qualities the creators of folklore have endowed the *mergens* with a power to perform actions beyond the possibilities of both ordinary men and shamans.

Nevertheless, the *mergen* is usually a generalized and concentrated personification of the qualities of real people. Naturally not all of them use their qualities for good purposes. The principal heroes who are always positive, often must come into collision with ill-willed *mergens* or *pudins* against whom they fight sometimes a long and stubborn battle. The final victory there, as also in the fight with the evil spirits and shamans, lies always with the positive hero. In his battles with the enemy the *mergen* is usually supported by a woman (*pudin*)—his chief wife, or elder sister, who often appears in the form of some animal. She encourages the *mergen*, guides his actions, heels his wounds. This role of hers is reminiscent of the function of the female spirits helping the shamans.

An almost compulsory element in the Nanai tale is the revenge on the evil spirits for the death of the father or, less frequently, of some kinsman of the same clan. This motif is usually the central idea of the composition, penetrating into the episodes, and serves as a motto for the hero in his actions. While the text is related—and it takes sometimes two or three evenings—this motif may fade into the background or disappear altogether; yet it reappears later to throw a light upon some earlier actions of the characters in the tale. In a great number of tales about male heroes the thirst for vengeance is associated with the *mergen*'s endeavour to acquire as many wives as possible. The *mergen* marries the lonely *pudins*, he encounters on his way, as well as the daughters of his friends and of his defeated foes. As a dowry he obtains whole villages, which he moves over to his homeland.

The heroine (*pudin*) is always a striking beauty, extremely skilful in sewing and decorating costumes and in making utensils and in cooking. In addition to this, she is very often represented as extremely bold, a good shot and deft in hunting and indefatigable in skiing. The *pudin* is practically always endowed with the gift of foresight, the faculty of transfiguring herself and the power of performing other supernatural actions.

Many a tale with a female creature as a leading character is founded on the motif of choosing a worthy husband. The *pudin* proposes to her wooers to compete with her in the most difficult arts of hunting, shooting with the bow, skiing and in other sports regarded by the contemporary Nanai as purely male activities. After many victories, the *pudin* meets a *mergen* who defeats her and becomes her husband. In some tales the *pudin*, like the *mergen*, fights a single combat with evil spirits to take vengeance for the death of her father. From among all the Tunguso-Manchurian peoples, the Nanai alone have preserved a greater number of heroic tales about female heroes and woman hunters.

In the tales with male leading characters the *pudin* plays the more modest role of a helper, sometimes tutoress, of the *mergen*, as has been said above.

The text here published is recorded in a transcription of distinctive sounds (phonemes) relying on the system of the International Phonetic Association. Punctuation marks are used for convenience in reading. Proper names are written with capital initials. The stress is not marked since it always falls on the last syllable of the word.

ńeŋmã.

1. *əm ʒōgdo gɔxi nī bālʒexã. təi naisal pulsīči, xaewadā wāri, n'əpultə, tō, bəjū wāri. totamdā ču pojaŋgon'i ču əgdi wāri. gə, təi pojaŋgon'i sīksə əuguxən'i — toeŋga āŋn'i buikiči. ʒea in'iwə sīksə əuguidun'i guči toeŋga āŋn'i buikin'i. tui burməri, toeŋga toeŋga buikiči, tui burməri, xəm buikiči. totapmāt xəm burbūčiči n'oan'i murčixən'i: ʒeaselbewa xəm wāxã, mimbǣdə̄ wāʒāma.*

2. *ʒōgdo kaɔka indəxən'i. əm dolbon'i aoŋgaxan'i, xaedā aba. totapmāt ʒea in'iwə tui bičində̄, ənūri xāʒombi, sōktaji xəm ʒapara, waesi əuxən'i sin'čijə̄də̄. totara elera, xəji ičəxən'i. n'oan'i əuxəmbi xəʒiələn'i ičəxən'i piurū-u bi poktowa taosi ənəpimə̄t ičəxən' naoto poktowan'i. čāwa ičəpi, murčixən'i: mī āŋnaewa wāxã ambã pokton'i biʒərə.*

3. *totara čāwa xoʒami ənəxən'i. baese daopi, boačando seo dōlan'i təi naoto aorewan'i bāogoxan'i. bāogopi, wāgoe daramadoji bī surəwə ačora, kərkiləgui əuriūčin'i, təi naoto nae xəsəʒīn'i uŋkin'i: aga, mimbǣwə əʒǣ wāra! mī simbǣ soŋgoe biʒə̄rəm ʒicimbi bičī. əci ənīwəsi sāra ənuji. simbǣ wāras xōntaori, nāŋgalami kaerã. naoto uŋkin'i: xə̄ŋguʒiji uirə, omoldolē uiru. mə̄rgə̄ uŋkin'i: maŋgagban'ǣ! simbǣwə xōn'e kutučiuri? kutučidujəwə čādo wāŋāstān'e.*

4. *təi xamealī uirə, mə̄rgə̃ māŋbowa ənəxən'i. kutučiurəsi. mə̄rgə̄ murčīn'i: māŋbodo semata oen'i bimi, min'ʒi gəsə ənəin'i. toetara n'aron'či tōxan'i.*

mə̄rgə̄ maselaxan'i. na͜ototān'i, n'o͜an'i xōldondolan'i dāremi, n'o͜amban'i kutučīn'i. mə̄rgə̄ murčīn'i: n'arondo semata n'o͜amban'i xətəmi, mimbə̄ewə kutučīn'i. to͜etara se͜o dōčen'i ənəxən'i. tādo maselōčan'i, na͜oto bāǯe maŋgaǯi n'o͜amban'i kutučīn'i.

5. to͜etara čexan'i osego͜anda, māŋboči ə͜uguxən'i. māŋboči ə͜ugupi, sīksəgu͜əndə a͜oŋgēči. a͜oŋgēdoje mə̄rgə̄ pimuwə, gi͜uwə wāxan'i. to͜etara selomba seločexan'i. na͜otoŋgoǯeje gəsə se͜axan'i selomba. ǯe͜a čeman'awa tə̄rə, susuxəči. n'aŋga ənəmi, xāme͜alē u͜ixən'i na͜otoŋgon'i u͜ičəwə čerambōčen'i, xamase ičəguxən'i. na͜oto n'ōŋge͜ā ǯukəwə pūr enə͜iǯə̄en'i turiŋkin'i. mə̄rgə̄ murčīn'i: čisən'iwə goroǯi ǯidə, ə͜in'ə̄ewə jadaxan'i. totara na͜otoči undīn'i: gə̄, əsikə̄ simbiwə wāwori! surəji u͜isi ə͜urixən'i. tādo na͜oto undīn'i: aga, mī čisən'ə̄ewə tu͜i goroǯi ǯimi jadāčendolaji əsi xa͜e jadaǯāmbe? mī mədəwə dōlǯexambi, sin'či gisurəguji ka͜oǯara͜oŋkečaji. xa͜e mədə̄? — undīn'i.

6. na͜oto undīn'i: mī, ə͜i čeman'ewa tə̄mi, tu͜i bi mədəwə dōlǯexambi. bu͜ə siũ tokando eseǯāpu əm dāji irgəmbə, bāldo͜e bajã irgəmbən'i. tə͜i bāldo͜e bajã asen'i əgdi. piktəgu͜i təŋ əmũ aselaji əm əktə piktəwə bāxan'i. tə͜i piktən'i ənusīn'i. ənūn'i tu͜i bīn'i: ujundulən'i bəgdin'i uliksəwən'i pe͜alačemi se͜arēn'i. xa͜edā se͜arēwan'i sāwase. bāldo͜i bajã, əgdi samamba gələrə, ǯōgdo͜e nəu͜čə͜in'i. tə͜i se͜are ǯaka se͜andaso͜ewan'i əm samandā sāraseči. se͜ara ənūčin'imət, samasal morēči. bu͜ə ǯakpalači ənə͜idupuwə na͜e simbiwə ičəpəri agdaǯāra: ə͜i na͜emāt tə͜i piktə ənūwən'i xorejam. aga, sī xa͜eǯi xorego͜e tīsi?

7. mə̄rgə̄ undīn'i: mī xa͜eǯi xore͜ori? si͜ur ənəǯə̄pumə. na͜oto undīn'i: aga, si͜ur ənəsisi. to͜edu əgdi na͜ji ətūri. čāwā xore͜orewa murči osen'i, mimbi͜əwə tuŋgələru. tu͜i tara na͜oto ǯelelaji pačelaxan'i. əm enda ǯelen'i očen'i. mə̄rgə̄ ǯapara tuŋgələxən'i. tu͜i tara ənəmi, si͜un tokondo irgəmbə esēdon'i to͜edo ǯu͜er na͜ji čome͜ačera, durbi͜əčirə ičəǯīči. tu͜i tara tə͜i na͜esal du͜isi morēči: bāldo͜e bajã, ǯōgdo͜e əǯ tə̄sirə! māŋbo sole͜aǯe͜an'i əmu na͜ji ǯidiji, tə͜i na͜ji sī piktəwəsi xoreǯāra.

8. mə̄rgə̄ to͜ewan'i esēn'i. bāldo͜i bajã, u͜ikələji ńi͜ərə, mə̄rgə̄ ŋāladon'i ǯapagora, ǯōkčeji tobogoxan'i. mə̄rgə̄, ǯōkčen'i īpi, ičəxən'i: xəǯəni saman'i to͜eŋga, mān'ǯo saman'i to͜eŋga, solõ saman'i to͜eŋga. xəm ǯo͜ato͜eŋga samã. tə͜i samasaldola əm solõ saman'i ču dāji. tə͜i dā͜e samã tə͜i ənusi piktə ǯampan dōlan'i bīdun'i laŋ bīn'i. bāldo͜e bajã mə̄rgəmbə tə͜i dāji saman ǯakpadon'i ose͜oŋken'i. bāldo͜e bajã se͜aworewa ase na͜ji aŋsewanda, se͜aredoji mə̄rgən'ǯi gəsə elã xōwa arakewan'i omēdoji mə̄rgən'či undīn'i: ām, ulən'ǯi təǯəməru! mə̄rgə̄ undīn'i: tu͜i dāji samasal čēlaxamban'i bu͜ə solge na͜ji xōn'e ta͜ori? bāldo͜i bajã undīn'i: ə͜idu bi samasal xa͜ewadā sārase. se͜are ǯaka, piktəjəwə se͜ara, ənūčin'imət, moramari səkpə̄čiči ba͜e abawa!

9. tu͜i tə͜idu͜əči sīksəguxən'i. xās əči͜ə ākpaŋgo͜ači bičī. tu͜i ta͜edoči tə͜i ǯo͜ato͜eŋga samasal po͜an'e ōn'ǯe, po͜an'e xuməsi tūlsixəči. bāldo͜e bajã undīn'i: ām, əsi ələ̄ esego͜e biǯə̄rə. esego͜edo͜an'i tultul samasal tu͜i osēči. tu͜i umi, bāldo͜e bajandā ōn'ǯi tūxən'i. təm mə̄rgəgdə̄ a͜ja.

10. mə̄rgə̄ ākpanda ičəǯə͜in'i: ǯu͜ər gōlǯõ palan tokondolan'i poŋtorčə̄e tami. əmũ mama arbeŋken'i bədurku, kučə̄ŋku, da͜osoŋgoko. tə͜i mama elera sirgundin'i tu͜i tara xacaŋsēn'i: ərdəŋgə̄, mīkə̄ əjəčimə̄ xāldā əčə̄e tāju. ə͜in'ə̄ewə ə͜i udən'i samã xərīdə̄, mindu͜i dāji samã xərixəm biǯə̄rə, ambã xərīdə̄, mindu͜i dāji ambã xərixəm biǯə̄rə. tu͜i tamdā ə͜i udədun'ikə̄ mindu͜i dāji samandā, ambandā aba. təm ə͜i māŋbo sole͜alan'i bi dəxi mə̄rgə̄ pojaŋgon'i də͜isuldə mərgə̄. tə͜i kere osen'i, tu͜ji ose͜ore. tə͜i mə̄rgə̄ ǯidīn'i mədən'i bičī. ǯeǯala biji biǯə̄rə. də͜isuldə mə̄rgə̄ osēn'i, gorokala ta͜okala pu͜ə!

11. *tui xasaŋsera, təi mama təi əktə piktə bīn'i ӡampan'či īxən'i. īrin'i — təi puӡ ataremir dərūxən'i: əmӡ·e bəgӡijəwə sealogoxã. xōn'e bālӡeore! tādomāt māī gəӡōŋgoxan'i: əi enda ӡelen'i min'či xaewa xasaŋsexan'i. tui tara enda ӡelewan'e arbembogora, təi mama ӡampan dōlan'i təsīwən'i ičəӡəkəp enda ӡeleӡen'e nāŋgalaxan'i. toekandon'i təŋ xaedā puikuin'imət seasī bičin'i. mama xuə pəgiəčin'i tūxən'i. tūmi kəukəgdə aba očogoxan'i.*

12. *aba osegoen'i — samāsal moramari səkpəčiči: ao, ao! bāldoe bajandā təguxən'i. suə, samāsal, bae abawa səkpəčīsu. mī piktəčī mədəsipmət sāӡāmbe. tui tara piktəčiji mədəsixən'i. piktən'i undīn'i: solo samã ӡakpaӡeaӡen'e nāŋgalaxã. təji ambã ӡiəkə maŋgalaxã biəsi.*

13. *tui tape bāldoe bajã samāsalba toekame puŋnəguin'i. puŋnəguiduji samāsalči undīn'i: xawoe samã piktəjə bəgӡidun'i uliksəwə bālӡegoandi? samāsal undīči: buə mutəsīpu. mərgəntən'i topen'ӡeji təi puӡī bəgӡičin'i pisixən'i. puӡī, elegora, palan'či pər puikuxən'i. təi puӡī — moča uləə puӡī. bāldoe bajã mərgən'či undīn'i: piktəji asegoase būrəmbi.*

14. *mərgə, asegoe bāra, aŋpaӡeji arakewa omemi, kaoka təi ӡōgdo indəxən'i. sīksəŋgiələ mərgə mədəsīn'i: aŋpa, əi dujələ pimusəl bīnu? bāldoe bajã undīn'i: biji. tui tame goroči əӡi tōra. əi dujə təŋ pamōre boa. əsimə ӡičī naji tərək pulsiəsi. mərgə undīn'i: aja, ӡeӡawa pulsī nītān'i xae pamara!*

15. *mərgə, ləkəi ӡapara, duisi tōxan'e. tōməe, tōməe, goroči tōxan'i. əm ӡōwa ičəxən'i. əlkə bəjičirə, pāwawan'i čumčukuən'ӡi əlkə bojalera, ičīn'i təi mama nakandoji puəsəmi bīwən'i. pāwala tan'e ičəučin'æ, əmӡimə nakandu təsī puӡī ičə tara xoӡexan'i. mərgə təi puӡimbə ičīn'i — asegoe bāxan'i puӡindui əmutunu, uləlō bīnu. tui tara təi puӡī xasaŋsēn'i: bumbæwə uidə xorease. bajam biuxə osendā aja naedā xoregoan'e.*

16. *mərgə uikələ čūl īxən'i. anda puӡī, ən'isi tui maŋga ənusīn'inu? tui tara, rakando tərə, mama ӡakpadon'e taočemi dərūxən'i. taočeməe, taočeməe, mərgə, elera, təi mama nuktədun'i ӡapara, bəsərəwə kəndəl kəndəl erčēmi, pulsimi taočeme, burbuəŋkin'i təi mamawa. burbūčən'i, undəxəmbə boače nān'čara, ӡirkəguxən'i.*

17. *ӡirkəguin'i — mama āŋmalan'e enda ӡelen'e arbeŋgoxan'e. enda ӡelen'e, arbeŋgora, mərgən'či undīn'i: anda mərgə, mī ambāpčekaje. tui ambāmba seaoxã ambāpčeorgoan'e. mī əsi ambāpčexambi ačoečendami ənəmbi. ənərə, ӡiӡujəwə xaedodā bīdə uləə ətuməčimi saečaro. mərgəntən'i, čəwə ӡapara, boala xumumbə xuləxən'i. tui tara, tei mamawa əmučkə n'iəurə, xumuxən'i.*

18. *īgupmət təi puӡin'či undīn'i: anda puӡī, mī maŋga ӡəmuluxəmbi. seagojawa aŋsero. tui bi ən'imbi xaewan'e kaerandēsi, ambāmba! tui tapmāt puӡī seaworewa aŋsexan'i. totara seaxači. mərgə təi puӡin'ӡi aselaxan'i. mərgə, asedoe bipæ, gewanēdoan'emāt əuguxən'i.*

19. *əuguiduji əm gormaxomba wāxan'i. ӡōwa esegora, īguxən'i. īgupi, ičīn'i — bāldoe bajã, ӡuə gōlӡə aldandon'i palando elegora, mamaji xulӡurīn'i kombowa ӡapagora, omenasēn'i. daramadon'etān'i penako. penadon'e seaore, kombo bičin'i. bāldoi bajã mərgə īguidun'i ičəuxən'i. ām, ӡiӡuxəsnu? mərgəntən'i undīn'i: ӡiӡuxəmbi. pamaxambi bicī. bāldoe bajã undīn'i: əi boa tui bijə, — undīn'i, — pamōr boa. tui tara xaӡōm ačogora, xoӡexan'e bāldoe bajã. mərgə, čeme seapi, susuxən'i.*

20. *tui ənəmæ, əm irgəndulə esexan'i. təi irgəndu nae kasa tēn'i. mərgəmbə ičəməri, jaoxemari, arakewa omeomari, etoandomāt turuguxəči. mərgəmbə kasa tae samã, sākse samã, ӡakpalan'i eseoaŋkeče. mərgə, əm tukuru arakewan'i gələrə, xərim təpčiuxən'i. sākse samandola tawaŋke xərimi mərgūsəldu etaŋgi*

pərəgʒiəʒin'i būrə, əlčiusəldu, kəkəsəldu etaŋgeʒi būrə, tui xə̄rimi, tukurui dolendola osēn'i, sākse samã piktən'i puʒin'či esexan'i. tui tara tī puʒindu guči guči etaŋgeʒeji būxən'i.

21. *sākse samã, ičəpi, undīn'i: ām, tui tame piktəiwə soktoandači. mə̄rgəntə̄n'i undīn'i: mī əsimə̄t būxəmbiwə tui goedame ǯapačēn'i n'emorea xaedā! tādotān'i sākse samã piktəčiji undīn'i: ām, nae būxəmbən'i turgə̃ omero! mə̄rgə̃ təi kalta tukuruji təi puǯimbə xəm omeoaŋken'i. puǯintə̄n'i, jəbə maŋga soktora, amemban'i jərgimələn'i əm tukuruwə ǯapara, xə̄rim tə̄pčiuxən'i amendolaji boŋgoǯe, tawaŋketān'i əmutu mə̄rgə̃ xərixən'imət. mə̄rgəndulə esedoan'i tukurun'i dolendola bičĩ. čawa mə̄rgəmbə xəm omeoaŋken'i.*

22. *mə̄rgə̃ esexam bəunduji aolekačēn'i. təi ǯōgdo uidə̄də̄ aba, xəm xaose āwandaxačenu. təm nū̆čku ārčokāragdā gōlǯõ keradoan'e tə̄sīn'i. ǯuə puǯĩ uikələ čūl īxəči. tui tara tī ārčokān'či mədə̄sīči: əin'ǣwə esexã mə̄rgə̃ xaedo bīn'i? ārčokā̃ undīn'i, tā malodo aorejām. buə mə̄rgə̃ maŋga soktoxã biǯə̄rəm gisurənduxəpu. amemban'e wāxã baegoan'e ākpalo piktən'i ǯidīn'i, ələ̄ ese. maŋga soktoxã osen'i, mə̄rgə̃ dolbaloǯāram.*

23. *tui tara mə̄nə dōla eremačera, mə̄rgə̃ aorečēn'i gəsə ənəxəči. tui tara xə̄sīčī: anda mə̄rgə̃, sənəru! mə̄rgəntə̄n'i aoxamāt i-i-i sənīn'i očen'i. anda mə̄rgə̃, bumbǣwə əǯi orkelara, sārejawa sāwandasoam. amembase wāxã baegoansi ākpalo piktən'i ələ̄ ələ̄ eseji. simbǣ maŋga soktoxam biǯə̄rəm sāwan'exapo. mə̄rgəntə̄n'i: nae sāwandēwan'i xaewan'e orkelōri. agdara xoǯeoregoan'i. tui tame təi ǯuə puǯĩ ənuxə̃.*

24. *ənū̆čiči, n'aŋga bimi, əm puǯĩ īxə̃. ojaǯen'i ičūrin'i — ambamāt učūri. təi ārčokān'či mədə̄sīn'i: əin'ǣ ǯičĩ mə̄rgə̃ xaedo bīn'ə̄m? təi ārčokāntān'i — sārasembǣ — uŋkin'i. sīdə̄ ələ̄ ǯajaese!*

25. *tei puǯĩ mə̄rgə̃ bāron'i undīn'i: anda mə̄rgə̃, aoredosi dolbalaxã un'ǯə̄čimə. sənəru! mī sin'ǯi sorendami ǯičimbi. mə̄rgəntə̄n'i undīn'i: mī suə bāroso ənī naji. eseočejawa soremi aja. əidu soremi, nae kaca tēwan'i waǯōleori. puǯintə̄n'i undīn'i: mī tui gorowa ǯobomi ǯičimbi. əidu soregoe tāmbe. mə̄rgə̃ undīn'i: gā, aja, soregoari.*

26. *tādotān'i puǯĩ čumčuən'ǯiji muju muju taxan'i. mə̄rgəndu xaedādā aba. puǯiŋ guči ŋālē ondar ondar taxan'i. mə̄rgəndu xāc xaedā aba. tōtaptān'i puǯĩ damaxewa omeore dajeǯeji mə̄rgəm bāron'i jupuələ̄xən'i. dajetān'i, ənərə, mə̄rgən' ǯuliələn'i tūxən'i. mə̄rgəntə̄n'i undīn'i: ələ̄n'inu? puǯĩ undīn'i: ələ̄n'i. mə̄rgə̃ undīn'i: taodagōreno? puǯĩ undīn'i: čexalē osēn'e, taodagoro. tādotān'i mə̄rgə̃ undīn'i: topembi gəriwən'i gəkpə toekoro! totara topeŋken'e. puǯĩ xotowan'i gəgbərəm toekoxan'i. puǯĩ buikĩ. mə̄rgə̃, əlčiusəlbə gələrə, təi puǯimbə xumuəŋkin'i.*

27. *totara mə̄rgə̃, boače n'iərə, duisi tōxan'e. duilə əm ǯō bičin'i. dōlan'i əgǯi puǯĩ ǯelgan'e čaō-o təin'i. mə̄rgə̃ dosoǯēn'i. poan' puǯĩ undīn'i: təi mə̄rgən'ǯi aoxã osen'i, buikində̄ aja. mə̄rgəntə̄n'i, soktoxã nī osera, uikələ̄ tuičəmi īxən'i. əgǯi puǯisə̄l xoldondolan'i xəm boači n'iəguxəči. təm ǯō ə̆ǯən'i puǯirə̄gdə dərəǯiguxən'i. mə̄rgəmbə ŋāladon'i ǯapara, tə̄wəŋkin'i. tui tara gisurəndurə, puǯĩ seaworewa aŋsexan'i. mə̄rgə̃ tādo aoŋgedoji tī puǯimbə asegoi bāxan'i.*

28. *ǯea čeman'awa mə̄rgə̃ asečeji undīn'i: əin'ǣwə arakewa omeori. čisən'iwə ǯipi, əčiədə̄ arakewa omeaje. ətəxi nī aba — xōn'e omeori? toetara aseǯeji gəsə kasa bāron'i əuguxəči. tādu arakewa omexači. sākse samã sīksəŋgiələ kasawa pan'endami n'iəxən'i. sākse samã maglakočeji əuxən'i. mə̄rgən'ǯi anōmačemare dərūxəči. mə̄rgə̃ sākse samāmba tūriǯin'i anaxan'i. tui tara sākse samāmba*

boa bāron′ tōrəǯen′i baŋsalaxan′i. tui tara duisi tōkoxači. bəktə̄ bipǣ, sākse samā̃ tūguxən′i nāče leparāmdā. čā̆watān′i mə̄rgə̄, ənərə, loap toaŋgoxan′i. toaŋgōčen′i, sākse samā̃ undīn′i: sī mimbǣ xətəkəsi. piktəji asegoase būrəmbi. mə̄rgə̄ sākse samā̃ piktən′i puǯin′ǯi aselaxan′i.

29. *toetara etoan′či mugdəwə sāwondagoxan′i. tādo sāwoedoje ičəxən′i mə̄rgə̄ əm mə̄rgə̄ ǯidīwən′i. təi mə̄rgə̄ eseočēn′i, buə mə̄rgəŋgupu n′exoraŋken′i: gučkə, bāčegoapo! sāomār xoǯepare, gusin′i ənuxən′i. ənuiduji undīn′i: kasawa xamarexambe xaewan′i biurǣm!*

30. *mə̄rgəndə̄, aseji ǯōkčean′e tōkora, susugujə bargečedon′i tī əmǯiŋə asen′i, sākse samā̃ piktən′i undīn′i: anda mə̄rgə̄, xaedā ərindun′i xamasi močogoedoje bundulə keregoxare. mī amembe — elā̃ axon dəū̃, xəm irgə̄ŋku. buə bīpu irgə̄ ču wawoeǯema. irgə̄mpu gərbun′i čečergerē̃.*

31. *mə̄rgə̄ tawaŋke susuxən′i. duəntə dōwan′i ənəmǣ, əm tōwa wāxan′i. tui tara əm on′ekāŋ keračen′i āwandagoxan′i. aoŋgagoe bargečēdon′i gusin′i ǯīčin′i, pondaǯoji olbeŋken′i. tādo gəsə aoŋgēči. mə̄rgə̄ gusin′čiji undīn′i: gučkə, tō pāwan′e, guǯə̄mbən′i searo!*

32. *gəsə aoŋgara, mə̄rgə̄ əmučə̄kə̄ čemi ərdə̄ gusimbi irgə̄n′čin′i ənəxən′i. gusimbi ǯōgdolan′i esexan′i. ǯōgdo əm mapa, əm mama bīči. mə̄rgəntə̄n′i tī gusim pondaǯon′i puǯī bəundulən′i esexan′i. in′ū-u gusin′i, pondaǯon′ puǯī esegoxači. ən′in′ boači n′iəxən′i. ən′in′čī mədə̄sīči: bundulə uidə̄ əciə eseano? mama undīn′i: čemi ərdə̄ əm mə̄rgə̄ esexam. mī xajǣ uŋkəji — gamasom bāogoream, pondaǯoje əǯi olbendam. əǯiku naewa xaegoe tui taore? nae tui taen′ osēn′i, buə xəm burburi goan′e! gusin′i, pondaǯon′i ǯōkče īguxəči. puǯin′ǯi mə̄rgə̄ əmutu asejemāt kupimi dərūxən′i. təi puǯī amen′i, āŋn′i, ən′in′i seaorǣdā searamari gasēči.*

33. *tui bidučiə siun tokondo ǯuə mə̄rgə̄ esexan′i. təi mə̄rgəsəl gərbun′i: sojala, gojala. īməri, mapači undīči: bundu būrisi piktəsi goenu, naedo būrisi piktəsi guči goenu? mapa xuməs tūxən′i. tūndu kol kol bīn′i.*

34. *tādotān′i bun′ǯǣ mə̄rgə̄ tī ǯicī mə̄rgəsəl bāron′i undīn′i: bue geagda nī əktə nī tui taewan′i xōn′e alecōri! sī aseǯeji aselagoro. mī bae andakačexankāčǣ oseŋātān′i. tī mə̄rgəsəltə̄n′i undīn′i: əmdə̄ ǯičimbi, xae tui xasaŋsēsi? əjəǯ ǯal, tajaǯi ǯal. tui tami bun′ǯǣ mə̄rgə̄, ǯuətuŋə mə̄rgəmbə nuktədun′i ǯapami, boači pāgǯalaxan′i geramardā korpeaseǯeači. tui tara boače n′iəurə, mapa sələm san′čen′i ǯuətuŋəsəwə dapsečemi wāxan′i.*

35. *tui tara īguxən′i ǯōkči. tādomāt tī mapa, mama, gusin′i tə̄guxəči, agdanasemari undīči: tui bi nae biəsilī tui taeno? gusimbi pondaǯoǯen′i aselaxan′i. tī ǯea čeman′ǣwa mə̄rgə̄ susuxən′i.*

36. *ənəmǣ, amembe wāxā̃ baegoan′čeji esexan′i. esera, təi baegoambi waesi, xə̄sirə, soreme wāra, irgə̄mbən′i dalegomi, ǯiǯumi, gusim irgə̄mbən′də̄ gəsə gaǯoen′i. tawaŋke ǯiǯumərǣ, sākse samā̃ irgə̄ndulən′i čečergerendola esegoraə, tādo əm xadowa aoŋgaxan′i.*

37. *əm dolbon′awa sākse samā̃ piktən′i asen′i jajame dolbo sənəxən′i. anda puǯī, sīdə̄ samanseno? xaedā mədə̄wən′i dōlǯeori biǯə̄rə. asen′i undīn′i: samaŋgola xae dāe samā̃! naotom asesi čemana siun tokondo əi māŋbola sologoǯāra xujū̃ dā aǯen′ǯi. cāwa siur ənuəndi osen′i, xoǯekase.*

38. *ǯea čeman′a tə̄pi, mə̄rgə̄ māŋbowa jap adoleǯe jakčera, ətūndəmi əurīdūn′i asen′i əm ǯogbowa būxən′i. siun tokō̃ esedōn′i xujū̃ dā aǯē̃, taja kalta sologoxan′-ǯeji adolea bāra. əjəči daogoedoan′i mə̄rgə̄ təŋ məipəndulən′i ǯogbolaxan′i. tui tara nāči xāboečami mə̄rgə̄ sorēn′i. aǯē̃, maselagope, mə̄rgəmbə muə bāron′i xawaŋela kutučīn′i. əlčiusəl bələčindəučən′imə̄t, nā bāron′i toaŋkeči. toampare,*

mɔ̄rgɔ̄, sələm sã torān'i loptorām toanda ažê želelan'i elã modã kuŋkulūčən'imɔ̄t, naotom asen'i ažê āŋmalan'i arbeŋgoxan'i. naotom asen'i xaoedā asedōndā bālžexan'i ču ulɔ̄, ču boŋgo asen'i.

39. *tādo əm xadolta bipərə̄e ənuguəri bargečegoedoači sākse samã, piktəji obogoan'i mɔ̄nɔ irgɔ̄mbi tokondolan'i obolara, gāžoandēn'i. tawaŋke susugurə žižumɔri, kasa irgɔ̄mbən'idɔ̄ gəsə gāžoe. bāldoe bajã irgɔ̄mbən'i esegopi, duil bi aseji, bāldoe bajã piktən'i aseji xəm gəsə gāžoen'i. bāldoe bajandā piktəduji kalta irgɔ̄mbi olbemboandēn'i.*

40. *tawaŋki susugupi, mɔ̄n žōgdolaji esegoen'i. āŋnān'e xəm ujŭ osegoxači. poan'e sagdaŋgoxan'i, tun'əpuŋku osegoxači. āŋnān'e xəm aseko, irgɔ̄kɔ̄ŋku. mɔ̄rgɔ̄, esegopi, gāžoxambi irgɔ̄səlbə əmutu xotomāt tɔ̄wəŋkin'i. duilə bi gurusəl mɔ̄wa waesi xodasendasoe, waela bi gurusəl duisi muəwə xodasendasoe. mɔ̄rgɔ̄ tui mɔ̄rgɔčimi bālžexan'i.*

Tale (translated)

1. In a house there lived forty people. These people went about to earn their living, they traded in fur animals, elk and meat animals. And the youngest brought down most of them. This young one comes home one evening—five of his elder brothers were dead. He comes home next evening—another five of his brothers had died. Dying like that, five by five, they died away, and they were all dead. Then, when all had died he thought, '[the evil spirit] has killed my comrades, it will surely kill me, too'.

2. He spent one day sitting at home, stayed up one night, but nothing happened. When also the following day passed in the same way, he took all his travelling things, his skis and went down to the riverside, straight to his icehole. He stopped and looked down the river. Somewhat lower than the place where he had descended he noticed a hardly visible trace. Only when he went nearer he saw the trace(s) of a washbear. Having caught sight of it he thought, 'surely this is the track of the evil spirit who killed my elder brothers'.

3. Then he followed it (the track). Having gone over to the opposite bank, he found a sleeping female washbear in the thicket of the island. He found it, pulled out the axe from behind his back to kill her, and when he raised the axe to hit, that washbear said in a human voice, 'Elder brother, do not kill me! I have come, thinking that you probably are crying. Now seeing that you have set out on your way, I'll go away'. 'Why should I not kill you? [It is] a pity (not to do so!)' The washbear said, 'Bind [me] with the packing strap and tie [me] to the belt'. The *mergen* said, 'Very difficult! How to drag you along?' 'When I shall be dragged after you, then you will kill me thereby'.

4. Having tied [her] behind himself, the *mergen* went along the river. [She] did not let herself be dragged (went by herself). The *mergen* thought, 'There is little snow on the river, that is why she comes with me'. Then they went up to a morass. The *mergen* strained every nerve. The washbear overtook him and pulled him. The *mergen* thinks, 'On the morass the snow keeps her up, that is why she is drawing me'. Then they entered a thicket of shrubs. There [he] strained every nerve, but the washbear pulled him even stronger. Then he waved his hand [gave it up as lost] and descended to the river.

5. Having come down to the river, [they] settled down for the night. In the night the *mergen* killed a hazel-grouse and a roe-deer, and cooked broth. Together with the washbear he ate the broth. Next morning they got up and set out on their way. Having walked a while [he] turned back when the washbear tied behind drew at her leash. The washbear, drawing her breath, scratched the blue ice. The *mergen* thinks, 'Yesterday we came a long way (walked much), today she is tired'. Then he spoke to the washbear, 'Well, it is time to kill you!' [He] raised his axe. Then the washbear said, 'Elder brother, since I was not tired yesterday, having walked so much, how should I get tired now? I heard news and I want to stop you to tell it'. 'What news?' he asked.

6. The washbear said, 'When I got up this morning I heard this kind of news. At noon we shall come to a large village, to the village of the rich man Baldoi. This rich man Baldoi has many wives. Also children—only from one woman he had one daughter. This daughter is ill. Her illness is like this: [though] she is alive, [somebody] tears the flesh off her feet and eats it, but who it is, is unknown. The rich man Baldoi has called in many shamans and keeps them in his house. None of the shamans knows when this devourer comes to eat. Only when it has eaten and gone, do the shamans shout. When we shall pass them, the people will see you and rejoice: only this man will heal the illness of that child. Brother, what are you prepared to heal with?'

7. The *mergen* says, 'What do I heal with? Let us pass them.' The washbear says, 'Brother, you will not pass. Many people are on guard by the river. If you think of curing her, put me in your bosom.' Then the washbear hit herself on the head and turned into a dog's head. The *mergen* took [her] and tucked [her] into his bosom. After that [he] went, and when he arrived at the village at noon [he saw] two people looking, screwing up their eyes and protecting them with their hands from the sun. Then these people started to shout toward the shore, 'Rich man Baldoi, do not sit at home! Up on the river comes a man. This man will cure your daughter'.

8. The *mergen* went ashore. The rich man Baldoi came out through the door, down to the river, took the *mergen* by the hand and led him into his house. The *mergen*, going into the house, saw five Nanai shamans, five Manchu shamans, five Solon shamans. Altogether fifteen shamans. Among these shamans the greatest is a Solon shaman. This great shaman sits by the bedside when the ill child is in bed. The rich man Baldoi set the *mergen* beside that big shaman. The rich man Baldoi ordered the women to prepare food. During eating he drank with the *mergen* three glasses of wine and said to the *mergen*, 'My dear, do your best!' The *mergen* says, 'How can we, simple people, do what such shamans could not?' The rich man Baldoi says, 'The shamans who are here know nothing. When the devouring creature has bitten my daughter and gone away, only then do they shout and wag their mouth'.[1]

9. Meanwhile (literally: while doing this) the night fell. [They] did not yet go to sleep. At this time those fifteen shamans fell some backwards, some prone. The rich man Baldoi says, 'My dear, now [it] will apparently

[1] When in trance, with a view to expelling the spirits of illness from the body of the diseased, the shaman shouts 'aot! aot!' as if trying to snap with his teeth at the escaping spirits, and meanwhile imitates the movements of the bear with mimetic gestures.

soon come. When [it] comes, this is what always happens to the shamans.' Saying this, the rich man Baldoi also fell back. To the *mergen* alone nothing happened.

10. The *mergen* lay down. He saw: between the two hearths, in the middle of the floor, raising a column of dust, an old woman made her appearance with a little table, a knife and salt. That old woman got up. She trembles, and says then, 'Strange, I have never felt like this. Obviously, today a shaman is flying in these parts, a stronger shaman than myself. Obviously, an evil spirit is flying, an evil spirit stronger than myself. But in these parts there is no shaman or evil spirit stronger than me. There is but one living up the river, the younger brother of the forty *mergens*, the *mergen* Deisulde. There was rumor about this *mergen*'s coming. Probably [he is already] nearby. If this is *mergen* Deisulde—away, pass by!'

11. Having said this, the old woman went to the bed curtain behind which the little girl was lying. [She] enters and that *pudin* starts moaning, 'One of my legs [she] has eaten all up, now she begins to eat my other leg. What can I do?' And here the *mergen* remembered, 'What did the dog's head tell me!' Then [he] produced the dog's head, looked where the old woman [was] sitting and threw the dog's head. When it fell down, a sound was heard as if something had been shot. The old woman fell to the edge of the plank bed. [She] fell and disappeared altogether.

12. When [she] disappeared, the shamans uttered a cry with their mouths, 'Ao! Ao!' The rich man Baldoi got up too. 'In vain you shamans shout with your mouths! I shall ask my daughter and shall know!' Then he asked his daughter. His daughter says, 'Near the Solon shaman [somebody] threw [something]. That evil spirit got it pretty well!'

13. After this the rich man Baldoi beat the shamans and drove [them] away. While driving away, he says to the shamans, 'Who among the shamans will grow flesh on the legs of my daughter?' The shamans said, 'We cannot'. But the *mergen* moistened the legs of the *pudin* with his saliva. The *pudin* got up and, as if nothing had happened, jumped to the floor. She was well, as if she had not been ill. The *pudin* is very beautiful. The rich man Baldoi says to the *mergen*, 'I give you my daughter to marry'.

14. Having married, the *mergen* drank wine with his father-in-law and spent the day in that house. When the night fell, the *mergen* asked, 'Father-in-law, are there hazel-grouse in this taiga?' The rich man Baldoi says, 'There are some. But do not go far. This taiga is a place where it is easy to lose one's way. A newcomer would not know the right way to find the road'. The *mergen* says, 'Alright, I shall go nearby, I'll not lose my way!'

15. The *mergen* took [a bow and] arrows and went to the taiga. He walked and walked and came a long way. He saw a house, silently sneaked up, carefully pierced the window with his fingers[2] and sees that the old woman, the evil spirit is tossing from side to side on the bedstead. No sooner did he look through the window than the *pudin* sitting on the other bed cast a glance and that was all. The *mergen* looks at that *pudin*—somewhat like that *pudin* whom he had married, if not prettier. After that the

[2] The windows of the Nanai houses were earlier panelled with oiled paper instead of glass.

pudin says, 'Nobody will save us. If we were rich, somebody would come to our rescue'.

16. The *mergen* entered through the door. 'Friend *pudin*, is your mother as ill as that?' Then he sat at the bed and started to do magic around the old woman. Bewitching, bewitching . . . The *mergen* stood up, grasped the old woman by the hair, walked and dragged her around the table doing magic, reduced the old woman to death. When she died, [he] went to the street to fetch a board and laid the deceased [on it].

17. He lays her there, and a dog's head appears from the mouth of the old woman. The dog's head made its appearance and says to the *mergen*, 'Friend *mergen*, I turned into an evil spirit. If you eat an evil spirit, you, naturally, turn into an evil spirit yourself. Now I shall go and get rid of this transformation. I shall go away, and you, wherever you will be, prepare yourself well and carefully for my return.' The *mergen* took a spade and dug a grave in the street. Then he carried that old woman out [alone] and buried [her].

18. As soon as he returned, he said to that *pudin*, 'Friend *pudin*, I am very hungry. Prepare me something to eat. And why do you deplore your mother, that evil spirit?' Then the *pudin* prepared the meal. They ate. The *mergen* married this *pudin*. Having been with his wife, the *mergen*, when it dawned, went away.

19. On his way he killed a rabbit. He came to the house and went in. He looks—the rich man Baldoi stands in the middle of the floor between the two hearths, grabs the scoop which the old woman had warmed up and drinks. Behind his back there is a knapsack. In the knapsack there is food and a scoop. When the *mergen* comes in, the rich man Baldoi turned back. 'My dear, have you come back?' And the *mergen* says, 'I have. I lost my way'. The rich man Baldoi says, 'This place is like that—[he] says—a place where it is easy to lose one's way.' Then the rich man took off his clothes and lay down. Having had breakfast in the morning, the *mergen* set off.

20. He went and arrived at some village. In that village somebody holds a great funeral feast.[3] Having caught sight of the *mergen*, they invite him, treat him to wine and lead him to the ritual show-booth. They take the *mergen* to the shaman celebrating the great feast, to Saksi.[4] The *mergen* asked for a bottle of wine and started to carry it around. He carries it around, starting from the shaman Saksi, gives each *mergen* two glasses of wine,[5] to the man servants and maids a glass each and carrying it around

[3] The great (third and last) feast (*kasa*), was celebrated by the shamans of the highest order one year after the death of the person in question. On these feasts which lasted from three to five days there were usually large gatherings of people, principally from the clan of the deceased. During the feast the family of the deceased richly regaled all those present. The purpose of these feasts was to assist the soul of the deceased out of the grave (*buni*). A wooden image of the soul of the deceased (*pańu*), as well as the representation of his body made of grass or carved from an aspen log (*mugdə*) about 1 meter long, were placed in a ritual booth erected especially for this purpose (*itoan*). The *pańu* is made immediately after death and remains in the house for a whole year. The *mugdə* is carved prior to the big festival. During the feast both the *pańu* and the *mugdə* are placed in the *itoan*. As a concluding act of the festival, the shaman burns the *pańu* and the *mugdə* on the funeral pyre.

[4] *saksi* 'magpie'.

[5] The home-made wooden cup (*itangi*) looks like an earthen saucer with a shallow deepening at the bottom where wine is poured.

like that when only half of the bottle was left, he went up to the *pudin*, daughter of the shaman Saksi. Then he gave her glass after glass.

21. Shaman Saksi saw this and says, 'My dear, this way you will not make my daughter drunk'. But the *mergen* says, 'What I have just given her, she has held it so long that it flowed on her hand'. The shaman Saksi says to his daughter, 'Darling, you had better drink what they have given you'! The *mergen* made the *pudin* drink that whole half of the bottle. But the *pudin* who had not become too drunk produced from the shaman box of her father another bottle and started to carry it around beginning with her father and so on, just as the *mergen* had carried it. When she came to the *mergen*, half of the bottle was left. She made the *mergen* drink all this. By that time night had fallen.

22. The *mergen* lies on the place which he had reached. In the house there is nobody. Everybody, it seems, had gone out somewhere to sleep. Only a little girl is sitting at the hearth. Two *pudin*s came in through the door. Then they ask that little girl, 'Where is the *mergen* who came today'? The little girl says, 'Over there, sleeping at the place of honour on the bed'. 'We thought that the *mergen* was very drunk and we have come to tell [him]. The elder daughter of the enemy who had killed his father coming here, soon she will come. We thought that the *mergen*, if he is very drunk, will be caught unawares.'

23. Then both, pushing each other, went to the place where the *mergen* was lying. They called [him], 'Friend *mergen*, wake up'! But the *mergen*, as if being asleep, began to wake up, 'I-i-i!' 'Friend *mergen*, do not blame us thinking that we have come to impart that which is known. Soon enough comes the elder daughter of the enemy who had killed your father. We have come to inform you, thinking that you are very drunk.' But the *mergen*, 'How can one blame for being warned? One can only be glad!' Then the two *pudin*s went away.

24. When they had gone out, somewhat later, another *pudin* came. From her appearance, she looked like an evil spirit. [She] asks the little girl, 'Where is the *mergen* who has come today'? But that little girl says, 'I really do not know'. 'You, too, are hiding him!'

25. The *pudin* says to the *mergen*, 'Friend *mergen*, you will probably say that in your sleep you have been caught unawares! Wake up! I have come to fight with you'. But the *mergen* says, 'I am just coming to you. I shall come, and then we shall fight. To fight here—[means that] we disturb the people who are celebrating the feast.' The *pudin* says, 'I have come with such difficulty from afar. I want to fight here.' The *mergen* says, 'Well, alright, let us fight!'

26. Then the *pudin* moved her finger. Nothing happens to the *mergen*. The *pudin* waved her hands. Again nothing happens to the *mergen*. Then the *pudin* threw (as a lance) her tobacco pipe to the *mergen*. The pipe flew and fell in front of the *mergen*. The *mergen* says, 'Enough, isn't it?' The *pudin* says, 'Enough'. The *mergen* says, 'Shall I reply?' The *pudin* says, 'Reply, if you want!' Then the *mergen* says, 'Spittle mine, fall and tear [her] scalp!' Then he spat. It fell so that her brains fell out. The *pudin* died. The *mergen* called the servants and ordered them to bury the *pudin*.

27. Then the *mergen* went out to the street and away into the taiga. In the taiga there was a house. Therein sounds of many *pudin*s were heard.

The *mergen* listens. Some of them say, 'If we had slept with this *mergen*, we might even die'. Then the *mergen*, pretending to be drunk, went staggering in through the door. Many *pudins* came out to the street past him. Only one *pudin* remained, the mistress of the house. She took the *mergen*'s hand and made him sit down. After this, when they had talked, the *pudin* prepared a meal. Passing the night there the *mergen* married that *pudin*.

28. Next morning the *mergen* says to his wife, 'Today one must drink wine. I came yesterday and have had no wine as yet.' 'There are no servants, how will you drink?!' Then together with his wife, he went to the feast. There they drank wine. When the night fell, all the people went out to celebrate the last rite of the feast.[6] The shaman Saksi went to his tower.[7] He and the *mergen* started to push one another. The *mergen* pushed the shaman Saksi so that he fell. Then he tossed him up with his foot so that he flew up into the sky. Then [all] went to the hills. A little later the shaman Saksi fell back to earth, and got stuck in it. The *mergen* came up to him and pulled him out. When he pulled him out, the shaman Saksi says, 'You have defeated me. I give you my daughter for wife'. The *mergen* married that *pudin*, the daughter of the shaman Saksi.

29. Then he went in to the ritual booth to say farewell to the *mugdə*. While bidding farewell, the *mergen* saw another *mergen* going. When that *mergen* came up to him, our *mergen* made a bow: 'Welcome, uncle!' Having finished leave-taking [bidding farewell to the *mugdə*], his mother's brother went out. Going out he says, 'I was late for the feast. Why should I stay here?'

30. The *mergen*, too, went to the house of his wife and when he prepared to leave, one of his wives, the daughter of the shaman Saksi says, 'Friend *mergen*, whenever you come back, come to us. My father has two brothers (literally: my father—three brothers). All have their villages. The village where we live is the closest to the river. Our village is called Čičirgirin.'

31. After this the *mergen* set out on his way. Going through the taiga he killed an elk. Then he went to the bank of a small brook for the night. While erecting his hut, his uncle came and brought along his [own] sister. They spent the night together. The *mergen* says to his uncle, 'Uncle, eat the liver and the stomach of the elk'.

32. After they had spent the night together, the *mergen* went alone early in the morning to the village of his uncle. He arrived at the house of his uncle. In the house there was an old man and an old woman. The *mergen* went up to the place of the *pudin*—the sister of that uncle of his. When the day came, the uncle and the *pudin*—his sister—also arrived. His (the uncle's) mother went out into the street. [Someone] asks the mother, 'Has nobody come to us?' The old woman says, 'Early in the morning there came a *mergen*. What did I say? You'll meet your nephew, as it is, do not take the sister! Why should one do that to a woman who has a husband! If one

[6] Those who take part in the festival bow to the *pańu* and the *mugdə* and bid them farewell before the shaman throws them into the fire.

[7] On the place where great feasts were held a special tower was erected (*maglako* 'place of throwing', from *maglaori* 'to throw, to fling'). The shaman climbed the tower, shot an arrow into the air and descended with another arrow. The participants of the feast endeavoured to keep him away with a stick but straining every nerve, he was meant to reach the ritual booth.

does like this, we all certainly perish!' The uncle and his sister ceme into the house. The *mergen* began to amuse himself with the *pudin* as with his wife. The father, the elder brother and the mother of that *pudin* do not eat anything, they are sad.[8]

33. At that time, at noon, two *mergens* arrived. Their names are Soyala and Boyala.[9] They come in and say to the old man, 'You will give us one of your daughters, and to him you will give the other (literally: your daughter to be given to us is another one, isn't she, and your daughter to be given to the man is again another one, isn't she?)'. The old man fell prone and did not move.

34. And then our *mergen* says to the *mergens* who had just arrived, 'How can we, married people, tolerate when a woman acts like that! Marry your own woman and let it be as if I had simply courted [her].' But those *mergens* say, 'Since you have come, why do you speak like this?' Bad language is heard from both parties. Then our *mergen* grasped both *mergens* by their hair and ran out into the street so [quickly] that they could not even keep pace. Then having gone out into the street, [he] killed both by beating them against the iron posts[10] of the old man.

35. Then again he went into the house. Then that old man, the old woman and his uncle got up and said with joy, 'If there were no such man, would this have happened?' [The *mergen*] married the sister of his uncle. Next day the *mergen* set off.

36. He goes and arrives to the enemy who had killed his father. Having arrived, he summoned his enemy to the river-bank, killed him in fight and now coming back, joins his village, and together with the village of his uncle drives it to his place. On his way back from there he arrived at the village of the shaman Saksi, Čičirgirin, and spent there a few days.

37. One night his wife—the daughter of the shaman Saksi—woke up and sings shaman-like, 'Friend *pudin*, are you also a shaman? Some news can be heard probably?' His wife says, 'Among the shamans, what a great shaman woman I am? Your washbear-wife will ride tomorrow at noon on this river on a nine-sagene long *kaluga*.[11] If you let her float by, it is the end of you!'

38. Having got up next morning, the *mergen* drew a net across the river and went on the look-out. His wife gave him a harpoon. When noon came, and the nine-sagene long *kaluga* swimming on that side of the river met the net and started to swim through to this side, the *mergen* threw his harpoon straight into its back. After that [he] fought [with it], trying to drag it out on the ground. The *kaluga* straining every nerve dragged the *mergen* into the water up to his armpits. And it was not until then that the servants came to help and pulled [it] out on the ground. When [they] pulled [it] out, the *mergen* pulled out an iron pole and after having hit the head of the *kaluga* three times, his washbear-wife appeared from the jaws

[8] This may be regarded as a memory of relatively recent times when the norms of polygamy started to slacken in Nanai society.

[9] From the noun *soja* = *goja* 'fang of wild-boar, musk, or wolf'.

[10] Posts for the drying of fish. They are usually of wood.

[11] A fish in the Amur river, of the species *Huso dauricus*, attaining sometimes a length of five to six metres and a weight of one ton.

of the *kaluga*. The washbear-wife, compared to any of his other wives in beauty, was the best and the first wife.

39. After he had lived there for a few days and started to leave, the shaman Saksi, to give his daughter a dowry, divided his village into halves and ordered to take [the half away]. From there they started on their way, they went and took along also the village where the celebrations were held. Having arrived at the village of the rich man Baldoi, [he] carried away with the others his wife living in the taiga and his wife, the daughter of the rich man Baldoi. And the rich man Baldoi ordered his daughter to take along half of his village.

40. Having started from there on their way, [he] arrives home. All his elder brothers are alive again. Some have become old, started to [walk] with crutches. All his elder brothers are married and have villages. The *mergen* arrived and settled the villages which he had brought, like a town. The setlement was so large that the people living on the elevated part [of the settlement] take wood to sell by the river, and the people living by the river, take water uphill to sell.[12] This is how the *mergen* lived like a hero.

[12] The settlement became so large that the people living at the far end from the river cannot go for water themselves but buy it from those living on the bank who, in turn, are so far from the taiga that they must buy wood from those living at the taiga end of the settlement.

The Three-Grade Amulets among the Nanai (Golds)

by

V. Diószegi, Budapest

In our study of the so-called *moi̯gańis*[1] (Nanai amulets against lung diseases), it has been pointed out that they belonged to three classes: (a) the *du̯entę moi̯gańi*, (b) the *ambanso moi̯gańi*, and (c) the *i̯arga moi̯gańi*, related to each other in so far as the 'efficacy' of the amulet is concerned. Benedek Baráthosi Balogh, a collector of these Nanai objects, says: 'The efficacy of the (different) *moi̯gańi* is not regarded as equal. At first the gentler and more benevolent *doonta (du̯entę) moi̯gańi* is used; if this is of no help, the *ambanso moi̯gańi*, and finally, the *i̯arga moi̯gańi* is applied.'[2]

Not only of the *moi̯gańi*, but scores of other amulets come under the *du̯entę, ambanso* and *i̯arga* classes.

Here are a few examples:

du̯entę ai̯ami	*ambanso ai̯ami*	*i̯arga ai̯ami*
du̯entę ai̯ami čāni	*ambanso ai̯ami čāni*	*i̯arga ai̯ami čāni*
du̯entę χermiktani	*ambanso χermiktani*	*i̯arga χermiktani*
du̯entę moi̯gańi	*ambanso moi̯gańi*	*i̯arga moi̯gańi*
du̯entę seu̯en	*ambanso seu̯en*	*i̯arga seu̯en*
du̯entę ungiptu	*ambanso ungiptu*	*i̯arga ungiptu*
kollkeru du̯əntę	—	*kollkeru i̯arga*
bui̯un dama du̯entę	*bui̯un dama ambanso*	*bui̯un dama i̯arga*
—	*ogǯimā ambanso*	*ogǯima i̯arga*
tomu du̯entę	*tomi ambanso*	*tomu i̯arga*

In this study we shall endeavour to solve the following three problems: (a) What are these three classes? (b) Why are just these applied for healing? (c) Why are they applied in the order of *du̯entę, ambanso* and *i̯arga?*

Comparing these three-grade amulets, we find that the *du̯entę, ambanso* and *i̯arga* refer to three different species of animals.

Take, for instance, the *moi̯gańi* amulets (Figs 1a, b, c). Cut out of a thin iron plate, they have the following features in common: They are kidney-shaped, their patterns consist of a certain number of snakes opposite and below each other. They differ in the animal figures which are cut below the snakes or hung on the brim. Those hanging on the *du̯entę moi̯gańi* (Fig. 1a) are thick and square-built and have no tails; the animals hanging on the *ambanso moi̯gańi* (Fig. 1b) or shown on the amulet itself under the snakes are thin and long with short tails. Finally the animals pictured on the *i̯arga moi̯gańi* (Fig. 1c) are slimmer and longer than those on the *du̯entę*

[1] Диосеги (1949) pp. 176–204.

[2] Description on inventory card, No. 136. 199 of the Ethnographic Museum of Budapest.

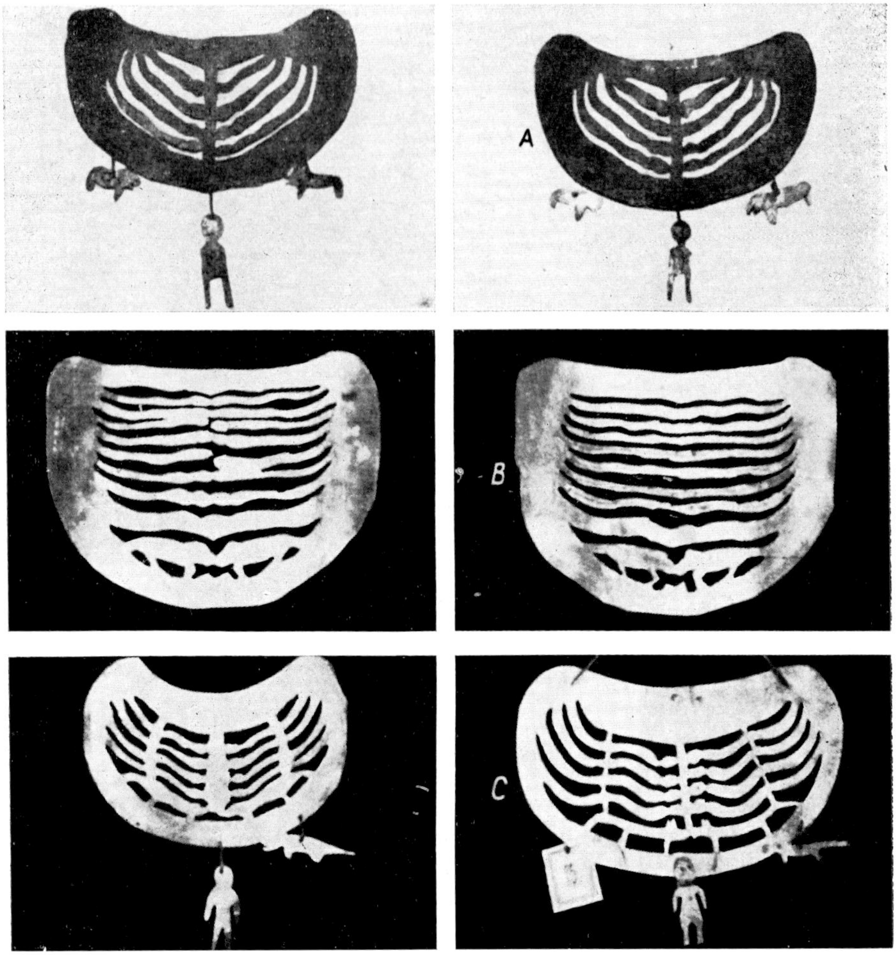

Fig. 1. Nanai *moi̯gańi* amulets: *A*) *dü̯ęntę moi̯gańi*; *B*) *ambanso moi̯gańi*; *C*) *i̯arga moi̯gańi*

moi̯gańi, but are smaller and more thick-set than those hung or pictured on the *ambanso moi̯gańi*.

Similar differences are also evident in the amulets of the *χermiktani* type (Figs 2a, b, c). These are made of dry grass and depict a fairly big four-legged animal to which five human and four bird figures are attached by a longer straw string. The *dü̯ęntę χermiktani*, *ambanso χermiktani* and *i̯arga χermiktani* differ from each other actually only in the shape of the quadruped. The principal figure of the *dü̯ęntę χermiktani* (Fig. 2a), a four-legged animal, is made of grass with a strip of willow-bark wound around its body to keep the grass together. Its legs, ears and tail are short, each being represented by a wisp of straw inserted into the body. Similarly, the main figure of the *ambanso χermiktani* (Fig. 2b) is a beast with short ears and

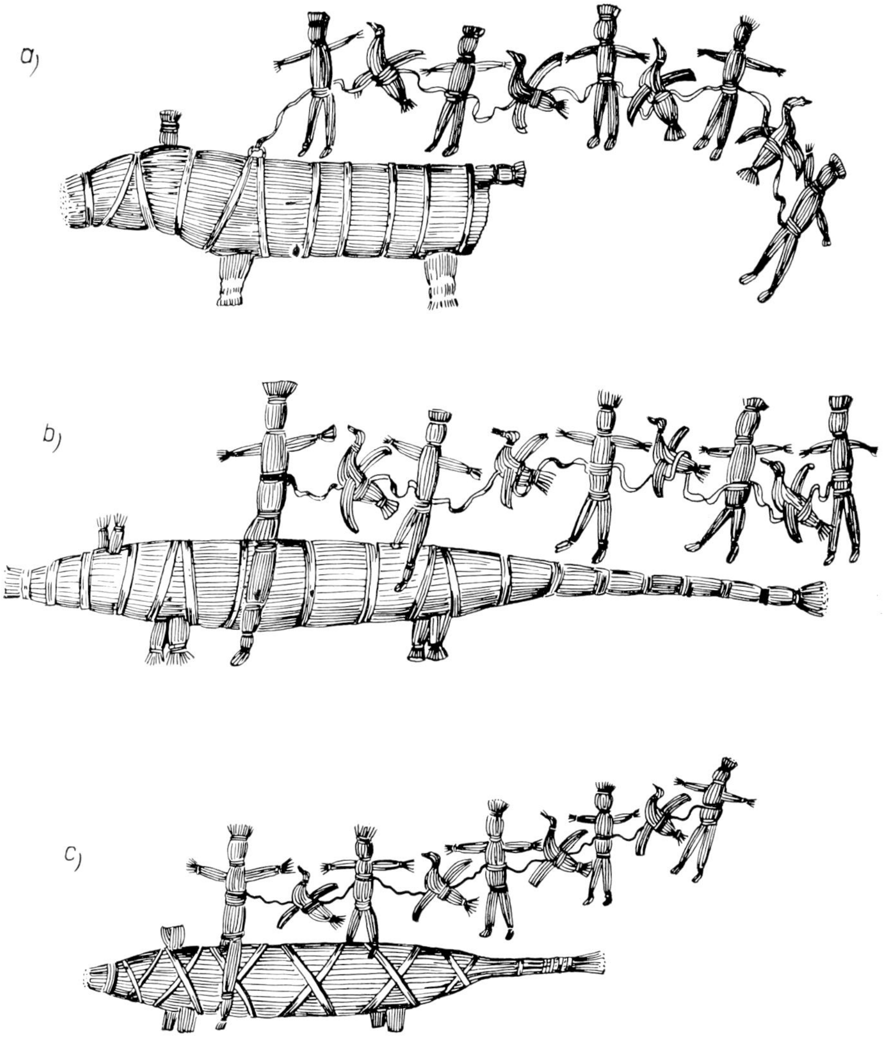

Fig. 2. Nanai *χermiktani* amulets: a) *d^{u}e̯nte̯ χermiktani;* b) *ambanso χermiktani*; c) *i̯arga χermiktani*

short legs, but with a long tail. Its body is decorated with willow bark in parallel strips. The chief figure of the *i̯arga χermiktani* (Fig. 2c) has also a long tail, but the whole animal is smaller than the *ambanso χermiktani*. Here too, the body is surrounded by willow bark, but the ribbons, instead of running in parallel stripes, crisscross each other producing a rather spotty effect.

It is quite difficult to describe accurately the features of the animals made of sheet-iron and dry grass. In fact they are hardly recognizable. More can be done with the amulets made of wood, which are painted and coloured in an effort to make the figures more life-like.

Hence, the anthropomorphic *ai̯ami* (Figs 3a, b, c) and *ai̯ami čāni* (Figs 4a, b, c) types of amulets, though not animal-shaped, make it easier for us to determine the three kinds of animals more accurately.

The *dŭ̯ente ai̯ami* and the *dŭ̯ente ai̯ami čāni* are human figures carved of flat wood. The head is oval and the surface of the body is flat. Since

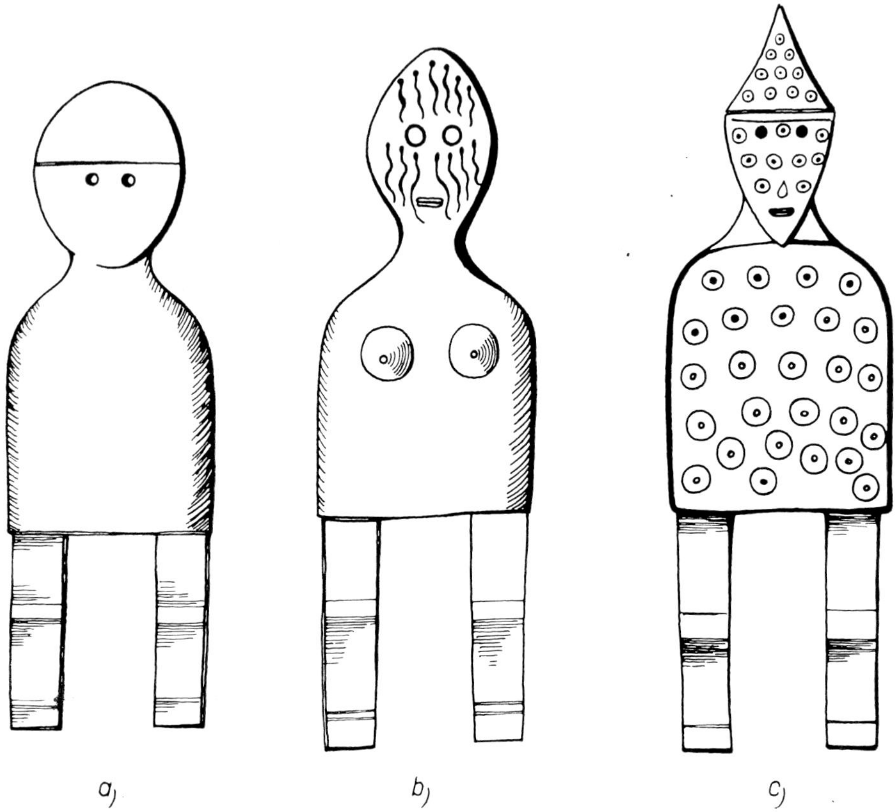

Fig. 3. Nanai *ai̯ami* amulets: a) *dŭ̯ente ai̯ami*; b) *ambanso ai̯ami*; c) *i̯arga ai̯ami*

the amulets now made are no longer coloured, we lack information as to the nature of the animal symbolized; the painted *ambanso* and *i̯arga* samples, however, are eloquent on this point. The *ambanso ai̯ami* and the *ambanso ai̯ami čāni* are human figures made of flat wood (the former without arms, the latter, without arms and legs). The body, and sometimes even the face, is covered with snakes either painted or burnt in. The *i̯arga ai̯ami* and the *i̯arga ai̯ami čāni* are also human figures carved of flat wood (the former without arms and the latter without arms and legs). The entire body is covered with small circles, each with a dot inside, burnt in, or painted. If several colours are used, the designs are black or red. Often the circles are red and the dots black, or the other way round. Baráthosi[3] says that according to the Nanai, 'the painting imitates the spots of the panther and the *ai̯ami*s of this type were originally lined with panther skin'.

The anthropomorphic amulets of the *ai̯ami* and *ai̯ami čāni* types have the same decorations as the designs on the main figures of the zoomorphic *χermiktanis*. The *ambanso*s are decorated with strips (in the case of the *ai̯ami*s parallel snakes and in the case of the *χermiktanis*, parallel bark ribbons), whereas the *i̯arga*s are covered with patches (in the case of the

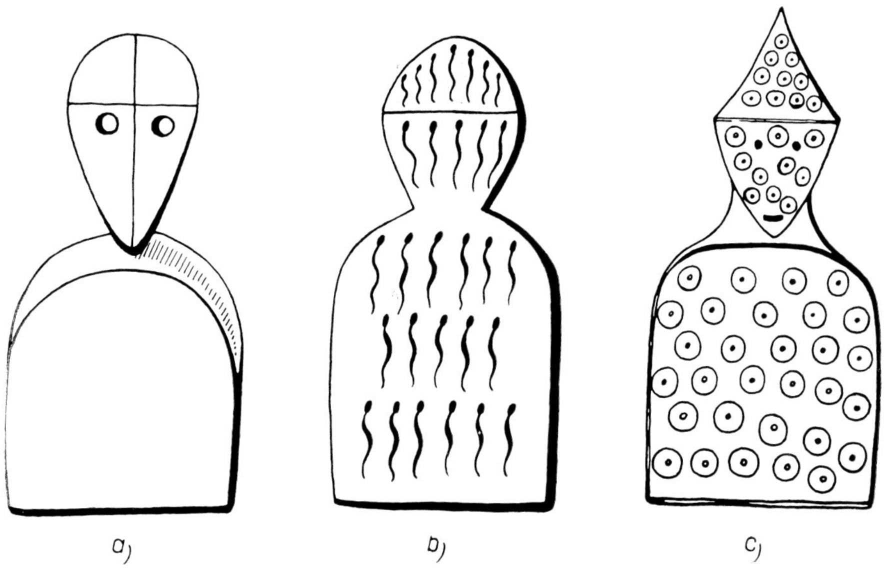

Fig. 4. Nanai *ai̯ami čāni* amulets: a) *du̯entę ai̯ami čāni*; b) *ambanso ai̯ami čāni*; c) *i̯arga ai̯ami čāni*

χermiktanis the strips of bark crossing each other in X forms, and in the case of the *ai̯ami*s red or black circles each containing a black or red dot), originally with panther skin.

Of the animals on the three-grade amulets the second is probably a tiger, judging by its striped 'skin', while the third, on account of its mottled body covered with patches, could hardly be anything else but a panther.

These assumptions are substantiated by the zoomorphic three-grade amulets of the Nanai, which give a sound basis for determining the species of the first animal. We present the *seu̯en* type as an example (Fig. 5). The *du̯entę seu̯en* is a plastically carved animal-figure from whose neck a *du̯entę moi̯gańi* is suspended. On its hind legs even the paws are elaborated, whereas its forelegs are incomplete. This animal figure is unmistakably a bear, namely a brown bear (Fig. 5a). The *ambanso seu̯en* is an animal figure made of wood with a snub nose, short legs and a slightly turned-up tail. The eyes are two concentric circles with a dot in the centre. On the nose, close to the mouth,

[3] Manuscript No. 975 in the Ethnological Files of the Ethnographic Museum of Budapest, p. 1.

three snakes looking toward the eyes are drawn one below the other. Beyond the eyes two cross-shaped mosquitoes can be seen, and under them two coiling snakes. The entire body and the tail of the animal are covered with black strips. The amulet represents, without any doubt, a tiger (Fig. 5b). The *i̯arga se̯u̯e̯n* is an animal figure, its head is elongated, its legs very short and its tail long. Its head, entire trunk and tail are covered with small circles almost every one with a dot in the centre. This animal is obviously a panther. The form and skin-colouring of the animals represented on the

Fig. 5. Nanai *se̯u̯e̯n* amulets: a) *du̯e̯nte̯ se̯u̯e̯n*; b) *ambanso se̯u̯e̯n*; c) *i̯arga se̯u̯e̯n*

three-grade healing amulets definitely suggest three different species: the *du̯e̯nte̯* displaying a bear, the *ambanso*, a tiger and the *i̯arga*, a panther (Fig. 5c).

This conclusion, drawn from the portrayal of the animals, is confirmed by the fact that all three species are indigenous in the area inhabited by the Nanai, that is to say the region of the Amur, Ussuri and Sungari rivers.

The brown bear (*Ursus arctos* L.) is common in Siberia, and, according to Maak is also well-known in the Nanai settlements: 'All along the Amur, and not only on the banks, but on the islands as well, bears are frequently seen. I saw the largest number of bear foot-prints around the mouth of the Ussuri river and farther down by the large lakes adjoining the Amur.'[4]

The existence of the tiger in this region is more questionable. We read in an old volume of the *Globus*: 'When we gave a report on Gustav Radde's

[4] Maak (1859) pp. 96–7.

trip to Dahuria a few months ago, we mentioned that there were tigers in that region. A short while ago we received an anonymous letter according to which tigers can hardly be found that far north. Nevertheless, this fact has been known and absolutely proved. In answer to this letter we are reprinting a section from A. Nordmann's zoological notes (*Erman's Archives* 1862, Vol. XXI, p. 348). "The tiger (*Felis tigris*) is commonly seen along the southern reaches of the Amur river and in the region of the Sungari and Ussuri, and occasionally turns up along the northern course of the Amur, too".[5] Also R. Maak asserts that tigers are seen here: "Along the middle course of the Amur river, on the area which begins at the Khingan Mountains and ends a little lower than the Ussuri river, tigers (*Felis tigris* L.) are frequently seen. According to the natives they are the most common along the Sungari and Ussuri rivers. Tigers may be seen, as they told me, along the Gorin river, too, and probably the eastern limit of the incidence of the Amur variety does not stretch much farther down than the river of that name".'[6]

In addition to the tiger, however, the panther is also known in the Nanai settlements. According to Maak: 'In the area which begins with the Sungari river and ends a little above the Gorin river, the natives know another feline species besides the tiger. All indications point to this being most likely *Felis irbis*,'[7] i.e. the panther. From the journey he made along the Ussuri, Maak receives a definite answer to the problem of the occurrence of the panther in the Nanai-inhabited regions. 'The panther or *irbis* occurs in the entire Ussuri valley, from the mouth to the upper course of the Ussuri and even around the source of the river. According to the natives, it is, however, by far not so common as the tiger, and in some districts it is completely unknown, as for instance, at the mouth of the Nor river in the area of the Kalang mountain and around the mouth of the Murena river. The natives maintain that it is not found generally along the left bank. The inhabitants of the right bank, however, say almost everywhere that the *irbis* has been seen there. The reason for the unequal distribution of panthers on the two banks of the Ussuri and its connection with the landscapes of the different parts of the Ussuri valley, cannot be settled today. At any rate, the *irbis* is well known to them. The way they described and drew it for me did not point to the tiger.'[8]

I. M. Prževalskij, who visited the Ussuri region almost a decade later than Maak, also testifies to the occasional appearance of panthers in the Nanai-inhabited areas. He writes: 'The panther (*Felis irbis* Müll; *Murtes zibellina*) is seen throughout the Ussuri region but it is incomparably rarer here than the tiger. Moreover, it hides in remote places and, except for very rare occasions, does not approach human habitations even in winter. It reaches the southern region of the Ussuri more often than the banks of the middle course of the Ussuri.'[9] The panther is known, moreover, not only to the Nanai living along the Sungari and Ussuri rivers, but also

[5] *Globus* (1863) p. 191.
[6] Maak (1859) pp. 102–3.
[7] Maak (1859) p. 102.
[8] Maak (1859) p. 115.
[9] Пржевальский (1947) pp. 193–4.

to those near the Amur. According to A. Normann: 'In the region of the Amur and the Ussuri another smaller variety of tiger occurs, the *Felis irbis*. It is not striped but spotted, and is much less frequent.[10]

Prževalskij also finds that 'the panther is found along the middle course of the Amur too, but very rarely'.[11]

The scientists who studied the fauna of the Amur, Sungari and Ussuri regions have contributed to the classification of the animals depicted in the three-grade amulets not only by confirming that bear, tiger and panther are indigenous to Nanai-inhabited areas, but also by informing us that they are the subject of a special cult here and that the idols of the region include these animals.

According to Maak, the Tungus (Nanai, Ulchi) and Nivkh (Gilyak) tribes regard the bear as a sacred animal: 'In their villages they keep them imprisoned in a place made especially for this purpose in order to have them close at hand for the great celebration at which they are offered as sacrifices and eaten'.[12]

The tiger is also honoured by the Nanai. Maak writes that it often attacks people and domestic animals and, therefore, is feared and even idolized by the people.[13]

'Tiger-heads carved of wood, crude but recognizable, are worshipped as idols by the Nanai, while the tiger itself is honoured in their shamanistic songs. They have a belief that, whatever his illness may be, a man recovers if he meets a tiger in the woods and escapes from being devoured.'[14] Maak also mentions certain tiger idols which were decorated with strips.[15]

The Nanai pay homage to the panther. Prževalskij writes: 'It seems that it is worshipped just as frequently as the tiger by the Nanai and the Orochi. Idols of panthers and tigers are often seen in the possession of these peoples.'[16] 'Among the natives living near the lower course of the Ussuri', Maak writes, 'I have seen idols representing panthers. Although they had the same shape as those representing tigers, their colouring was quite different. The figures of panthers were covered with black and red or only red spots.'[17]

The conclusion that the *d̯ṵe̯nte̯*, *ambanso* and *i̯arga* are in fact bear, tiger and panther amulets, respectively, drawn from the form of these animal amulets and from the observations of the naturalists who had visited the Nanai, is fully supported by the language, since the meanings of the above three Nanai words are 'bear', 'tiger', and 'panther', as can be seen from the following examples:

d̯ṵe̯nte̯ 'bear, *Ursus arctos* L.': in Nanai (Baráthosi)[18] *doonta* (ɔ: *d̯ṵe̯nte̯*) 'bear, *Ursus arctos* L.' in Nanai (Lopatin)[19] *doonte* (ɔ: *d̯ṵe̯nte̯*) 'bear'.

[10] *Globus* (1863) p. 191.
[11] Пржевальский (1947) p. 193.
[12] Maak (1859) p. 97.
[13] Maak (1859) p. 102.
[14] *Globus* (1863) p. 191.
[15] Maak (1861) p. 115.
[16] Пржевальский (1947) p. 194.
[17] Maak (1861) p. 115.
[18] Benedek Baráthosi Balogh, *Vocabulary*, in manuscript.
[19] Лопатин (1922) p. 225.

amban 'tiger, *Felis tigris* L.': in Ussuri Nanai *ambán* 'tiger'.[20] In Amur Nanai *ambán* 'tiger'.[21] In the Samar dialect of the Nanai *amba* is 'tiger'[22], and in the language of the Samagir[23] living in the neighbourhood of the Nanai, *amba* is also 'tiger'.[24]

i̯arga 'panther, *Felis irbis* Müll.': in Amur Nanai *i̯arga* 'panther'.[25] In Lower-Ussuri Nanai *jargá* '*Felis irbis* Müll'.[26] In Ussuri Nanai *jerga* '*Felis irbis* Müll'.[27] In Amur *jerga* 'panther'[28] and '*Felis irbis* Müll'.[29] In the Ulchi dialect *jerga* '*Felis irbis* Müll'[30] and in literary Manchu *yarγa* ~ *yarχa* 'panther'[31].

This, we believe, is the answer to the first of our questions raised in connection with the Nanai three-grade amulets. The meaning of the words *du̯ente̯*, *ambanso* and *i̯arga* is now clear.

In the second part of this paper we wish to attempt to clarify why just these animals are pictured on the three-grade healing amulets.

It has been shown in the quotations from naturalists that the bear enjoys special respect among the Nanai. Ethnographers insist on this respect for the bear. It is important from our point of view first to deal with the belief prevalent among the Nanai people that man is descended from the bear.

According to a Nanai legend, the bear visits a woman in the yurta, and from their union children, the *nani χala* 'the Nanai clan' are begotten. When these children grow up, the woman goes to the woods and changes into a female bear. Finally the bear is the victim of a hunter, he disobeys his bear-mother who had asked her children (i.e. the Nanai) not to kill any bears encountered in the taiga.[32]

The same belief is prevalent among the closest relatives of the Nanai, the Udegeis and Orochi.

According to the Udegeis there lived once on earth a man called Egda and his sister. Not knowing that she was his sister, he married her. A little boy and a little girl were born to them. The little boy, however, found out that his parents were brother and sister and betrayed this secret to his father, who therefore sent his wife to certain death and took his son and daughter into the woods. Here he found a bear track and left the little girl there. Then he saw a tiger track and left the boy there. Then he drowned himself in a river. A bear picked up the little girl and lived with her as her husband; the little boy was found by a tigress who lived with him as his wife. The Udegeis stem from the former marriage, from the latter marriage no progeny were born. The legend goes on to relate that when the little boy

[20] Maak (1861) p. 11.
[21] Пртотодьяконов (1901) p. 4.
[22] Schmidt (1928a) p. 21.
[23] For the problem of the Samagir language see Diószegi (1953) pp. 33–44.
[24] Schmidt (1928a) p. 21.
[25] Протодьяконов (1901) p. 52.
[26] Maak (1859) p. 102.
[27] Maak (1861) p. 115.
[28] Maak (1861) p. 14.
[29] Grube (1900) p. 56.
[30] Grube (1900) p. 56.
[31] Gabelentz (1864) p. 225.
[32] Анисимов (1950) p. 312.

grew up, he began to hunt. Once when hunting he came across a bear and mortally wounded it with his arrow. The dying bear first told him that he was his sister's husband, then began to make his will. In the future the boy should never permit his sister to eat bear meat, women should never lie on his skin, and his *os penis* should be preserved and passed down through the female line. All these rules are strictly obeyed by the Udegeis to this day.[33]

An informative Oroch legend is quoted by I. A. Lopatin from V. K. Arsenev's diary, which he read in manuscript form. Once a hunter went out hunting and his sister was left alone at home. While the brother was away, a bear came and took away the girl. The bear married her. The brother kept looking for his sister for a long time and finally found her in the forest. The woman told him that she had married and that her husband was not at home. The brother went out to hunt again. Not far from the sister's yurta he saw a bear and killed it. He dissected it and presented the meat to his sister. When he took the skin and head of the bear to her, she recognized her husband and died of the shock, while the brother lost his mind.[34]

The northern relatives of the Nanai, the Evenki (Tungus) and the Evens (Lamuts) are also familiar with the legend of the marriage between humans and bears. From the Evenki living by the Stony Tunguska river A. F. Anisimov has recorded the following story: 'The bear kidnaps a girl from the settlement (according to one version the first woman living on earth, and according to a different version, one of several girls who were picking lilies on the river bank), makes her his wife, then meets her brother in the forest by the fire, who mortally wounds him with an arrow. As he is dying, he confesses, in a song, that he is the brother-in-law of his killer and that he is to be buried (flayed) not by the hunter but by someone from the bear clan, from the clan of the hunter's brother-in-law.'[35]

The Yakutized Evenki, the Dolgans, have also a traditional story of a girl who became the wife of a bear.[36]

The Even (Lamut) traditions regard the bear's sister as their ancestress. Hence, when they surround the bear's den during the hunt to drive him out, they join hands and sing: 'Father Bear! Our grandmother who is your sister *dantra* has told you not to frighten us but to die.'[37]

The Nanai language provides important information on the native name for the bear. We have so far come across two names: the Nanai *dŭ̯e̯nte̯* and the Even *dantra*. It must be pointed out that these two words are phonetically related: the Even *-ndr-*, *-ntr*, *-nr-* ~ Evenki *-nd-*, *-nn-*, ~ Nanai *-nt-* consonant variations are common in the Tunguso-Manchurian languages.[38] The Even *dantra* (or, according to the more accurate recording, *dū̆ntre̯*) and the Nanai *dŭ̯e̯nte̯* being the two extremes of this Tunguso-Manchurian phonetic correspondence.

The word *dū̆ntre* ~ *dŭ̯e̯nte̯* is a euphemism for bear, but the original meaning of the Even or Nanai word is not 'bear' but 'taiga'. Examples:

[33] Арсеньев (no year) p. 174.
[34] Лопатин (1922) p. 206.
[35] Анисимов (1950) p. 312.
[36] Анисимов (1950) p. 312.
[37] Богораз (1926) p. 9.
[38] Цинциус (1949) pp. 195–203.

Even *dúnre* 'taiga'[39], *dōnrē* 'taiga'[40]; Evenki *dúndę* 'land'[41], *dūnne* 'land'[42]; Nanai *dúęnta na* 'plain covered with pine woods'[43], *dų̂ęntę* 'taiga'[44]; Ulchi *dúęnte na* 'plain covered with pine woods'[45], *dúęntę* 'taiga'[46]. The euphemistic name of the bear, i.e. *dų̂ęntę ęję́ńi* 'the lord of the taiga' goes back to the word meaning 'taiga'. The second part of the expression was gradually dropped, leaving the first word *dų̂ęntę* to stand for 'bear'.

Ethnographic observations, as well as an analysis of the words denoting bear, suggest that the Nanai bear associations are of totemistic origin. The healing power of amulets representing bears should be ascribed to this. The animal ancestors, the totems, occupy an important position in healing among the Siberian peoples, including the Nanai. 'The vast majority of the *ongon* spirits, particularly those which have an animal shape, control diseases, sending them upon people and then recalling them. The *ongons* obviously have inherited this function from their predecessors, the totem-animals. In the history of religion the medical role of the oldest totem and of the *ongon* spirits is regarded as primary, for it was the combat against illness that led man to the cult of the totems, and later to the cult of the *ongons*. In this way primitive people hoped to find protection against disease.[47]

The *tiger*—as I. A. Lopatin points out—probably enjoys an even greater respect among the Nanai than the bear.[48] N. A. Bajkov's collections, on the other hand, show that not only pictures of tigers but even certain parts of the tiger may be used as amulets: tiger claws are worn as amulets offering protection against evil men and the evil eye; tiger skins are believed to destroy all kinds of bad intentions operative from outside and to calm the nerves of the person who is resting on them; the toe-bones of the forelegs bring success in every kind of enterprise; whereas dried tiger-eyes make possible an insight into things normally concealed from the human eyes.[49]

The degree of respect shown to the tiger by the Nanai is evident from the ethnographic descriptions given below.

'A Nanai will never shoot a tiger. If somewhere in the taiga he steps on the footprints of a tiger, he will immediately move off the path. If during the hunt he happens to come across a tiger, he will throw away his gun, bow to the ground and say: *ęję, mafa! ul'andi fuliwanu, mumbiwa lak-lak bauno,* which means: "Well, old man, give us a good hunt, give us food, let the animals come in front of my weapon".'[50]

According to A. N. Bajkov, the Tunguso-Manchurian inhabitants of the Amur region, the Nanai, the Orochi, the Solon, the Manegir and the Evenki

[39] Василевич (1958) p. 126.
[40] Цинциус–Ришес (1952) p. 602.
[41] Benedek Baráthosi Balogh, *Vocabulary*, in manuscript.
[42] Василевич (1948) p. 103.
[43] Grube (1900) p. 86.
[44] Петрова (1935) p. 32; Оненко (1959) p. 323.
[45] Grube (1900) p. 86.
[46] Grube (1900) p. 86.
[47] Зеленин (1936).
[48] Лопатин (1922) p. 206.
[49] Байков (1925) p. 17.
[50] Лопатин (1922) p. 207.

not only honour the tiger as a deity, but present sacrifices to it from their prey. When a tiger happened to kill a man, it was customary in the not too distant past to sacrifice a man tied to a tree if the shaman thought this advisable. When it tore a man to pieces, they said that the victim must have been a pig or a dog in his earlier life, and that the tiger recognized him by the smell, for otherwise it would certainly not have done him any harm. The native fur-hunters refuse to hunt tigers and, in fact, do their best to protect them from recent settlers such as the Chinese and the Russians.[51]

Lopatin writes that if a Nanai kills a tiger by accident, the 'unlucky' hunter feels infinite sorrow and fear. In such cases, as a Nanai related, a richly clad, white-haired old man appears to the hunter in his dream saying to him: 'Why did you kill my son; you should not have crossed his tracks and gone the way he walked. You should not have raised your eyes to him, and if you happened to see him from the distance, you should have touched the ground before him.' Then the white-haired old man scolds the hunter for some time, while the offender must keep quiet, with an air of resignation. When the hunter wakes up, he goes to a near-by settlement where a Nanai of the *aktenka* clan is living. There he reports to the head of the clan what has happened and asks for help. The chief summons all the members of the *aktenka* clan for a conference. Those assembled question the hunter about the circumstances of the offense and carefully think over the seriousness of his guilt. Then they decide what food and drinks will have to be sacrificed by the hunter. Usually from half a pail to three pails of brandy, 2–3 puds (35–50 lb) of meat, dried fish, etc. are allocated, which are taken by the hunter to the same settlement before a fixed time. The head of the clan makes a small model boat of wood and puts in it half a bottle (or less) of brandy, some lard, dried fish, etc. Then the whole party, together with the penitent hunter, go out into the woods to the clan tree of the *aktenka*s. There they all bow. The chief makes the following prayer: 'Old man, be not angry! He killed your son accidentally. The next time he will be more careful. Here is his offering to you. Drink and eat of it and forget the past.' The model boat containing the offering is hung on the tree and the men return to the settlement, where they arrange a feast and eat and drink the hunter's whole sacrifice. They eat the meat with thin porridge made from *buda* and more food prepared by the women from the stored reserves. All through the feast they drink great quantities of brandy. The penitent hunter takes a glass of brandy first to the head of the clan and kneels before him. The latter accepts the glass and says: 'Don't do more harm to the old man'. They all drink, big and small, but the hunter himself must not swallow a drop. Because of the extra brandy, the feast lasts one, two, or three days and sometimes even a whole week. After the completion of the rite, the Nanai who killed the tiger considers himself absolved, feeling that he has atoned for his guilt and has satisfied the old man who disturbed his nights. Should the dream recur, he knows that his sacrifice was insufficient, and repeats the ceremony.[52]

[51] Байков (1925) p. 17.
[52] Лопатин (1922) pp. 207–8.

Actually, the taboo in respect of shooting a tiger extends to any animal wounded or pursued by the tiger. If it later transpires that an animal killed by the hunter was fleeing from a tiger, nothing can force a Nanai to take its meat or skin, or even to touch it. In such cases the hunter must turn towards the disappearing tiger and say: '*mafa*, the animal belongs to you. It was not I who killed it but you.' On the other hand, if the hunter does take the tiger's prey, the tiger will appear to him in his dream and claw him. Here again, he will visit the *aktenka* clan and take them the sacrifice asked for by the clan council.[53] The Nanai are not satisfied merely with sparing tigers and relinquishing any booty it has pursued, but they try to assist it in all possible ways.

Once a Nanai went out hunting in the spring. He left his boat and advanced a few steps when he heard the terrible roar of a tiger. He crawled towards the place whence the roar was coming and noticed that a tiger was trying to crush a giant snake coiled around him against a tree. The Nanai pulled out his hunting knife, the *saamu*, and cut the snake in two with a single blow, thus delivering the tiger from its embrace.[54]

Not only do the Nanai help the tiger, the tiger also helps the Nanai, as is evident from the following tale.

This happened a very long time ago. A Nanai went to hunt in the taiga with his wife and child. They went far up a small river, left their boat on the bank and went deep into the taiga. Soon a storm rose and swept away their boat. They were left without arms and food in the taiga. They wandered about for a long time and nearly starved. The poor things were already preparing for death when it occurred to them to pray to the tiger. They knelt down and cried out loud: 'Our father, seest thou not that we are dying of hunger? Bring us some food!' And the branches of the copse moved and a tiger appeared with a deer in its jaws. It threw its booty at the feet of the Nanai and retreated into the forest. The Nanai, however, did not want to eat the meat of the deer given by the tiger, but the tiger shouted to them from the woods: 'Eat it quickly lest you die'. When they came home, they wanted to make a propitiatory sacrifice, but an old white-haired man appeared to the head of the *aktenka* clan in his dream and told him that no offense had been committed and that he was not at all angry.[55]

It is quite clear from what has been said that the Nanai regard the tiger as a friend, and never kill it. At the same time the tiger is also strikingly 'friendly' toward the Nanai.

The relationship of the tiger to the Nanai and the Nanai to the tiger is characterized by an almost 'fraternal' readiness to help. The existence of this 'fraternal' relationship is supported by the fact that an offense against a tiger (e.g. killing it or taking its prey) can be redeemed only by a great food sacrifice, in the rituals of which the *whole clan* participates. The clan tree, the head and all the members of the clan have to benefit from this sacrifice. Hence, the killing of a tiger is an injury not only to the tiger, but also to one of the Nanai clans, the *aktenka* clan, and the offender must propitiate the clan.

[53] Лопатин (1922) pp. 207–8.
[54] Лопатин (1922) p. 208.
[55] Лопатин (1922) p. 209.

Thus, the relationship of the Nanai to the tiger seems to be based on the assumption of kinship with the tiger, which, like the bear, is considered the ancestor of certain clans. That the *aktenka* clan consider the tiger their ancestor is proved by the following legend about their origin.

This happened long, long ago, before the Russians had burnt down the woods, and the woods still had an abundance of all kinds of animals. Once the Nanai went to hunt on the high Meke hill, taking along with them a girl so as to have someone to cook their meals in the camp. They hunted for a long time and bagged many animals. Once they returned to camp and saw a tiger at the entrance. They were frightened, but the tiger silently retreated to the woods. After that the hunters often found the tiger at the entrance, but the animal never did them any harm. The hunters began to question the girl asking her whether she was not troubled by the tiger. She did not complain and said that it came to see her every day. Once the hunters returned earlier than usual, and when they went into the camp, they found the tiger lying with the girl. The girl pretended to be asleep, whereas the tiger jumped out and ran into the forest. The hunters then knew that the girl was living with the tiger. They questioned her about the tiger's visits. She told them that one morning when nobody was at home except herself, the tiger appeared. She was very much frightened, but the tiger calmed her in a human voice and asked her to be his wife. Soon the girl gave birth to a little boy who was named *aktenka*, meaning 'born of a tiger'. He grew up quickly, became a skilled huntsman and married. His wife gave birth to many children. The *aktenka χala* or *aktenka* clan, and all those who now belong to the *aktenka* clan, descend from him. As Lopatin writes, the Nanai who told this story added: 'Well, that is why we, *aktenka*s, are not afraid of the tiger and that is why he does us no harm. When we happen to meet him during the hunt, we must stand up and show our face to our father, so that he may recognize us. Then he immediately leaves us alone and withdraws into the woods.'[56]

The name of the *aktenka* clan reveals another important point. The tiger is not only ancestor but also ancestress in the beliefs of the Nanai. According to the Udegei legends in connection with the bear origin, Egda's daughter lived with a male bear, and his son with a tigress. The etymology of the clan name *aktenka* (ɔ: *e̮kte̮nke̮*) proves that the Nanai, like the Udegei believed not only in marriage between women and male tigers, but also between men and tigresses. Like the bear, the tiger was regarded not only as an ancestor but also as an ancestress. The Nanai word ends in a variant of the suffix *-ka, -ke̮, -ku, -kŭ* attached to a strong stem (i.e. a stem ending in *-n*) and meaning 'belonging to something'. E.g. *namu-ka, lamu-ka* 'name of an Oroch clan', (belonging to the sea) < *namu, lamu* 'sea'.

Similarly the root of the Nanai clan name is *akte* (ɔ:*e̮kte̮*). This word is in fact known in the sense of 'tiger' in the Tunguso-Manchurian languages, e.g., *etti* (ɔ: *e̮tte̮* < *e̮kte̮*) 'tiger' in Evenki.[57]

The Nanai name for the tiger *akte* (ɔ: *e̮kte̮*) is, however, again a euphemistic name: the original meaning of the word is not tiger, but 'female animal'.[58]

[56] Лопатин (1922) p. 209.
[57] Василевич (1948) p. 296.
[58] Grube (1900) p. 13; Оненко (1959) p. 218.

For instance, Ulchi *ekta* (ɔ: *e̜kte̜*) 'female animal'.[59] The word *e̜kte̜* is also used for 'woman, wife' in Nanai and Ulchi.[60] Therefore the evidence furnished by the Nanai language is in harmony with the totemistic legend attached to the tiger, since the name of the *aktenka* clan really means etymologically 'belonging to the tiger', though according to the testimony of the language the animal ancestor was not a male, but a female tiger. The euphemistic name given to the tiger, and the fact that it is taboo to pronounce the real name of the animal ancestor, again proves that the tigress is a totemistic animal in the Nanai religion.

Apart from the word 'tiger' contained in the Nanai *aktenka* clan-name, the general Nanai name for tiger—*amban*—is also euphemistic. It exists in the Nanai language in two forms: *amban* and *pữre̜n ambańi*. Originally neither of them designated the tiger. The meaning of *amban* is 'great, powerful.' Cf., for instance, Nanai *amba* 'great',[61] Ulchi *amba* 'great',[62] Jurchin *amba* 'powerful',[63] Manchu *amba* 'great'[64]. Thus the original meaning of the *amban* name for tiger is 'great', 'powerful' or 'possessed of power'.

The term *pữre̜n ambańi* means 'the great, the master of the remote places of the taiga'; cf. Nanai *pữre̜n* 'remote place in the taiga'.[65] The *-ńi* ending of *ambańi* is a suffix indicating the third person possessive.

An analysis of ethnographic descriptions and of the names for tiger leads us to conclude that the ideas of the Nanai associated with the tiger, as those which are associated with the bear, are of a totemistic nature. This explains why the tiger—like the bear—has come to be regarded as an animal endowed with the power of healing.

Nothing similar to the bear or tiger cults is known in connection with the panther. The observations made by naturalists and ethnographers on the spot only prove the existence of idols or amulets depicting panthers. No traces of totemistic associations with this animal have been found so far in ethnographic collections. The *i̯arga* 'panther' is not a totem animal, nor is its name euphemistic. How does it then have the power to cure diseases, and how has it come to figure on the healing amulets?

The Nanai believe that illness is caused by some evil spirit which robs its victim of its life spirit and keeps it in captivity. In the meantime it inflicts different kinds of tortures on the life spirit. For instance, when it dips the life spirit into cold water, the owner gets the shivers; when it dips the life spirit into hot water, he develops a temperature. Only the shaman, with the assistance of the *ai̯ami* (or assisting spirit) is able to free the robbed spirit.

We are of the opinion that the healing function attributed to the panther is rooted in this conception of the cause of illness entertained by the Nanai. The names of the life spirit *erga* (ɔ : *e̜rge̜* ∼ *e̜rge̜n*)[66] and of the panther (*i̯arga*) in Nanai are so close phonetically that popular etymology might

[59] Grube (1900) p. 13.
[60] Grube (1900) p. 9; Петрова (1935) p. 36; Оненко (1959) p. 64.
[61] Grube (1900) p. 9.
[62] Grube (1900) p. 9.
[63] Grube (1900) p. 90.
[64] Haneda (1937) p. 19.
[65] Петрова (1935) p. 66.
[66] Grube (1900) p. 11.

have easily linked them together. The *i̯a-* ~ *i̯e-* ~ *e* sound alternation is known in Nanai, e.g.:

i̯aχa[67]	*i̯eχχa*[68]	*eχa*[69]	'cow, ox'
i̯asa[70]	*i̯esa*[71]	*esa-l*[72]	'eye'
i̯anda-ku[73]	*i̯endo-la*[74]	*enda*[75]	'dog'

Thus, owing to this phonetic alternation, the form of the Nanai *i̯arga* 'panther' grew more and more similar to *erga* meaning 'life spirit' and the word *i̯arga*—as we have seen—does actually occur in the variant *i̯erga* and even in the form *erga*, as in the Nanai *erga* ('panther').[76] As a result of word magic prevalent among the Nanai, the accidental resemblance between the two names may have given rise to the idea that the *erga* or panther could help them in regaining the kidnapped *erga*, or life spirit.

At the same time, endowing the panther with healing power also may have been assisted by other facts. For one thing, for instance, other animals are included in the medicinal practice of the Nanai, and one of them is related to the panther, namely the tiger. Then again, its strong, ruthless and terrifying nature makes the panther fit for overcoming the evil spirit.

We may, therefore, safely conclude that the answer to the second question, namely 'Why have the Nanai chosen the figures of the bear, the tiger and the panther for healing?' is that the first two are totem animals and the third became a curing animal as a consequence of word magic.

Now why are the three-grade amulets used in the order bear—tiger—panther? The answer seems to be an obvious one. The degree of fear elicited by the three animals is reflected in the order of the application of the three amulets. Naturalists say that the bear is less dangerous than the panther or the tiger and the panther is more terrifying than the tiger. Prževalskij, for instance, writes: 'The natives here are more afraid of the panther than of the tiger, because they cannot escape from a panther by climbing a tree since it is an excellent climber'.[77] The order of the three classes of the amulets does correspond to the ferocity of the animals they represent and to the degree of fear inspired by each. Fear, then, played its part. But we are of the opinion that it contributed to the preservation of a custom rather than being its origin. The real causes of this established order have deeper roots.

We should attempt to solve this problem by examining the nature of the conceptions associated with the animals, their range among the Nanai, the peculiarities of the names of the animals and their geographical distribution.

[67] Grube (1900) p. 55.
[68] Grube (1900) p. 55.
[69] Grube (1900) p. 55.
[70] Протодьяконов (1901) p. 52.
[71] Grube (1900) p. 51.
[72] Grube (1900) p. 51.
[73] Протодьяконов (1901) p. 52.
[74] Grube (1900) p. 15.
[75] Grube (1900) p. 15.
[76] Benedek Baráthosi Balogh, *Vocabulary*, in manuscript.
[77] Пржевальский (1947) p. 194.

Totemistic concepts associated with the bear occur among all the Tunguso-Manchurian peoples—among the Evenki and Evens of the northern branch, just as well as among the Nanai, Ulchi, Udegeis, or those Evenki who have become closely assimilated with the Yakuts—the Dolgans. As to the time of the origin of these bear concepts it may be assumed that, since all the ideas about bear ancestors are found in very similar forms both among the northern and southern groups, the belief in the bear totem must have arisen before these peoples broke up into the Tungus and Manchu groups.

On the other hand, the tiger has a totem value only among the southern Tunguso-Manchurians:—the Nanai and, perhaps, the Udegei. Hence, there are two possible conclusions: Either the idea existed among all the Tunguso-Manchurian peoples, but the northern group forgot it, or it arose only among the clans of the southern branch as a separate development. The first possibility is not very likely, because it is difficult to imagine that concepts associated with a totem animal could disappear without any trace. Reference to the totem animals should be found at least in some totemistic clan names. For instance, the *djuku-git, djuke-il, juke-il* clan of the Evenki has preserved the memory of the otter (cf. *djukun* 'otter') in its name, the *kukti-gir, kuχti-gir* clan, the name of the cuckoo (cf. *kukti* 'cuckoo'), the *sili-gir* clan, the name of the stork (cf. *sili* 'stork'), the *kongnoko-gir* clan, that of the elk (cf. *kongnoko* 'elk'), etc. as animal ancestors.[78] Far more acceptable is the theory that the tiger as a totem animal developed only among the southern Tunguso-Manchurian peoples. It is strongly supported by the geographical distribution of the tiger—the Amur, Sungari and Ussui regions. It is only natural that the group of the Tunguso-Manchurian peoples who had never lived in these areas (i.e., the clans of the northern branch) could not have honoured this animal as a totem, since they did not know it, and if the tiger became a totem animal only after the southern branch had moved south, then the belief in a tiger ancestor is certainly from a later date than the belief in a bear ancestor.

Certain conclusions in connection with the temporal sequence of the development of the ideas associated with the two animal ancestors may also be drawn from their names. Whereas the names of the amulets representing the bear and the panther are *dʉ̯̑ente̯* ('bear') and *i̯arga* ('panther'), respectively, the name of those depicting tigers is not *amban* ('tiger')—as one would expect—but *ambanso* ~ *ambanseo*. It is this *-so*, *-seo* ending that provides us with a clue to the element of time. As is evident from other forms, as *amban sewo* (cf. *amban sewo muχani*[79]) it corresponds to the words *sewo* ~ ~ *seʉ̯̑en* ~ *sùn* and means 'animal-shaped spirit'. This ending was affixed to the word *amban* (tiger) to distinguish the tiger from a different *amban*—the evil spirit. The name for evil spirit is also *amban* in Nanai: The Nanai *amban* ~ *amba* means 'evil deity, devil'[80], the Ulchi *amban* 'evil spirit, devil'[81], the Oroch *amban* 'evil spirit, devil'[82], etc. Again this name is euphemistic, for etymologically it means 'great' and, hence, is derived from the word *amba* meaning 'great' as does the name of the tiger. The phonetic

[78] Анисимов (1951) p. 194.
[79] Poniatowski (1923) p. 3.
[80] Протодьяконов (1901) p. 4.
[81] Grube (1900) p. 9.
[82] Schmidt (1928b) p. 24.

identity of the names for tiger and evil spirit was, however, in the way of thinking of the Nanai, a source of great danger because of word magic. Whenever the Nanai uttered the name of the tiger or of the tiger spirit—that is to say, the word *amban*—the danger always existed that not the helpful tiger spirit, but the harmful evil spirit would appear at the call. In order to avoid this, when they spoke of the tiger spirit, they did not say 'tiger' or *amban* as in the case of the bear or the panther, but 'tiger spirit', i.e., *amban sewo*～*ambanseo*～*ambanso*. From the chronological point of view, important for us, it is a significant fact that the modification took place in the name of the tiger spirit and not in that of the evil spirit *amban*. This makes it obvious that the name *amban* for evil spirit—and with the name also the notion—existed already prior to the development of the name and concept of the tiger spirit. It must have been the very existence of a concept for the evil spirit that necessitated the addition of the tiger spirit to the bear spirit in healing. The Nanai probably believed, on the basis of word magic, that the healing of an illness caused by the evil spirit, the *amban*, would be much more successful through the spirit of the *amban* than through that of the *dų̈ente̦*.

The belief in the healing power of the panther spirit must have developed still later. Its later origin compared to the bear is proved by its geographical distribution. It occurs in the very same southern region (occupied only by the clans of the southern group at a later date) where the tiger is indigenous. That it developed after the belief in the healing power of the tiger spirit, is proved by the mere fact that it became a healing animal at all. It did not originate in the ancient totemistic notion of the cause of diseases (by a totem animal: hence only a totem animal can drive it out), but in a belief of much later origin, (that diseases are the consequence of the kidnapping of the life spirit or *erga*). The healing power of the panther is due to word magic. We have demonstrated in the present study that the *dų̈ente̦*, *ambanso* and *i̯arga* amulets are used successively in the given order for the treatment of the same disease and that they represent the spirits of three animals: the bear, the tiger and the panther. These three animals were given a place in Nanai healing practice on account of their notions of the cause of disease. According to them men descend from certain animal ancestors and these send diseases upon them or call them back. The spirit and image of the bear became a healing spirit or amulet, because the bear is the totem animal of numerous clans of the Tunguso-Manchurian peoples.

'Healing' with the bear amulet was, however,—we may well imagine—not always successful. The cause of failure came to be explained by supposing that in addition to the animal ancestors known until then, other disease-causing factors had developed. This prompted the development of the belief in the evil spirits of *ambas*. The spirit of the bear, the bear amulet, of course, could not be expected to counteract these with the same efficacy.

The changes in beliefs connected with the causes of diseases and the increase in the number of accepted causes went (necessarily) hand in hand with modifications among the healing spirits. A new healing spirit and amulet developed: the tiger spirit and its representation. As an animal ancestor it could successfully counteract diseases caused by totem animals, and because of its name *amban* it could be counted on being effective against the evil spirit or *amban* through word magic.

But then the remedy by means of the tiger amulet often failed, too. Hence a new belief developed: Diseases were caused by the disease spirit (the animal ancestor or the evil spirit) which kidnaps the life spirit of the person who becomes ill, and inflicts different kinds of torture on it. This change was shown once more in the concept and representation of the healing animal spirits. The panther, that is the *erga*∼*i̯erga*∼*i̯arga* was included among them, for its spirit and representation were believed to be capable of successfully combating the disease-causing spirits which had kidnapped and held captive the life spirit or *erga*.

The bear, tiger and panther on the three-grade amulets constitute proofs of changes that have occurred in the notions of disease and healing among the Nanai over a period of thousands of years.

REFERENCES

(Anisimov, A. F.) Анисимов, А. Ф. (1950) Культ медведя у эвенков и проблема эволюции тотемистических верований (The Bear Cult of the Evenki and the Problem of the Evolution of Totemistic Beliefs), in *ВИРА*.

(1951) Шаманские духи по воззрениям эвенков и тотемистические истоки идеологии шаманства (Shaman Spirits in the Belief of the Evenki and the Totemistic Sources of Shamanistic Ideology), in *СбМАЭ*, Vol. XIII.

(Arsenjev, V. K.) Арсеньев, В. К. (no year) Лесные люды, удехейцы (Forest People, the Udegei), in *Сочинения*, Vol. 5.

(Bajkov, N. A.) Байков, Н. А. (1925) *Маньчжурский тигр* (The Manchu Tiger), Kharbin.

(Bogoraz, V. G.) Богораз, В. Г. (1926) Миф об умирающем и воскресающем звере (The Myth of the Dying and Resurrected Beast), in *ХФ*, Vol. I.

(Cincius, V. I.) Цинциус, В. И. (1949) *Сравнительная фонетика тунгусо-маньчжурских языков* (Comparative Phonetics of the Tunguso-Manchurian Languages), Leningrad.

(1952) *Русско-эвенский словарь* (Russian–Even Dictionary), Moscow.

(Diószegi, V.) Диосеги, В. (1949) Развитие одного вида целительных амулетов у гольдов (The Development of One Form of Curative Amulet among the Golds), in *FE*, Vol. I.

Diószegi, V. (1953) Le problème de l'ethnogenèse des Samaghirs, in *AO*, Vol. III.

Gabelentz, H. C. (1864) *Mandschu–Deutsches Wörterbuch*, Leipzig.

Globus (1863) Vol. IV.

Grube, W. (1896) *Die Sprache und Schrift der Jučen*, Leipzig.

(1900) *Goldisch–Deutsches Wörterverzeichnis*, Saint Petersburg.

Haneda, T. (1937) *Manwa jiten* (Manchu Dictionary), Kyoto.

(Lopatin, I. A.) Лопатин, И. А. (1922) *Гольды амурские, уссурийские и сунгарийские* (The Golds of the Amur, the Ussuri, and the Sungari), Vladivostok.

(Maak, R.) Маак, Р. (1859) *Путешествие на Амур* (Journey to the Amur River), Saint Petersburg.

(1861) *Путешествие по долине реки Уссури* (Journey along the Valley of the Ussuri River), Vol. I, Saint Petersburg.

(Onenko, S. N.) Оненко, С. Н. (1959) *Русско-нанайский словарь* (Russian–Nanai Dictionary), Leningrad.

(Petrova, T. I.) Петрова, Т. И. (1935) *Краткий нанайско-русский словарь* (Short Nanai–Russian Dictionary), Moscow-Leningrad.

Poniatowski, St. (1923) Materials to the Vocabulary of the Amur Gold, in *BUP*, Vol. 10.

(Protodjakonov, Pr.) Протодьяконов, Пр. (1901) *Гольдско-русский словарь* (Gold–Russian Dictionary), Vladivostok.

(Prževalskij, I. M.) Пржевальский, И. М. (1947) *Путешествие в Уссурийском крае 1867–1869 гг.* (Journey in the Ussuri Region in 1867–1869), Moscow.

Schmidt, P. (1928a) *The Language of the Samagirs*, Riga.

(1928b) *The Language of the Oroches*, Riga.

(Vasilevič, G. M.) Василевич, Г. М. (1948) *Русско-эвенкийский (русско-тунгусский) словарь* (Russian–Evenki, i.e. Russian–Tungus Dictionary), Moscow.

(1958) *Эвенкийско-русский словарь* (Evenki–Russian Dictionary), Moscow.

(Zelenin, D. K.) Зеленин, Д. К. (1936) *Культ онгонов в Сибири* (The Cult of the Ongons in Siberia), Moscow–Leningrad.

Features of the Ancient Religious Rites and Taboos of the Nivkhi (Gilyaks)

by

Č.M. Taksami, Leningrad

The present article is based on the personal observations of the author and on material he collected from old Nivkhi during a stay in the Amur district (1956) and on the Sakhalin Island (1957) for the purpose of examining religious rites and taboos.

According to the ancient belief of the Nivkhi, every settlement had its own 'master' (host): *mif yz* 'master of the earth', *tol yz* 'master of the water' and *kur* 'master of heaven', on whom life itself and the health and welfare of the entire settlement depended. In this connection, the Nivkhi observed a ceremony ('the feeding of the master spirits') the aim of which was to secure successful fishing, lucky hunting and good health.

Unlucky events, such as the death of a family member (especially of a child), unsuccessful fishing, etc. were considered evidence of malice borne to them by the master spirit of the place. When moving to new living quarters, they offered sacrifices to the master spirit of the new place within nine days after arrival. A redhaired dog was needed for this purpose, which was either to be carried there, or procured on the spot. The sacrifice was offered in the forest, while the head of the family uttered the following prayer, *mif yza! ńyŋnamagur χumguja. olagu morǩagur χumguja. ńyŋaχ čoǧo naǧo malǧogur ŋyŋguja!*[1] 'Master of the earth! Give us a good life! May our children live and be healthy. Help us to kill much fish and game!'

The new settlers arranged feasts in their houses in which the whole population of the village participated. This custom of offering sacrifices was called 'feeding the master of the place' (*mif yz ard'*). Sacrifices were offered in order to protect the family against diseases and calamities and to obtain, as a reward from the master spirit, lucky fishing and hunting. It was believed that, if the entire family remained in good health and hunted and fished with success for 1–3 months after the change, the master spirit had admitted the newly installed family. If, however, members of the family began to fall ill or died, and if fishing and hunting were bad, it was believed that the master spirit had rejected the newcomers, and the family moved to another place.

If somebody (especially a child) was about to leave his village, the Nivkhi offered a sacrifice to the master spirit of the place. This was called 'looking for luck' (*kyz ŋyŋd'*). During the feeding, the Nivkhi appealed to the 'master of the earth', asking him to protect the child from diseases and to grant him a good journey and good health. They believed that if this 'feeding' was omitted, the person who left might get sick or even die. The sacrifice was generally offered by the mother of the departing person. Two or three days before the departure the women prepared meals; the night be-

[1] The following phonetic symbols are used in this paper for the transcription of the Nivkhi words: *p*ʿ for aspirate *p*, *ť* for aspirate *t*, *ǧ* for uvular *g*, *ǩ* for uvular *k*, and *η* for the velar nasal.

fore the departure, the traveller took a bit from every dish, then his mother took a spoonful of everything in a dish and carried it to the forest near the village. On her way, she was accompanied by her son who was leaving. Once there, she sprinkled food on the earth with a small spoon: the first spoonful was meant for the 'master' on whose ground the village was standing, the second one, for the 'master' of the mother's native place, and the third one, for the 'master' of the place where the son was going to. The feeding was accompanied by the following prayer: *ty miva kinŋygur oǧlagu nyŋguja ťa mujnyguja* 'Earth! Make it so that my children should have a good journey and have no illness'.

A similar 'feeding' took place, when somebody returned to his native village after a long absence. The traveller was given leaf-tobacco, which he had to bury in the ground. This was the 'feeding of the earth', and it was accompanied by the following prayer: *miva ńaχ namagur χumguja ťa mujnyguja* 'Earth! Make it so that I should live well and not become ill!'

The Nivkhi had also major religious ceremonies attended by the whole population of the settlement. Besides the well-known bear festivities, there were religious feasts like the 'freezing of the Amur', 'the thawing of the Amur', 'the feeding of heaven'. The aim was always to secure good fishing.

The feast of the frozen river was called *toľ fynď* which means literally 'to leap into the water' or 'to throw into the water'. Sometimes the Nivkhi called it simply 'the feeding of the water' (*toľ arď*) and celebrated it immediately after the water was covered with ice. The day of the feast was fixed by the oldest man in the settlement. After he had issued his orders, every family prepared good meals: mush (*rak̆*), jelly made of fish-skin (*mos*), jam (*χolod̆*), sun-dried fish (*ma*), etc. Late in the evening of the chosen day, the head of each family took the dishes and a salmon tail dried in autumn, and went to the Amur, every man to his own ice hole. First, the salmon tail was stuck into the hole with the words *k̆a-k̆a-k̆a*, whereby the Nivkhi believed to have given food to the draught-dog (*tol k̆anarď*) of the 'master of the water'. Then, with appropriate imploring words, a bit of food from every dish, meant for the 'master of the water' himself, was thrown into the hole. After this, everybody went home and treated the neighbours to the remaining food. Only men were allowed to attend the ceremony; women, sitting on benches during that time, were forbidden even to walk on the floor. The next day, a great popular feast was celebrated in the village, with dancing, dog races, etc.

The thawing of the Amur ice was celebrated in the same way as the freezing, only the dog race was substituted by a boat race. Before the 'feeding' of the Amur, the Nivkhi did not fish, since they believed that the master spirit of the water would not give them fish anyway.

The third common feast of the village was the 'feast of heaven' (*člyraχ čyunyd*), celebrated simultaneously with the opening of the winter path along the Amur. Three larch trees were erected in a triangle in the middle of the village. First the old people sat down around them, behind them the youth. The trees were smeared with food, with the blood of a pig and a cock that had been killed there, and sprinkled with vodka. The oldest Nivkhi implored the master of heaven to protect the inhabitants against various misfortunes. After having 'treated' the larch trees, those present at the feast ate the remaining food, and everybody went home and continued the cel-

Fig. 1. Ritual cups (*tolŋir*) for 'feeding the river' at the freezing and thawing of the Amur

Fig. 2. Dwelling-house with carved wooden figures of the sun and the moon and shavings hanging from the roof

ebration. On that day nobody worked or lent anything, because it was believed that luck and happiness would go with the lent object. According to the Nivkhi, the master called *t'ly-yz* or *kur* lives in heaven. It is he who takes care of all men, gives the new-born to the parents and also fixes the date of their death. In order to obtain the favour of this mighty 'master', the Nivkhi regaled him as well.[2]

Originally, religious feasts were celebrated within the clans; but as the clans broke up, other villagers, irrespective of the clan they belonged to, began to attend the ceremonies.

The building of houses and the conducting of indoor life were accompanied by various rites and taboos. The place of the winter residence was generally chosen by the oldest man of the village. A place was selected, where no building of any kind had ever stood before. Sometimes the shaman was invited to choose the place for the house. He 'made inquiries' whether or not it was inhabited by evil spirits. If he 'noticed' the presence of such spirits, he designated another place and the house was eventually built there. Before the construction was started, slabs of stone (or, more frequently, of flint), as well as eglantine twigs, were put into the holes dug for the corner props. According to the Nivkhi, it was necessary to do this in order to prevent the 'evil spirit' (*kins*) from boring a hole into the timber and entering the house through it. The eglantine twigs were supposed to safeguard the logs from the evil spirits. The same result could be obtained by placing big, dried pike heads with sharp teeth under the timber. Sometimes, on the advice of the shaman, flour was dusted on the construction site before placing the first beams. It was considered to be a favourable omen if the thin sheet of flour was untouched or showed human footprints the next morning and the building operations continued. But if the flour showed the prints of a human hand or of the paw of a wild animal, the house had to be built in another place, as this was a bad omen. When the work was finished, a stone was fixed under the threshold. The Nivkhi had a special word for it: *nit'—vaχ*. This stone then remained there for good. It was believed that the master who protected the door against the evil spirits lived under the threshold. The Nivkhi of the western shores and of the northern part of Sakhalin Island, put an axe on the threshold for the night, after the building operation had been completed. This was to prevent the evil spirit from entering the house during the night in the absence of its inhabitants. At sunset of the day when the building was finished, the master went into the house to find out whether there was any light to be seen or noise to be heard. A noise or creaking was thought to be a bad omen, proving that the evil spirit had settled down in the house.

When a new house was completed, the owners immediately moved in, so that the evil spirits should not get in before them. A special ceremony *tyv-ard* 'the feeding of the house' was performed on that occasion. Four dogs were strangled, one in each corner of the house. Sometimes only two dogs were killed in the two front corners. (The number of the killed dogs depended apparently on the wealth of the family.) Every corner was smeared with their blood; their meat was given as a treat to the guests. But other tasty Nivkhi dishes were offered as well: meat jelly (*mos*), mush

[2] Cf. the story about the 'master' of heaven. See *Text* 1 on p. 419.

(rak̆), berries mixed with seal-fat *(als)*, a special kind of berry jam *(χolad')*. It was not permitted to serve roasted food. The master of the house took a piece from each dish with which he smeared the various parts of the house: the main girder, its supports, the joints where these meet, the little hole in the front wall, etc. The scraps of food were thrown into the fire. During this ceremony, the master addressed the following words to each part of the house: *urla dyf muja. ńōlagu kinηygur χum-guja. t'a sid'χagin kolaftaχumguja.* 'Be a good house! Grant my children a happy life! Let them not be sick!' Then, the honoured guests, the old men, sat to table and began to eat and to drink vodka. Every guest felt obliged to repeat the prayer of the host. The Nivkhi believed that special 'master spirits' live in these parts of the house, hence they were always fed when there was a special treat.

Fig. 3. The master spirit of the house (*tyv yz*)

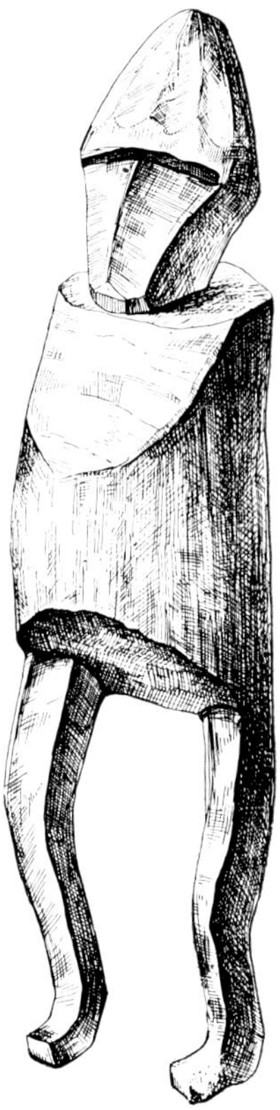

Fig. 4. The master spirit of the house (*tyv yz*)

The couch *(p'onaχ)* in the rear was considered the most distinguished part of the house. It was here that the guests of honour were received and the sacrifices offered. Dogs' skulls, bound in old nets, hung above the door of old houses. In Russian-type log houses, the dogs' skulls were suspended in the loft, on the front wall. Dogs' skulls also hung above the entrance of the summer houses. The Nivkhi believed that they prevented the evil spirits from entering the house.

After moving into the new house, the master planted two long stakes near the building, one at the front and the other at the rear traverse wall *(čχyu ziğr)*. A rope was stretched between the stakes with wooden figures of the sun *(keη)* and the moon *(loη)* and wood shavings *(čaχ)* hanging on it. Sometimes figures of axes and knives were also hung up. According to the religious beliefs of the Nivkhi, these objects protected the house and its inhabitants against misfortunes. The name of this rite was *tyv ziğd'* or *tyv-źak̆aod'* 'the strengthening of the house'.

Life inside the house was governed by numerous taboos strictly observed by the Nivkhi.

Every clan had its own firestone. In the Nivkhi house, the fire was subject to several prohibitions. The first to light a fire or a candle in a new house was the oldest member of the family, and he had to use the tribal firestone. Only members of the family or of the tribe were allowed to light a fire or a candle in the house. The fireplace could be cleaned only in the morning. To clean the fireplace and to carry out the ashes in the evening meant to bring down disaster and death on a family member.

During the bear festivities, ashes and charcoal were removed from the fireplace, the earthen lining was replaced and fire was kindled with the tribal firestone. Before lighting the fire, a small feast was arranged in honour of the 'mistress' or the 'master' of the fireplace. If there were two fireplaces in the house, bear meat was cooked in the one nearest to the forest (*palerk̄t̄ur*). In the evening or at night, when a member of the family went out of the house, it was forbidden to put the lamp out, or else the person who went out, would die. It was considered to be a grave offense if the fire or the lamp was extinguished by somebody belonging to another clan. The offender had to pay a fine (*nit' lytt'*), such as a dog, a trough, etc.

When members of a clan, or of a family belonging to the clan, moved to a new place, the oldest member of the clan broke the firestone in two and gave one piece to the oldest man in the leaving family. It was to be used to make fire at the new place and had to be passed on the other members of the clan for the same purpose. Hence originated the phrase, 'to break the fire', meaning the breaking up of a clan.

The Nivkhi thought that the master or mistress of the fire lived in the fireplace. He or she had great power, supplied men with warmth, cooked the meat and the fish, boiled the water and protected men against misfortunes. The master of the fire was also considered to be a mediator between mankind and nature, able to offer help during the hunt. While the hunter was asleep, the master of the fire hunted for him; if the spirit had not killed anything for him, the hunter would return with an empty bag. Therefore, when the hunter offered a sacrifice to the fire before going hunting, he not only felt that he had strengthened the 'master' of the fire, but he had also assured himself successful hunting. As an outcome of this belief, wherever they happened to be, the Nivkhi always fed the fire during meals, casting a morsel of the food they were eating into it. For instance, when arriving at a hunting ground, the hunter portioned out some food for the master of the fire on the very first day and threw its share into the fire before each meal. Since the master of the fire was supposed to live in the fire, the Nivkhi never touched the fire (now the stove) with a knife, nor did they spit into the fire. When somebody, by mistake, dropped his knife into the fire or pricked the stove with its point, the Nivkhi sprinkled the spot with tobacco. They believed that a person who spat into the fire, would get ulcers on the tongue and face or, if she were a woman, her breasts would begin to hurt.

Fig. 5. Amulet for curing children

When the men were out hunting, those who remained at home had to observe strict taboos handed down from one generation to the other, aiming at the protection of the hunter from a tragic death and at the ensuring of a successful hunt. The violation of these was considered a great sin. For instance, it was strictly forbidden for women to embroider, since it was believed that the hunter might get lost in the taiga, or walk in circles, along the same path, like the ornaments of the embroidery under the skilful hands of the women. Nor was it permitted to invite the shaman during that time, because the sound of the drum would bounce back from the traps set by the hunter and the game would run away. For the same reason it was forbidden to sing. Dough could not be stirred as it might stick to the eyes even from far away. If an object had been in the house during the night, it could not be carried out or given even to a member of the clan. Fire was not to be carried out of the house, and it was not allowed to receive guests for the night. The women could not comb themselves at home but had to go to a house whence nobody went out hunting, for else they would dirty the floor and the hunter would fall on impassable ways.

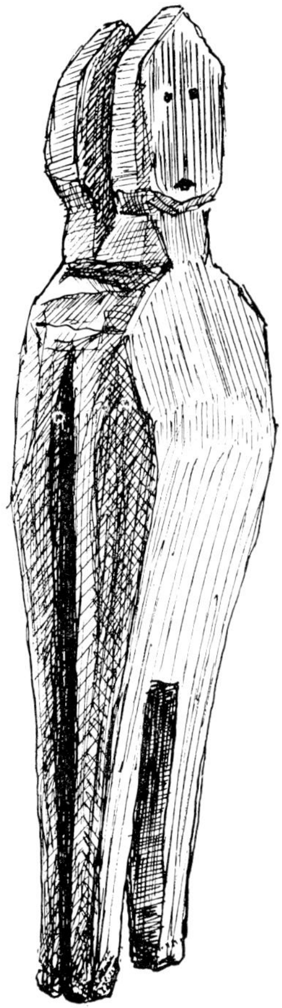

Fig. 6. Amulet for curing children

Many taboos had to be observed at home when the men were hunting sea-animals. It was strictly forbidden for the women to wash in the house of the man whose boat the hunters were using. To reach something over the fire to another person was equally prohibited, because it meant that the seal caught in a net could swim through it. In the house the guest was not supposed to touch the handle of the kettle or of the teapot; if he did so accidentally, he had to tear off a piece of his sleeve and put it on the fireplace.[3]

The Nivkhi were very much afraid of female 'impurity'. Women were considered to be particularly dangerous during pregnancy, delivery and menstruation. This, of course, did not apply to little girls and old women. Menstrual blood was considered so impure that even bears were afraid of it. It was thought to be most inadvisable, and to involve all kinds of troubles and misfortunes for a woman to touch a bear. Women, being impure, were not allowed to bear their children at home. When the time came, irrespective of the season or the hour of the day, they had to go to a tent that had been built and furnished specially for that purpose. Nobody was allowed to enter except the midwife. Only after special purifying ceremonies could a man have intercourse with a woman after confinement.

[3] Крейнович (1934) p. 85.

The purification ceremonies was performed as follows: The infant was carried home immediately after the birth, but the mother remained in the tent for four days if the child was a girl, and for three, if it was a boy. Then a clean dress was put on her and her hair was combed. Since her impurity was believed to pass into the dresses she had on during the delivery, they were rolled together and preserved in a barn until the next birth. Once in the house, the woman was moved near to the fireplace and laid on the floor on some straw which had been spread there for her. The woman had to stay there for three or four days depending on whether the child was a boy or a girl. On the last day she got up early in the morning, changed clothes and carried the old ones and her 'bed' into the barn. She washed and combed herself, and only then was she allowed to lie on the couch. After delivery, when the woman went in or out of the house, the fire was protected from her either by covering it, or by putting it out. This went on for three days. The Nivkhi thought that, if they failed to do this, they would catch no fish. For five days following the birth no guests or foreign tribesmen were received in the house. The menstruating woman was not allowed to touch the fireplace or to prepare the food for the people living in the house. The cleansing ceremony was called 'self-purification' (*psarnd*).[4]

From the very moment the child was born, until it grew up, the chief concern of the parents was to isolate it from the evil spirits that penetrated into the house. If a child died shortly after its birth—and this was by no means a rare occurrence—the parents, when having a new child, performed a special ceremony. This was described to me, near the mouth of the Amur, as follows:

The new-born baby was taken into the house and put down near the threshold. The midwife swept the floor and scattered the sweepings onto the child, which was then lifted up. The sweepings were shaken off it, and the child was carried to the couch. Sometimes it was seated on a shovel on which it was pulled to the couch. The natives believed that, in this way, the evil spirit which had entered with the infant, could be expelled.

On the island of Sakhalin it was customary, if newborn children had died one after the other in a family, that the last one—so as to be saved from death—was not taken into the house through the door, but through a hole that had been cut under the lowest rafter of the rear wall. The father passed the snow-shovel through the hole he himself had cut; the mother came out of the hut, carried the infant to the hole and placed it on the shovel. The father then pulled the child into the house and carried it on the shovel to the door. Now a girl, specially invited, entered the house and stepped over the infant lying on the threshold. This was called the 'stepping over' (*jalimynd*) ceremony. E. A. Krejnovič thinks that the idea of pulling the baby in through the hole was to make the evil spirit lose its way.

When a death occurred in a house, a special rite was observed. The corpse was instantly turned with the head toward the west. The Nivkhi believed that the 'realm of death' (*mly-vo*) was where the sun sets and that the dead had to go there.

A dead man was laid out in the house for three days, a dead woman for four. As soon as the corpse had been carried out of the house, a piece of iron

[4] Крейнович (1930) p. 93.

was driven into the earth where the dead had been lying and another one near the threshold. The Nivkhi said that while the iron was being driven into the earth, the whole house resounded from the noise and the evil spirit (*kins*) who had killed the man felt that he too would be nailed to the floor together with the iron, and ran away after the dead. The iron driven into the threshold was supposed to prevent the evil spirit from re-entering the house. After the corpse had been carried out, the house was surrounded by a string under the eaves, so as to prevent the evil spirit from discovering that someone had died in the house. If this had not been done, the evil spirits would crowd in the house to kill the remaining members of the family. The string served as a sanitary cordon, so to speak, separating the house from the surrounding area where the evil spirits were still making mischief. It was never removed but, in course of time decayed and fell off by itself.

The Nivkhi term for the ceremony of driving iron into the floor and into the threshold, and for enclosing the house with a string, is *tyv-źaǩoď* (literally: 'to strengthen the house'). Its main purpose was to protect the remaining members of the family from death.

If somebody lay dead in the house, the Nivkhi never raised their voice above a whisper; this custom can be observed even in our days. It is strictly forbidden to speak loud, to make noise or to laugh, because that would hurt the dead. The author frequently attending funeral rites had the opportunity to observe the prevalence of this custom. L. Ja. Šternberg's statement, according to which tradition demanded that in the presence of the dead one had to laugh and make fun and that silence in the house of the dead was a sin,[5] does not correspond to the facts today.

When somebody dies in a house, two candles or lamps are lit, one near the head of the deceased, the other, in its usual place. In the morning, the candle near the head is first extinguished, or else the other member of the family will die soon. A woman told me the following story. In the summer of 1954 her elder brother died in the village of Vaida. His relatives lit up one lamp instead of the two and failed to carry it to the burial ground with his other belongings. Because of this, the widow was obsessed by visions for a long time: she saw her husband coming and stopping near the window. After much hesitation, the lamp was taken to the grave of her husband and, from that time on, he no longer returned. This narrative emphasizes the reason behind the Nivkhi custom of not leaving the belongings of the dead in the house but of carrying them to the burial ground together with the corpse.

During the bear festivities, the Nivkhi observed various taboos relating to behaviour in the dwelling house. Since it was believed that bears could not stand menstrual blood and even died from it, women and girls were allowed to touch only certain objects and to walk only in certain parts of the house. They were forbidden to walk between the rear couch and the stake to which the bear was bound, or to touch the pillars.

During the bear festivities, the bear's head and skin were carried through the upper smoke-hole into the house if it was built half dug into the ground, and through the window of the rear wall if it was built on ground level. According to L. Ja. Šternberg this ceremony is a survival of the days when the Nivkhi entered their homes through the smoke-holes. Nowadays, the

[5] Штерноерг (1933) p. 80.

Fig. 7. Amulet for curing children

Fig. 8. Amulet for curing children

Nivkhi living in the Amur district and on Sakhalin account for not carrying the bear's head and skin through the door by the fact that menstruating women cross the threshold when entering the house. The winter houses of the Nivkhi (*to ryf*: the house dug into the ground, *čad ryf* or *čand ryf*: the house built on the surface of the ground) were generally occupied by several (3 to 5) families, including often as many as 15 to 30 persons. As a matter of course, it was most probable that one of the girls or women, was menstruating. Therefore we may assume that the Nivkhi avoided the door while carrying the bear's head in and out. If the window was used for this purpose, the window ledge was stained with blood that could not be washed off. If a person did so, he was sure to die.

An interesting ceremony was performed in the house during the wedding. Before the bride left the house, her parents placed a big cast-iron kettle (*la vyń*) in the street, near the threshold. The parents of the bridegroom placed a similar kettle near the threshold inside the room. The newly married couple crossed the threshold, stood with both feet in the kettle and only then started on their way. The parents of the husband took the kettle and carried it home, leaving their own to the parents of the bride. If the young woman's parents wished their daughter to give birth to a boy, they put a fork in their own kettle, and a spoon, if they wanted a girl.

The exchange of the kettles between the families of two clans had obviously an economic significance. It confirmed, so to speak, the economic links between the two families. We have observed the exchange of kettles in the village of Kalma near the Amur river in 1946. The Nivkhi said that the ceremony was also performed after the bear feast. During the festivities, the representatives of the bridegroom's clan placed the kettle and an axe near the threshold, inside the door, while the bride's family left the kettle outside, near the threshold. Then the exchange ceremony was performed. It may be assumed that the members of the two clans confirmed the economic relations between each other by this act.

Various wooden idols were kept on the couches in a Nivkhi house. Among these, the *tyv yz* (literally: 'the master of the house', *tyv* 'house', *yz* 'master') was remarkable because of its dimensions. It was a human figure life-size or half-life-size, carved from a cylindrical piece of wood. The features and the hair were indicated schematically. It was rolled up in white-coloured fabrics and surrounded by shavings. Such idols were kept behind the rear couch, on special shelves. The Nivkhi believed that they were alive and possessed super-human power which they could use against the evil spirits. According to the old Nivkhi, it was the *tyv yz* that protected the house and its inhabitants against the evil spirits. The 'masters' of the houses were always 'fed' at meals (their lips were smeared with food). If the hunt was particularly successful, a great banquet was arranged in their honour.

The *tyv yz* was made only on the advice of the shaman, or the decision of the oldest and most honoured member (*pila nivχ* 'great man'; see: *Text 3* on p. 421) of the family or the clan.

The *tyv yz* was the clan protector of the house and was handed down only through the male line. He was generally called for aid when a new-born infant died or a child was ill. Since infant mortality was extremely high among the Nivkhi, almost every family had recourse to him. This is the reason why every explorer who visited the Nivkhi, encountered these wooden figures in every house.

The existence of the *tyv yz* among the Nivkhi was first reported by L. Šrenk: '. . . When I asked them to show me the idols in the yurta, I saw on the shelves above the rear couches three big human figures carved of wood, blackened by smoke and soot around their neck, in the same way as was customary among the Ulchi'.[6]

L. Ja. Šternberg has also records of the *tyv yz* he observed among the Nivkhi.[7]

Nowadays there are only a few old Nivkhi who still keep the *tyv yz* in their lofts. At the end of the present article a text will be found, describing the wooden figure of the *tyv yz*.

Besides the *tyv yz*, the Nivkhi had other wooden idols, named *ken*, which were made when children were ill. They believed that every idol was capable of 'curing' this or that disease.

The following is a description of some of these idols which we have collected for the Anthropological and Ethnographic Museum of the Academy of Sciences of the USSR.

1. atk̆an. Tiger figure carved of a single piece of wood. Stylized representation. The body is extended, the head is joined to the trunk without a neck, the snout is blunt, a cut running crosswise indicates the mouth. The legs are small rectangular extensions: the tail is straight and pointed. Painted streaks represent the skin. Such idols were made upon the advice of the shaman in order to cure children. According to the Nivkhi, the *atk̆an* absorbed the child's disease. The idol was kept in the house until the child had grown up; then it was put into the barn. It could not be thrown away.

2. šajká. Two human figures as it were entwined, carved out of a single piece of wood. The bodies are flat, the heads are pointed, the legs are marked by two rectangular extensions. They have no hands. Their eyes and

[6] Шренк (1903) p. 112.
[7] Штернберг (1908) p. 127.

mouth are marked on the face. The *šajká* was made when the child was ill or frightened; it was put under the pillow and kept there until the child was cured.

3. umgu and *utku.* Both are anthropomorphic figures, each carved out of a separate piece of wood. The representation is stylized.

The *umgu* is a male figure, with a big conical head and a flat face with eyes, nose and mouth. A piece of cloth is stitched on the short, cylindrical body. The figure has neither hands nor feet.

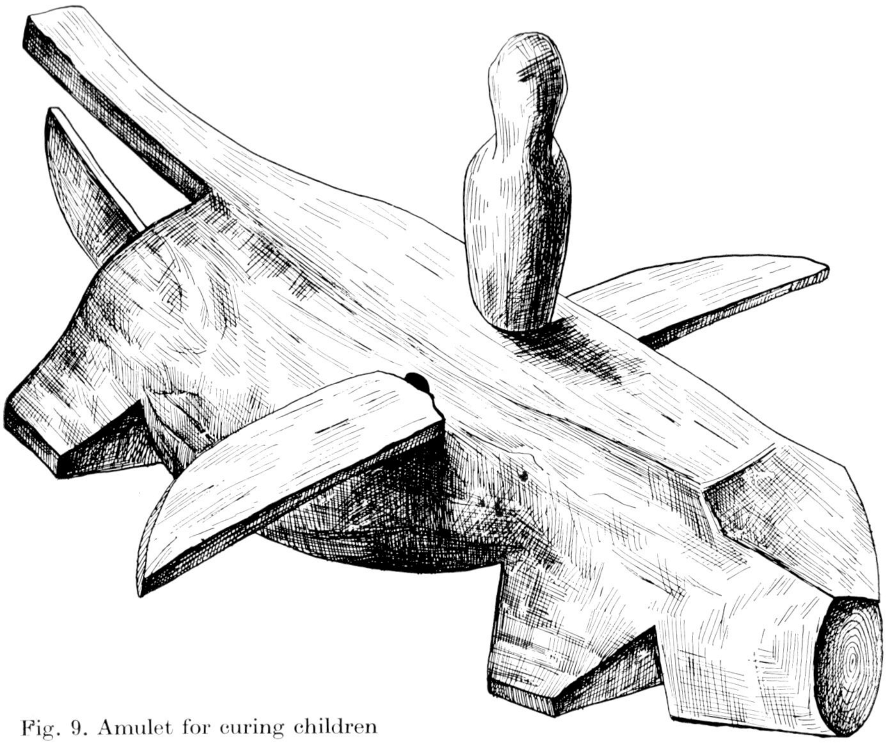

Fig. 9. Amulet for curing children

The *utku* is a round-headed female figure with the eyebrows, eyes, nose and mouth marked on the face. On the backside, the plaits are represented by cuts. The body is cylindrical, with arms bent at the elbows carved on it. A protruding transverse hoop represents the belt.

These idols were made when the children cried much. It was thought that both girls and boys cried to show that they wanted to marry. Therefore male idols were made for the girls and female ones for the boys. These idols were kept in the house until the girl or the boy was married.

4. pyj χapa. A winged tiger, with a human figure sitting on it. The tiger has an animal shape with a blunt snout, a double tail and short legs. A transverse slot was cut in the trunk to hold the wings, and another in its back for the human figure. The *pyj χapa* was made when the children were ill, but sometimes it was also made for adults. The body of the ill person was rubbed with shavings. This ceremony was called *psutat* ('to wash'). The shavings were put on the tiger figure, carried eastwards early in the morning and left in the forest. It was believed that the *pyj χapa* takes the illness with itself. It was given wings to be able to fly away, while the human figure was supposed to take over the disease.

From the foregoing it may be concluded that these numerous ritual ceremonies and taboos were mainly aimed at protecting the Nivkhi—and especially their children—from diseases and death, and at securing luck during fishing and hunting.

APPENDIX

The following texts concerning the ceremonies and taboos dealt with above have been recorded on the spot and translated into Russian by the author of the present article.

Text 1.

Two young Nivkhi from the same village, went far away hunting. On the hunting ground they stopped near a stately larch tree which was so large that it could only be embraced by four men. They decided to spend the night there. They made a fire, prepared their meal, then ate and lay down to sleep.

One of the hunters was a bachelor, the other a married man whose wife was pregnant. During the night, the bachelor heard a conversation from inside the tree, but he did not pay any attention to it and, in the morning did not say anything about it to his companion. After breakfast, each of them went hunting separately. The bachelor returned around noon with an empty bag. The married man returned later with a rich quarry.

After supper they lay down to rest. At the same time they heard the following conversation from inside the tree: 'Ugh, how I feel my feet! I am tired'.—'Where have you been?'—'Everywhere along the Amur and the sea.' 'Were many children born?'.—'Only one child was born, the son of the hunter who is next to us. I wanted this child to be provided with meat, therefore I arranged that his father should kill much game.' The hunters wanted to know who was talking and began to search in the hollow of the tree. They even climbed up to the top of the tree but found nothing.

They went home and told the old men everything they had heard. The old men explained to them that the master of heaven was very mighty; it is he who decides when the children should be born and appoints also the time of men's death. When this time comes, people die. If anyone be dissatisfied with anything, he has to die earlier than his appointed time. That is why the Nivkhi arrange feasts for this master to reconcile him.

Text 2.

A long time ago, two Nivkhi went out hunting from a village. One of them was a good man, the other one wicked. When they arrived at the hunting ground, each built a hut for himself and made a fire nearby. They sat down to eat together and then each went in his own hut. Before eating, the good man always threw a bit of food into the fire (*p'ind nigryχku*).

Every day he had a good bag, killing elks, reindeer, sables, etc. The wicked man did not believe in anything, nor did he throw food into the fire; so he had no luck in hunting.

One evening, the two hunters sat down together to eat. The good man took the food, threw a bit into the fire and only then began to eat. The wicked hunter asked the good one, 'Why do you throw food into the fire? I have observed you doing so before every meal.' The good man answered, 'If I give the fire to eat, it goes before me and detains the game until my arrival, and I kill the game. If the fire is hungry, it cannot detain the game'.

The wicked one said, 'You are lying! The fire eats wood!' Later, the good hunter asked him whether he had fed the fire before eating. The wicked hunter replied, 'Why should I feed the fire, since it eats wood. I feed it with wood and if it is not enough, I get some more.' The good man replied, 'If you don't believe me, don't feed it'.

The wicked man said, 'Tomorrow I shall go home, since I have run short of provisions'. The good man decided to stay another week. Next morning, the good hunter went out hunting. The wicked one gathered his own belongings together, packed them on the dog sledge and made a fire. He picked up the scraps of food, halved them and put his axe near him. Half of the food he ate himself and the other half he threw into the fire, saying: 'Here, eat now!' At the same time he snatched up his axe and hit in the middle of the fire. Then he packed up his things and off he went.

In the evening the good hunter returned from hunting to the hut and heard human voices. He went to the hut of the hunter who had left and there he found, at the fireplace, a naked man, nothing but skin and bone, with blood running from his broken skull. The weary man called the hunter: 'Come, oh come!' and he added, *ynyja ni mynyďra, ńğonuja* (*ńmorkuja*) 'I am dying, revive me, (cure me!) My master hit me with the axe and went home. He never fed me but forced me to hunt.' Then he asked the hunter, 'Go and fetch pine resin and put it on the cut on my skull; then put there two or three twigs and bandage the whole lot with shavings. And if you have something, feed me.'

The good hunter did everything the poor man asked and brought him some meat and other victuals he had there. In reply the exhausted man said to the hunter, 'Next morning when you get up, put your things on the dog sledge and sit on top of it yourself. I shall give you my dogs, they will pull you home. It took you five days to come here, but you will be home tomorrow before sunset. When you arrive at the fields of your village, stop the dogs and give them something to eat. In the village you have to pull the sledge by yourself'. Hardly had the hunter looked around, when the lean figure disappeared together with the food brought for him.

Next morning the hunter did as he was told and, indeed arrived home at the said time. There he was met by his relatives, the villagers, and also by the wicked hunter. They started to ask him all about the hunting. He told them in detail about the man in distress—the master of the fire. Next morning they learned that the wicked hunter and his whole family had died; they had not woken up any more.

Text 3.

I have recorded this in the village of Kalma, near the Amur, as related by a Nivkhi, Jachko. His parents had five children all of which had died within 1–2 months after their birth. Following the advice of the shaman, in 1906 or 1907 the family moved from the village of Ust-Amgun to the village of Khok. In addition, the shaman said to his parents: 'make a *tyv yz* ("master"). For this purpose, go to the lake Dalzhinskoe whence a little brook springs, follow its course until it reaches a lonely fir grove. The *tyv yz* must be carved out of one of these fir trees. The journey must be made in the summer together with the whole family. The lake must not be crossed, you should go along its left shore. Take one rest on the way. If you have luck, after the halt a reindeer will come running at a gallop from the mountains straight toward you. Kill it and dissect it at once. Cook its meat near the fir wood, have a good rest, carve the *tyv yz* and return home the next day along the right shore of the lake. Henceforth the children will not die, but all will grow up.' His parents did as the shaman told them. At home, the carved figure was put on the shelf above the rear couch. After this, they had five more children none of whom died, but all had a long life. One of those surviving is Jacko. After the death of their parents, the *tyv yz* was taken care of by the eldest son. Since his death it has been kept by Jacko. The *tyv yz* can be kept only by the menfolk, therefore his sisters did not take it.

REFERENCES

(Krejnovič, E. A.) Крейнович, Е. А. (1930) Рождение и смерть человека по воззрениям гиляков (The Birth and Death of Man in the Beliefs of Gilyaks) in Э, Nos 1–2.

(1934) Морской промысел гиляков дер. Куль (The Maritime Economy of the Gilyaks of the Village of Kul), in СЭ, No. 5.

(Šrenk, L.) Шренк, Л. (1903) *Об инородцах амурского края* (The Native Inhabitants of the Amur Region), Vol. III, Saint Petersburg.

(Šternberg, L. Ja). Штернберг, Л. Я. (1908) *Материалы по изучению гиляцкого языка и фольклора* (Materials for the Study of the Gilyak Language and Folklore), Saint Petersburg.

(1933) *Гиляки, орочи, гольды, негидальцы, айны* (Gilyaks, Orochi, Golds, Negidals and Ainu), Khabarovsk.

Remarks on Nivkhi (Gilyak) Proper Names (Anthroponyms)

by

V.Z. Panfilov, Moscow

Until recently the Nivkhi did not use surnames, unless we consider the clan names as coming under that heading. Each member of a family had his or her individual name or nickname which could by no means be passed on either to the other members of the family, or to the children.

The proper names of the female members of a family are usually formed by means of suffixes other than those attached to the names of the male members.[1] Most of the suffixes used for men's names have their counterparts in women's names, differing only in their last components.

Such twin suffixes are: *-kun* ~ *-gun* ~ *-gʿun* ~ *(xun)*[2], *-tun* ~ *-run* ~ *-dun*, *-tan* ~ *-ran* ~ *-dan*, *-tin* ~ *-rin* ~ *-din*, *-lin*, *-man*, *-zin* ~ *-sin*, *-kin* ~ *-gin* ~ *-g'in* used in men's names, corresponding to *-kuk* ~ *-guk* ~ *-gʿuk* ~ *(-xuk)*, *-tuk* ~ *-ruk* ~ *-duk*, *-tak* ~ *-rak* ~ *-dak*, *-tik* ~ *-rik* ~ *-dik*, *-lik*, *-mak*, *-zik* ~ *-sik*, *-kik* in women's names.

It can readily be seen that all these suffixes are compounds and consist of two components, the first one being the same for both men's and women's names. The second component in all suffixes for men's names is identical, i.e. *-n*. This also applies to the second component of the suffixes for women's names which is *-k*. The component *-n* recurs in most of the other names of men, and the women's names also contain the component *-k*.

In addition to this, as has been said previously, a large group of men's names is formed with the suffix *-n* alone, and many of the women's names with the suffix *-k*. The component *-n* in the names of men is considered identical with the suffix *-n* found in a number of nouns denoting living beings and kinship, such as for instance, *ok̆on* 'son-in-law', *kʿiun* 'younger brother'; *axmaln* cf. *axmalk* 'father-in-law'; *yzn*, cf. *yzik* 'master, owner'; *von* 'villager', cf. *vo* 'village, settlement'; *k̆an* 'dog', cf. *k̆ax̆* 'leading dog in a race', and also in the numerals *ńen* 'one', *men* 'two' in the sextodecimal system used for counting people.

The suffix *-k* also occurs as a fossil formative component of some terms of kinship, as for instance: *ytyk* 'father', *ymyk* 'mother', *atak* 'grandfather, father of the husband, father of the mother', *yťik* 'grandmother, old woman', *apak* 'father-in-law', *atk* 'wife's father', *ačk* 'wife's mother', *aťik* 'younger brother'.

Nowadays the Nivkhi have started using proper names for surnames, and so, each family member receives a name common to all members. The difference between names of men and those of women is beginning to disappear since both are used irrespective of sex.

[1] Krejnovič wrote: 'most of the women's names end in *-guk*, *-uk*, *-k*, and the men's names in *-gun*, *-un*, *-in*, *-n*.' Крейнович (1937) p. 84, note.

[2] In this paper we use the material of the Amur dialect and the related dialects spoken on the western shores of Sakhalin collected by the author during the expeditions of 1950, 1953, 1954 and 1957 on the spot.

The etymology of many proper names is not quite clear; hence it is often impossible to tell to what part of speech the word from which they are derived belongs, or what formative suffixes they contain. Yet, based on the well-established etymologies of proper names, it may be concluded that many proper names in the Nivkhi language are derived from verbal stems (including words denoting properties) and from nouns.

Suffixes forming names applied to persons of the male sex are as follows:

1. -kun ~ *-gun* ~ *-gʻun* ~ (*-xun*). Examples:

Rajgun (< *rajuď* 'to write', cf. *Rajlik* woman's name),
Efkun (< *ev'ď* 'to hold, to take'),
Mygun (< *myď* 'to listen'),
Sevgun (< *espť* (*sevd'* ~ *čevď*) 'to stab, to prick'),
Yzgun (< *jyzď* 'to call, to shout'),
Vigun (< *viď* 'to go'),
Xitkun (< *xituď* 'to avoid danger by running', cf. the woman's name *Xitkuk*),
Lezgun (< *lezď* 'to carry something under the arm'),
X̌ovgun (< *x̌ovď* 'to turn, roll together'),
Ozgun (< *ozď* 'to stand up'),
Mugun (< *muď* 'to become, to die'),
Togun (< *tod'* 'to cry, whine'),
Pʻxygun (< *pʻxyď* 'to return'),
Jorgun (< *jorď* 'to face somebody'),
Vorgun (< *vorď* 'to meet one another'),
Pilgun (< *pilď* 'to be big'),
Maŋgun (< *maŋ(g)laď* 'to be strong, powerful'),
Palgun (< *pal* 'hill, forest'),
Vezgun (< *ves* 'crow'),
Taligun (< *tali* 'bag, sack') and so on.

Among these the suffix *-kun* ~ *-gun* ~ *-gʻun* ~ (*-xun*) is the most productive. Etymologically it can be compared with the plural suffix in the north-eastern dialect: *-kun* ~ *-gun* ~ *-gʻun* ~ (*-xun*) and in the Amur dialect *-ku* ~ *-gu* ~ *-gʻu* ~ (*-xu*)[3].

The same suffix can be detected in some nouns denoting kinship (e.g. *targ'un*).

2. *-tun* ~ *-run* ~ *-dun*. Examples:

Zarun (< *zaď* 'to hit'),
N'urun (< *ńuď* 'to look'),
Firun (< *fid'* ~ *pʻiď* 'to be somewhere' or *fiď* ~ *pʻid* 'to bore, to drill'),
Smedun (< *ezmuď* (*smod'* ~ *čmod*) 'to love'),
Vadun (< *vaď* 'to fight'),
Zdortun, *Ksyvrun*, *Plarun*, *Nantun*, and so on.

[3] For more details see Панфилов (1958) and (1962).

This suffix is less productive than the former.

3. -kan ~ *-gan*. Examples:

Evgan (< *evď* 'to hold'),
Irgyn (< *irď* 'to keep, to preserve'),
Vetkan (< *vetaď* 'to dress'),
Tajgan (< *taj* 'patch, to blot'),
Vyčkyn (< *vyť* 'iron'),
Čafkan (< *caf* 'whetstone'),
Makan (< *maď* 'near, close'),
Onigan, *Čergan*, *Lezgan*, *Nytkan*, *Zajgan*, and so on.

4. -tan ~ *-ran* ~ *-dan*. Examples:

Xoran (< *x̌oď* 'to take off the net'),
Lodan (< *loď* 'to tie, to pack, to tighten something; to be tight'),
X̌atan (< *x̌aď* 'to shoot'),
Pajdan (< *paj* 'in vain'),
Kistyn (< *kis* 'mud'),
Lovrťan (< *lovr* 'armpit'),
Vaŋdan, and several others.

The suffix *-tan* ~ *ran* ~ *dan* is also used to form words denoting clans, for instance: *čoran* (< *čo* 'fish'), *lezgran* (< *lezŋg* 'nook where the killed bear and also its bones are kept'). This suffix is identified with the word *-tan* ~ *-ran* ~ *-dan* 'household, members of a family, generally of some closed circle'. (Cf. north-eastern dial. *taf* 'house' in which *-f* is probably a formative suffix meaning 'place'.) See, for instance: *N'yŋ dangu alz ŋat viďg'u* 'our people went to pluck berries' (*-ku* ~ *-gu* ~ *-g'u* ~ *-xu* is the plural suffix).

5. -kin ~ *-gin* ~ *-g'in* ~ *-xin*. Examples:

P'imgin (< *p'imď* 'to run'),
Purgin (< *vurď* ~ *purď* 'to make the bed'),
Syukin (< *syuď* 'to take off'),
Čaŋgin (< *čaŋď* 'white'),
Tanskin (< *tans* 'patch'),
X̌arxin (< *x̌ar* 'wasp'),
Mitrkin (< *mitr* 'scoop'),
Čajgin (< *čaj* 'tea'),
Patkin (< Amur dial. *pyt*, east-north. dial. *pat* 'tomorrow'),
Surgin, *Sazgin*, *Lolskin*, and others.[4]

6. -tin ~ *-din* ~ *-rin*. Examples:

Udin (< *uď* 'to burn'),
Ğ'odin (< *ğ'od'* 'to string'),
Ovrin (< *jovď* 'to mend the net'),

[4] The suffix *-kin* ~ *-g'in* ~ *-gin* ~ *-xin* is used in the word *yg'rkin* (<*yg'rd'* 'black') a name for a dog.

Tutin (< *tu* 'lake'),
T'odin (< *ťo* 'bank, beach'),
Čaudin, *Kostin*, and several others.

7. *-lin*. Examples:

Ŗajlin (< *rajuď* 'to write', cf. *Ŗajlik*),
V̦alin (< *vaď* 'to fight'),
X̌ajlin (< *x̌aj* 'wild pigeon'),
Igulin, etc.

8. *-lun*. Examples:

Tulun (< *tu* 'lake'),
Frolun.

9. Many men's names are formed with the suffix *-n* from widely different parts of speech. Examples:

Kʿyryn (< *kʿyrď* 'to starve'),
Vynun (< *vynuď* 'to close, to block'),
ŋarǩon (< *ŋarǩaď* 'to click, to smack'),
ŋyryn (< *ŋyrď* 'to lay under, to put to'),
Pʿivryn (< *pʿivryď* 'to tidy up the house'),
Otain (< *otaď* 'to row backwards'),
Vyskyn (< *vyskť* 'to fight'),
Jylkun (< *jylkť* 'to stammer'),
Kʿegʿŋain (< *kʿegʿŋaiď* 'to stand aghast'),
Ŗukun (< *rukť* 'to hit'),
Irmyn (< *irmyď* 'to cross, to go through something'),
Jalin (< *jaliď* 'to have no time, not to succeed'),
Vyksun (< *vykzď* 'to throw out'),
Ityn (< *itť* 'to speak'),
Pʿolǧʿon (< *pʿolǧʿď* 'to pierce'),
Ķeŋran (< *keŋrad'* 'to enjoy oneself, to be happy'),
Ķaman (< *ǩamaď* 'to run'),
Čox̂tyn (< *čox̂tť* 'drunk'),
Paǩan (< *paǩť* 'about eyes red with trachoma'),
Ǩʿavrin (< *ǩʿaur* 'obstacle, brake'),
Putin (< *puť* 'sea cabbage'),
T'aŋgin (< *ťaŋgi* 'boss, chief'),
Tʿykyn (< *tʿyky* 'edge of the plank-bed'),
Plityn (Russian плита 'plate, slab'),
Hagʿin (< *hagʿi* 'bird-cherry'),
Čurin (< *čur* 'straight'),
Jyfťun (< *jyfťu* 'deliberately'), etc.[5]

Besides the above-listed suffixes, several other formative suffixes can be deduced from men's names, used for this purpose in sporadic cases:

[5] The vowels in some of these words between the root and the formative suffix are obviously epenthetic and do not belong organically to the suffix.

1. the suffix *-man* (e.g. *Itman* < *itt'* 'to speak', *Pańman* < *p'ań* 'knee-cap' ?), 2. the suffix *-zin* ~ *-sin* (e.g. *Parkizin, Vyrzin*); 3. the suffix *-vin* (e.g. *Čutvin, Udvin*), which apparently is derived from the verbal stem *fiď* ~ *p'iď* 'to be somewhere' and can be deduced also from some tribal names as, for instance, *Pyrkivin*, literally: 'staying in *Pyrki*'.

It should be mentioned that a large number of common nouns are also used, unchanged, for proper names in the Nivkhi language. Sometimes even other parts of speech are used in this sence. (Cf. for instance, *Ǩ'arx̌* literally: 'pole, column', *Čeǩaj* literally: 'snaring loop', *Tyf* literally: 'house', *Tapat* literally: 'ear muffs', *Čaf* literally: 'whetstone', *P'op'o* literally: 'the sound produced by a dolphin when exhaling', *Aŋa* literally: 'who' and many others, *Luŋ* literally: 'driller'; *T'yxt* literally: 'fastidious', etc.)

Suffixes forming names applied to persons of the female sex are as follows:

1. -kuk ~ *-guk* ~ *-g'uk* ~ *-xuk*. Examples:

Xitkuk (< *xituď* 'to avoid danger by running'),
Jarguk (< *jarď* 'to feed'),
T'yrguk (< *t'yrď* 'to look'),
Tykrguk (< *t'ykrď* 'break loose from the hook, etc.'),
Moguk (< *moď* 'to try'),
Myiguk (< *myiď* 'to fall to the ground'),
Porguk (< *porď* 'to lie'),
G'ezguk (< *g'ezď* 'to strain, to filter'),
Vakuk (< *vaď* 'to fight'),
Leguk (< *leď* 'to be effaced'),
Čryguk (< *čry* 'bank, beach'),
Čaŋguk (< *čaŋ* 'peg', possibly *čaŋ(la)* 'white'),
Yzguk (< *yz* 'master, host', possibly < *jyzď* 'to call'),
X̌arguk (< *x̌ar* 'wasp'),
Kipkuk (< *kip* 'handle'),
Liskuk (< Russian Лиза),
P'ńag'guk (*p'*-prefix of possessive pronoun, *n'ax* 'eye') and some others.

2. -tuk ~ *-ruk* ~ *-duk*. Examples:

Feruk (< *feď* ~ *p'eď* 'berry-plucking'),
Umruk (< *umď* 'to be angry'),
Ovruk (< *ovď* 'to fight'),
Sambruk.

This suffix is less productive than its counterpart *-tun* ~ *-run* ~ *-dun* used for the formation of men's names.

3. -tak ~ *-rak* ~ *-dak*. Examples:

Oftak (< *ovď* 'to fight'),
Lavťak (< *lavď* 'to go by, to pass'),
Amtyk (< *jamaď* 'to look').

Unlike its counterpart *-tan* ~ *-ran* ~ *-dan*, this suffix occurs sporadically only, as the previous one.

4. -tik ~ *-rik* ~ *-dik*. Examples:

Vidik (< *viď* 'to go'),
Tʿatik (< *ťʿaď* 'to breathe'),
Morik (< *moď* 'to try'),
Pağrik (< *pağʿ(la)* 'red'),
Lyxtik (< *lyx* 'weather'),
ŋoxtik (< *ŋox* 'fat'), etc.

5. -lik. Examples:

Rajlik (< *rajuď* 'to write'),
Vojlik (< *vojiď* 'to appear'),
N'avlik (cf. *N'avan*), *Oblik*, *Kuklik*, etc.[6]

6. Numerous women's names are formed by adding the suffix *-k* alone. Examples:

Jazruk (< *jazruď* 'to respect, to honour'),
Jan'uk (< *jan'uď* 'to thrash, to give a thorough trash to sby'),
Jox̌tik (< *jox̌tť* 'to heal'),
Rorik (< *rorď* 'to miss somebody, to pine for sth'),
Induk (< *indyď* 'to look'),
Jurik (< *juruď* 'to read'),
Tulmuk (< *tulmuď* 'to mutter, to grunt'),
T'olak (< *ťolaď* 'to be satisfied'),
Hek (< *jeď* ~ *heď* 'to cook'),
Līrk (< *līrt'* 'to harden with cold, to freeze'),
Ršysk (< *ršyzď* 'to put underneath, to substitute'),
Pʿerk̆uk (< *pʿerk̆ť* 'to bend, to be curved'),
Ajzik (< *ajz* 'gold'),
Vyťik (< *vyť* 'iron'),
Kʿuksk (< *kʿuks* 'crumb, pigmy'),
Čvark (< *čvar* 'chain'),
Pulkuk (< *pulk* 'little loaf'),
Vamk̆uk (< *vamk̆* 'mitten'),
Čŋyruk (< *čŋyr* 'grass').[7]

Apart from the above suffixes some other formative suffices can be deduced from women's names, used for the same purpose in individual cases: 1. the suffix *-zik* (cf. e.g. *X̌ovzik* < *x̌ovď* 'to turn, to roll up', *Kʿuvzik* < *kʿuva* 'thread'), 2. the suffix *-kik* (cf. e.g. *Mūfkik* < *mūf* 'day'), 3. the suffix *-pak* (cf. e.g. *Nelpak* < *nelaď* 'to show one's teeth', *San'pak*), 4. the suffix *-mak* (cf. e.g. *Semak*, *Kojmak*).

[6] Compare also the word *čaŋlik* (< *čaŋ)la* ('white'), a name for a dog.
[7] The additional vowels inserted between the root and the suffix *-k* are epenthetic.

It should be mentioned that some nouns, as well as words belonging to some other parts of speech are indifferently used both for men's and women's names. (Cf. for instance, *Malk̄* literally: 'knot', *Matak̄* literally: 'vessel of birch-bark for gathering berries', on the one hand, and *Jyks* from the verb *jykz-ď* 'to catch hold of', *Pyby* from the verb *vypy-ď* 'to lull', on the other hand, *Kʿay* from the verb *k̄ʿay-ď* 'no, it is missing'.)

Our linguistic analysis shows that proper names have been formed from words differing widely in meaning. Any action, property or object may serve as basis for the formation of proper names which, at first glance, might seem to have been given by chance.

It would indeed be misleading to say that all these proper names are accidental in character. Many proper names were given because of some characteristic trait of the person or because of some property the parents wished the child to possess, e.g.

Togun < *toď* 'to cry, to whine',
Umruk < *umď* 'to be angry',
Pilgun < *pila* 'big',
Rajgun < *rajuď* 'write',
Vadun < *vaď* 'to fight', 'wrestle', etc.

Other proper names can be traced back to circumstances, events, etc., which took place when the child was born. Thus, for instance, the name *I(gʿ)run* shows that the child was born on the fourth day of the dog race, when the dogs rest (cf. *Var ǧan i(gʿ)ryuď* 'the racing dogs, the dogs that have taken part in the race rest'); the name *Myiguk* < *myiď* 'to go to ruin, to fall' was given if the child was born when the dog sledge broke down.

Some proper names are obviously nicknames which the bearer received after growing up. See, e.g.

Čoxtan < *čoxtť* 'drunk',
Vyxtun < *vyxtuď* 'to fail to finish something, not to accomplish',
Jalin < *jaliď* 'to have no time to do something', etc.

It should also be remembered that in primitive communities naming was influenced by the general belief prevailing in these societies that there were unbreakable links between the name and the phenomenon or object denoted by it, and especially between the proper name and the person who bears it.

We have in mind the well-known fact that people in primitive societies used to take care not to divulge the name of any member of the community to any stranger for fear of giving the latter power to cause harm to the person.

The proper names, as discussed and stated above seem to have been given to the children because of some characteristic feature, action or phenomenon which the parents wished their child to partake of. Yet opposite considerations might also have played a part in giving a name: If, for instance, the children had died in a family the parents tried to give the new-born baby a name which had not been used before and had an incidental character.

According to their belief, this excluded any possibility of anybody doing any harm to the child by magic manipulation with his or her name. (See, for instance, such proper names as

Vigun < *vid'* 'to go',
Vorgun (*Jorgun*) < *jord'* 'to meet',
Syukin < *syud'* 'to take off',
Mitrkin < *mitr* 'little scoop', etc.)

The origin of the Nivkhi (Gilyak) people is still an open question. Their language occupies an isolated place among the other languages of Siberia.

On the other hand, one of the greatest experts in this question, L. Ja. Šternberg has shown that close tribal links exist between the Nivkhi and the neighbouring Tunguso-Manchurian peoples and the Ainus. 'While making a census of the Gilyaks—he writes—I took care to register every piece of information and legend on the origin of each tribe. When studying these data I came to the surprising conclusion that the founders of the overwhelming majority of the clans that now exist had been representatives of neighbouring tribes and not Gilyaks.'[8] In this context the complete coincidence, or close relationship, of many formative suffixes in Nivkhi proper names and clan names with those of the neighbouring Tunguso-Manchurian peoples is of particular interest.

For example, the Nivkhi suffix *-kan* ∼ *-gan* can be collated with the Evenki suffix *-gan* used in words denoting the inhabitants of a region (cf. e.g. *ngēgan* 'riverbank inhabitant', *dīgen* 'mountaineer',etc.) and with the Manchu suffix *-gan* ∼ *-gon* ∼ *-gen* in clan names[9]; the Nivkhi suffix *-kin* ∼ *-gin* ∼ *-xin* can be compared with the Evenki suffix *-gir* (plural), *-gin* (singular), 'indicating the affiliation of women to the clan organization'[10]; further, compare the Nivkhi suffixes *-rik* and *-lik* with the Evenki suffix for women's names *-rik* ∼ *-lik;* the Nivkhi suffix *-k*, with the Evenki suffix *-k* for women's proper names; the Nivkhi suffixes *-tu-n, -ru-n, -du-n, -li-n, -lu-n, -mu-n, -ma-n, -zi-n,* with the suffixes in Manchu clan names *-du (* ∼ *-dun)* ∼ *-tun; -li, -lu, -mu, -ma* ∼ *-me* ∼ *-mo, -si,* respectively[11] and also the Nivkhi suffix *-man,* with the Chinese suffix *-myn* > *mn* used for the plural of nouns denoting persons.

These and similar linguistic relations undoubtedly corroborate the assumption that the Nivkhi (Gilyak) people have mingled intimately with the neighbouring Tunguso-Manchurian peoples.

[8] Штернберг (1933) p. 20.
[9] Лебедева (1958) p. 106.
[10] Лебедева (1958) p. 107.
[11] Лебедева (1958) pp. 102–4.

REFERENCES

(Krejnovič, E. A.) Крейнович, Е. А. (1937) *Фонетика нивхского (гиляцкого) языка* (Phonetics of the Nivkhi [Gilyak] Language), Moscow–Leningrad.
(Lebedeva, E. L.) Лебедева, Е. Л. (1958) Лингвистический анализ родовых названий маньчжуров (A Linguistic Analysis of the Clan Names of the Manchus), in *ВЯ*, No. 3.
(Panfilov, V. Z.) Панфилов, В. З. (1958) Грамматическое число существительных в нивхском языке (The Grammatic Number of Nouns in the Gilyak Language), in *ДСИЯ*, Vol. XI.
(1962) *Грамматика нивхского языка,* часть I (The Grammar of the Gilyak Language, Part I), Moscow–Leningrad.
(Šternberg, L. Ja.) Штернберг, Л. Я. (1933) *Гиляки, орочи, гольды, негидальцы, айны* (Gilyaks, Orochi, Golds, Negidals and Ainu), Khabarovsk.

Popular Conceptions, Religious Beliefs and Rites of the Asiatic Eskimoes

by

G.A. Menovščikov, Leningrad

The Asiatic Eskimoes, representing the most ancient branch of the Eskimo tribes scattered over wide arctic areas, have lived for a long period of time in close vicinity with the Chukchi, Koryaks (Nymylans), Yukagirs (Oduls) and other Paleo-Asiatic peoples. Hence their spiritual and material culture offers a considerable resemblance to the culture of those neighbouring peoples. Such similarities between several northern tribes have been studied and described in great detail by V. G. Bogoraz,[1] the well-known Russian ethnographer, who has also made interesting observations concerning the religious conceptions of the Asiatic Eskimoes. The data on the religion of Paleo-Asiatic peoples, collected and published by him, are still of great scientific value.

The modest aim of the present study is to add supplementary information and new contributions to the knowledge of the spiritual culture of the Asiatic Eskimoes. The material in this article has been collected by the author himself during observations on the spot over several years. The circumstance that some of our facts agree with the religious concepts and rites observed by Bogoraz among the tribes of the Chukchi Peninsula, confirms the assumption that in the past they existed also among the Asiatic Eskimoes and the Chukchi. The author has made use of some of E. S. Rubcova's observations concerning the way of life and the religion of the Asiatic Eskimoes.[2]

Not considering himself a specialist in this field, the author of the present article does not intend to make a comparative analysis of the vast data existing on the religious concepts of the Eskimoes. In any case, this could not be undertaken safely within the scope of a short study such as this. We believe it very useful, however, to publish direct, personal observations which may later serve as factual raw material for scientific studies on the spiritual evolution of this underdeveloped people. The readers may be particularly interested in the new material submitted here, since it contains some elements of the folklore of the Asiatic Eskimoes which have not been so far observed or taken into consideration by anyone dealing with the religious beliefs of this people.

As demonstrated by many scholars by means of research into the spiritual culture of other peoples, the religious beliefs of the Asiatic Eskimoes contain many features in common with the beliefs of the Chukchi, Koryak,[3] Nivkhi (Gilyak)[4] and other smaller peoples of north-eastern Asia and north-western America.[5]

[1] Богораз (1908, 1939, 1910, 1932) and (1930) p. 63; Анисимов (1958) p. 234.
[2] Рубцова (1954).
[3] Jochelson (1905).
[4] Шренк (1883–1903), Штернберг (1933).
[5] Boas (1888), Birket-Smith (1929), Nelson (1899), Dall (1870) and many others have treated the problems of Eskimo spiritual culture.

*

Depending on the natural environment and climatic as well as geographic conditions, the developmental stage of the forces of production represents the source and basis of the popular ideas and religious beliefs of peoples. For many long centuries the given material foundation of an Eskimo society had remained almost completely unchanged. Among the Eskimoes any evolution pointing toward a well-established knowledge or any development in their spiritual and material circumstances had always been stunted by their severe living conditions, the unfavourable climate and the rather scanty and unreliable physical resources in the arctic region. In spite of all inconveniences, the Eskimoes living farthest north have become, with their harpoons, kayaks and boats, the creators of a unique arctic culture. Ages ago they learnt how to kill gigantic sea animals such as whales from kayaks or boats by skilfully using their remarkable bone- or iron-headed harpoons. They improved their tools constantly, and in the course of many centuries they learned how to cope with the cruel conditions of the arctic climate. But available natural resources were so limited that fins, bones and stones remained the only material for manufacturing these tools until the representatives of this remote arctic culture came into contact with the Europeans.

Any well-established knowledge gained from centuries of experience is to be ascribed partly to the observation of natural phenomena, and partly to the productive work of everyday life. The success of a hunting party —in fact, often the hunter's life itself—depended on the weather, the state of the ice, the sea currents and many other similar phenomena. Therefore, arctic man had always been a keen observer, storing up numerous practical experiences in his mind.

For instance, when the sea is covered with a thick sheet of ice and there is no wind, the Eskimo hunters do not go to the distant corners of the ice, because if a floe were to break off, the currents would carry it to the high sea, and the men on it would never return. Only when the south wind is blowing is it possible for them to hunt on the ice. Hence the judicious Eskimo saying: 'Don't hunt at sea in calm weather: the ice breaks off and the current carries it away. If the south wind blows, hunt, then you are not driven away.'

Some precepts for judging the weather or the most propitious time for the summer sea hunting have been derived from observations of the behaviour of birds. For example, if in a long south wind the wild ducks remain on the water or fly with the wind, it means: 'Don't expect good weather for hunting on boat, a storm is coming!' If, on the other hand, the ducks start to fly against the wind, the experienced hunters say: 'Tomorrow there will be north wind, the waves will get shorter, out we go for walrus hunting!'

The beginning of walrus hunting in the summer (walruses usually come at that time from the south-west) is timed by the migration of the snipe. The Eskimoes call them *antam suğmyği* 'slops snipe' because, when they arrive, at the beginning of July, they fly to the settlements and rummage in the rubbish heaps, looking for worms. The arrival of the snipe was the starting signal for the summer walrus hunting. The Eskimoes waited impatiently for this season, when they were able to kill the walruses near the settlement and to skin them on the shore, and the entire population could eat plenty of sea crabs from the stomach of the walrus, and fresh guts, too.[6]

During the cold winter months when the sea is covered with ice to the line of the horizon, permitting seal hunting on foot, the Eskimo hunters scan the wast ice sheet for an ice hole. If they cannot find one on the surface of the frozen sea, they direct their eyes towards the sky, where a dark spot among the white misty clouds is considered to be the reflection of a distant ice hole. In this way they locate a spot good for seal hunting on the ice.

In the summer, the floes retreat far out to the sea and cannot be seen from the shore, but the experienced hunters find them by their reflection in the blue sky where they appear as white spots. The men put to sea and sail in the direction of these spots; on the big ice floes they kill the walruses and seals.

If, on a bright winter day, a black streak appears all of a sudden above the sea in the direction of distant mountains, it is not advisable to make a long journey with the dogs because there will be a terrible blizzard.

In the life of the Eskimoes there are periods when sea game is lacking for months: the dreaded days of hunger set in. Then any edible thing that the inhospitable arctic environment will yield, is devoured. In summer and autumn the Eskimoes gather cockles, crabs and various seaweeds which are washed ashore. In time, the cockles will deteriorate, and eating them involves the danger of being poisoned. Autumn is known to be one of the dangerous periods 'when the mountain peaks are covered with snow and the cockles become poisonous'.

Unlike other peoples who live under more favourable climates where conditions are suitable for the evolution of practical therapeutics and the use of numerous medicinal plants against diseases, the Eskimoes have never been in a position to develop this branch of popular science. They did, however, possess some knowledge of the field, as is proved by many of our records. Thus, for instance, they treated tubercular patients by improving their meals with whale ham and warmed seal fat. More than once this method led to satisfactory results. It was also customary to cure tuberculosis with the blood of a freshly killed walrus. Physical exercises were also recommended for the treatment of consumptive persons. When the case was not too serious, the patient had to take long walks in the mountains or run some distance at a moderate speed.

Profusely bleeding wounds were cured by spreading burnt reindeer hair on them.

Nausea which was caused by diseases or bad food was cured with the smoke of an umbelliferous plant called *typʟuk* (*Umbelliferae, Coalolerum gmelinii*), which seems to have a disinfectant effect. Disagreeable smells in the dwelling houses were also suppressed by this smoke.

The practical knowledge of the Eskimoes included a special terminology for hunting and other productive works. Big-game hunting at sea requires signals which are in every respect exact and reasonable for quick orientation and for aiming with the harpoon or spear. The Eskimoes used more than

[6] The sea molluscs, found in the stomach of the walrus, were looked upon as delicacies by the Eskimoes.

fifteen different expressions for the walrus, depending on its position in the water, its age and the direction in which it was swimming. These terms not only named the animal, but also drew the hunter's attention to the direction. Here are a few, which I have recorded from the Chaplino Eskimoes:

ajvyk̆	'walrus',
putukak̆	'walrus floating on the water with its head bent down' (< *putu* 'noose', *putukak̆* 'noosed'),
ilğax̆tyk̆	'walrus sleeping in the water with only its nose visible',
ak̆yluğrak	'walrus swimming without any particular aim' (literally: 'quiet guest'),
arguğak̆	'walrus swimming westwards' (derived from the verb *arğuğak̆uk̆* 'to swim against the current'),
k̆avryk̆	'walrus swimming northwards',
manilğuk̆	'walrus flock swimming towards the shore' (from the verb *manik̆uk̆* 'approaching'),
uxsilğak̆	'walrus lair',
nunavak	'walrus (or other game) on the ice floe' (< *nuna* 'mainland', 'earth'),
unkuvak	'one year old walrus',
unkuvagağaxpak	'young walrus with small tusks',
miksagak̆	'young walrus',
aŋyjyğak̆	'rapacious walrus that has grown up and is living alone' (< *aŋy* 'big'), and so on.

The Eskimo language uses more than twenty demonstrative pronouns for quick and precise orientation, each indicating the exact place of the object, the side and the distance; for example:

tana	'this here, at hand',
iŋna	'that one, somewhat further away, opposite to the speaker',
samna	'that one, near the sea, on the southern side',
amna	'that one, beyond the horizon',
aŋna	'that one passing by here',
pikna	'that one above the speaker, opposite to the sea, on the elevation',
ukna	'that one approaching',
ikna	'that one on the other side of something',
k̆agna	'that one there outside', and so on.[7]

About forty terms are known for straps for various purposes and of various sizes and shapes. They clearly show the rational application of linguistic means by the Eskimoes to the concrete names for a great many homogeneous objects, important for productive work. Compare:

k̆ašk̆ak̆	'strap made of the skin of a young walrus' (< *k̆asik̆ak̆* 'young walrus'),
makʟingak̆	'strap made of the skin of a bearded seal' (< *makʟak* 'bearded seal'),

[7] See Меновщиков (1954) p. 93 and (1955) pp. 26–41.

tapx̌aǩ	'strap (general term)',
kalgun	'broad strap for hauling a whale' (from the stem *ka-* 'to approach'),
iǩyğutaǩ	'strap for hauling a walrus' (< *iǩytutala* 'breadth'),
sirak	'supplementary strap attached to the harpoon strap for holding a wounded or killed walrus above the water',
tamlagutaǩ	'strap for pulling canoes and whale boats ashore' (< *tamlagaǩuǩ* 'moored'),
tanŋynilga	'all sorts of twisted ropes',
pix̌aǩ	'rope braided like a plait' (< *pix̌aǩuǩ* 'to braid'),
niğran	'thin strap for sledges' (< *niğraǩ* 'cage'),
uliǩ	'broad strap',
nağuʟx̌un	'harpoon strap' (< *naguʟxağaǩuǩ* 'to harpoon'),
wapa	'warp (end)',
sux̌utaǩ	'strap for hauling a killed seal',
ujataǩ	'strap with one end fastened to something' (strap with dog-collar, headstall; derived from *ujaǩuǩ* 'neck'),
kamutaǩ	'special strap for hauling seals' (< *ǩamuǩ* 'that which is hauled'),
mağjax̌ǩux̌sin	'pulling strap, strap for harness of dog-sled, draught' (< *mağjax̌ǩuǩ* 'to guide'),
pytugutaǩ	'strap, chain or rope for fastening something' (< *pitugaǩ* 'fastens', 'ties'),
pytugvik	'strap or rope for loading', 'dog-harness',
nyx̌singğaǩ	'strap made of seal skin' (< *nyx̌saǩ* 'seal'),
aŋaʟǩatasiǩ	'strap for casting' (from *aŋax-* 'to carry, to cast'),
ǩuvrax̌ǩaǩ	'strap for net' (from *ǩuvraǩ* 'strap net'),
magnyğraǩ	'thick strap of walrus skin for whale harpoons',
tafsi	'waist-belt',
tafsin	'strap for drawing together the *yaranga* (dwelling house)',
amağan	'strap for the shaft of the harpoon',
ukamataǩ	'strap for boats hauled by men, girth' (< *ukamaǩ* 'haulage', 'pulling the barge ashore'),
asiğutaǩ	'strap for fastening the load',
tasiǩ	'girth of towing rope',
akmagin	'strap for fastening loads on one's back' (< *akmagin* 'to haul on one's back'),
ujğisiǩ	'anchor line, chain' (< *ujğak* 'stone'),
pajğaǩ	'thin strap made of sealskin for tying and fastening' (< *pajği-* 'to meet'),
aŋjax̌paǩuvingğaǩ	'hempen rope, rope' (literally: 'part of the barge') (< *aŋjax̌pak* 'bark'), and so on.

The Eskimoes have a deep knowledge of anatomy. They know perfectly well not only the structure of the skeleton and the external tissues of the body, but also all the intestines in great detail. Every part of the organism has its own name. It must be assumed that anatomic knowledge associated with the main occupation of the Eskimoes—hunting—developed in ancient times, since many of the terms for the objects of material culture are copied from names of the different parts of living organs, either formally or through their functional resemblance. Also the numerical system of the Eskimoes points to an ancient origin of their anatomic knowledge since it contains notions about the human body and its parts. Thus for example the number 5 (*taʟimat*) derives from *taʟik* 'hand', and the number 20 (*juginak̆*) from *juk* 'man'; *juginak̆* means literally 'the whole man'.[8]

Names of parts of the body are frequently used for different parts of a boat or of a framework for a house or again for some topographical terms. In such cases the words indicating parts of the body are used either in their original or in a modified form; cf.:

iğu	'foot',
iğulut	'rafter of the framework for a house',
iğuŋak̆	'sledge runners' and, a later derivation, *iğuŋağyk* 'forked support for gun',
k̆ukak̆	'waist', and 'border of the barge near the stern',
kanagak̆	'knee', hence *kanagajagyk̆* 'summer thighboots for men',
kujak	'backbone' and 'keel' or 'bottom of a boat',
sanik̆ak̆	'side' and 'gunwale',
tulimak̆	'rib' and 'rainbow',
saja	'chest' and 'outer part of prow',
suŋak̆	'bile' and 'pearl',
patyk̆	'spinal cord' and 'pith',
xata	'back' and 'tree stump',
tykyk̆	'forefinger', hence *tykyğak̆* 'promontory, point, peninsula',
k̆yzik̆	'lips', hence *kyzilyk* 'drinking vessel, jug' and *k̆yziğin* 'edge of the sole', and so on.

Accumulated practical knowledge in one special field was transferred by the Eskimoes to objects and phenomena of other spheres in their environment which habit resulted in the thoroughest possible utilization of the resources of their native tongue.

The practical knowledge which had been gathered by many generations undoubtedly covered all the fields of their vital functions. However, much of that knowledge has disappeared owing to changed living conditions, migrations and the dissolution of formerly coherent ethnic units into minor local groups.

*

The harsh environment and the inability of the Eskimoes to understand many of its phenomena formed an excellent background for the birth of erroneous and fantastic ideas, which served as a basis for religious belief.

When faced with a natural phenomenon beyond the scope of their comprehension or experience, they 'transanimated' it and attributed it to supernatural forces. They believed in various evil and good spirits (*tungak*) and in the master spirits of the sea and the tundra. The religious ideas of the Eskimoes are of great historic interest, as they reflect archaic forms of social and religious conditions.

The basis of the religious conceptions of the Eskimoes was animism, which endows natural forces and surrounding objects with human qualities. It is

[8] Меновщиков (1956).

a state in which 'man does not separate himself from nature but identifies himself with its phenomena and forces'.[9] According to the animistic concepts of the Eskimoes, objects and phenomena had a 'soul' and hence could think and act like men.

Their mythology also endowed the representatives of the animal world with animistic characteristics. One of their beliefs was that man was bound by family ties to an animal or a lifeless thing. This was evident in celebrations in honour of the family divinity. Their religious attitude is clearly reflected in their folk-poetry: the tale 'The whale born from a woman'[10], the myth of the kingfisher, which was transformed into a wolf, etc.

The Eskimoes believed in the power of magic. A man could influence nature and his fellow-beings or, the other way round, natural forces could influence man. Such phenomena as death and diseases, winds and storms, were attributed by them to supernatural forces, because they could not be explained logically. Ideas connected with the effects and the actions of supernatural forces gave rise to the development of religious manifestations, such as incantations, omens and prohibitions (taboos). Magic was the underlying principle of shamanism (see below) among the Eskimoes, as well as among other northern peoples.

They believed that the whole world was populated by good and evil spirits and that man's life was completely dependent on them. The good spirits were generally identified with the hunted animals and birds. In the imagination of the Eskimoes, the evil spirits (*tuğnyğat* or *tunğat*) were beings of fantastic shapes and the *tuğnyğaks* caused all the evil which was inflicted on mankind. To avoid their dangerous witchcraft, men had to perform a long row of ceremonies and to observe numerous prohibitory taboos.

The endowment of animal with human characteristics, based on realistic biological details, was clearly evident in the ceremonies consecrated to the killing of big game. Killing a wild animal was not considered to be a murder: the verbs 'to kill' and 'to murder' were only used in connection with man. In the hunting terminology it was enough to say 'he reached him', 'he brought him down', 'he got hold of a seal, a walrus, a bear', etc. According to the ideas of the Eskimoes, the game was not killed, it came spontaneously to the man as a guest. This guest, however, had to be brought down with the help of a harpoon or a spear. The killed game was highly praised and persuaded not to be offended but to return again to the hunters. To assure this they offered sacrifices: they cut off small pieces of the animal's nose and threw them back into the sea, if it was a sea animal, or on the tundra, if the animal came from there. While casting the sacrificial morsels into the ocean, the hunter would express his gratitude to the animal, 'You have come to me as a guest; I thank you. I allow you to go back now. Return next time with several others!' If a brown bear was killed, his head was placed with the snout northwards, and the pieces cut from the nose were thrown with the same words on the tundra.

E. S. Rubcova, who lived for a long time among the Asiatic Eskimoes, describes the reception ceremonies of a whale as follows. 'The killed whale is a grand guest. He does not go to stay with everybody, only with the ones he likes. The hunter who has killed many whales has a great reputation

[9] Косвен (1957) p. 157.
[10] Меновщиков (1958) p. 67.

and prestige far outside his own native village. Considering that the killed whale was the dearest of all guests, the hunter who killed him did his best to receive him as kindly as possible. For a few days the hunter did not leave his house, because, as they explained it, "If a dear guest visits one, one doesn't leave him alone either. The whale, being the dearest of all guests, cannot be left alone or else he will be offended".'

Whales which were killed in the spring were solemnly seen off in the autumn, while those caught in the autumn, had their 'farewell parties' in the spring. To celebrate the farewell festival of the whale was mainly necessary because he would have been sorry if he had seen his comrades leaving. Until that time, they tried to please the whale in every possible way. The killer of the animal was not allowed to provoke anybody to a quarrel, neither could he refuse a request. Since the whale had to be merry, his host beat the drum and sang frequently to make him content, so that he should return there again. Ceremonies consecrated to the whale varied according to the different families, but they all had the motif of the treating of the guests and the giving of gifts in common.

'After each ceremony'—concludes E. S. Rubcova—'members of the feasting family, sitting according to their age, carried dried whale meat in a sacred vessel to the sea and threw the morsels into the water. By this act, the whale was supposed to return to the sea.'[11]

V. G. Bogoraz draws attention to the fact that it was generally a woman who offered the sacrifice to the sea, whereas a man—the head of the family—at the same time presented his equipment to the master spirit of the sea, soliciting success for the next hunt.[12]

The Asiatic Eskimoes considered the raven and the spider to be sacred beings. Both of them were representatives of age and wisdom, and killing them was taboo. Living for decades at the same place, the ravens were well known by the Eskimoes, in fact, some of them were called by their name. These intelligent and longeval arctic birds of prey were aware of the friendly disposition of the Eskimoes and were permanent guests in the rubbish heaps, near the *yarangas*. In the 1930's, in the village of Sireniki, I often saw ravens snatching up pieces of meat which the Eskimoes had given to their dogs. The dogs snarled and went for the ravens, but the Eskimoes only smiled indulgently and looked pleased to see the courageous thieves coming to the village for their breakfast.

The raven cult in the Eskimo mythology and religion is vague; we do not know whether it has totemistic roots or not. In folk tales, the raven appears as a sage who gives good advice to people (see the tale 'They have Reached the Sun'[13]), or as a learned magician who makes people happy (see 'The Man and the Raven'[14]) or again, as a cunning creature who punishes the wicked and the stupid (see 'The Raven and the Wolf'[15]).

The cult of the spider is more specific. The Eskimoes firmly believed that the spider lived the same life as man: it spun a net with which it caught food. In many Eskimo myths and tales the spider appears as the rescuer

[11] Рубцова (1954) pp. 218–9.
[12] Богораз (1939) pp. 85–6.
[13] Меновщиков (1958) p. 40.
[14] Меновщиков (1958) p. 60.
[15] Меновщиков (1958) p. 19.

and guardian of distressed people in need of help. For example, in the tale about the girl who did not want to marry, the spider appeared as her rescuer; the brothers of the man she had killed wanted to give her as food to gigantic worms which were specially bred for the purpose.[16]

The kingfisher (*mysuŋysak̆*) was also regarded as a holy being and, hence, could not be hunted. Among the Asiatic Eskimoes there was a widespread belief that the kingfisher and the wolf were one and the same animal. In the summer the kingfisher lives near the sea, causing terror to all coastal inhabitants, whereas in the winter it dons a fur, roams about the tundra and ravages the reindeer of rich nomads.

Formerly, neither wolves nor kingfishers were killed. The Eskimoes regarded them as their guardians. This is reflected in their folklore. In a tale called '*kikmirasik*'[17] the wolves ravage the whole reindeer herd of the wicked nomad because he illtreats the orphan girl whom they finally rescue from the persecution of her master. Here, the belief in the strength of good, helpful spirits appearing in the shape of an animal is linked up with the idea of sufferings resulting from social inequality.

Explanations for incomprehensible natural phenomena were found in everyday life and in beliefs. The wonderful glitter of the northern lights was thought to be the souls of dead children playing ball in heaven. The rain was interpreted as a girl washing her clothes in the sky, or as the tears of the dead whose souls went to heaven. The wind was believed to be the angry breathing of the master spirit of the earth (*kijağnyk̆*), etc.

The belief that the souls of the dead continue their life in the 'upper' or 'lower' worlds to return as infants to the earth once more, is reflected in the habit of giving new-born children the name of the dead.[18] The child who received the name of a dead person had to try to acquire all the good qualities of the dead, which were taught by the parents or relatives with great severity. Eskimo folklore reflects primitive conceptions of the survival of the human soul after death.

Among the Asiatic Eskimoes, the belief in *tuğnyğak̆*s was common. They were usually considered to be evil spirits who brought diseases, death and a long list of troubles upon men. In the imagination of the Eskimoes, they assumed the most fantastic shapes: giants resembling men, old men and women or balls with gigantic mouths and eyes swinging on straps. They could also be half-men, with one hand, one foot and half a head. They were often animals, (mainly bears and foxes) who talked to each other like human beings. In some cases they even took a normal human shape. These were the most dangerous *tuğnyğak̆*s because they could not be recognized. In many mythological tales, when pursued by courageous men, the *tuğnyğak̆*s sank under the ground only to re-appear at another place for breath, then slipping again under earth like sea animals which are able to spend a long time under water. Not being immortal, the *tuğnyğak̆*s were generally defeated by men. In the tales, as V. G. Bogoraz remarks, the shamanistic spirits of the Chukchi were also thought to have the shape of real things — wild animals, birds, men (giants and dwarfs), house tools, stones, etc.[19]

[16] Рубцова (1954) pp. 183–91.
[17] Меновщиков (1947) pp. 200–5.
[18] Богораз (1939) pp. 179–80.
[19] Богораз (1939) p. 19.

The mediator between the *tuğnyğaǩ*s and men was the shaman (*aliğnalğit*) who was able to communicate with them by means of 'magic acts'. Nansen gives a fine description of shamanism among the Greenland Eskimoes, 'They (the shamans) exercised an influence over their fellow-countrymen, acquiring their ascendancy mostly by their mysterious capacity for magic, which they usually performed in the winter when people lived indoors. The lamps were turned out and the windows covered with hides. Complete darkness prevailed. The *angakok* (shaman) squatted on the floor and produced a horrible noise which made the whole house tremble. He changed his voice, growled, shouted, ventriloquized, sighed, groaned, beat the drum, whined, burst out laughing satanically, and performed all kinds of tricks to make the others believe that he was visited by various spirits and that it was they who really made the noise.'[20]

As I myself more than once observed in Sirenik and in Chaplino, the shamanistic ceremonies of the Asiatic Eskimoes were very similar. In the dark, with the lights turned out in the *yaranga*, the shaman could make such impressive noises that those present, seized with horror, prostrated themselves on the ground and covered their eyes—although it was impossible to see anything even without their doing that! The shaman had to possess considerable knowledge, skill and experience to make those present believe that he was communicating with the *tuğnyğaǩ*s. This behaviour of the Chukchi shamans is well illustrated by V. G. Bogoraz.[21]

Among the Eskimoes no one could become a shaman who had not heard—asleep or awake—the call of his 'image' or 'helper' (in the Eskimo language: *jugwa* 'his man, his master'). Then the shaman-candidate who was subjected to many taboos, prescriptions and precautionary measures, had to meet alone the *tuğnyğaǩ*s with whom he made an alliance, thus becoming an intermediator between spirits and men. Bogoraz says that the Asiatic and other Eskimo tribes have a so-called shaman language (language of the spirits), which is used in conferring with the *tuğnyğaǩ*s. It contains no direct terms for objects and phenomena, but only periphrastic substitutions.[22]

The shamans' influence on their surroundings was immense: the Eskimoes believed that with their help diseases could be driven away, a hunter who was lost in the tundra or carried off by detached floes could be brought home, fine weather could be obtained at sea, game could be forced to appear during unsuccessful hunting, etc. To fulfil these demands, the shaman began his dramatized performance in the darkened house. The ceremonies remained ineffective unless gifts were bestowed on him. Afterwards the sick man's health was restored only if he himself, or his relatives, took a munificent present to the shaman, who had cured him, in the shape of skins, meat, fat or clothes.

As we may see, communication with the spirits rested on an economic basis from the very beginning and payment was essential for any mediation between them and the Eskimo people. During the ceremonial performances the shaman addressed himself to the *tuğnyğaǩ*s, promising them the future gifts from the people for his assistance. Clever shamans acquired wealth in this manner and often became well-to-do landlords (*nunalyxtaǩ*). The

[20] Нансен (1937) p. 344.
[21] Богораз (1939) pp. 105–44.
[22] Богораз (1919).

religious creed of the tribe was abused by the shamans for money-making and boundless exploitation.

Shamanism failed to take firm roots among the Asiatic Eskimoes; in fact, owing to the steadily growing network of Soviet cultural and medical institutions, it definitively lost the ground from under its feet.

Fetishes play an important role in the religious concepts of the Eskimoes, according to which some protected the hunters from the dangers of the sea and the tundra, others brought luck in hunting and still others safeguarded them from the *tuğnyğak's* who had the power of cursing mankind with dreadful diseases. Minor services, such as protection against evil spirits, ensuring the friendly dispositions of the good spirits of the family hearth, or personal welfare, were achieved by female amulets.

These included family and personal amulets. The former, associated with the festive ceremonies, were carefully guarded and inherited from one generation to the other. They included parts of animals, birds or other things regarded as divinities. Among personal amulets there were all kinds of items, ranging from human, animal, bird or fish figures made of bone or wood to bunches of flowers. They served also for decorative purposes and were worn on dresses, belts, or head-bands and on the neck, hands or feet.

Nansen, the famous Norwegian arctic explorer, who made a thorough study of the cultural activities of the Greenland Eskimoes, gives, in his work *The Life of the Eskimoes*, an interesting description of the religious beliefs connected with amulets. They are almost identical with those of the Asiatic Eskimoes. He writes as follows, 'The Eskimoes believe in amulets which all of them wear and which consist of certain objects related in one way or another to animals or to men . . . They wear them throughout their life, mostly on their body or on their clothes. Men frequently sew them into their leather braces, on their chest, whereas women put them into their hair. Others are kept on the roof, in the tent, or in the kayak to prevent it from capsizing. Everybody has usually several amulets, which, they believe, protect them against bewitching, safeguard them from the injury of evil spirits, assist them in dangers, and bestow special qualities on them.'[23]

I was able to ascertain personally that this multiplicity of amulets, and ceremonies and beliefs attached to them, was widespread among the Asiatic Eskimoes as late as the 1930's. Sometimes the belief in the magical force of a given amulet spread among a large group of hunters. If, for instance, a man who wore a raven figure had luck during a whole season, while the others returned with empty bags, raven-shaped amulets or parts of the raven itself (beak, head, feathers) were chosen to ensure successful hunting for the whole community.

At the beginning of the 19th century, Ju. Lisjanskij, a Russian explorer, collected valuable material on the amulets and their importance among the Kadyak Eskimoes in Alaska.[24]

Besides the amulets which protected the people from diseases and various ailments and also brought them luck in hunting, the incantations of the shamans occupied an important place in the religious practices of the Asiatic

[23] Нансен (1937) p. 347.
[24] Лисянский (1812).

Eskimoes. They were used to expel diseases, to attract the wild tundra and sea animals when the hunting was unsuccessful for a longer time, to ensure the safe return of hunters carried off by floes driven away by the wind, etc.

E. S. Rubcova[25] gives us an interesting description of the preparations that preceded the incantation among the Chaplino Eskimoes. The aim of the ceremonies, consisting of two parts, was to drive away a disease. Before performing the incantation, the shaman had to replace his 'dirty' mouth by a clean one, which upon his request 'the old woman living in the upper heaven' (*ḱilagym ḱulim ağnaʟḱwağa*) lends him for that time. The first part is consecrated to the replacement of the dirty mouth, which is spat in the middle of a sea current, while the clean one is borrowed from the old woman in heaven. Diseases can be driven away only by incantations performed with the clean mouth. During the second part of the ceremony 'the heavenly daughters of the old man who does not lift his head' are asked to come down from heaven, 'the younger daughter with gloves of the skin of an Arctic hare and the elder one with gloves of ordinary hare skin'. The younger one comes and caresses the face of the old man, the elder one his nape. The old man lifts his head and begins to smile. According to Rubcova, the Eskimoes considered the 'old man who does not lift his head' to be the 'chief spirit of life to whom they appealed through the intermediary of his daughters'.[26] Thus the people are rescued by 'the old one who does not lift his head'. He is just like ordinary people as far as appearance and habits are concerned.

The burial of the dead was performed in the following manner: First the corpse was washed and dressed in a new fur suit, then it was tied together with straps so that there would be 8–10 strap ends by which to carry it. A long stick was fastened to the body to prevent it from bending. The head was covered with the skin of a young reindeer to prevent the soul of the deceased from seeing the road, so that he should not be able to return. Before the corpse was carried out of the house, the relatives and the guests sat to a funeral feast, as this was the last occasion for the dead to take a meal together with the others. Family members were not allowed to carry the corpse to the burial place. The dead could not be taken out through the door and a special opening had to be cut for this purpose in the back part of the house. Once the corpse had been removed, the opening had to be carefully sealed up so as to hinder the soul of the dead from finding its way back. If the death of a family member was soon followed by that of another member of the same family, it was evident that the first soul had found the way back and had taken the other man with him. In such cases the house was torn down, its parts scattered to the winds or burnt, and a new house built at some other place. It was not a custom among the Eskimoes to bury or cremate the dead. They were left on the ground, with small stones around them to symbolize the fence that was to prevent them from returning. The clothes and straps of the deceased were cut into small strips at the place of burial, and the corpse was surrounded by personal belongings. If the deceased was a man, his hunting

[25] Рубцова (1954) pp. 290–1.
[26] Рубцова (1954) pp. 290–1.

equipment (his gun, knife or harpoon, or sometimes even his boat or kayak) was placed near him, it if was a woman, she was surrounded by her household utensils (a knife, a wooden dish called *kajutak̆*, a cup, a teapot, a sewing machine, needles, furs, etc.). The objects left near the dead were usually destroyed so that nobody should be able to use them except the soul of the dead. Bogoraz thinks that the reason for this was to separate these objects from their own souls so that the deceased could take them with him.[27]

According to ancient customs, persons who returned from the funeral were dusted with ashes and had their hands washed. For one day afterwards it was forbidden to speak aloud, because the dead man's soul could hear it and might return.

On the day of the funeral the Eskimoes avoided tying knots, sewing, and fastening things with straps, because the ends of the threads, or of the straps, might pull the soul of the dead person back and attach illness to the house.[28]

If the deceased had friends with whom he had grown up, played and hunted, one of them scattered residue from a pan of fat around the *yarangas* of the others on the evening of the funeral so as to fence them off from the soul of the dead. Persons of the same age as the dead tied their little finger and their third finger together with grass to escape the necessity of having to leave their family.

The first night after the funeral, the 'image' (*pistyŋa*, 'his active force') of a dead person goes about visiting the houses of his contemporaries to see them play and work, just as he had seen them when he was alive. This 'going about' endangered their lives, because the dead man had become an evil spirit.

If, during a hunt or a journey by dog sledge, a corpse was found lying on the road, the hunter or traveller had to step over it or kick it with his foot, lest the dead return and take his life too. The invisible image that accompanied and safeguarded the man, was called *pista* 'active force' and *tağnyk̆* 'shadow'.

Five days of deep mourning followed the death of a woman, whereas the mourning for a man lasted only four days. During this time nothing was done and complete silence was observed in the house. After this period of time, and when the deceased had taken leave of his relatives, all traces leading to the funeral place were carefully destroyed so that the dead, transformed into an evil spirit, could not return.

The Asiatic Eskimoes had beliefs, charms and taboos; the origin and existence of these were more or less connected with their everyday life, many aspects of which they could not understand.

A considerable part of these beliefs were associated with the sea and the wind. Strong sea winds caused much trouble for the hunters. Light boats and kayaks were driven out to the high sea, floes with hunters on them were detached and driven to unknown places, and hunting on the sea was often stopped for a long time by storms—and famine was the inevitable result.

[27] Богораз (1939) p. 192.
[28] Рубцова (1954) pp. 78–9.

During my frequent meetings with Asiatic Eskimoes I was able to collect many of their beliefs, prophecies and taboos. Here I shall record only a few of them, as was told to me by Numylen, the fearless hunter, in 1954, in the village of Sirenik:

'In the summer you must not carry an uncovered drum in the street, for if you do, a strong wind will arise and the sea will be stormy. The drum must be covered by a leather coat.'

'Don't burn seaweed on a camp fire, for bad weather is sure to follow, and there will be a big surge.'

The majority of the beliefs, prophecies and taboos of the Eskimoes were related to their struggle for existence and for the hunting of animals living in the sea and on the tundra. Some revealed their natural desire to preserve game for the future, since the haphazard and ruthless exploitation of nature would necessarily lead to famine and ruin. We wish to stress the following Eskimo maxims:

'If you meet five identical animals: five wild reindeer, five wolves, five hares, five foxes or other animals—you must not kill all of them. One is taboo, it belongs to the creator (*kijağnyk̆* "creator, life"). If you kill all of them you must pay with the life of your relatives, your clan will die out. Neither must you kill all the sea animals: if there are five together, one must be spared even if it does not go away by itself.'

'If you kill an animal in the tundra—hare, polar fox, fox, wild reindeer or mountain ram—touch its ear. If the ear is torn or there is a cut in it, you must not take the animal home but turn it with its head to the east and say the following words, "I have killed you out of ignorance, go back home to your master".'

'When fishing or angling, you must not keep the biggest fish. The biggest fish belongs to the master spirit of the sea (*myğym jugwa*). If you keep it, you will have no luck.'

'If on the fins of the sea animal you have killed there are six phalanges instead of five, you will be a lucky hunter all your life and so will your posterity. If, however, you find less than five phalanges, you will never have luck in hunting.'

The beliefs referring to women reflect in a most striking fashion the dominating role which men played in production.

'A married woman must not wear her husband's clothes because her weight would be transmitted to him and he would drown if his boat capsized and would fall into the water while jumping from one floe to another.'

'When giving meat to her husband, a woman must place the dish before him so that it should not tilt, for if it does, his boat will rock on the sea, or the ice will crush it and he may get drowned.'

'A woman must not hang up her clothes to dry above the head of her husband, or else he will have no luck in hunting. If he stalks a sleeping animal, the ravens, owls and gulls will fly there and awaken the animal which will run away.'

'If, after a husband has eaten his meal, pieces of meat are left over on the dish, his wife must collect them in a small heap and eat them all up at once; then animals will come flocking to her husband when he is hunting.'

The hunting prophecies and prohibitions clearly reflect the everyday life and hunting habits of the Eskimoes.

'Don't roast walrus meat on the camp-fire, because the walruses will get wild and will turn over your boats when you are hunting.'

'You must not bend over the kettle during the meal, or else the harpooned walrus, or whale, will swim in different directions for a long time.'

'Small children and lads must not be trashed either with gloves or with other clothing, or else the hunted game will become wicked and will hide.

'A young man must not eat the scraps on the dish, for if he does, he will kill only poor game.'

'When hunting on the ice, don't shout or call your comrades at the top of your voice, even if it is to catch up with the game or to tell them that the current is carrying you far out, because you will frighten the game away. You may shout only when you are in real danger and need help. In case of danger you must shout one single word: *gys*! This will be repeated by the echo, above the ice, and will be heard by the others.'

The prophecies reflect the cosmogonic ideas of the Eskimoes in connection with the sun and the moon as sources of life, controlling warnish, light and weather.

'When there is an eclipse of the sun or the moon, the oldest men and women must come out of their *yarangas* and sing old songs to the sun or moon, with the shoulder-blades of a reindeer in their hands. A dog must be killed and its blood sprayed in the direction of the eclipse; this brings luck to the hunters.'

Similar examples of beliefs, forecasts and witchcraft have been recorded by Bogoraz among the Chukchi.[29]

Eskimo notions of the world and the earth are connected with a mythical being called *sila*, *siʟa* or *sʟa* ('world, universe, outer space, weather, divinity'). The master spirit of the universe, the creator of every living and lifeless thing on earth was called *siʟam jugwa*, the man of the universe. It was his duty to follow with attention whether men were observing the old customs and traditions. In the Eskimo mythology he was represented as a simple man, with a dog sledge, kayak, harpoons, paddles, leather straps, a wife and sometimes even with children. He killed sea animals and wild reindeer just like earthly men. He was, however, superior to mankind in so far as he possessed the supernatural force of carrying men to the 'upper world' and there of teaching the evil-doers earthly virtues (see the tale called 'The Alaikhpagmits').[30]

The ideas of the Eskimoes about the creator of the world, the master spirit of the universe, the divinity living in heaven, were not clearly defined. This dignitary, as opposed to the devils (*tuǧnyǧak̆*) and 'images' (souls), had no definite shape and name—as I have discovered—but assumed one or another name according to the situation in which he was mentioned (in a tale, song, incantation, belief, prohibition, etc.). Among these appellations, in addition to *siʟam jugwa* or *sʟam jugwa*, the following were used by the Asiatic Eskimoes: *kijaǧnyk̆* 'who gives life' (or 'life', from the stem *kijaǧ-* 'to live'), *kynlyk* 'which has limits' (from *kynla* 'border', 'limits'). Cf. the prayer to the killed animal: *k̆ajux̆ʟax̆k̆un naʟunk̆usin, tyǧikumamkyn, aga agla ʟpyk kynlyxpynun, juxpynun.* 'Because I have killed you by

[29] Богораз (1939) pp. 145–65.
[30] Меновщиков (1958) p. 125.

error, go to the one having limits, to your man (meaning: 'your master'), *almysim jon* 'man of the customs' or 'master spirit of the custom' (from *almysin* 'custom', Naukan dialect); *tynagyrgyn*, the name of the divinity of the upper world who watches over the adherence to the old customs. (The name of this deity is borrowed from Chukchi mythology.) The Chaplino Eskimoes generally used the expression *kijağnyk̃* for it, whereas Naukansk used, in addition, *almysim jon*. This supreme being was invested with human features just like *siʟam jugwa*: he went hunting, possessed a *yaranga*, a boat or a kayak, a wife (sometimes even two) and children. He took the guilty hunters with himself in order to instruct them and then let them descend to earth by a strap of seal skin. He chose his wives from among the seaside hunters and also let them come down to earth when he was tired of them. This realistic background of the life of the Eskimo protector deities reflects their immature and primitive ideology which was based on a distorted version of their own life.

From what has been said, we see that monotheism remained at a low stage of development among the Asiatic Eskimoes. Their religion was founded on a belief in various spirits and contained elements of ancient animism. This can be observed among many other peoples at a lower stage of development. It was only under the Soviet system that the Asiatic Eskimoes started to discard, step by step, their distorted views concerning natural and social phenomena. Now their life has become imbued with the materialistic ideas of Marxism–Leninism and, governed by technology and science, they have ceased to be dependent on the whims of nature.

The religious conceptions and rites described in this article are only my personal observations and do not give a full picture of the spiritual culture of the Asiatic Eskimoes. In their evolution and in material and spiritual culture of the neighbouring tribes, there are still many blanks to be filled up by historians, ethnographers and linguists.

REFERENCES

(Anisimov, A. F.) Анисимов, А. Ф. (1958) *Религия эвенков* (The Religion of the Evenki), Moscow–Leningrad.

Birket-Smith, K. (1929) *The Caribou Eskimoes*, Copenhagen.

Boas, F. (1888) *The Central Eskimo*, Washington.

(Bogoraz, V. G.) Богораз, В. Г. (1908) Религиозные идеи первобитного человека (The Religious Ideas of Primitive Man), in *З*.

(1910) К психологии шаманства у народов северовосточной Азии (Contribution to the Psychology of Shamanism among the Peoples of North-Eastern Asia), in *ЭО*.

(1919) О так называемом языке духов (шаманском) у различных ветвей эскимосского племени (About the So-called [Shaman] Language of Spirits among the Different Branches of the Eskimo Tribes), in *ИАН*, series VI, Vol. XIII, Nos 8–11.

(1930) Чукотский общественный строй по данным фольклора (The Structure of Chukchi Society on the Evidence of Folklore Data), in *CC*, No. 6.

(1932) Религия как тормоз соцстроительства среди малых народностей севера (Religion as a Hindrance to the Building of Socialism among the Small Tribes of the North), in *CC*, Nos 1–2.

(1939) *Чукчи* (The Chukchi) Vol. II, Религия (Religion), Leningrad.

Dall, W. H. (1870) *Alaska and its Resources*.

Jochelson, W. (1905) *The Koryaks*, Vol. I, Religion and Myths, The Jesup North Pacific Expedition, in *MAMNH*, Vol. VI, No. 1.

(Kosven, M. O.) Косвен, М. О. (1957) *Очерки истории первобытной культуры* (Essays on the History of Primitive Culture), Moscow.
(Lisjanskij, Ju.) Лисянский, Ю. (1812) *Кургосветное путешествие* (A Journey Round the World), Moscow.
(Menovščikov, G. A.) Меновщиков, Г. А. (1947) *Ун ипаг' атыпут* (Our Tales) Leningrad.
(1954) *Эскимосско-русский словарь* (Eskimo–Russian Dictionary), Leningrad.
(1955) Указательные местоимения в эскимосском языке (Demonstrative Pronouns in the Eskimo Language), in *ВЯ*, No. 1.
(1956) Из истории образования числительных в эскимосском языке (From the History of the Development of Numerals in the Eskimo Language, in *ВЯ*, No. 4.
(1958) *Эскимосские сказки* (Eskimo Tales), Magadan.
(Nansen, F.) Нансен, Ф. (1937) *Собрание сочинений* (Collected Works), Vol. I, Moscow.
Nelson, E. W. (1899) *The Eskimo of the Smithsound*, New York.
(Rubcova, E. S.) Рубцова, Е. С. (1954) *Материалы по языку и фольклору эскимосов. Чаплинский диалект* (Materials on the Language and Folklore of the Eskimoes, Vol. I, Chaplino Dialect), Leningrad.
(Šrenk, L. I.) Шренк, Л. И. (1883–1903) *Об инородцах амурского края* (On the Minorities of the Amur Region), Saint Petersburg.
(Šternberg, L. Ja.) Штернберг, Л. Я. (1933) *Гиляки, орочи, гольды, негидальцы, айны* (Gilyaks, Orochi, Golds, Negidals and Ainu), Khabarovsk.

The Preservation of Animal Bones in the Hunting Rites of Some North-Eurasian Peoples

by

I. Paulson Stockholm

In his excellent study on the hunting rites of the northern peoples of Asia and Europe, Uno Harva[1] has devoted a long chapter to the preservation of bones of wild animals.[2] The present summary, which has been based on the extensive material collected there, and on the data available in subsequent publications by other authors who dealt with the problem,[3] is intended to be a preliminary report on a research field the author has been interested in for a long time.[4]

The preservation of single bones or even of the whole skeleton of the game animal is a most common and widespread custom throughout North-Eurasia (Northern Europe and Siberia). It may be observed not only in the hunting rituals, i.e. magic rites aimed at preserving and increasing the stock of game, but also in animal sacrifices, mostly of domestic animals.[5] As regards the purely hunting aspects of the custom, the author has referred to its occurrence also among northern peoples in North America.[6]

To sum up, the following conclusions may be drawn in connection with the preservation of bones in North Eurasia.

1. In the records about animals whose bones are preserved or saved in one form or another, the following are mentioned: bear, reindeer, elk, deer, fox, hare, sable and other, not specified, fur animals, wolf, lynx, wolverine, walrus, seal, swordfish, as well as various kinds of birds and fish.

The bear occurs most frequently, and we possess the most detailed data about the preservation of bear bones. It must be emphasized, however, that rites performed with bear bones are by no means fundamentally different from those where other animals are involved, the difference being only one of degree and not of manner. This may partly be ascribed to the circumstance that the rites concerning bear bones are, in many places—for instance among the Saams (Lapps), the Mansi (Voguls), the Khanty (Ostyaks) the Nivkhi (Gilyaks) and the Ainu—included in major bear festivals, and therefore appeal to the natives, as well as to scholars and reporters, to a far greater extent than do the more common and simple hunting rites connected with other animals. As a matter of fact, these minor rites become more prominent whenever bear festivals are either missing or they are less developed, as among the Evenki (Tungus), the Yakuts, the Yukagirs (Oduls), etc. It is the rites with bear bones that offer the greatest insight into the mental processes and religious conceptions

[1] Cf. Harva (1938) pp. 434 ff.
[2] Holmberg (1925) pp. 34 ff.
[3] Hallowell (1926); Gahs (1928); Friedrich (1941, 1943); Lot-Falck (1953) pp. 205 ff; Nachtigall (1953); Zélénine (1952) pp. 60 ff; Edsman (1958).
[4] Cf. Paulson (1961).
[5] See Holmberg (1925) pp. 45 ff; Harva (1938) pp. 446 ff; cf. also Gahs (1928).
[6] Paulson (1959).

underlying the bone-preserving rites as a whole. Also everywhere the bear is marked by individual traits, unlike the other animals which readily lose their character in the great collective of their species. However, on a basis of comparison, we may say that some useful game such as the wild reindeer, the elk and the deer, or some beasts of prey such as the wolf, and in the coastal area, some of the big sea animals (walrus, seal, whale), are found to occupy a more prominent place than others.

2. The customs connected with the preservation of animal bones may be classified into four groups, depending on how the bones are preserved or buried. (a) *Burial in the ground.* This method is customary mainly in Lapland in respect of the bear, though it has been reported sporadically from all over North Eurasia also in connection with other animals. (b) *Burial in trees* or *on platforms.* A method especially popular in Siberia with various tribes and different animals. (c) As a transitional form we may mention the costume by which the bones are simply *laid out on the earth* with or without a cover of stones, wood and the like. In the final analysis, this class includes most of the cases in which reporters mention only the mere fact of bone preservation, without any particulars as to the method. (d) *Burial in water.* This method is often used when the animal's natural habitat is water (in cases of aquatic animals of both inland waters and sea mammals, fish, etc.). There is no evidence of bones of land animals ever having been sunk in water in northern Eurasia, whereas information of this kind is available from North America.[7] Since there is only very vague reference (e.g. in the Khanty material) to the ritual burning of animals' bones, it is out of place to speak of cremation in connection with beasts of prey.

From the form of burial some scholars (e.g. Gahs, Nachtigall) have tried to draw far-reaching conclusions, concerning the meaning of bone preservation in general. The author thinks he cannot agree with them. In spite of local differences, the form of burial hardly allows any inference as to the underlying motives of bone preservation. Harva, who has also dealt with the problem pointing out the difference between animal burial in Lapland (in the ground) and in Siberia (burial in trees and on platforms), is of the opinion that the rites of animal burial are closely connected with the forms of human burial.[8]

3. The quantities as well as kinds of bones preserved seem to supply more important implications than what can be derived from the different forms of burial. The material contains many direct indications pointing to the preservation of the whole animal, i.e. the skeleton, and probably also the soft tissues. Of course, a great number of the references on the subject do not specify at all whether the entire animal or only certain parts were involved. The skull is mentioned most frequently. Special ritual preservation of bear skulls, as well as skulls of other animals, can be observed throughout the whole area, right from Lapland and Finland to the Far Eastern regions where the Nivkhi and the Ainu live. Sometimes, but less frequently, other bones are mentioned as being preserved either together with the skull, or more sporadically, alone. Occasionally, the skull of the bear was preserved in a tree, while the other bones were either buried or even burnt (Khanty). Beside the skull, mostly long bones are mentioned in this connection.

[7] Paulson (1959).

[8] Holmberg (1925) p. 40; Harva (1938) pp. 440 ff.

In a wider scope, the rites of bone preservation include cases where some of the soft parts (flesh, hide, etc.) were preserved together with the bones. Small animals, with or without their fur, or big beasts of prey, not used for food, were often preserved entirely. This is interesting in so far as it reveals the intention of preserving as much as possible of the parts of the animal which man did not want for himself. Certain parts (as, e.g., the skull), however, were considered as substitutes for the whole animal.

Finally, the prohibition, often referred to, of breaking or injuring the bones of animals should be mentioned here. It can occur also independently of the actual preservation.[9]

4. Most of the data, however, say nothing about the reasons or motives underlying bone preservation, but merely describe the ritual proceedings or mention the occurrence of the custom, without giving any explanation. Whenever an account by the natives themselves is given to motivate the act, it usually implies that by preservation rites the bones are safeguarded from dogs and from the wild animals. Sometimes they maintain that the killed animal, or its 'soul' develops an interest in the fate of its earthly remains and even demands that special rites should be performed for them. These should be carried out so as not to rouse the animal's wrath, to incur its vengeance, etc. At the same time, another motive usually emerges: By saving and burying the bones of the killed animal in strict accordance with ritual prescriptions, and by observing certain laws and taboos, the natives not only aim at future luck in hunting, but also wish to avert any danger that might threaten the hunter from other animals of the same species. They hold, for instance, that a bear permits itself to be caught more readily by a hunter who has always treated bear bones correctly, whereas it will attack and kill one who has omitted to pay these last honours.

Quite a number of data have been gathered from Lapps, Khanty, Nivkhi and Ainu which are highly instructive in respect of the conceptions underlying this custom. The bear, after its bones had been paid due honours, is requested to visit other animals of its kind and tell them about the fair treatment and the honours received from men. The bear is also supposed to ask other bears to let themselves be caught willingly by these hunters. Occasionally, a special place is named where the animal (namely the bear) is sent after its death and after the appropriate rites have been performed. In this connection the Finns speak of a 'forest home', the Nivkhi, of the 'Master of the forest', of the 'Master of the mountain'. The Tungus send the 'souls' of the animals 'toward the West', i.e. to a land situated in the west and considered to be the land of the dead for the wild animals.

There exists another interesting group of data, according to which the killed animal is not only supposed to continue life in some simple, unspecified form, but is also believed to be revived or resurrected. Such beliefs among the Lapps included the 'slow resurrection' of the killed bear, or 'new bears' and 'new birds' to be formed of the bones of the killed animals. Elsewhere it is said that an animal will resume its life if its bones are properly and correctly preserved (Yukagirs).

The concept, according to which the soul or spirit (or the dead body) of the animal had had a particularly important role in the bone-preserving

[9] Cf. for North-America: Paulson (1959).

ceremonies, can be fairly well demonstrated by data from the Khanty. A description of Khanty bear festivals explicitly states that the earthly remains of the bear must be paid all honours, 'because his shadow (i.e. his so-called "free soul", *is* which is represented as a shadow) demands that this should be done'.[10] Bear skull preservation, referred to by several authors, is said to have as its aim the 'sending home of the shadow soul', since the skull is regarded as the residence of the soul for all living beings, men as well as animals.[11] After the songs have been chanted during the bear festivities, the shadow soul of the bear rises to the heavenly god *num torem* who is believed to have once sent the bear down to earth.[12] It was a rather familiar view among some North-Eurasian peoples (Lapps, Mansi, Khanty, Nivkhi, Ainu, and others) that, during the lifetime, the immortal soul (free soul) of the bear and of other animals had its seat in the skull. Hence the skull had to be treated with particular reverence when the 'sending home' ceremonies took place after the animal had been killed.

*

Manifold attempts have been made by various scholars[13] to explain the custom of preservation of the bones and the hunting rites related to it. The problem has been sufficiently elaborated and does not need a lengthy repetition. Nevertheless, we think it worthwhile to summarize some of the most important results.

All scholars agree that the bone rites are performed for the sake of the soul or spirit of the dead animal (Frazer, Harva, etc.). In this rite the 'soul' stands for the 'animal itself', is its form of appearance after death, and corresponds genetically to the 'free soul' of the living animal.[14] Although the preservation and ritual treatment (burial) of the whole animal or its skeleton is by no means general, such honours being awarded sometimes to one part only, it is beyond doubt that the animal is always considered as a whole, on the principle of *pars pro toto.*

As a rule the primitive hunter sees in an individual animal a being representing the whole species or genus. For this reason, the individual soul of the animal mostly dissolves into the collective soul of the animals. Recently Hultkrantz has demonstrated for North America[15] how easily the idea of the soul of an animal, or even of a species, can merge with that of a collective guardian spirit (*animal guardian*) of a given genus. These data concerning the guardian spirit of animals, i.e. animal species can be supplemented with a range of other data from North Eurasia.[16] Of course, there are considerable differences in this respect among the various genera. The soul, or dead spirit, of a powerful bear, for instance, stands out much more as an individual figure than the souls of smaller animals which are more likely to amalgamate in their genus taken as a collective entity. Among the Siberian tribes, the

[10] Patkanov (1897) pp. 127, 130; Karjalainen (1921) pp. 14 ff.
[11] Paulson (1958a).
[12] Patkanov (1897) p. 147; Karjalainen (1921) pp. 386 ff.
[13] To be added to those in Note 3: Frazer (1949) pp. 518 ff.
[14] Paulson (1958a).
[15] Hultkrantz (1953) pp. 497 ff.
[16] Paulson (1961).

Yukagirs seem to have had particularly well developed concepts of individual and collective animal guardians.[17]

Some of the data show that bone ceremonies were performed to honour the guardian spirit of the species. Apart from the protective spirits of the individual species, there had come up also a general guardian spirit of the entire animal world. The Yakuts call him the 'Master of the forest' (forest spirit and protector of the forest animals), the Nivkhi the 'Master of the forest' or of the 'mountain' or the 'sea' (for sea animals). It is he who demands that bones of animals should be ritually treated—or, at least, it is for him that they are carefully preserved. Thus, Friedrich has linked up the idea of revival from bones with that of the 'Master of the forest' who plays a leading part in it. He,[18] however, admits that this cannot be proved on the basis of reports. In contrast, Harva, who has also investigated the role the forest spirit played in the custom of bone preservation, is not willing to acknowledge such a function.[19]

The theory (Gahs, Schmidt and Nachtigall) which holds that the preservation of the bones of game animals had originally been offerings to the heavenly god or to the so-called supreme being is erroneous and unacceptable. The so-called heavenly sacrifice (Gahs) cannot be more ancient than the hunting magic (Gahs 'totemistic magic') which is declared to be a secondary development by the school of the so-called cultural history. The latter exists virtually everywhere in the circumpolar and subpolar zone, from Lapland to Labrador, whereas the former—that is to say the real sacrifice to the heavenly god—can be found only among some nomad tribes of Central Asia, and even there only domestic animals, and never wild animals, are used for the purpose.

The idea of the *revival* of the bones can be found in many places (Friedrich). It is, however, important to note here that the natives themselves think rather of the animal's continued existence than of a revival in the proper sense of the word. The question has been asked by many scholars in their analyses[20] as to what mental processes are behind the rites with game bones, and the belief of the animal's continued life attached to them. In answer to this question, it must first be emphasized that no view maintains the subsistence of the so-called life soul[21] in the bones, either in the skeleton as a whole or in one of its important parts, such as the skull.[22] It is not the life soul, or the qualitative principle of being alive that survives in the durable parts of the body (bones, etc.) of men and animals (in fact, this escapes when death sets in)—but a quantitative substance, called life stuff.[23] Different hunting magic rites have resulted from this, first of all, the preserving of bones. Religious motives were added for sacrificial animals, and eschatological concepts and speculations for human beings. Undoubt-

[17] Jochelson (1926) pp. 144 ff.

[18] Friedrich (1943) p. 203.

[19] Harva (1938) p. 445.

[20] See in particular Friedrich (1943).

[21] For conceptions about the soul among the peoples in the North of Europe and Asia, see Paulson (1958b), and for animal spirits, Paulson (1958a).

[22] A view advocated, e.g., by Pettersson (1957) pp. 54 ff.

[23] For the difference between these two conceptions, see Hultkrantz (1953) pp. 160 ff, 180 ff, 190 ff; and Paulson (1958b) p. 232.

edly, however, the idea of the immortal soul, the surviving dead spirit of animals and men must also be regarded as one of the further reasons for the preservation of bones.

Both the spirit or 'soul' and the body (i.e. its durable remains such as the bones, the entire skeleton or parts of it, like the skull), represent the creature *in toto*, in a closed personal identity after death. For the primitive imagination, dead animals, as well as men, are always doubly present, as corpses and as surviving souls or spirits. These are in a mystic relation with each other, so that an injury done to the bones, for example, would cause a similar damage to the soul. Thus, the belief in an animal's soul surviving, and representing it after death, had important consequences in the evolution of hunting and religious rites among primitive peoples.

Finally, let us refer to another important aspect of the problem. With Harva we may conclude, 'In all likelihood, the wild animals themselves had originally been the subject of all customs and beliefs connected with hunting'.[24] The basic reason for reviving animals is the same as the reason for the beliefs connected with death: the general conception of soul. This is also proved by the fact that a deeper veneration is shown to the game animal after it has been killed. Exceptions, such as the worship of the living bear or other big game, do not affect the general principle. Animal worship is a cult of the dead, says Karjalainen[25]; it is the veneration of the animal spirit that has outlived death.

However, as everywhere else in animal worship (animalism), the basic issue is the disposition of man—especially of the primitive hunter—who feels as if he were linked up in a mystic relationship with the animals. At the end of his study on the representations of the animal and plant souls, as found among the North-American tribes, Hultkrantz says, 'The hypothesis we have advanced above also gives us an insight into a piece of human psychology. In the evolution of the belief in souls among animals, plants and objects, we find the pure expression of the fellow-feeling of the primitive hunter and collector with Nature.'[26] In his work about the beliefs of the Dolgans, the famous Russian explorer A. A. Popov has called this attitude of primitive men toward Nature and, above all, toward animals—*assimilatism*.[27]

REFERENCES

Edsman, C. M. (1958) Studier i jägarens religion (Summary: Studien zur Religion des Jägers), in *AARS*, Kungl. Vetenskapssamhällets i Uppsala Arsbok, No. 2, Stockholm.

Frazer, J. G. (1949) *The Golden Bough*, A Study in Magic and Religion, New York.

Friedrich, H. (1941) Die Forschung über das frühzeitliche Jägertum, Mitteilungen zur Kulturkunde, in *P*, Vol. 2, Fasc. 1–2.

(1943) Knochen und Skelett in der Vorstellungswelt Nordasiens, in *WBKL*, Vol. 5.

Gahs, F. (1928) Kopf-, Schädel- und Langknochenopfer bei den Renntiervölkern, in *Publication d'hommage offerte au P. W. Schmidt*, Wien.

[24] Harva (1938) p. 394.
[25] Karjalainen (1921) p. 48.
[26] Hultkranz (1953) p. 497.
[27] Попов (1958) pp. 80 ff.

Hallowell, Z. B. J. (1926) Bear Ceremonialism in Northern Hemisphere, in *AA*, Vol. 28.
Harva, U. (1938) Die religiösen Vorstellungen der altaischen Völker, in *FFC*, Vol. 125.
Holmberg, U. (1925) Über die Jagdriten der nördlichen Völker Asiens und Europas, in *JSFOu*, Vol. XLI.
Hultkrantz, A. (1953) Conceptions of the Soul among the North American Indians, in *EMSMS*, No. 1.
Jochelson, W. (1926) The Yukaghir and the yukaghirized Tungus, in *MAMNH*, Vol. IX, The Jesup North Pacific Expedition.
Karjalainen, K. F. (1921) Die Religion der Jugravölker, Part I, in *FFC*, Vol. 41.
Lot-Falck, E. (1953) *Les rites de chasse chez les peuples sibériens*, Paris.
Nachtigall, H. (1953) Die erhöhte Bestattung in Nord- und Hochasien, in *A*, Vol. 48, Fasc. 1–2.
Patkanov, S. (1897) *Die Irtysch–Ostjaken und ihre Volkspoesie*, Part I, Saint Petersburg.
Paulson, I. (1958a) Die Vorstellungen von den Seelen der Tiere bei den nordeurasischen Völkern, in *E*, Nos 2–3.
(1958b) Die primitiven Seelenvorstellungen der nordeurasischen Völker. (Eine religionsethnographische und religionsphänomenologische Untersuchung), in *EMSMS*, No. 5.
(1959) Zur Aufbewahrung der Tierknochen im Jagdritual der Völker des nördlichen Nordamerika, in *MMVH*, Vol. 25. ('Miscellanea Americana'.)
Pettersson, O. (1957) Jabmek and Jabmeaimo. (A Comparative Study of the Dead and the Realm of the Dead in Lappish Religion), *LUÅ*, N. F. Avd. 1, Vol. 52, No. 5.
(Popov, A. A) Попов, А. А. (1958) Пережитки древних дорелигиозных воззрений долганов на природу (The Survivals of Ancient Pre-Religious Conceptions of Nature among the Dolgans), in *СЭ*, Vol.2.
Zélénine, D. (1952) *Le culte des idoles en Sibérie*, Paris

Mother Cult in Siberia

by

O. Nahodil, Prague

The time-honoured force of tradition is one of the most active factors in the history of religion. It is responsible for many elements in contemporary religions, which connect the beliefs of the people of the twentieth century to those of the ancient representatives of mankind, the anonymous creators of primitive cultures. It is owing to tradition that we are still able to discover in the different religions of the world various historical stages and causes of the vitality of ancient religions, often concealed under a more or less transparent cloak of conceptions of later origin. By lifting this, we can understand the transformations which some ancient elements of religion have undergone in the course of history. The dignified beauty and nobility of the Renaissance Madonna, manifested through the sublime art of the Italian painters of the 14th–16th centuries, meant something more than an artistic expression of the religious concepts of that period. In that same sense we may say that the cult of the Holy Virgin was not, in itself, a prerogative of Christianity, but indicates an intimate emotion which has always been, and still remains, the most important foundation of all beliefs—I refer to the universal respect of woman and motherhood, ancient as religion itself.

In Christianity these primeval ideas are only a reflection and the completion of ancient traditional beliefs which can be traced to the dawn of religion, right back to the Aurignacian period.

The cult of woman and mother is practised all over the world. It appears in many forms, some of which are vague and can only be treated correctly by examining (a) their mutual relationship, as was suggested by Edward B. Tylor[1], and (b) their relation to social and economic evolution, as was done by J. J. Bachofen, Th. Achelis, N. Kharuzin, and recently also by G. Thomson and others.[2]

Nowadays, however, the former methods of research are not fully satisfactory any more. The thousand years old traditions can only be evaluated if we do away with the idealistic interpretation in which this cult has been wrapped lately by R. Thurnwald and K. Dittmer (i.e. that matriarchate is only an exceptional state[3]), and by adhering to the opinion, based on the investigation of J. J. Bachofen and L. H. Morgan, that matriarchate is a universal, historical stage through which the whole of mankind had to pass at the beginning of the era of clan organization before patriarchal relations evolved. One either accepts the theory of the priority of matriarchate

[1] Tylor (1881).

[2] Bachofen (1861, 1927); Харузин (1905) pp. 289–9 ff; Achelis (1908) p. 121; Thomson (1949).

[3] Thurnwald (1958) p. 173; Dittmer (1954) p. 53. Here it is impossible not to remember the splendid work in which M. O. Косвен (1948) summed up data concerning the history of matriarchate. For some contributions from Czech ethnography see Находил (1958c) pp. 101–9.

over patriarchate as generally valid, or one must consider matriarchate as a non-typical aspect of social evolution. The first theory seems to afford a possibility of discovering the religious conceptions reflecting or containing traces of matriarchate, whereas the second puts an end to exact historical research and to the understanding of the connections of certain cults with the social organization of ancient times. *Tertium non datur.*

For several years, the author has studied the problems of the cult of woman, and has succeeded to synthetize the fragmentary details in one general picture which also shows a systematic classification of the relevant data. The wide complex of the phenomena in question links the Later Paleolithic period, (for which a fundamentally correct solution was given by P. P. Efimenko, in spite of the relative scarcity of ethnographic parallels cited[4]) with the Neolithic and Eneolithic Periods and with very early civilizations. Female statuettes from the Later Paleolithic period, the so-called Venuses, known in vast territories extending from Western Europe all the way to Siberia[5], show striking resemblances to one another both in artistic execution and in symbolic meaning. The present state of research seems to justify the assumption that the numerous Neolithic and Eneolithic female figurines[6] represent not only the Mother Earth cult (known from A. Dietrich's monumental work[7]) but also the cults of the mother of grain, water, mountains, etc., derived from the Paleolithic mother-proto-parent cult (protectress of the clan and its household and of the fire). Since this cult can often be attested also in the subsequent ages, it may be considered a connecting link—however sceptical J. Filip[8] may have been in this respect.

The archaeological material from pre-Columbian America consists mainly of female sculptures distributed over several strata. These 'mysterious clay figurines', as H. D. Disselhoff[9] calls them, display a degree of uniformity in both form and meaning. This is perfectly evident when we compare the material from almost the whole of America and from related regions with the cults known in the Old World.

Finally the cults of the oriental 'Great Mothers' (the interrelatedness of which has been demonstrated by B. Noss and A. P. Každan[10]) definitely represent a culminating point in the evolution of earlier mother cults. It is therefore not surprising that, for instance, the cult of the old Egyptian goddess, Eset (Isis), should have been acceptable to the ancient Greeks, Romans and Germans[11] who themselves had a number of expressions for autochthonous cults of the female principle.[12] Nor is it surprising that the cult of great mothers penetrated into the mythology of Buddhism, Mithraism, Christianity, Islam and Shintoism, since the tradition originating from the period of matriarchate,

[4] See Ефименко (1953) pp. 382–94.
[5] See E. Saccasyn-Della Santa (1947); Absolon (1949) pp. 201–20; Klima (1955); Всемирная история Vol. I, Moscow, 1955, pp. 70–2.
[6] See Neustupny (1956).
[7] Dietrich (1905).
[8] Filip (1948) p. 109.
[9] Disselhoff (1953) pp. 30–2, 42–4.
[10] Noss (1956); Каждан (1957).
[11] Salac (1915).
[12] Müller and Trathnigg (1954) pp. 168, 186, 192–5, 197, 202–3, 259–61, 263–4, etc.; Vries (1957) Vol. II, pp. 288 ff.

practically the longest period of pre-history, was still very strong and vital in early oriental agricultural civilizations.

The author has published a few chapters[13] on the problems touched in short notes on mother cult in his previous papers.[14]

*

The actual character of the mother cult, particularly as far as ethnographic parallels with the female figurines of the Paleolithic Period are concerned, is plastically evident in the fragmentary, yet extremely valuable material from Siberia. The matriarchal system existed in some parts of this territory not so long ago and some vestiges of it have been preserved up to the present day among the indigenous nationalities. This material is characterized by a few fundamental types which recur also in other territories and ethnic communities and which, therefore, may be considered as typical features of cults, formerly probably no less universal than the matriarchal order of primitive society at its earliest stages.

Particular interest has been attached to certain remains of mother cult in the West-Siberian communities of Finno-Ugric tribes (Khanty and Mansi), where they are rather distinct, although it is difficult to throw light on their original ethnographic function. The above ethnic groups are of common origin, descending from two phratries both having (if we analyse retrospectively the ascendant lines) mythical feminine characters as ancestors. In the phratry *por*, the first woman, obviously the mythical founder of the female line, appears as a she-bear.[15] In the phratry *moš* the matrilinear filiation is headed by the '*kaldaš* woman' or the '*kaldaš* mother'[16] represented as a female hare or as a butterfly.[17] While in the phratry *por* the first woman appears as the daughter of the totemic ancestor, in the phratry *moš* the image of the female founder of the maternal line corresponds to the conception of totemic ancestors, such as the female hare or the butterfly. According to the ideas of the Northern Khanty recorded by K. Karjalainen[18], the '*kaldaš* mother' gives souls to the children and determines the life story of every child at its birth, recording it on a wooden plate. At the same time she records whether the child will live on earth in wealth or poverty, whether it will be happy or unhappy.[19] In the folklore tradition the '*kaldaš* mother' takes care of the people of her phratry and brings aid to them when they are in need.[20]

[13] Nahodil (1959) pp. 1, 3 ff.

[14] Nahodil (1954) pp. 87–95, (1956) pp. 22–8, (1957) pp. 122–6, (1958a) pp. 584–8.

[15] Чернецов (1939) p. 23; Прокофьева (1956a) p. 593.

[16] Чернецов (1939) p. 30.

[17] Прокофьева (1956a) p. 593.

[18] Karjalainen (1921–1927) Vol. III, p. 177.

[19] The analogy with the Slavic *рожаницы* or *sudičky* ('Fate Women'), is certainly not accidental. See Grohman (1864) p. VII; Máchal (1891) pp. 76–7 and (1907) pp. 55, 57–8; Brückner (1918) pp. 104, 106–7; Mansikka (1921) pp. 164, 246, 250, 307; Vakarelski (1958) pp. 160–2, 166. Also the ancient Celtic and German conceptions of the so-called 'matres' and 'matronae' show certain analogies in this respect. See Müller–Trathnigg (1954) p. 168; Vries (1957) p. 296.

[20] Чернецов (1939) p. 31.

Moreover, in the ancient traditions of the Khanty and Mansi there appear a number of female personalities, such as 'seven women with stone eyes',[21] the tribal 'earthy mother', 'mother of the lower world (earth)', '*torum*'s mother' or '*torum* mother'[22] or the 'clan mother', appearing as a totem, such as in the case of the clan called the 'Frog's People' tracing their origin to the 'Great Frog Woman'.[23]

The West-Siberian Finno-Ugrians show quite clearly the correspondence of the image of some sort of a mythical woman—the original mother of the clan—with the image of the totem animal. They appear in figures which are, according to W. Steinitz[24], understood in a complex way, that is to say, both in their animal and human forms. This image shows that the mother cults have a common base with totemistic ideas. We may regard these phenomena as original forms of one and the same matriarchal cult of ancestors,[25] which is obviously ancient—as can be clearly seen from the comparative material.

We meet with analogous material in the religious ideas of the Samoyeds, where we find the good goddess *ja-neb'a* 'mother earth'.[26] A whole number of early religious ideas of the Selkups (Ostyak Samoyeds) concerning ancestors also refer to mother animals. In these conceptions, often concealed under recent strata, we can readily discover old evidence of the former existence of matriarchate.

Among the Selkups all major spirits, masters of the house, appear as feminine beings. Of great importance is the heavenly mother with her daughters, the giver of life, the old woman *ylynda kota*, who sends souls to newly born people in the form of birds. As among several other Siberian nationalities, in the religious ideas of the Selkups appears the Mistress of Fire, who is the 'grandmother to fire and wind'. Among the female supernatural beings they have also the Ruler of Snakes, called *selči šÿt emysyt*, which means 'mother of all snakes'. The ruler of the forests is the female sable, and also the sun is a female being.[27] The installation of a shaman of the Selkups could be performed only with the previous consent of the giver of life, the mother.[28]

The above data give us the first concrete pictures of the mother cults on Siberian soil. The material collected is, however, far more voluminous extending over the South-Siberian territory, the eastern part of Siberia and the Far East (including Sakhalin). The data on the whole show not only certain surprising common features, but also prove various historical stages in the development of these religious ideas.

The South-Siberian Sayan-Altai Turkic nationalities offer rich and interesting material relating to an ancient mother cult, which can be traced right

[21] Steinitz (1938) p. 128.
[22] Чернецов (1939) p. 31.
[23] Чернецов (1947) p. 162.
[24] Steinitz (1938) p. 134.
[25] Nahodil (1956) pp. 34–5, (1957) p. 128. Mother cults, the nature of which is probably very similar, can be found also among the Estonians and Saams (Lapps). See Токарев (1958) pp. 124 and 140.
[26] Прокофьева (1956b) p. 627.
[27] Прокофьева (1952) p. 102; similarly, the sun cult can be encountered among the Udegeis. See Иванов–Левин–Смольяк (1956a) p. 837.
[28] Прокофьева (1952) p. 102.

back to the period of ancient Turkic according to the testimony of the Orkhon-Yenisei inscriptions from the 6th to the 8th centuries of our era. We refer to the cult of the goddess *umaj*, protectress of children and fertility, worshipped as early as the time of the Kaghanate[29], when it presumably already had a long tradition behind it. Remnants of this cult are found not only in the religious ceremonies of the Shors, where the *umaj änä* 'mother *umaj*' appears again as the protectress of children,[30] but also in those of the Southern and Northern Altai, the Tuvas and Khakass, as well as among the Central-Asian Kirghiz and Uzbeks, whose ancestors lived in the lands of the ancient Kaghanate.[31] Data concerning the nature of the cult of this goddess are few. In general she is characterized as a protectress of children. Among the Khakass, for instance, she manifests herself clearly as a mother, as the feminine principle of life, as a protectress of children. The Khakass see her as the placenta or the umbilical cord of a new-born child, the essence of life, the human soul. In the not very remote past they organized every month a special ritual of feeding new-born children in honour of the goddess *umaj*. In the course of this ceremony a special broth was prepared which was given to the children from ritual cups by means of special spoons.[32]

Traces of the mother cult are discernible also in the mountain cult, which existed among the Altais until recent times.[33] It was a pre-shamanistic tribal or clan cult, connected with a number of taboos valid for women who were taken from other tribes, in compliance with the strict principle of exogamy. The tribal mountains were represented in the anthropomorphous form of domestic spirits—the rulers of the mountains, who were imagined to be either men or women. According to L. P. Potapov, these rulers usually appeared in the shape of naked young women or girls, usually with large breasts.[34] Altai traditions have it that these supernatural feminine mountain rulers gladly had sexual intercourse with hunters, particularly of the unmarried ones whom, in return, they endowed with luck in hunting.[35] This legend is in full agreement with the contents of the prayers to the mountains and waters recited by the Altais for luck in hunting and fishing.[36]

The matrilinear origin of the mountain cult of the Altais is also evident in shamanism. For instance, the Kumand shaman received the magic drum from the matrilinear spirits through the mediation of the tribal mountain of his mother's tribe.[37] Elements of ancient mother cults which eventually penetrated into late shamanism manifest themselves in the contents of the prayers of the Khakass shaman during an offering to the tribal mountain,

[29] See Потапов (1953) p. 86, (1956) p. 10.

[30] Потапов (1936) p. 150.

[31] Potapov (1957) p. 142. See also Дыренкова (1928),

[32] Шибаева (1948) p. 204. As far as the preparation of the ritual broth is concerned, the analogy with Slavic customs of ancient origin, connected with the cult of the *рожаницы* or *sudičky* ('fate woman'), recorded among the Great Russians, White Russians and Czechs, is conspicuous. See Mansikka (1921) p. 307; Brückner (1918) pp. 106–7.

[33] See Потапов (1946) pp. 145–60.

[34] Потапов (1946) p. 153.

[35] Потапов (1946) p. 159.

[36] Потапов (1946) p. 149. The Shors have a very large number of different 'masters' of mountains and rivers, represented as naked women. See Потапов (1936) p. 151.

[37] Потапов (1936) p. 151.

when he prays, above all, for success in life and for the fertility of women.[38] Research work conducted by L. P. Potapov in this field deserves attention. His comparative analysis of the place-name *ötükän* in Orkhon inscriptions and documents, based on Uigur, Mongolian and Old Turkic material, proved that the soundest interpretation of the term *ötükän*[39] was 'the goddess or divinity of the earth'.[40]

Apart from old legends about the origins of the individual *sök*'s (clans), the Altais have preserved certain objects which are related to cults of woman.[41] In the same way as the Teleuts and the Shors, they worshipped supernatural family protectresses, whom they represented as dolls made of cloth, called *ämägändär* 'grandmothers'. The women brought offerings to them and fed them with specially prepared dishes.[42]

The most extensive and reliable treatment of this material is contained in a book by N. P. Dyrenkova, whose detailed explanation is so convincing that we cannot share S. A. Tokarev's[43] doubts regarding her conclusions.

The term *ämägändär* applied by the Teleuts to these dolls means 'old mothers, grandmothers'.[44] In the language of the Kumandins it also means 'old women, women, wives' and in Mongolian 'old women, ancestral mothers'.[45]

The dolls are, however, known also by other names. The Shors call them *öräkkännär*, meaning 'old women, esteemed women' or *qurtujaq*, (*qurtyjaq*) 'old women, women, housewives'.[46] Not only the semantic development of these terms is characteristic. Their plural form is in full harmony with the familiar and matrimonial relations of the matriarchal period, in respect of which the existence of group marriage is a necessary assumption.

The cult of these dolls representing *ämägändär* female spirits contain a number of features that eloquently testify—as far as known comparative material goes—to the indubitable manifestation of an old feminine cult which once existed, long ago, in the days of the matriarchal organization of society. It comes under a category of cults which W. E. Peuckert called 'Geheimkulte'.[47] They are not general cults, but survive in feminine circles only, representing a permanent supernatural connection of a married woman with the house of her parents, her mother, her line of ancestors, her uncle, etc.[48] It is not without interest to note that the dolls were usually made by old women from materials provided by the bride's uncle.[49]

[38] Потапов (1947) p. 114.

[39] Ancient Turkic name of a mountain.

[40] Потапов (1956) pp. 5–6.

[41] See Токарев (1931) p. 17.

[42] See Ефимова (1926) pp. 225–46; Каруновская (1927); Дыренкова (1937) pp. 123–44; Токарев (1947) pp. 143–4, and (1958) p. 441.

[43] Токарев (1947) p. 144.

[44] Дыренкова (1937) p. 127.

[45] Дыренкова (1937) pp. 128–30.

[46] Дыренкова (1937) p. 130.

[47] Peuckert (1951).

[48] See Дыренкова (1937) p. 132 ff. It is necessary to mention that, in the case of the West-Siberian Kets, the object of the family cult was also a special spirit, protectress of the family, called *alalt*. Her picture was to be found in every family. Токарев (1958) p. 496 states that it was a being of the feminine sex, probably constituting a vestige of matriarchate.

[49] Дыренкова (1937) p. 132.

Comparative ethnographic data confirm our interpretation of these representations of female spirits which served to protect women in labour, to make births successful, and to safeguard the new home of a woman, her hearth and her property.[50] N. P. Dyrenkova correctly asserts that these dolls personified the original protectresses of the family, who saw to its welfare, assisted multiplication, and watched over the family hearth.[51]

Nor are we far from the truth, I believe, when we assume that this cult continued to survive after the transition to patrilocal matrimony, for even then, it still constituted an ideological link between married women and their matrilinear ancestors.

Finally, the results of research on the vestiges of totemism in Siberia[52] point to the fact that even later shamanism absorbed elements from the strongest of the earlier cults and was based on older traditional conceptions. There is little doubt that the manufacturing of *ämägändär* dolls passed over to a modern period characterized by shamanism just because of its long tradition in society. Similar dolls with feathered heads can be seen on the ritual cloaks of Altai shamans. Here N. P. Dyrenkova very aptly draws attention to an analogous situation among the Yakut shamans, who called similar dolls attached to the ritual clothes *ämägät*, which is the plural of the Mongol word *ämägän* and means 'ancestral mothers'.[53]

Data from the Turkic nationalities of Southern Siberia, which are far from being exhausted in this study, show that ethnographic parallels have a definite function in the interpretation of archaeological relics.

In the preceding pages it has been mentioned more than once that mother cults were connected with the cult of fire and of the hearth. Recent investigations show that also fire was a feminine principle—a kind of ancestral mother. This connection is brought out more forcibly by sculptures of ancient Venuses that have been found in the vicinity of hearths or right in them, as was already mentioned.[54]

[50] Дыренкова (1937) pp. 133–5.

[51] Дыренкова (1937) p. 139.

[52] Nahodil (1958).

[53] See Дыренкова (1937) pp. 142–3. Analogous material is contained in the study of K. V. Vjatkina (1946), who shows that the Buriat Mongol ancestor *ebüge* is always closely connected with the maternal line and with the name of the totem. Hence, it is in the author's opinion comprehensible that the same clan cult was practised also by the representatives of various *jasun* (i.e. new clans originating by means of division of older clan units), since the cult was connected with the idea of the ancestress of the clan, protectress and ancestral mother of the Buriat clans, who was a female shaman (p. 138). Here we must also mention the inspiring and precious study of Иванов (1951), in which (pp. 119–31) the author (a) emphasizes the importance of the *ongons* of matriarchal origin, (b) points to the names of *ongons* with female images ('mother', 'daughter', 'wife'), (c) analyses vestiges of the matriarchal clan ideology, (d) introduces new material dealing with the problem of identifying the image of the 'mother' with that of the totemic animal (swan), (e) analyses the prayer addressed to the mythical ancestral mother of the Buriats and (f) gives theoretical generalizations. Again we note the important fact that also these vestiges of the matriarchal cult were preserved among the *women*. Ethnographic material from the South-Siberian Turkic peoples is thus supplemented and elucidated by means of ethnographic data referring to the cults of the Buriat Mongols, which—considering the ethnogenetic connections—is understandable.

[54] Nahodil (1958a) p. 586.

Also in this respect the abundant South Siberian material from the Sayan-Altai nationalities is valuable; it not only helps us in finding old traces of mother cults, but also gives us a better understanding of the ethnographic material from other territories of Northern Asia, particularly from its eastern part and from the Far East.

L. P. Potapov, the eminent specialist in Altai ethnography, is fully justified when he stresses that fire worship had been very widely spread among the Altais. Every day, when they sat down to eat, the head of the family or his wife, poured some tea on the fire which was burning in the middle of their hut, or threw into it a few morsels of the food when they took their meals. The first part of every meal always belonged to 'mother-fire' (*ot änä* in Altai) as protectress of the family. On a long voyage when an Altai stopped to rest or sleep, he first lit a fire and, before eating, fed his protectress with the food prepared for himself. If the rest was only a short one, he fed the fire with a pinch of tobacco at least. This custom of feeding fire was strictly adhered to by Altai hunters during a hunt, at which time they associated the practice with their prayers for a plentiful booty.[55]

It was a custom among the Southern Altais, particularly among the Kachins and the Sagais that a young bride, when entering her husband's house for the first time, bowed first of all before the fire and offered food to it. After the wedding, when visiting her husband's relatives, it was her duty to bow to the fire in every yurta and to throw a morsel of food into it. When they received the young wife into their house, they fed the fire, praying at the same time that it should take the new member of the family under its protection. After the birth of the first child, the wife organized a special feast in honour of the fire and, with her child in her arms, bowed in front of it. When entering a strange house with her infant, the woman immediately addressed a supplication to the fire, beseeching it to take her under its protection, while the head of the family fed it with victuals. It was believed that if a woman failed to observe this custom, she would contract a disease in the breasts and lose her milk.[56]

Fire as a feminine principle appears also among the Shors, as may be seen from the religious epithets *otuz paštu ot änä* 'thirty-headed mother fire' and *altyn tondu ot änä* 'mother fire in a golden fur coat'.[57]

It is a custom among the old men of the Khakass tribe to squirt some vodka or throw morsels of food into the fire during their meals. The women usually do the same. The Khakass believe that Mother Fire, an old woman with many braids, resides in the hearth of every dwelling. Finally, the fact that it was the woman, the mistress of the house, who was the first to bow before the fire in the course of the shamanistic ceremony, proves that fire worship as practised by the Khakass goes back to a dim and distant past.[58]

Much valuable material on the mother-fire cult of the Altai Kizhi and Teleuts has been collected by N. P. Dyrenkova.[59] She points out that these ethnic groups consider fire as the cleanest and most widely worshipped of all spirits. Apart from the *ot änä* 'mother fire' there exists a very archaic

[55] Потапов (1957) pp. 135–6.
[56] Потапов (1957) p. 136. See also Дыренкова (1927) pp. 67–8.
[57] Потапов (1936) p. 150.
[58] Шибаева (1948) p. 204.
[59] Дыренкова (1927) pp. 63–78.

conception of a supernatural mistress of fire, represented by a white fire girl called *kyz änä* 'virgin mother'.[60] It is referred to by L. Ja. Šternberg as parthenogenetic, anthropomorphous mother fire.[61]

The prayers of the Teleuts often contain invocations to the 'thirty-headed mother fire' and the 'forty-headed mother fire' as well as to the 'mother raker of the stone hearth', the 'mother fire with the decorated head' or the 'mother fire on a cream-coloured hearth'.[62] These terms prove the great respect paid by them to the fire as a feminine principle. Birket-Smith's assertion[63] that the hearth is the sacred centre of the primitive dwellings and that fire was one of the oldest forces to be worshipped[64] by man is proved to be true, time and again, in examining the customs of the Turkic population of Southern Siberia and of many North-Asian nationalities.

In the eyes of the Altai Kizhi and Teleuts fire is primarily the protectress of life in the *sök* (clan), and is a member of the clan as a whole as well as of every individual family from whose house it must not be removed.[65]

According to the beliefs of the Altai Kizhi, 'mother fire' helps all the inhabitants of a house. She assists in the preparation of meals, blesses the house with children, ensures the multiplication of the cattle, etc. Mother fire watches over the souls of the deceased, lighting up their way to the other world with a bright flame.[66]

Since fire is anthropomorphic and is considered to be the living mother, or the ancestral mother, the behaviour of human beings in its presence is governed by a number of taboos, which are met with in one form or other also among the other Siberian peoples. The Altais treat fire with particular circumspection: it is forbidden to behave disrespectfully towards it, to quench it unnecessarily with water, or to throw garbage into it. Fire must not be touched with an iron knife lest its head be injured.[67] According to their religion, the fire is a mediator between man and spirits and deities. It consumes the offerings and forwards them to the gods. Therefore no shaman (*qam*) visits any spirit without the special permission of 'mother fire'.

A fire is a place of sacrifice in itself. It was a custom among the Altai Kizhi and Teleuts, in the past, to see that the fire received a part of every meal.[68]

Finally in Altai Burkhanism, (a shamanistic religion modified under the influence of Lamaism) the old custom of worshipping mother fire survived with some very strict taboos concerning relations towards fire.[69]

As can be seen from the preceding paragraphs, material on the mother cults of the Sayan-Altai peoples is surprisingly abundant. A comparison of

[60] Дыренкова (1927) p. 64.
[61] Штернберг (1936) p. 368.
[62] Дыренкова (1927) pp. 74–7. In this connection it would be possible to consider other explanations of the ethnic name of the Kirghiz. (According to popular etymology the word *kyrk kyz* means 'forty virgins'.) See Хайтун (1956) p. 94; Токарев (1958) p. 386.
[63] Birket-Smith (1946) p. 233.
[64] Дыренкова (1927) p. 66.
[65] Дыренкова (1927) pp. 66–7.
[66] Дыренкова (1927) p. 69.
[67] Дыренкова (1927) p. 69.
[68] Дыренкова (1927) p. 71.
[69] Дыренкова (1927) pp. 72–4.

the local manifestations of these cults with other Siberian and Central Asian material permits us to come to certain general conclusions.

Before passing on to a study of the ethnographic material of the so-called minor nationalities of Eastern Siberia and the Far East, where we meet with a number of obviously related customs, let us first consider vestiges of mother cults among the Yakuts—in so far as they are recorded at all.

Reconstructing the ancient clan order of the Yakuts, we see that, as late as the 17th century, certain relics still survived of an ancient matriarchal clan organization which had existed among them in times prior to their transition to pastoralism. Traces of the old matriarchal order are evident, for instance, in the legends about the filiation of the individual clans from women.[70] In the religious conception of the Yakuts the strongest reminiscence of former matriarchate is the cult of a blonde deity called *ajȳsyt*, the giver of fertility, protectress of women in labour, the same goddess from whom mothers receive their children and in honour of whom they bring offerings after the birth. It was assumed that, after the birth of a child, for three days she was a guest in the house in which the birth took place. Hence, during those three days a special women's ceremony for the purpose of accompanying *ajȳsyt* was held, characterized by the fact that men were not permitted to take part in it.[71] Nor is the existence of female shamans among the Yakuts unimportant in this connection.[72]

A very interesting picture of the remnants of mother cults, with rich comparative material, can be seen in the case of the minor nationalities of Eastern Siberia and the Far East, whose conception of ancestral mothers of the clans is almost invariably connected with fire. Similarity among them amounts almost to uniformity.[73]

This applies first of all to the Tunguso-Manchurian peoples of Eastern Siberia. The Evenki (the Tungus proper), as proved by A.F. Anisimov, worshipped as the chief spirit of the clan a female being called *togo mušun* which, translated literally, means 'mistress of the fire'. The Evenki imagined her to be a half-blind old woman, a grandmother (*ənəkə*), regarded as the spirit of fire, the mother of the clan, the mistress and head of the tent, and also the protectress of the souls of the members of the clan and the family. The Evenki made offerings of pieces of meat or fat to *togo mušun*, which they threw into the fire, uttering prayers usually connected with their wish for a successful hunt.

Fire was considered a symbol of the unity of the clan, every one of which, among the Evenki, had one common hearth, which was sacred. The fire which cemented together the whole collective of the clan consisting not only of the living, but also of the deceased members and those about to be born, was not permitted to pass into the hands of a stranger. It is no mere accident that fire, which was the symbol of collective material life in the clan was represented as a woman, as the ancestral mother. It was probably due to the fact that this ancestral mother was more or less identical with fire (as can be assumed from a comparison with material known from other

[70] Токарев (1940) p. 29.
[71] See Попов (1928), (1949) p. 268 and (1956) p. 302
[72] Бахрушин–Токарев (1953) p. 176.
[73] Nahodil (1958a) p. 586.

Tunguso-Manchurian tribes) that we do not find any pictures of this holy old woman.[74]

According to the religious conceptions of the Evenki, *bugady mušun* was another being belonging to this class. She was a female spirit, who was the protectress of the clan, the head of the whole husbandry of the clan, the mother of the animals, the mistress of the taiga, of the world and of the universe. The Evenki imagined her to be a very sensible, old, yet strong woman. It is again typical that the representations of this supernatural, ancestral mother of people and animals are rather varied. Sometimes she is anthropomorphous, at others zoomorphous. Once the *bugady mušun* is an old woman, then again she is a big female elk or a wild female reindeer. However, also in her zoomorphous, totemic form she was considered *enin-te*, i.e. a woman who is 'considered a mother'.

The Evenki believed her to be the guardian of dwellings and of people, as well as the protectress of supplies, of reindeer herds and of the whole husbandry of the clan which she defended against the activities of evil spirits. Of great importance is the fact that in every clan the shaman preserved *bugady mušun* as a common idol and that every family had in its tent a small image of this idol.

Long ago the *bugady mušun* was the centre of attention and respect in everyday life. From time to time the women brought home a picture of her, threw pieces of fat into the fire and let the smoke pass over it. This was the customary way of making offerings to the *bugady mušun*. When sitting down to eat, the Evenki used to cut off pieces of meat and fat and throw them into the fire, saying '*togo mušun*, eat, be satiated, *bugady mušun*, eat, be satiated! Eat to satiety, do not be angry with us and protect us' or 'make our hunting successful', or 'make our herd bear more reindeer', etc. In the case of accidents and difficulties, the Evenki considered their female spirits to be their best friends and protectors.[75]

In connection with the cult of the *bugady mušun* D. Khaitun informs us, on the basis of material collected by A. P. Okladnikov, of a very interesting fact: The Evenki once worshipped sacred places of the clan, called *bugady*. These were rocks, which the Evenki regarded as living beings — animals, most often elks. The *bugady* animal was recognized as the ancestor, totem and protector of the clan.[76]

Due to lack of space, in this article we cannot go into greater detail in connection with the two ancestral mothers worshipped by the Evenki. However, they are of primary importance for a proper understanding of mother cults not only in Siberia but, we may say without exaggeration, also all over the world. Let us extend the picture of these Evenki cults by comparing them with some other Evenki cults which have been brought to our attention lately by A. F. Anisimov. I refer to the cult of the mistress of the earth, called *dunnə mušun*,[77] the cult of old women, protectresses of

[74] Анисимов (1936) pp. 47–8; see also his new work (1958a) pp. 93–8, 101–2.
[75] Анисимов (1936) pp. 56–7 and (1958a) pp. 29, 32, 78–9, 83, 103–5.
[76] Хайтун (1956) p. 124. See also Анисимов (1951) pp. 192–3 and (1958a) pp. 29, 82–3; Токарев (1958) p. 504. For the totemistic worship of a rock among the Orochi, see Лопатин (1925) p. 33; Nahodil (1958) pp. 146–7.
[77] Анисимов (1958a) pp. 25, 31, 83.

the roads to the realm of the dead[78], and finally the cult of the old woman, who is the mistress of the *buni* (the world of the dead).[79]

Here it is necessary to note that the Evenki conception of the *bugady mušun* corresponds more or less to the Nanai (Gold) cult of the female being called *sengi-mama*—the ruler of the earth and of the world, the ruler of the taiga, of the game and, hence, of successful hunting. The Nanai hunters addressed a prayer to this transcendental female being whenever the time of hunting approached. Ceremonies of this type were divided into two cycles: the first one took place in every village near the sacred tree of the clan, the *tutga,* before the hunters departed, the second one, in the forest at the hunting place. In the first case the Nanais addressed their prayers to *sengi-mama,* the ruler of the taiga, and in the second case to *boa,* 'mother of the nature', called the 'Great Mother'.[80]

A similar cult existed also among the close relatives of the Nanais—the Udegeis. The Nanai *sengi-mama* is analogous to their *sangia-mama,* who is again a female mother spirit, supernatural ruler of the taiga, mother of animals (and people) and mistress of the earth and of the world. As the first Udegei writer, Džansi Kimonko[81], writes there existed certain idols of this mother and ruler of the taiga, who at the same time was the spirit of success in hunting. These idols were 'fed' with the blood of the sacrificed animal. The idol was brought offerings consisting of the head and heart of a reindeer, accompanied by the following prayers:

> *sangia-mama! sangia-mama!*
> Hear our prayer,
> Accept our offering with the blood of our animal!
> For the elk which you sent us,
> We bring you its head.
> We all kneel
> And humbly bow to you, *sangia-mama*![82]

To the offering made to the supernatural ruler of the taiga and successful hunting, a pious supplication was added calling for further 'assistance' in hunting. D. Kimonko writes about his father, who cut off a piece from the tongue of an elk with his knife and threw it down the stream of the river, and then cut a piece from the nostrils of the elk, throwing this up the stream of the river. Part of his prayer runs as follows:

> This is for thee, *sangia-mama*!
> Accept it from us, please!
> Send us another elk.
> May it come to the river bank
> In front of my boat
> And may it neither hear nor see me . . .![83]

[78] Анисимов (1958a) p. 65.
[79] Анисимов (1958a) p. 78.
[80] Анисимов (1951) p. 192.
[81] Kimonko (1954) p. 19.
[82] Kimonko (1954) p. 20.
[83] Kimonko (1954) p. 20.

Here, we find a very close analogy of the Evenki *bugady mušun* with the idols of the Tunguso-Manchurian tribes near the Amur river. Should more detailed data referring to the cult of the feminine spirits of *sengi-mama* and *sangiamama* be recorded, it is highly probable that further common features come to light.

With this, however, the analogy between the religious ideas of the Evenki and the Amur tribes is not exhausted. Here the Evenki *togo mušun* 'old woman of the fire' is even more characteristic.

The attitude of the Nanais towards fire was a very respectful one. According to an ancient belief of theirs, fire had a supernatural mistress: the spirit *fadźa-mama*, an old woman.[84] I. A. Lopatin[85] mentions that the Nanais always kept a glowing fire alive, from which the hunters, when leaving for long trips into the taiga, kindled a wick which they carried along and from which they lit their temporary fires in the open air, away from their homes. Another characteristic document says that old men used to come in the winter to the fire which they had lit in the taiga asking the *fadźa-mama* for a place. The attitude of the Nanais towards fire and its supernatural mistress is finally shown by the fact that children and young people were not allowed to dash to the fire suddenly lest that hurt the old woman *fadźa-mama*.

As the Evenki and the Nanais, so also the Amur Orochi believed that the mistress of the fire had to have offerings brought to her. According to A. M. Zolotarev, also in the eyes of the Amur Ulchi the common fire was the first and foremost sign of fellowship in the clan. In the fire the 'mistress of the fire' called *poza-mama* lived. Of obviously later origin (when patriarchate had arisen) is the idea of a *poza-mafá* 'old man of the fire'. This supernatural mistress or master of the fire is identical with the supernatural master of the taiga and the mountains, viz. with the natural sustainer of the people. In the old Ulchi houses there used to be two hearths, one of which belonged to the supernatural master of the taiga and was considered sacred. Whenever sitting down to eat, the Ulchi threw into this hearth, which was originally only the seat of the feminine spirit, the first mouthful of food, so as to ensure the success of the next hunt.[86]

Undoubtedly the Negidals had very similar ideas, for they kept near the fire a picture of the mistress of the fire called *kutug: á*, as well as a picture of the master of fire, *myjkudý*, as is recorded in the notes of L. Ja. Šternberg.[87]

From what has been said we can see that the worship of fire as a feminine principle, the supernatural ancestral mother, was of great importance among the Tunguso-Manchurian tribes and nationalities.

Also in the Far East, we find the remnants of an analogous cult, among the Paleo-Asiatic nationalities. It manifests itself in the most striking

[84] Иванов–Левин–Смоляк (1956d) p. 804.

[85] Лопатин (1922) p. 107.

[86] See Иванов–Левин–Смоляк (1956a) p. 851; Золотарев (1939) pp. 38–9. We encounter the fire cult also among the Udegeis who pictured the spirit of the fire as a being of the female sex. See Токарев (1958) p. 519.

[87] Штернберг (1933) pp. 534, 544. In connection with the Sakhalin Oroks, we are informed that every element had its own master spirit—to whom offerings were brought. It remains uncertain, however, whether the fire had alsoits mistress in their mythology. Иванов–Левин–Смоляк (1956b) p. 858.

manner among the Nivkhi (Gilyaks) for whom the communal character of the fire was an important symbol of the unity of the clan. Its fire and the behaviour of the people in the vicinity of the hearth were connected with a number of taboos. The spirit of the fire—its supernatural mistress—was imagined as an old woman residing in every hearth, who was also the mythical ancestral mother of the clan. According to another conception, evidently of a later origin, the hearth was the seat of an old man, of an old woman and of their descendants. Šternberg stressed that the conception of the old woman-mistress of the fire of a clan was an allusion to its distant feminine origin.[88]

Fragmentary information of this kind exists also in connection with the Paleo-Asiatic Koryaks and Chukchi. The former use a special instrument for lighting the fire; it is a wooden plate having the shape of a human body. This very important item among the household utensils was called the 'fire-woman'.[89] According to V. G. Bogoraz-Tan, the Chukchi made offerings to the fire which were sometimes called 'the feeding of the fire'—a special ceremony which took place during the journey from the winter to the summer pastures. It is important to note here that every evening, after the tent had been set up for the night, it was the woman who had to light a small fire in front of the entrance and give it an offering of fat, lard, bone marrow and similar foodstuffs.[90]

The picture is completed by ethnographic material from the Sakhalin Ainu, among whom the ancient cult of fire is connected with the cult of sacred beings called *kamui* and also with the cult of the *ináu*. Among the *kamui* of the clan an important place was always occupied by the old woman (mistress) of the fire. The *ináu* itself was nothing but a stick, cut smoothly upwards with a knife so as to form a head consisting of thin, spirally wound shavings at the top. It, thus, resembles the simplified picture of a man of the Ainu tribe, remarkable in anthropology for his great quantity of hair. Usually stuck into the ground, it was originally the effigy of a supernatural woman—a mother, the ancestral mother of the clan. This is also proved by the fact that for the mistress of the fire the Ainu used to stick the so-called *unti-ináu* into the hearth.[91]

Ancient mother cults have survived, sometimes in small fragments, in all native Siberian nationalities. Here are some more illustrations from the northernmost part of Siberia (tundra).

The Taimyr Nganasans (Tavgi Samoyeds) never used matches but a flint and fire-steel for lighting a fire when celebrating their chief religious festival, the so-called 'festival of the clean tent'. The fire was always lit by women, every day by three different ones. Young women were preferred as they were said to bring luck.[92]

The south-eastern neighbours of the Nganasans, the Dolgans, preserved remnants of matriarchate right up to the arrival of the Russians. Thus, for example, a community of several families living in one *golomo* elected

[88] Штернберг (1933) pp. 87–8. See also Иванов–Левин–Смоляк (1956c) p. 874.
[89] See Ефименко (1953) p. 402.
[90] Богораз-Тан (1939) p. 79. For the role of the Chukchi women in the magic hunting ceremonies see also Анисимов (1958a) pp. 47–8.
[91] Штернберг (1935) pp. 52–3; (1933) pp. 131 (Fig. 3, Table I), 572–3, 614, 617.
[92] Попов (1936) pp. 64–5.

a common mistress to whose decision even men submitted. Still recently the Dolgans considered the matrilineal relationship closer than the patrilineal. It was the woman who kept the fire and 'fed' it and supervised also the sacred places of the clan.[93]

It is not impossible that further information in this connection concerning the North Asian countries will be found. Even what we have presented here is sufficient to bring out clearly many common features existing in the religious ideas of the nationalities of the North Asian territory. The very fact that fire was worshipped as a mother, old woman, mistress, etc. over such a vast territory is of great importance.[94] We could quote analogous cases from other Asiatic countries and from the ethnography of other continents. This, however, is not the purpose of our article.

By summing up all the small details concerning mother cults in Siberia we may reach certain generally valid conclusions.

1. In most cases the material itself shows that, in the ideology of the autochthonous population of Siberia, pronounced vestiges of a former matriarchal order of society were preserved. This constitutes—not for the first time—a proof of the fact that matriarchate is not a form of social organization which is common only among tribes engaged in agrarian husbandry, as assumed by P. Wilhelm Schmidt and his school[95] in their thesis which was not in accordance with historical truth. On the contrary, matriarchate is the oldest and most universal form of clan organization in society which existed even among the hunters of the Siberian taiga, the Siberian reindeer breeders and sea fishermen.

2. Ethnographic material from Siberia offers a splendid opportunity for proving the old theoretical assumption that matriarchate had its characteristic reflection in the religious ideology of that time. This assumption, formulated more than once in earlier works,[96] has been confirmed lately also by the new results of research work done by A. F. Anisimov.[97]

3. Our material on the mother cult in Siberia goes back in many cases to remote ages and permits us to see the connections with other cults, particularly with the totemistic cults. The oldest seems to be the conception of the mythical parthenogenetic mother, (ancestral parent of the clan, the ancestral mother, old woman, mother, etc.), to whom the individual clans traced back their origin, in matrilinear filiation. This supernatural mother often helps women in labour, gives children their souls and sometimes even predetermines the life of a newborn child; she is represented as the protectress of children.

The image of the ancestral mother is sometimes anthropomorphic, at others, zoomorphic. In certain cases it is clear that she is the mother of men and animals, and her human figure assumes also the shape of the

[93] Попов (1956a) p. 750.

[94] According to this, the ancient Greek goddess of fire, Hestia, a virgin worshipped by an eternal flame kept burning by unmarried women, and the ancient Roman goddess Vesta, whose fire was kept burning by vestal virgin priestesses, are not goddesses of Greek and Roman Antiquity, but plainly the descendants of those primitive old women, mothers and mistresses of the fire from the age when matriarchal society was born. If we went back far enough, we would reach the paleolithic Venuses.

[95] See, for instance, Schmidt (1946) pp. 6–7.

[96] Nahodil (1954) pp. 88 ff and (1957) p. 126.

[97] See Анисимов (1958b) pp. 261–82.

totemic 'animal'. It has thus been demonstrated that, in the genealogy of the religious ideas of mankind, mother cults and totemistic cults had the same basis.[98] Through the mother cult, totemism itself becomes more intelligible. Totemism, and conceptions that of the mythical ancestral mother, as a supernatural protectress of the clan and of its household, the mistress of the fire, the eternal fire, as the centre and the symbol of the community of the clan, are very ancient ideas. We can assume that they go back to a period which may be called 'Early Matriarchal Society', or in terms of archaeology, the 'Aurignacian Period'. Hence, it is obviously this fragmentary material which is the most suitable for drawing a parallel with the late Paleolithic cult of woman evident in the Venus sculptures.[99]

4. In a far later period, most probably in the Neolithic, when matriarchate flourished, further cults originated, such as the cult of the earthly mother (of the clan), ruler of the earth and the world; and even the deity of the earth. Of relatively later origin are the cosmological ideas of supernatural mothers of heaven, mothers of the lower world, world of the dead, and the conception of the mother deity *torum*. Also the female shamanistic ceremonies are obviously a vestige of matriarchal ritualism. The sun as a female principle should be explained rather as the result of the influence of a foreign, probably Chinese, religion.

REFERENCES

Absolon, K. (1949) The Diluvial Anthropomorphic Statuettes and Drawings, Especially the So-called Venus Statuettes, Discovered in Moravia. (A Comparative Study), in *ArtA*, Vol. XII, 3.

Achelis, Th. (1908) *Abriss der vergleichenden Religionswissenschaft*, Leipzig.

(Anisimov, A.F.) Анисимов, А. Ф. (1936) *Родовое общество эвенков (тунгусов)* (The Clan Society of the Evenki [Tungus]), Leningrad.

(1951) Шаманские духи по воззрениям эвенков и тотемистические истоки идеологии шаманства (The Shaman Spirits in the Beliefs of the Evenki and the Totemistic Sources of the Ideology of Shamanism), in *СбМАЭ*, Vol. XIII.

(1958a) *Религия эвенков в историко-генетическом изучении и проблемы происхождения первобытных верований* (The Religion of the Evenkis Studied Historico-genetically, and the Problems of the Origin of Primitive Beliefs), Moscow–Leningrad.

(1958b) Религия эпохи расцвета матриархата (Religion in the Climaetic Age of the Matriarchate), in *ЕМИРА*, Vol. II.

Bachofen, J.J. (1861) *Das Mutterrecht.* Eine Untersuchung über die Gynaikokratie der alten Welt nach ihrer religiösen und rechtlichen Natur, Stuttgart.

(1927) *Mutterrecht und Urreligion*, Leipzig.

(Bakhrušin, S.V.–Tokarev, S. A.) Бахрушин, С. В.—Токарев, С. А. (1953) *Якутия в XVII веке* (Yakutia in the 17th Century), Yakutsk.

Birket-Smith, K. (1946) *Geschichte der Kultur*, Zürich.

(Bogoraz-Tan, V. G.) Богораз-Тан, В. Г. (1939) *Чукчи II. Религия* (The Chukchi. II. Religion), Leningrad.

Brückner, A. (1918) *Mitologja slowianska* (Slavic Mythology), Cracow.

(Černecov, V.N.) Чернецов, В. Н. (1939) Фратриальное устройство обско-югорского общества (The Phratrial Organization of the Ob Ugrian Society), in *СЭ*, No. 2.

[98] Let us mention here that these supernatural mothers who were at the same time totems, were also, in certain cases, sacred mountains or rocks.

[99] See also Хайтун (1955) pp. 129–42.

(Černecov, V.N.) Чернецов В. Н. (1947) К историю родового строя у обских угров (On the History of the Clan Organization of the Ob Ugrians), in *СЭ*, Vols VI–VII.
Dietrich, A. (1905) *Mutter Erde.* Ein Versuch über Volksreligion, Leipzig–Berlin.
Disselhoff, H. D. (1953) *Geschichte der altamerikanischen Kulturen*, Munich.
Dittmer, K. (1954) *Allgemeine Völkerkunde*, Braunschweig.
(Dyrenkova, N. P.) Дыренкова, Н. П. (1927) Культ огня у алтайцев и телеут (Fire Cult among the Altai Kizhi and the Teleuts), in *СбМАЭ*, Vol. VI.
(1923) Умай в культе турецких племен (The Umaj in the Religion of Turkish Tribes), in *КПВ*, Vol. III.
(1937) Пережитки идеологии материнского рода у алтайских турков (Survivals of the Ideology of the Maternal Clan System among the Altais), in *Памяти В. Г. Богораза* (Memorial Volume of Bogoraz), Moscow–Leningrad.
(Efimenko, P. P.) Ефименко, П. П. (1953) *Первобытное общество* (Primitive Society), Kiev.
(Efimova, A. E.) Ефимова, А. Е. (1926) Телеутская свадьба (The Teleut Wedding), in *Материалы по свадьбе и семейно-родовому строю народов СССР* (Contributions to the Matrimonial and Familiar Order of the Peoples in the U.S.S.R.).
Filip, J. (1948) Pravěké Československo (The Prehistory of Czechoslovakia), in *USP*.
Grohmann, J. V. (1864) *Aberglauben und Gebräuche aus Böhmen und Mähren*, Vol. I, Prague–Leipzig.
(Ivanov, S. V.) Иванов, С. В. (1951) Происхождение бурятских онгонов с изображением женщин (The Origin of Buryat *Ongons* with Female Images), in *ТрИЭ*, Vol. XIV.
(Ivanov, S. V.–Levin, M. G.–Smoljak, A. V.) Иванов, С. В.–Левин, М. Г.–Смоляк, А. В. (1956a) Орочи (The Orochi), in *Народы Сибири* (The Peoples of Siberia), Moscow–Leningrad.
(1956b) Ороки (The Oroks), in *Народы Сибири* (The Peoples of Siberia), Moscow–Leningrad.
(1956c) Нивхи (The Nivkhi), in *Народы Сибири* (The Peoples of Siberia), Moscow–Leningrad.
(1956d) Нанайцы (The Nanais), in *Народы Сибири* (The Peoples of Siberia), Moscow–Leningrad.
Karjalainen, K. F. (1921–1927) Die Religion der Jugra-Völker, Vols I–III, in *FFC*, No. 44.
(Karunovskaja, L. Ê.) Каруновская, Л. Э. (1927) Из алтайских верований и обрядов, связанных с ребенком (Some Altai Beliefs and Rites Associated with the Child), in *СбМАЭ*, Vol. VI.
(Každan, A. P.) Каждан, А. П. (1957) *Религия и атеизм в древнем мире* (Religion and Atheism in the Ancient World), Moscow.
(Khajtun, D. E.) Хайтун, Д. Е. (1955) К вопросу об интерпретации женских изображений ориньякской эпохи (On the Interpretation of Female Images from the Aurignacian Period), in *Учение записки VII. Труды историко-филологического факультета.* (Scientific Papers. Communications of the Historico-Philological Faculty.)
(1956) Пережитки тотемизма у народов Средней Азии и Казахстана (Survivals of Totemism among the Peoples of Central Asia and Kazakhstan), in *УЗТГ*, Vol. XIV.
(Kharuzin, N.) Харузин, Н. (1905) *Этнография* (Ethnography), Vol. IV, Moscow.
Kimonko, Dz. (1954) *Tam kde teče Sukpaj* (Where the Sukpaj Flows), Prague.
Klíma, B. (1955) Výsledky archeologického výzkumu na tábořisti lovců mamutů v Petřkovicích, okr. Ostrava v roce 1952 a 1953 (Results of the Archaeological Investigations in the Camping Place of the Mammoth Hunter in Petřkovic, Ostrava District in the Years 1952–1953, in *ČSM*, Vol. IV.
(Kosven, M. O.) Косвен, М. О. (1948) *Матриархат. История проблемы* (The Matriarchate. The History of the Problem), Moscow–Leningrad.
(Lopatin, I. A.) Лопатин, И. А. (1922) *Гольды амурские, уссурийские и сунгарийские* (The Golds of the Amur, Ussuri and Sungari Rivers), Vladivostok.
(1925) *Орочи — сородичи маньчур* (The Orochi, Kinsmen of the Manchus), Kharbin.
Máchal, J. (1891) *Nákres slovanského bájesloví* (Outline of Slavic Mythology), Prague.
(1907) *Bájesloví slovanské* (Slavic Mythology), Prague.
Mansikka, V. J. (1921) Die Religion der Ostslawen, in *Quellen*, *FFC*.

Müller, W.–Trathnigg, G. (1954) *Religionen der Griechen, Römer und Germanen*, Wunsiedel–Wels–Zürich.
Nahodil, O. (1954) *O povodu náboženství* (On the Origin of Religion), Prague.
(1956) *Vznik náboženství* (The Birth of Religion), Prague.
(1957) *O kulture prvobytné spolecnosti* (About the Culture of Primitive Society), Prague.
(1958a) Praveke Venuse a posvätny ohen (Prehistorical Venuses and the Sacred Fire), in *Priroda a solocnost*, Vol. XI, Bratislava.
(1958b) Überreste des Totemismus bei den tungusisch-mandschurischen Völkerschaften, in *AE*, Vol. VII, 1–2.
(Nahodil, O.) Находил, О. (1958c) Из истории проблемы матриархата (On the History of the Problem of the Matriarchate), in *СЭ*, No. 3.
(1959) Materske kulty (Mother Cults), in *ČE*, No. 1.
Neustupný (1956) Studies on the Neolithic Plastic Arts, in *Acta Musei Nationalis Pragae*, Vol. X-A-Historia Nos 1–2.
Noss, J. B. (1936) *Man's Religion*, New York.
Peuckert, W. E. (1951) *Geheimkult*, Heidelberg.
(Popov, A. A.) Попов, А. А. (1928) Материалы по шаманству. Культ богины Аисит у якутов (Contributions to Shamanism. The Cult of the Goddess Aisit among the Yakuts), in *КПВ*.
(1936) *Тавгийцы* (The Tavgy Samoyeds), Moscow–Leningrad.
(1949) Материалы по истории религии якутов б. Вилюйского округа (Contribution to the History of the Religion of the Yakuts Living in the Former Vilyui District), in *СбМАЭ*, Vol. XI.
(1956а) Долганы (The Dolgans), in *Народы Сибири* (The Peoples of Siberia), Moscow–Leningrad.
(1956б) Якуты (The Yakuts), in *Народы Сибири* (The Peoples of Siberia), Moscow–Leningrad.
(Potapov, L. P.) Потапов, Л. П. (1936) *Очерки по истории Шории* (Essays on the History of Shoria), Moscow–Leningrad.
(1946) Культ гор на Алтае (The Mountain Cult on the Altai), in *СЭ*, No. 2.
(1947) Этнический состав сагайцев (The Ethnic Composition of the Sagais), in *СЭ*, No. 3.
(1953) *Очерки по истории алтайцев* (Essays on the History of the Altais), Moscow–Leningrad.
(1956) *Применение историко-этнографического метода к изучению памятников древне-тюркской культуры* (Historical and Ethnographical Methods Applied to the Study of the Monuments of the Ancient Turkic Culture), Moscow.
Potapov, L. P. (1957) Šamanismus u národnosti Sajansko–Altajské vysočny jako historický pramen (Shamanism among the Peoples of the Sayan–Altai Upland as a Historical Source), in *ČE*, No. 2.
(Prokofjeva, E. D.) Прокофьева, Е. Д. (1952) К вопросу о социальной организации у селькупов. (Род и фратия) (About the Social Organization of the Selkups [Clan and Phratry]), in *ТрИЭ*, Vol. XVIII.
(1956а) Ханты и манси (The Khanty and the Mansi), in *Народы Сибири* (The Peoples of Siberia), Moscow–Leningrad.
(1956б) Ненцы (The Nenets), in *Народы Сибири* (The Peoples of Siberia), Moscow–Leningrad.
Saccasyn-Della Santa (1947) *Les figures humaines du Paléolithique supérieur eurasiatique*, Anvers.
Salač, A. (1915) *Isis, Sarapis a božstva sdružená dle svědectví řeckých a latinských nápisů* (Isis, Sarapis and Associated Deities in the Light of the Greek and Roman Inscriptions), Prague.
Schmidt, P. W. (1946) *Rassen und Völker in Vorgeschichte und Geschichte des Abendlandes*, Luzern.
Steinitz, W. (1938) Totemismus bei den Ostjaken in Sibirien, in *E*, Nos 4–5.
(Šibaeva, Ju. A.) Шибаева, Ю. А. (1948) Пережитки эпохи матриархата у хакасов (Survivals of the Matriarchal Epoch among the Khakass), in *СЭ*.
(Šternberg, L. Ja.) Штернберг, Л. Я. (1933) *Гиляки, орочи, гольды, негидальцы, айны* (Gilyaks, Orochi, Golds, Negidals, Ainu), Khabarovsk.
(1936) *Первобытная религия в свете этнографии* (Primitive Religion in the Light of Ethnography), Leningrad.

Thomson, G. (1949) Studies in Ancient Greek Society, in *The Préhistoric Aegean*' London.
Thurnwald, R. (1958)] Das Gesellungsleben der Naturvölker, in *Lehrbuch der Völkerkunde*, Stuttgart.
(Tokarev, S. A.) Токарев, С. А. (1931) *Докапиталистические пережитки Ойротии* (Pre-Capitalistic Survivals in Oirotia), Leningrad.
(1940) *Очерк истории якутского народа* (Outline of the History of the Yakut People), Moscow.
(1947) Пережитки родового культа у алтайцев (Survivals of the Clan Cult among the Altais), in *ТрИЭ*, Vol. I.
(1958) *Этнография народов СССР* (The Ethnography of the Peoples of the U.S.S.R.), Moscow.
(Tokarev, S. A.–Gurvič, I. S.) Токарев, С. А.–Гурвич, И. С. (1956) Якуты (The Yakuts), in *Народы Сибири* (The Peoples of Siberia), Moscow–Leningrad.
Tylor, E. B. (1881) *Anthropology, an Introduction to the Study of Man and Civilization*, London.
Vakarelski, Chr. (1958) Kultúrne prežitky (Cultural Survivals), in *Slovenský národopis*, Vol. VI, 1–2.
(Vjatkina, K. V.) Вяткина, К. В. (1946) Пережитки материнское рода у бурят-монголов (Survivals of the Maternal-Clan System among the Buryat Mongols), in *СЭ*, Vol. I.
Vries, Jan de (1957) *Altgermanische Religionsgeschichte*, Vol. II, Berlin.
(Zolotarev, A. M.) Золотарев, А. М. (1939) *Родовой строй и религия ульчей* (Clan Organization and Religion of the Ulchi), Khabarovsk.

About the Art of the Ancient Tribes of the Lower Tomi

by

V.I. Matjuščenko, Tomsk

During the archaeological expeditions conducted under my guidance in the years 1953–1957 for the Historical Museum for Material Culture, State University, Tomsk, a large number of archaeological finds were exposed from the basin of the Tomi and Yai rivers. The sites dating from the neolithic period (3rd Millennium and the beginning of the 2nd before our era) and from the Bronze Age (2nd millenium before our era) have been subjected to scientific examination.

Here we wish to deal with the following sites:

1. A neolithic burial ground near the settlement of Samus, at a distance of 40 kilometres from Tomsk down the Tomi river. Contained 16 tombs.

2. A neolithic burial ground in the old Moslem graveyard at the southern outskirts of Tomsk, examined by Prof. A. P. Duljzon in 1955–56.[1] It contained 20 neolithic tombs.

3. A neolithic burial ground on the right bank of the Yai river, near the settlement of the same name in the Kemerov district. Two undisturbed tombs were found here.

4. A Bronze Age settlement of the Turba period situated at the northern end of the settlement Samus, where the Samus burial ground was investigated. This Bronze Age settlement has so far been insufficiently excavated.[2]

Let us now analyse the monuments of art exposed from these burial grounds.

The statuette of a bear carved in sandstone was found in the Samus burial ground. Its snout is sculptured in the most realistic way. Its lips, nostrils and sharply curved forehead are only outlined. The eyes are points made with some sharp object, possibly a stone. The ears are reliefs (Fig. 1). The front paws are shaped in the form of small cylinders, while the hind ones are faintly indicated by a curvature at the ending of the trunk, dissected by a transverse hollow in the belly.

The statuette of a bear found in a tomb of the Moslem graveyard is made of muddy yellow sandstone shale. The head is well formed, the snout is somewhat long for a bear. The ears are marked by projections. This same solution is used to represent the legs. Judging by its posture, the figure is the representation of a running bear. One receives this impression from the hind legs which are markedly thrown back. Length 10·8 cm (Fig. 2).

The same tomb contained a burned clay object representing a fish with one side flat and the other in relief. Judging from its form, it is a young fish. There are transverse grooves on its head, probably for the purpose of suspending it. Length 6·1 cm.

In a tomb of the Yai burial ground two bone figures representing birds were excavated. They were part of a necklace composed of the teeth of

[1] Дульзон (1956) pp. 113–4.
[2] Матющенко (1956).

Fig. 1. Statuette of a bear

Fig. 2. Statuette of a bear

wild animals and of plates made of bone and stone. The necklace was found on the belt of a skeleton.

The first one represents a flying waterfowl with outstretched body. It gives a perfect impression of an impetuous flight, though no wings can be detected on the statuette. Length 7 cm (Fig. 3).

The other bird is less artistic as regards execution. The parts of its body are protruding, and, as the first figure, it has no wings. The details are coarse. Length 5 cm (Fig. 4).

Further similar finds representing bears and birds dating from the Siberian Neolithic Era are of no particular interest since they come up frequently in this territory, especially along the Yenisei river (Afontov Hill, Bazaikba).[3]

In a settlement from the Turba Period near Samus two objects were found: the head of a bear carved in stone and a human face.

The bear's head is made of fine-grained sandstone. The snout is elongated, the eyes salient, the forehead straight. (The nape is flat and slanting. The mouth is marked by longitudinal incisions.) Length 7·2 cm (Fig. 5).

The human face was sculptured, by all likelihood, in the round, and from flint. A part of the face chopped from the forehead was of great use in reconstructing the traits of the man represented. He is Europeoid having a straight nose, large eyes, and sharply outlined and salient chin. Size 7·3 cm (Fig. 6).

[3] Окладников (1949) p. 10, Figs 1–3 and (1951) Figs 10 and 13–9.

Fig. 3. Figure of a waterfowl

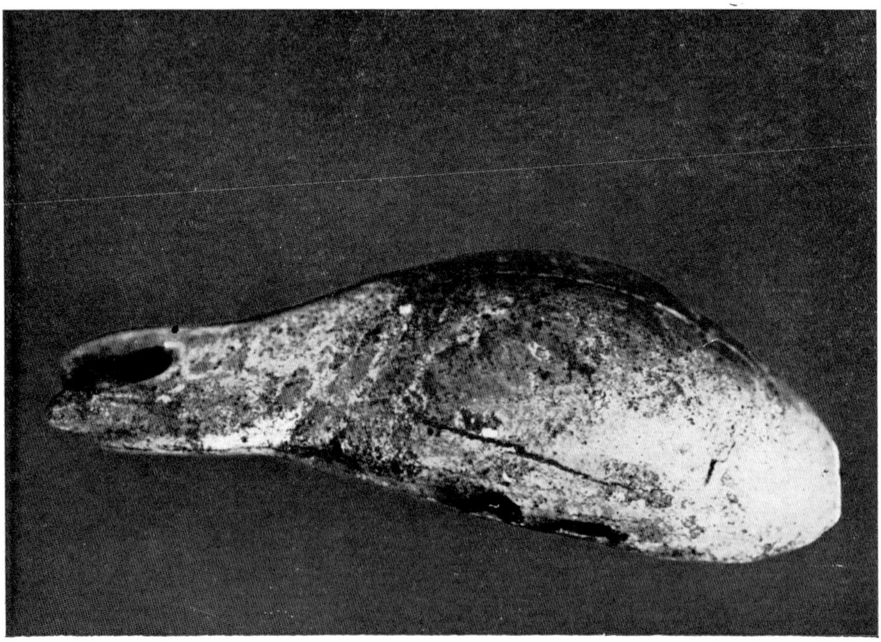

Fig. 4. Bird

Fig. 5. Bear's head

Fig. 6. Man's face

Besides these stone figures, stylized human figures have been found painted on vessels. Here the arms and legs are marked by broken carved lines, and the head is represented by two or three short vertical cuts. The trunk is shown by two lines connected by two shorter transverse lines.

In the present paper we are not concerned with the significance of these representations. There is still much to be done in this field. One thing, however, is quite certain: all these works of art are proofs of the rich spiritual content of the life of the ancient inhabitants of Tomsk.

REFERENCES

(Duljzon, A. P.) Дульзон, А. П. (1956) Археологические памятники Томской области (The Archaeological Monuments of the Tomsk District), in *ТТОКМ*, Vol. IV.

(Matjuščenko, V. I.) Матющенко, В. И. (1958) К вопросу о бронзовом веке в низовьях р. Томи (On the Problem of the Bronze Age on the Lower Tomi River), in *CA*.

(Okladnikov, A. P.) Окладников, А. П. (1949) Неолитические погребения на Афонтовой горе (Neolithic Graves on the Afontov Hill), in *КСИИМК*, Vol. XXV.

(1951) Из истории этнических и культурных связей неолитических племен Среднего Енисея (From the History of Ethnical and Cultural Relations of the Neolithic Tribes along the Middle Reaches of the Yenisei), in *CA*, No. 1.

Two Asiatic Sidereal Names

by

L. Mándoki, Pécs

Valuable ethnic and genetic inferences can be drawn from the names of constellations and from the study of the conceptions and beliefs associated with them. Also results obtained from other sources can be corroborated by a thorough analysis of these names.

Research has so far been concentrated on the names of constellations of the northern hemisphere as handed down by classical traditions.[1] Very little has been written[2] about such terms in Asiatic languages and their documentation is scanty.[3] The aim of the present paper is to stress the importance of exploring this rather neglected territory, by showing the connections between the European and Siberian names of two constellations.

THE PLEIADES

Outstanding among the works devoted to the investigation of sidereal denominations is Lehmann-Nitsche's treatise on the Pleiades (referred to as a *sieve*) based on profuse though not complete material. It ends with the following interesting remark: 'Future research may possibly get more knowledge about this from other Siberian tribes. It would be worthwhile to examine seemingly unimportant facts which, however, may reveal ancient connections between peoples who today live far apart.'[4] With this thought in mind we shall attempt to throw new light upon the type represented by the names for the Pleiades of the Taurus constellation.

Without going into the voluminous material contained in Lehmann-Nitsche's well-known treatise, we wish to add some recent Livonian data to it, such as *səggəl*, Liv. W, *si'glə, sü'gle*[5] and a Hungarian form *szitáslyuk* 'sieve-hole'[6] recorded in the Hungarian language area in several cases during the past century as a name for the Pleiades. The Hungarian word *szita* is not related to the Finno-Ugric words meaning 'sieve'; but is of Slavic origin.[7] Yet the Slavic peoples do not think of the Pleiades as a sieve, no such denominations can be found in their languages. What has happened here is that the ancient Finno-Ugric conception has survived in Hungarian but it is expressed by *szita* of Slavic origin, meaning also 'sieve'.

[1] Volpati (1932); Lehmann-Nitsche (1933a); Schmidt (1954); Mándoki (1958), etc.
[2] Holmberg (1927) pp. 417–48; Lehmann-Nitcshe (1933b) pp. 224–32.
[3] Золотницкий (1875) pp. 142–61; Vámbéry (1879); Потанин (1881) pp. 124–7, (1893) pp. 315–55; Jochelson (1905) pp. 122–3; Bogoras (1904) pp. 307–14; Гордлевский (1909) pp. 95–7; Kannisto (1958) pp. 88–90, etc.
[4] Lehmann-Nitsche (1933b) p. 232.
[5] Kettunen (1938) p. 360a.
[6] Ipolyi (1929) Vol. II, p. 22; Toroczkai–Wigand (1916) p. 26.
[7] Bárczi (1941) p. 291.

Among the non-Finno-Ugrian peoples, Lehmann-Nitsche quotes the name of this constellation from the Lithuanians and from the Koryaks living on the Kamchatka peninsula, after Jochelson[8] and Bogoras[9].

To the versions quoted by Lehmann-Nitsche, we wish to add a few Chuvash data, connected with the Hungarian *szitáslyuk*. The problem arises whether these are borrowed from Udmurt, since the 'sieve' type of naming of the Pleiades is unknown in other Turkic languages.

The Chuvash words are: *àlā śe̯lɒe̯r*[10], 'sieve star', *ala śəlɒər*[11] 'sieve star'— Pleiades. Zolotnickij also mentions that the Chuvashes in some places call the Pleiades *ala śuldi̯r* 'sieve stars'[12]. Asmarin gives a more detailed account of the use of this term. Under the entry ала сăлтăр 'Vergiliae, Pleiades' in his dictionary he quotes sayings associated with the Pleiades from the following villages:

Berezovoe (Chuvash Autonomous Socialist Soviet Republic, Syndyrak district)
Novye Karmaly (former Ufa province, Belebeev district)
Srednie Yumashi (= Buinsk, Chuvash ASSR, Ur. Marsk district)
Staroe Chekurskoe (former Simbirsk province, Buinsk district)
Cherepanov (former Simbirsk province, Buinsk district)
Novoe Izambeevo (Chuvash ASSR, Jylchik district).[13]

The most recent dictionary of the Chuvash (literary) language has the entry ала сăлтăр for the name of the Pleiades[14], which shows that the name meaning 'sieve' is still used in Chuvash side by side with the usual Turkic term *ülger* 'Pleiades'.

This proves that the 'sieve' idea is fairly widespread among the Chuvashes, but does not exclude the possibility of its being borrowed from Udmurt.

As Paasonen points out in his dictionary, the word *ala* 'sieve' is the Chuvash equivalent of the Kazan Tatar *iläk* 'sieve'.[15] Radloff quotes the word *iläk* 'sieve' from Tobolsk and Kazan Tatar, from Baraba and Chagatai.[16] Evidently both members of the Chuvash compound *ala-śəlɒər* 'sieve star' (Pleiades) are of Turkic origin (since *śəlɒər* $\sim$ *yüldüz* is obvious). Nevertheless, the Turkic origin of the idea itself has still to be proved.

Among the very scarce data on the sidereal names in Asiatic languages we find one reference to the 'sieve' idea among the Kirghiz living in North-Western Mongolia: *ilek* 'sieve' (Pleiades) recorded by Potanin[17] from the Kirghiz Musa Čermanov in the Bayan Aul district. This to me seems sufficient to cast doubt upon the Udmurt origin of the Chuvash word.

Nevertheless, we do not want to draw hasty conclusions from the data available at present. Grimm has published Finnish, Estonian, Lettish and Lithuanian data.[18] Uno Holmberg refers to similar Udmurt ideas:

[8] Jochelson (1905) p. 122.
[9] Bogoras (1904) p. 313.
[10] Mészáros (1909) p. 84.
[11] Paasonen (1908) p. 11.
[12] Золотницкий (1875) p. 161.
[13] Ашмарин (1928) Vol. I, p. 107.
[14] Дмитриев (1951) p. 479b.
[15] Paasonen (1908) p. 11.
[16] Radloff (1888–1906) Vol. I, p. 1479.
[17] Потанин (1881) p. 126.
[18] Grimm (no year) Vol. II, p. 608.

'The Votyaks and even the Lithuanian and the Baltic Finns called this constellation "the sieve".'[19] Beside Lehmann-Nitsche's quotation and our own examples from Livonian and Hungarian, we have demonstrated the occurrence of this type of term and idea also among Turkic peoples. These sporadic words, registered from so vast a territory, must be complemented by further research; then we shall have a firm and sound basis for comparative studies.

VENUS

The name of Venus, the closest planet and the brightest star on the firmament, exists in the languages of all peoples. It is visible twice a day and is called morning or evening star. In ethnical astronomy it raises many problems[20] with which we do not intend to deal here. One of its names, widespread over a large part of Asia in territories inhabited by Altai peoples, will be the subject of our present investigations.

In the Turkic, Mongolic and Tunguso-Manchurian languages the name *solbon* ∼ *čolbon* is used.[21]

In Osmanli Turkish we have *čulpan*[22], as well as corrupt forms such as *čobanyïldïzï* 'the shepherd's star'[23], etc.; Radloff *čolpan* (چولپان)[24]. Gordlevskij has published a legend collected by him in Constantinople from Kara-Khisar of the Sivas Vilayet.[25] The legend contains a corrupted form of the name *čolbon*, 'Once upon a time a merchant ordered his servants to wake him up at dawn. At night the servants, seeing the shepherd's star (*čoban jyldyzy*)[26] mistook it for the yellow star (*šary jyldyz* or by its other name *kervan kyran jyldyz*, "caravan-leading star") and thinking that dawn was approaching, saddled the horses and set off. On the way they were attacked by highwaymen and robbed.'[27]

In the Bashkir language Venus is termed *sulpan*. A distinction is made between the evening star *ėŋėr sulpanė* and the morning star *taŋ sulpanė*.[28]

[19] Holmberg (1927) p. 418.

[20] The morning star and the evening star have not been identified everywhere, yet they have for long been connected on account of their identical light and size: one is regarded as a male, the other as a female, or—as also in the Hungarian material (cf. Hegedüs [1952] p. 296)—they are considered to be brother and sister.

[21] These words come from the verb *čol-* 'to shine'. See Vámbéry (1879) p. 7.

[22] TDK (1945) p. 132a.

[23] TDK (1945) p. 132a.

[24] Radloff (1888–1906) Vol. III, p. 2025.

[25] Гордлевский (1909) pp. 96–7.

[26] This star is called *sabah jyldyzy* 'morning star'.

[27] On the evidence of the material at our disposal it is, unfortunately, not possible to tell to which star the *šary jyldyz* and the *kervan kyran jyldyz* refer. Vámbéry also quoted a name in this connection: *kervânkuš* 'caravan killer' and writes: '*Sekiz jolduz*' are the eight stars that appear only on the eastern horizon. They are visible for eight days, disappear on the ninth, reappear on the twelfth, i. e. they can be seen three times within one month. The star of misfortune, the *zejan-jolduz* 'the Scorpion' is one of them. (See Будагов [1871] Vol. II, p. 363. This star is known also in Persia by the name of *kézdum* or '*akrab*'). In Central Asia and Persia it is referred to by another name *kervânkuš* 'caravan killer', because the caravans, mistaking it for the Orion, and believing that dawn is approaching, drive their animals to death in order to arrive at their station in time. See Vámbéry (1879) p. 7.

[28] Pröhle (1904) p. 263.

In his great work on the Bashkir popular culture Rudenko gives *sulpan* for 'Venus' and *taŋ sulpany* for the 'morning star'.[29]

We find a Kazan Tatar form *čulpan* in Pröhle's 1901 collection, registering a Kazan Turk form beside the Bashkir forms.[30] Karakalpak *solpan*[31], Kirghiz *čolpon*[32], Kirghiz in North-Eastern Mongolia[33] and in the Tarbagatai mountains *čolpan*[34], Sart in the Khami oasis *čolpan*[35], Khoton and Tuva (Tangnu-Uryankhai) *čolpan*[36], Salar *čolman, čolľan*[37].

From the languages of the Altai Turks we can produce many examples. In Radloff's dictionary we find the following forms: Teleut *čolbon, čolmon*[38], Taranchi *čolpan*[39]; Sagai and Koibal *šolban*[40]. The Teleuts have separate names for the evening star and for the morning star: *iŋir čolmony* 'evening star', *taŋ čolmony* 'morning star'[41]: we have the same in Sagai and Koibal *īr solbany* and *tan solbany* respectively[42]. Verbickij quotes the following data: Kondoma dial. *šolban*[43], Abakan dial. *šolban*[44], Teleut *čolmon*[45], Lower Biya dial. *čolpon*[46], Palmbakh's Tuvan dictionary has *šolban*[47] (the definition Stella Polaris is an obvious error or a very recent shift of meaning, since the Turks call the polar star *temyr kazyk* 'iron column'); Khakass (Abakan Turkic) *solban* is registered by Baskakov. Also the Khakass distinguish between evening star (*īr solbany*) and morning star (*tan solbany*)[48]. In Altai Kizhi (Oirot) *čolmon* is the name for Venus, although in the women's language it means 'star' in general.[49]

In describing the representations on a drum of a Khakass (Minusinsk Tatar) shaman, Klemenc remarks that the star on the drum, marked No. 1, is called *solban-īr*, which the shaman interpreted as 'dawn'.[50] (Fig. 1.)

In Pekarskij's Yakut dictionary we find the form *čolbon* interpreted as '1. morning star, 2. proper name in tales'.[51] Potanin mentions that in the Verkhoyansk Collection of Articles[52] there is a tale about a man whose father had been killed and who married three girls one after the other, the daughter

[29] Руденко (1955) p. 315.
[30] Pröhle (1904) p. 263.
[31] Баскаков (1947) p. 226.
[32] Абрамзон (1946) p. 65.
[33] Потанин (1881) p. 127.
[34] Потанин (1881) p. 124.
[35] Потанин (1881) p. 127.
[36] Потанин (1881) p. 127.
[37] Потанин (1893) p. 338.
[38] Radloff (1888–1906) Vol. III, p 2025.
[39] Radloff (1888–1906) Vol. III, p. 2025.
[40] Radloff (1888–1906) Vol. IV, p. 555.
[41] Radloff (1888–1906) Vol. III, p. 2025.
[42] Radloff (1888–1906) Vol. IV, p. 555.
[43] Вербицкий (1884) p. 449.
[44] Вербицкий (1884) p. 303.
[45] Вербицкий (1884) p. 431.
[46] Вербицкий (1884) p. 431.
[47] Пальмбах (1953) p. 179.
[48] Васкаков (1953) p. 192.
[49] Баскаков–Тощакова (1947) p. 180.
[50] Клеменц (1890) p. 26.
[51] Пекарский (1917) p. 3635.
[52] Потанин (1891) p. 45

Fig. 1. The drum of a Khakass (Minusinsk Tatar) shaman. From Клеменц (1890)

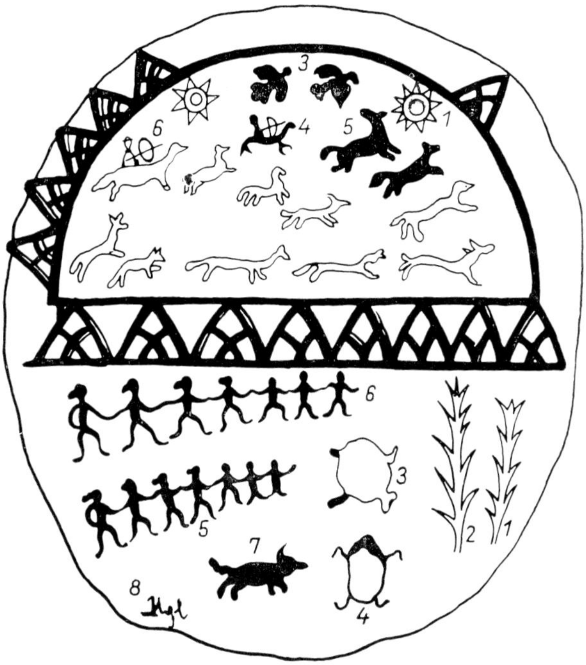

of the *čolbon* star (Venus), the daughter of *ürgel* (Pleiades) and the daughter of the Moon so as to bring his father back to life.[53]

We know two Yakut legends about Venus, both recorded by Seroševskij, the first in 1866 in the Namsk *ulus*: '*čolbon*, though sometimes occurring as a proper name, usually refers in most cases to Venus. It is said to be "the daughter of the Devil and to have had a tail in the early days". If it approaches the earth, it means destruction, storm and frost, even in the summer; "Saint Leontius, however, blessed her and thus her tail disappeared".'[54]

According to another legend, recorded in 1885 at Aldan Bayag '. . . *čolbon*, the daughter of the Devil is a beautiful girl; "there, behold, her light sparkles", the Aldan Yakuts say pointing toward Venus. "That is the brilliance of her eyes; she is the bride and the sweetheart of Satan's son—*ürgel* (Pleiades). When these two stars come close to one another, it is a bad omen; their eager quivering, their discontinuous panting cause great disasters: storms, blizzards, gales. When they unite, fathomdeep snow will fall even in the summer, and all living beings, men, animals and trees will perish . . ." '[55]

In 1921 Ksenofontov recorded a tale in the 2nd Odunin *nasleg* (West Kangales district) from Ilin Nikolaj, a seventy-three-year-old Yakut, entitled 'The Fight of the Shaman with the Star Bringing Frost'.[56] The name of the star

[53] Верхоянский сборник Vol. I. 3: pp. 110–1.
[54] Серошевский (1896) p. 668.
[55] Серошевский (1896) p. 668.
[56] Ксенофонтов (1929) pp. 94–6.

in question is *čolbon*. When commenting on this tale Ksenofontov writes in a note: 'The Yakuts have two words for the "star": *sulus* and *čolbon*. The first means simply "star", the second refers to stars that change their place in the sky, sometimes appearing and disappearing. Nowadays, however, it no longer—or very seldom—refers to other planets than Venus and has almost become its name. Yet, as we have seen, in legends also other "*čolbon*s" (i.e. planets) are mentioned.'[57]

Whether Ksenofontov's hypothesis is right, cannot be judged on the basis of the data available so far. Before proceeding, we wish to quote the known Mongolian vocabulary.

Kowalewski *čolman, čolmon, čolma* and *čolbon*[58], Kalmyk *tsolwn̥, tsolwr̥*[59]; *örīn tsolmn̥* is the name of the morning star[60]. In Podgorbunskij's Buryat dictionary we find *odon solbon* and *uχa solbon* from the Alar dialect, and *sagan solbo* and *mečin solbo* from the Tunka dialect.[61] In Buryat ethnographic literature further material is found. In Buryat mythology the personification, i.e. the master spirit of this planet (*ežin*) is a high-ranking deity '. . .*solbon-sagan-tengeri* or *uχa-solbon* . . . the *ežin* of the *solbon* star or planet is a celestial deity, according to a shaman of the Olkhon island, the son of *esege-malan-tengeri* . . . *solbon* had three wives, the last was a simple Buryat woman, Kherkhen-Saralgan daughter of Ukhandaj, a Buryat of the Olzon clan in the Verkholensk district. Though engaged, she was carried off by *solbon* . . . when going to her bridegroom amidst the guests of the wedding feast, and was taken up to heaven . . .'[62]

The name *solbon* for Venus occurs in a tale about the first shaman of the Barguzin Buryats: 'The shamans of Barguzin say that they had obtained their ability at shamanizing and divining from heaven. This is what they mean when they "relate" the story of Dogedoj, the shepherd of the *solbon* star. When the master spirit of the cattle, *solbon* had gone to the western *teŋgeri*s, Dogedoj decided to go for a walk with his dog Burto. He left the herd and did not return for three days. When he finally came back to the animals, he found them elsewhere and many of them had been devoured by wolves. He started to gather the animals together when master *Solbon*, who had come back from the western *teŋgeri*s, appeared. Noticing the disorder, *Solbon* gave Dogedoj a beating with a hook which he had at hand. He injured one of his eyes, his mouth and crippled his legs. The shaman, when performing, imitates all these situations: he jumps about, dances, skips, calls his dog Burto by shouting *vtu, vtu,* he pretends to have a deformed mouth, distorts his lips as if they were wounded, utters unintelligible words, feigns a lame foot, etc. He walks about among the audience, telling how he had become so deformed . . .'[63]

The Khalkha Mongols call Venus *čolmon*.[64] In the last quarter of the 19th century Potanin recorded the names *čolman, čolby, urin čolbyn* among

[57] Ксенофонтов (1929) p. 102.
[58] Kowalewski (1844–1849) Vol. III, p. 2206b.
[59] Ramstedt (1935) p. 429b.
[60] Ramstedt (1935) p. 298a.
[61] Подгорбунский (1909) p. 97.
[62] Агапитов–Хангалов (1883) p. 6.
[63] Гомбоев (1890) pp. 122–3.
[64] Дамдин-Сурэн–Лувсан-Вандан (1942) p. 30b.

them.[65] In another paper of his we find names of different stars, including some referring to Venus, which the Kara Yogurs and the Šira Yogurs call *čolbon*[66], the Shirongols of San-Tsuan *čolbon χotu* and *čorbon χotu,* and the Torguts of Kobok Žair *čolbon*[67]. The Khalkha Mongols living in this area use the following terms: *čolmon* in the *hošun* of Naiman-Van; *čolmon* in Ongiin, the *hošun* of Tushegun; *urin čolmon* in the *hošun* of Chokur-Van and *čolmun*[68] in Gurban-Saikhan, the *hošun* of A-Gun. According to Mastaert the Ordos Mongols use the names *tsʿolmon* and *tsʿu̯lmu̯n*[69].

In the Tunguso-Manchurian languages we have the following forms: Solon of the Khailar region *čolbon*[70] Evenki *čolbon* and *soldon*[71]. In Vasilevič's Evenki dictionary we find the name: *χōlban* 'Mars'[72]. Since Vasilevič also quotes another name for Mars, *χulānde,* it may be assumed that the identification of the name *χolban* with Mars is erroneous.

As we have said in connection with the information given by Klemenc about the Khakass (Minusinsk Tatars), representations of Venus are found on some shamans' drums. We have other evidence showing that in the regions of the Altai and the Baikal—an area that should be regarded as the cradle of the name type *solbon* ~ *čolbon*—Venus is often represented on shaman's drums and, consequently, that this star has a significance in shamanism as practised by these peoples.

In her paper on Teleut shamanism Dyrenkova mentions that the evening Venus (*äŋir čolmon*) and the morning Venus (*taŋ čolmon*) occur among the representations on Teleut shamans' drums.[73] (Fig. 2.) Potapov analysing the drawings on the drum of Marfa Todyševa (*jutty* clan), a Teleut female shaman living in the Cholukhoi *ulus,* writes that the shaman called the picture of the morning Venus *kün čolmon* 'sun *čolmon*' and that of the evening Venus, *ai čolmon* 'moon *čolmon*'.[74]

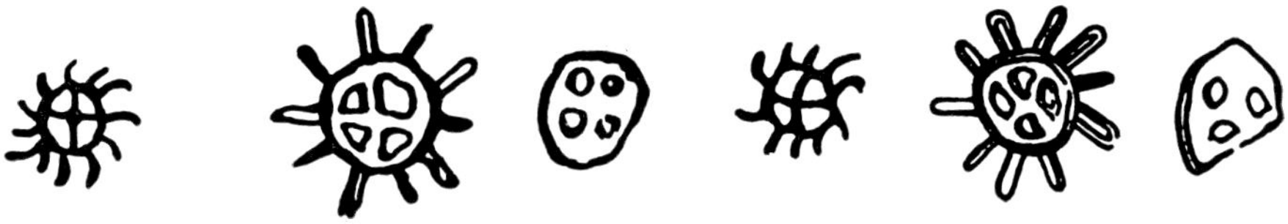

Fig. 2. Morning Venus and evening Venus

Ivanov has published statistics showing the frequency of objects represented on fifty Altaic shaman's drums. The main celestial bodies (Sun, Moon, Venus) occur on 38 of them.[75]

[65] Потанин (1881) p. 127.
[66] Потанин (1893) p. 338.
[67] Потанин (1893) p. 337.
[68] Потанин (1893) p. 337.
[69] Mostaert (1941–1944) Vol. II, p. 711.
[70] Потанин (1893) p. 337.
[71] Василевич (1948) p. 33.
[72] Василевич (1948) p. 141.
[73] Дыренкова (1949) pp. 118–9.
[74] Потапов (1949) p. 196.
[75] Иванов (1955) p. 189.

Fig. 3. A Teleut zodiac. From Ivanov (Иванов [1954] p. 651)

The following figures give a few characteristic examples:

In the third figure, the sign near the Sun is the morning Venus, i.e. *küniŋ čolmoni* 'the *čolmon* of the Sun', and the one near the Moon is the evening Venus, i.e. *ajdyŋ čolmony* 'the *čolmon* of the Moon'.[76] (Fig. 3.)

[76] According to Potapov, the name *taŋ čolmon* 'morning star' and *äŋir čolmon* 'evening star' published by Dyrenkova, and the names *kün čolmon* 'sun *čolmon*' and *ajdyŋ čolmon* 'moon *čolmon*' all refer, to Venus. So far we can agree with him, but we cannot accept his statement (Потапов [1949] p. 196) that 'this makes no difference in the meaning'. As we have already seen, these are the names of Venus not only on the drum of the Teleut shaman Marfa Todyševa but also on the Teleut zodiac published by Ivanov. Hence we may infer that this type of name, revealing a special idea, is quite common among the Teleuts. Examples like this can be quoted also from the Hungarian material. In several parts of Hungary, e.g. Göcsej region, the evening star is called *holdvezető csillag* 'moon-guiding star' the morning star, *napvezető csillag*

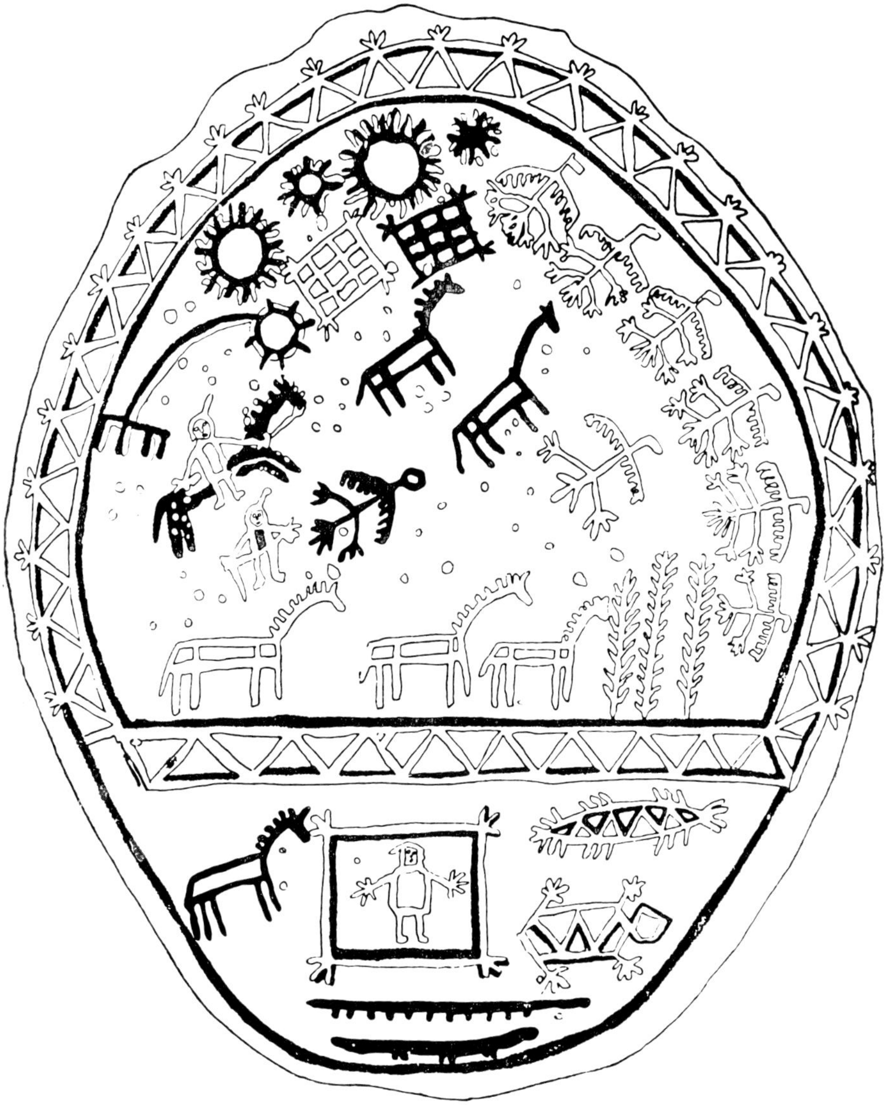

Fig. 4. A Teleut shaman's drum. From Ivanov (Иванов [1954] p. 655)

'sun-guiding star'—cf. Gönczi (1914) p. 190. The former has been recorded from Balatonszentgyörgy and Osztopán, Somogy county; Ohat, Hajdú county. The idea of Venus 'guiding the moon, resp. the sun in the sky' is about to fade away in Hungarian, though we have found corrupt variants of this type (*vezércsillag* 'leading star' from Rigács, Ódörögdpuszta, Veszprém county; Nagyréde, Heves county; etc.) which the informants associated with the above interpretation. It would not be appropriate to suggest a connection between the Teleut and the Hungarian variants (referring, in our opinion, undoubtedly to the same idea), but the necessity of further research into this problem seems to be obvious.

Dyrenkova has published separately drawing of Venus as seen on Teleut shaman's drums.[77]

As to the manner of representation, we quote Ivanov's words: 'The celestial bodies are usually painted in white, rarely in red and in exceptional cases in black. When the three principal celestial bodies, the sun, the moon and Venus, are represented near one another, the first two are drawn in white and the third in red, or the other way round. On most drums only two celestial bodies are represented—the Sun and the Moon.'[78]

Venus is represented in the upper zone of the drum, symbolizing the sky. (See Fig. 4.) We must ask ourselves the question, why Venus is drawn on shamans' drums. Ivanov, an outstanding specialist of the field of shamanism and the ornamental art of the Siberian peoples, writes: 'It is not very clear what significance the shamans attributed to Venus (*solbon*) which so often figures next to the Sun. It seems to have lost its symbolic value, and its presence on shamans' drums is simply due to tradition. Sometimes two Venus stars (evening star and morning star) are depicted. The shamanistic songs refer to Venus as the "prince of heaven". Ideas connected with it are expressed most clearly by the Buryats who associate the idea of fertility with this planet.'[79]

In order to make the function and the representation of this planet in the 'sky' of the shamans' drums quite clear, let us quote another of Ivanov's statements, based on Mikhajlov's manuscript[80], referring to the Buryats, 'According to the explanation given by some shamans, Venus, i.e. *uχā-solbon* "morning star" has nine daughters who have the power to work miracles, to bring the dead back to life, to make the poor rich, etc. The souls of the young candidates who are training to become shamans, come in contact, it seems, with the daughters of *Solbon* when they rise into heaven.'[81]

We have very little material referring to the sex of the personified Venus.[82] As we have seen, the Yakuts look upon Venus as a woman, the daughter of the devil. On the other hand, the Buryats consider it to be a man, saying that he is the son of *esege-malan-tengeri*. Potanin has recorded from the Kirghiz in the Tarbagatai mountains the following saying: *ürker ajdyŋ oly, čolpan ajdyŋ kyzy* 'the Pleiades are the moon's son, and the evening star the moon's daughter'.[83]

Summing up, we may come to the following conclusions:

1. The words used by the Turkic and Mongolian and most of the Tunguso-Manchurian peoples to denote Venus (*solbon* ~ *čolbon*) go back to one common root. The Yakuts alone are known to apply the word *čolbon* to planets in general (Ksenofontov), although earlier sources (Seroševskij, Pekarskij) do not support this. Thus the interpretation 'Mars' of the Evenki word *χōlban* must also be considered as erroneous, since it originally referred to Venus.

2. The names of Venus, as used by the Altai peoples, were borrowed by the Udmurts (Votyaks) along the Volga from the Kazan Tatar language,

[77] Дыренкова (1949) p. 110.
[78] Иванов (1955) p. 191.
[79] Иванов (1955) p. 192.
[80] Михайлов (1912).
[81] Иванов (1954) p. 730.
[82] Holmberg (1927) pp. 433–4, also touches upon this problem.
[83] Потанин (1881) p. 124.

in the form of *čulpon*. Munkácsi[84] has recorded forms like *čukna-čulpon* 'morning star' and *žit-čulpon* 'evening star'. The latter occurs also as *čulpon-kiźil'i* '*čulpon* star'.

3. Among the stars' names in the languages of the Altai peoples, this name has spread over the largest territories and the planet it designates plays an important part in the religion of the Altai peoples, and in shamanism.

We hope that we have proved the importance of the conclusions that can be drawn from sidereal names in the elucidation and corroboration of ethnic relations, and have shown the necessity of studying them without relying on classical (Graeco-Roman) or Arabo-Persian traditions.

REFERENCES

(Abramzon, S. M.) Абрамзон, С. М. (1946) *Очерки культуры киргизского народа* (Outlines of the Culture of the Kirghiz People), Frunze.

(Agapitov, N. N.–Khangalov, M. N.) Агапытов, Н. Н.–Хангалов, М. Н. (1883) Материалы для изучения шаманства в Сибири. Шаманство у бурят Иркутской губерни (Material for Studying Shamanism in Siberia. Shamanism among the Buryats of the Irkutsk Province), in *ИВСОРГО*, Vol. XIV, Nos 1–2.

(Ašmarin, N. I.) Ашмарин, Н. И. (1928) *Словарь чувашского языка* (Dictionary of the Chuvash Language), Kazan.

Bárczi, G. (1941) *Magyar Szófejtő Szótár* (Hungarian Etymological Dictionary), Budapest.

(Baskakov, N. A.) Баскаков, Н. А. (1947) *Русско-каракалпакский словарь* (Russian–Karakalpak Dictionary), Moscow.

(1953) *Хакасско-русский словарь* (Khakass–Russian Dictionary), Moscow.

(Baskakov, N. A.–Toščakova, T. M.) Баскаков, Н. А.–Тощакова, Т. М. (1947) *Ойротско-русский словарь* (Oirot–Russian Dictionary), Moscow.

Bogoras, W. (1904) The Chukchee, Vol. II, Religion (The Jesup North Pacific Expedition), in *MAMNH*, Vol. VII, No. 2.

(Budagov, L.) Будагов, Л. (1871) *Сравнительный словарь турецко-татарских наречий* (A Comparative Dictionary of the Turkish–Tatar Dialects), Saint Petersburg.

(Damdin-Surên, C.–Luvsan-Vandan, Š.) Дамдин-Сурэн, Ц.–Лувсан-Вандан, Ш. (1942) *Орос-монголь толь* (Russian–Mongol Dictionary), Ulan-Bator.

(Dmitriev, N. K.) Дмитриев, Н. К. (1951) *Русско-чувашский словарь* (Russian–Chuvash Dictionary), Moscow.

(Dyrenkova, N. P.) Дыренкова, Н. П. (1949) Материалы по шаманству у телеутов (Material on Shamanism among the Teleuts), in *СбМАЭ*, Vol. X.

(Gomboev, D. G.) Гомбоев, Д. Г. (1890) Сказания бурят, записанные разными собирателями (Legends of the Buryats Recorded by Different Collectors), in *ЗВСОРГОЭ*. Vol. I, No. 2.

Gönczi, F. (1914) *Göcsej s kapcsolatosan Hetés vidékének és népének összevontabb ismertetése* (A Concise Survey of the Region and People of Göcsej and the Adjacent Hetés), Kaposvár.

(Gordlevskij, V. A.) Гордлевский, В. А. (1909) Представления османцев о небесных телах (The Notions of the Celestial Bodies among the Osmanli Turks), in *ЭО*, Vol. LXXXIII.

Grimm, J. (no year) *Deutsche Mythologie*. Fourth ed. by E. H. Meyer, Vols I–III.

Hegedüs, L. (1952) *Moldvai csángó népmesék és beszélgetések* (Moldva Csángó Folk Tales and Conversations), Budapest.

Homberg, U. (1927) Finno-Ugric, Siberian, in *The Mythology of All Races*, Vol. IV, Boston.

Ipolyi, A. (1929) *Magyar Mythologia* (Hungarian Mythology), 2nd ed., Budapest.

(Ivanov, S. V.) Иванов, С. В. (1954) Материалы по изобразительному искусству народов Сибири в XIX-начала XX в. (Material on the Fine Arts of the Siberian Peoples in the 19th and at the Beginning of the 20th Centuries), in *ТрИЭ*, Vol. XXII. New Series.

[84] Munkácsi (1896) p. 329.

(Ivanov, S.V.) Иванов, С.В. (1955) К вопросу изображений на старинных предметах культа у народов Саяно-Альтайского нагорья On the Significance of the Representations on Ancient Objects of Worship among the Peoples of the Sayan-Altai Upland), in *СбМАЭ*, Vol. XVI.

Jochelson, W. (1905) The Koryak, Vol. I, Religion and Myths (The Jesup North Pacific Expedition), in *MAMNH*, Vol. VI, No. 1.

Kannisto, A. (1958) Materialien zur Mythologie der Wogulen, in *MSFOu*, Vol. 113.

Kettunen, L. (1938) Livisches Wörterbuch, in *LSFU*, Vol. V.

(Klemenc, D.) Клеменц, Д. (1890) Несколько образцов бубнов минусинских инородцев (Some Examples of Drums Used by the Natives of the Minusinsk Region), in *ЗВСОРГО*, Vol. II, No. 2.

Kowalewski, J. E. (1844–1849) *Dictionnaire mongol–russe–francais*, Vols I–III, Kazan.

(Ksenofontov, G. V.) Ксенофонтов, Г. В. (1929) *Хрестес* (Cross), Irkutsk.

Lehmann-Nitsche, R. (1933a) Das Sternbild der Mäher, in *ZfVk*, Vol. V.

(1933b) Das Sternbild des Siebes, in *MSFOu*, Vol. 67.

Mándoki, L. (1958) Az Orion csillagkép a magyarságnál (The Constellation Orion among the Hungarians), in *NÉ*, Vol. XL.

Mészáros, Gy. (1909) *Csuvas Népköltési Gyűjtemény* (Collection of Chuvash Folk Poetry) Vol. I, A csuvas ősvallás emlékei (Survivals of the Primitive Religion of the Chuvashes), Budapest.

(Mikhajlov, V. A.) Михайлов, В. А. (1912) Религиозная мифология бурят (The Religious Mythology of the Buryats), ms., in *Архив АН СССР* фонд 282, опись I, No. 70.

Mostaert, A. (1941–1944) *Dictionnaire Ordos*, Vols I–III, Peking.

Munkácsi, B. (1896) *A votják nyelv szótára* (Dictionary of the Votyak Language), Budapest.

Paasonen, H. (1908) *Csuvas szójegyzék* (List of Chuvash Words), Budapest.

(Palmbakh, A. A.) Пальмбах, А. А. (1953) *Русско-тувинский словарь* (Russian–Tuva Dictionary), Moscow.

(Pekarskij, Ê. K.) Пекарский, Э. К. (1907–1930) *Словарь якутского языка* (Dictionary of the Yakut Language), Petrograd.

(Podgorbunskij, I. Ja.) Подгорбунский, И. Я. (1909) *Русско-бурят-монгольский словарь* (Russian–Buryat–Mongolian Dictionary), Irkutsk.

(Potanin, G. N.) Потанин, Г. Н. (1881) Названия созвездий и небесных явлений и поверия о них (Names of Constellations and Celestial Bodies and the Belief in Them), in *Очерки Северо-западной Монголии*, Saint Petersburg. Vol. II.

(1891) Ставр Годинович и Гэсэр (Stavr Godinovič and Geser), in *ЭО*, Vol. X.

(1893) Звездный и животный эпос (Sidereal and Animal Epos), in *Тангутско-Тибетская окраина Китая и Центральная Монголия* (The Tangut-Tibetian Border of China and Central Mongolia), Saint Petersburg.

(Potapov, L. P.) Потапов, Л. П. (1949) Бубен телеутской шаманки и его рисунки (The Drum of a Teleut Female Shaman and the Designs on It), in *СбМАЭ*, Vol. X.

Pröhle, V. (1904) Baskir szójegyzék (List of Bashkir Words), in *KSz*, Vol. V.

Radloff, W. (1888–1906) *Versuch eines Wörterbuches der türkischen Dialekte*, Vols I–IV.

Ramstedt, G. J. (1935) Kalmükisches Wörterbuch, in *LSFU*, Vol. III.

(Rudenko, S. I.) Руденко, С. И. (1955) *Башкиры* (The Bashkirs), Moscow–Leningrad.

Schmid, T. (1954) Milch von den Sternen, in *Orientalia Suecana*, Vol. III.

(Seroševskij, V. L.) Серошевский, В. Л. (1896) *Якуты (опыт этнографического исследования)* (The Yakuts [a Study in Ethnographic Research]), Saint Petersburg.

TDK (1945) *Türkçe sözlük* (Turkish Dictionary), Istanbul.

Toroczkai-Wigand, E. (1916) *Öreg csillagok* (Old Stars), Budapest.

Vámbéry, Á. (1879) *A török-tatár nép primitív culturájában az égitestek* (The Celestial Bodies in the Primitive Culture of the Turko-Tatar Peoples), Budapest.

(Vasilevič, G. M.) Василевич, Г. М. (1948) *Русско-эвенкийский (русско-тунгусский) словарь* (Russian–Evenki [Russian–Tungus] Dictionary), Moscow.

(Verbickij, V. I.) Вербицкий, В. И. (1884) *Словарь алтайского и аладагского наречий тюркского языка* (Vocabulary of the Altai and Aladag Dialects of the Turkic Languages), Kazan.

Volpati (1932) Nomi romanzi degli astri Sirio, Orione, le Pleiadi e la Iadi, in *ZfRPh*, Vol. III.

(Zolotnickij, N. I.) Золотницкий, Н. И. (1875) Названия бога, неба и светил небесных (Names of God, Sky and Celestial Bodies), in *Корневой чувашско-русский словарь* (Chuvash–Russian Dictionary of Roots), Kazan.

Abbreviations

A = Anthropos
AA = American Anthropologist
AARS = Annales Academiae Regiae Scientiarum Upsaliensis
AE = Acta Ethnographica
AfR = Archiv für Religionswissenschaft
AfV = Archiv für Völkerkunde
AL = Acta Linguistica
AMNP = Acta Musei Nationalis Pragae
AO = Acta Orientalia
ArtA = Artibus Asiae
AUFA = Annales Universitatis Fennicae Aboensis
AUS,SSCR = Acta Universitatis Stockholmiensis, Stockholm Studies in Comparative Religion
BUP = Bibliotheca Universitatis Polonae
ČE = Československá Ethnografie
ČSM = Časopis slezkeho muzea
DDMÉ = A debreceni Déri-Múzeum évkönyve
E = Ethnos
EMSMS = The Ethnographical Museum of Sweden (Statens Etnografiska Museum), Monograph Series
Ethn = Ethnographia
FE = Folia Ethnographica
FFC = Folklore Fellows Communications
FUF = Finnisch-Ugrische Forschungen
H = Hermes
I = Irodalomtörténet
IHUHT = Instituti Hungarici Universitatis Holmiensis, Thesis
IRHGS = International Review of the Hungarian Geographical Society
ISK = Instituttet for Sammenlignende Kulturforskning
JAF = Journal of American Folklore
JMVL = Jahrbuch des Museums für Völkerkunde zu Leipzig
JSFOu = Journal de la Société Finno-Ougrienne
KSVK = Kalevala Seuran Vuosikirja
KSz = Keleti Szemle
KV = Kalevalaseuran Vuosikirja
LSFU = Lexica Societatis Fenno-Ugricae
LUÅ = Lund Universitets Årsskrift
MAES = Monographs of the American Ethnological Society
MAG = Mitteilungen der Anthropologischen Gesellschaft in Wien
MAMNH = Memoirs of the American Museum of Natural History
MMVH = Mitteilungen aus dem Museum für Völkerkunde in Hamburg
MNGy = Magyar Népköltési Gyűjtemény
MNy = Magyar Nyelv
MNyr = Magyar Nyelvőr
MNyTK = A Magyar Nyelvtudományi Társaság Kiadványai
MSFOu = Mémoires de la Société Finno-Ougrienne
MSOS = Mitteilungen des Seminars für orientalische Sprachen zu Berlin
MTANyIOK = A Magyar Tudományos Akadémia Nyelv- és Irodalomtudományi Osztályának Közleményei
NÉ = Néprajzi Értesítő
NT = Néprajzi Tanulmányok
NyF = Nyelvészeti Füzetek
NyK = Nyelvtudományi Közlemények
P = Paideuma. Mitteilungen zur Kulturkunde
S = Sociologus

SN = Slovenský národopis
Sz = Századok
SzF = Szegedi Füzetek
T = Terra
USP = Uvod do studia pravaku
UJ = Ungarische Jahrbücher
V = Virittäjä
VD = Verba Docent
WBKL = Wiener Beiträge zur Kulturgeschichte und Linguistik
ZfE = Zeitschrift für Ethnologie
ZfRPh = Zeitschrift für romanische Philologie

ВИРА = Вопросы истории религии и атеизма
ВРГО = Вестник Русского Географического Общества
ВЯ = Вопросы языкознания
ДСИЯ = Доклады и Сообщения Института языкознания
ЕМИРА = Ежегодник Музея истории, религии и атеизма
ЖМНП = Журнал Министерства народного просвещения
ЖР = Живописная Россия
ЖС = Живая Старина
З = Землеведение
ЗВСОРГОЭ = Записки Восточносибирского Отдела Русского Географического Общества по этнографии
ЗЗСОРГО = Записки Западносибирского Отдела Русского Географического Общества
ЗПОРГО = Записки Приамурского Отдела Русского Географического Общества
ЗРГООЭ = Записки Русского Географического Общества по Отделению Этнографии
ЗРГООГ = Записки Русского Географического Общества по Отделению Географии
ИАН = Известия Академии Наук
ИВСОРГО = Известия Восточносибирского Отдела Русского Географического Общества
ИГ = Иллюстрированная газета
ИКЕ = Известия Казанской Епархии
ИОАИЭКУ = Известия Общества археологии, истории и этнографии при Казанском университете
КПВ = Культура и письменность Востока
КрС = Краткие сообщения Института Этнографии
КСИИМК = Краткие сообщения Института истории материальной культуры АН СССР
МЗП = Магазин землеведения и путешествий
МИА = Материалы и исследования по археологии СССР
НИАИНС = Научно-Исследовательская Ассоциация Института Народов Севера, Труды по фольклору
СА = Советская Археология
СбМАЭ = Сборник Музея антропологии и этнографии
СВ = Сибирский вестник
СЖС = Сибирская Живая Старина
СС = Советский Север
СЭ = Советская Этнография
ТОИТ = Труды Общества Изучения Татарстана
ТрИЭ = Труды Института этнографии
ТТОИС = Труды Томского Общества изучения Сибири
ТТОКМ = Труды Томского областного краеведческого музея
УЗИФФВ = Ученые записки историко-филологического факультета во Владивостоке
УЗКУ = Ученые записки Казанского университета
УЗТГ = Ученые записки Таджикского госуниверситета
ХФ = Художественный фольклор
ЧОИДМУ = Чтения в Обществе истории древности при Московском университете
ЧОИДРМУ = Чтения в Обществе истории древности России при Московском университете
Э = Этнография
ЭИ = Этнограф-исследователь
ЭО = Этнографическое обозрение